The Cultural Feast

An Introduction to Food and Society

Second Edition

Carol A. Bryant
University of South Florida

Kathleen M. DeWalt
University of Pittsburgh

Anita Courtney
Lexington-Fayette County Health Department

Jeffrey H. Schwartz
University of Pittsburgh

THOMSON

WADSWORTH

Australia • Canada • Mexico • Singapore • Spain • United Kingdom • United States

THOMSON
WADSWORTH

Executive Editor: Peter Marshall
Development Editor: Elizabeth Howe
Assistant Editor: Madinah Chang
Technology Project Manager: Star McKenzie and
    Travis Metz
Marketing Manager: Jennifer Somerville
Production Project Manager: Belinda Krohmer
Print/Media Buyer: Doreen Suruki
Permissions Editor: Kiely Sexton

Production Service: Publishing Support Services
Composition: BookComp, Inc.
Interior Design: Hespenheide Design
Photo Researcher: Sarah Evertson
Copy Editor: Margaret Moore
Cover Designer: Bill Stanton
Cover Image: GettyImages
Text and Cover Printer: Transcontinental Printing, Inc.
    Louiseville

Printed in Canada
1 2 3 4 5 6 7 06 05 04 03 02

For more information about our products,
contact us at:
Thomson Learning Academic Resource Center
1–800-423-0563
For permission to use material from this text,
contact us by:
Phone: 1–800-730-2214
Fax: 1–800-730-2215
Web: http://www.thomsonrights.com

Library of Congress Control Number: 2002116414

ISBN 0-534-52582-2

Wadsworth/Thomson Learning
10 Davis Drive
Belmont, CA 94002–3098
USA

Asia
Thomson Learning
5 Shenton Way #01-01
UIC Building
Singapore 068808

Australia
Nelson Thomson Learning
102 Dodds Street
South Melbourne, Victoria 3205
Australia

Canada
Nelson Thomson Learning
1120 Birchmount Road
Toronto, Ontario M1K 5G4
Canada

Europe/Middle East/Africa
Thomson Learning
High Holborn House
50/51 Bedford Row
London WC1R 4LR
United Kingdom

Latin America
Thomson Learning
Seneca, 53
Colonia Polanco
11560 Mexico D.F.
Mexico

Spain
Paraninfo Thomson Learning
Calle/Magallanes, 25
28015 Madrid, Spain

This book is dedicated to:
Sterling, James, Laurie, and Whit McLamore
Billie, Saara, and Gareth DeWalt
Delaney, Delia, and Jake Gibbs
Lynn Emanuel

BRIEF CONTENTS

CONTENTS

Part II Food and Culture   83

4   Eating Is a Cultural Affair   84

5   Food Technologies: How People Get Their Food in Nonindustrial Societies   119

6   Food Technologies: How People Get Their Food in Industrialized Societies   157

7   Food and Social Organization   190

8   Worldview, Religion, and Health Beliefs:
The Ideological Basis of Food Practices   221

Part III Strategies for Addressing
Nutrition Challenges   257

Chapter 9 Hunger in Global Perspective   258

You are invited to join us around the table of *The Cultural Feast,* Second Edition, to savor the tantalizing topic of food and culture. As anthropologists and nutritionists, we will act as your guides to explore the deep and far-reaching roots of the eating habits of people around the world. We'll investigate the complex matrix of technological, social, and ideological factors that influence human food systems and the impact that diet has on our social, political, and economic structures. And we will introduce the most promising tools for influencing dietary behavior change and creating a world in which everyone has enough to eat. *The Cultural Feast* blends theoretical and practical information in a readable format that we hope will leave the reader feeling satisfied, though perhaps hungry for more.

Audience

This book is intended for a variety of students. Undergraduate students who study Anthropology and Food, the Ecology of Human Nutrition, and Food and Society will find a comprehensive and current treatment of those subjects in *The Cultural Feast.* Graduate students can use this book as an excellent foundation for related courses; it will prepare them for the more detailed readings they will encounter in these courses. Nutrition students who want to examine the complex factors that influence what we eat will find *The Cultural Feast* a refreshing and useful resource. Those who study or work in community nutrition can use this book as a supplemental text to enhance their appreciation of the intricacies of facilitating dietary behavior change. *The Cultural Feast* also provides nutritionists with useful tools for developing community nutrition programs. As our culture becomes more diverse, it is imperative that those in the field of nutrition have a working knowledge of the broad range of eating practices that exist in the world today. Hopefully, this book can serve as a handbook for those who intend to implement nutrition programs.

Organization and Scope

The Cultural Feast is presented in three main courses. The introductory chapter describes the biocultural, theoretical framework which is the foundation for the text. This useful model provides an understanding of dietary practices as they fit within the biological, sociocultural, and environmental systems that influence food production, distribution, and consumption. With the biocultural model as the starting point, each chapter builds on this theme as a holistic view of food and culture fully emerges. As the reader is confronted with questions about why people eat as they do, this theory will provide a reliable tool for finding the most well-reasoned answers.

In Part 1, readers will look at evolutionary and historical roots to learn the origins of today's human dietary practices. Chapter 2 asks the question "Are we what they ate?" An exploration of primate and hominid diets and their implications for today's nutrition and health will help readers discover the answer. The history of food-related practices is covered in Chapter 3. The focus is on the series of dramatic changes that encouraged the evolution of humans relying on hunting and gathering food to our current, complex, global food systems.

In Part 2, the focus shifts to modern populations and how they use the major components of culture—technology, social organization, and ideology—to create food systems. Chapter 4 introduces the concept of culture, using colorful and sometimes exotic examples to describe its influence on food practices. Chapter 5 describes the technologies people use in nonindustrialized societies to obtain food: foraging, horticulture, agriculture, and pastoralism. Chapter 6 examines the technologies industrialized societies use, discussing the strengths and drawbacks of these systems. Food and social relationships is the subject of Chapter 7. Readers will learn the connections among food and kinship and familial alliances, friends and neighbors, and economic and political alliances. This chapter also expands on the link between one's diet, gender, and position in socioeconomic terms. Chapter 8 uses the lens of worldview and religion to bring food practices into clearer focus.

In Part 3, readers will see beyond the concept of culture to explore the need for, and ways to make, changes in contemporary food systems. The extremes of overnutrition and starvation that exist in our world today need significant adjustment if people are to live to their fullest potential. Chapter 9 takes the complex subject of world hunger and sums it up in a manner that is clear and readable, without being simplistic. Chapter 10 goes a step further to address the factors that contribute to food security at the household, regional, national, and global levels, and it introduces the current and most promising solutions to this grave problem. In Chapter 11, the reader will look at individual dietary behaviors and some of the most respected theories for helping individuals make health behavior changes. Chapter 12 discusses the environment in which individual behavior change occurs by covering large-scale nutrition intervention approaches, including community-based and social marketing techniques.

Each chapter concludes with a Highlight that covers a related topic which warrants in-depth examination. This allows students to delve into interesting topics such as body image and eating disorders, vegetarianism and cultural competence. These highlights are a springboard for lively classroom discussions on topics to which students can easily relate. Focus boxes within chapters provide more detail on subjects to illuminate ideas. Margin definitions aid comprehension.

New to this Edition

A great deal has changed in the study of food and food culture since the first edition of *The Cultural Feast* was written in the early 1980s. This book reflects those changes and includes a wealth of current, more detailed knowledge about food and a fuller array of examples from cultures around the globe.

As described earlier, this second edition is built on the *biocultural* model of understanding food and culture. This model, combined with a generous helping of information about practical applications, provides readers with a more expansive way to see the world of food culture than was available in the first edition.

Our hope is that after you read this book, you will never look at your dinner in the same way—that you will have a deeper appreciation of the complex network of factors that influence what's on your plate. We also hope that you will be motivated to go back for another helping, continue to expand your understanding of the cultural context in which food habits exist, and proceed responsibly and effectively as you take on the nutrition challenges of our time.

Acknowledgments

Carol Bryant: I would like to thank Jessica Lowy and Arlene Calvo for assistance in compiling the bibliography and Moya Alfonso for her efforts in preparing a summary of the composition of U.S. households for Chapter 7. Best Start Social Marketing provided copies of the Loving Support and Texas WIC Marketing Program materials featured in Chapter 12. I would also like to thank Jim Lindenberger for his steadfast support and encouragement during the revision of *The Cultural Feast.*

Kathleen DeWalt: I would like to thank the reviewers for their helpful comments.

Anita Courtney: My gratitude to my husband, Jake Gibbs, for helping to update Chapter 3 and for tracking down numerous references for other parts of the book. My thanks to Carol Bryant for coordinating many of the details of the book and doing so with kindness, intelligence, and perseverance. Thanks to my sisters, Mary Anne Courtney and Patty Caudill, for caring for their nieces so that I could write. I am grateful to my parents, James Courtney and Edwina Peterson Courtney, for instilling in me a love of words. And, finally, I'd like to thank my daughters, Delia and Delaney Gibbs, for providing me with encouragement, laughter, and perspective.

Jeffrey Schwartz: I would like to thank the many curators who made possible my study of the human fossil record, and my students, who continue to ask the important questions.

Collectively, the authors would like to thank the following professionals for their thoughtful advice for this revision of *The Cultural Feast.* Mia Moore Barker, Indiana University of Pennsylvania; Sharon Davis, University of Nebraska at Kearney; Jan Goodwin, University of North Dakota; Brian Stross, University of Texas; Ann Sulivan, St. Francis Xavier University; John Van Willigen, University of Kentucky; William Witt, Grand Valley State University.

Dr. Carol Bryant is a Professor in the Department of Community and Family Health at the University of South Florida College of Public Health and Co-Director of the Florida Prevention Research Center. Dr. Bryant received her doctorate in Anthropology from the University of Kentucky in 1978 and an M.S. in Clinical Nutrition from the same institution in 1984. Before joining the faculty at the University of South Florida, she served as Deputy Commissioner for Nutrition and Health Education at the Lexington-Fayette County Health Department and co-founded Best Start, Inc., a nonprofit social marketing organization. Dr. Bryant has directed social marketing research on a wide variety of public health topics, including breast and cervical cancer screening, prenatal care, immunizations, early childhood intervention services, underage drinking, and pesticide exposure. She has directed five state WIC marketing projects and recently co-directed the development of the National WIC Breastfeeding Promotion Project. In addition to research, Dr. Bryant teaches a variety of graduate-level social science and social marketing courses and coordinates the National Social Marketing and Public Health Conference. She is also editor of the *Social Marketing Quarterly.*

Dr. Kathleen M. DeWalt is Professor of Anthropology and Public Health at the University of Pittsburgh. She currently serves as Director of the Center for Latin American Studies. She received her doctorate in anthropology in 1979 from the University of Connecticut, with a specialization in medical and nutritional anthropology. She has over 30 years of experience in researching food-security issues in Latin America, with work in Mexico, Honduras, Ecuador, and Brazil. Dr. DeWalt has also carried out research on nutritional issues in families with young children and older adults in rural communities in Kentucky. In her international work, she has focused most closely on the impact of agricultural research and development on the food security of small farmers. Recently, she has investigated the nutrition and health impacts of policies promoting nontraditional agricultural exports in Latin America. She is currently working on a study of the impact of development projects for women on women's role in household and community decision making in southern Ecuador and on the welfare of their children; and a study of the impact of urban sprawl on food security of vulnerable populations in the Brazilian state of São Paulo. Dr. DeWalt is the author of over 70 books, articles, and technical reports that deal primarily with issues in health, nutrition, food security, and agricultural research and development programs.

Anita Courtney is the Director of Health Promotion at the Lexington-Fayette County Health Department, where she has worked in public health nutrition for 22 years. She directs programs in community and clinical nutrition, physical activity, diabetes, and wellness. She is a member of the Executive Steering Committee of the Lt. Governor's Task Force on Childhood Nutrition and Fitness and has worked extensively on legislation to address childhood obesity. She is the statewide coordinator of the Kentucky *5 A Day Challenge* campaign. She received both her B.S. and M.S. in nutrition from the University of Kentucky and is a Registered Dietitian.

Jeffrey H. Schwartz is Professor of Physical Anthropology and of History and Philosophy of Science at the University of Pittsburgh. He is also a Research Associate in the Department of Anthropology at the American Museum of Natural History and in the Department of Vertebrate Paleontology at the Carnegie Museum of Natural History. His research spans topics in evolutionary theory, paleontology, developmental biology, and forensic anthropology and osteology. He has traveled extensively throughout the world on archaeological and paleontological expeditions as well as for the study of the remains of fossil and living primates. For almost 20 years his services as forensic osteologist have been sought by coroners' offices and police departments throughout western Pennsylvania. The author of 4 books and editor of another, Schwartz has also published over 100 articles on topics in evolutionary and developmental biology, primate systematics, and human and comparative skeletal anatomy. With co-author Ian Tattersall, he has written a general book on human evolution. He and Tattersall are currently engaged in the first study of the entire human fossil record, which is being published by Wiley-Liss as a three-volume work. He has recently developed a new theory of evolution, which he presented in a feature article of *The Anatomical Record (New Anatomist)* (1999, volume 257) and in the book *Sudden Origins: Fossils, Genes, and the Emergence of Species* (1999, John Wiley & Sons), that for the first time reconciles genetics and the fossil record.

PART I

Evolutionary and Historical Roots of Human Dietary Practices

CHAPTER 1

Setting the Table for a Cultural Feast

We are about to embark on an exploration of the diets and nutritional behaviors of human beings to examine why people eat what they do. Why do teenagers in the United States like fast food? Why is obesity on the rise worldwide while people starve in the same countries? What role does the prohibition of eating beef play in Indian society? Why do some people enjoy bugs while others wouldn't consider eating them? These and many other questions will be addressed as we open the cupboards of cultures around the world. More importantly, you are about to expand your understanding of peoples throughout the world and their most important resource—food—in ways that will be beneficial to you for the rest of your life. You will learn how you fit into the complicated and challenging circumstances that influence the exploitation of food resources. By gaining an understanding of why people eat as they do, you'll come to understand many other things about them as well. And you will see how different, and how similar, you are to the vast array of humans and food cultures that populate our ever-shrinking planet.

We're on the trail of understanding the interconnectedness of many facets of life, using the scent of food as our guide. *The Cultural Feast* lifts the lids off of cooking pots, snoops around food markets, and digs into the dirt of the gardens and farms of many cultures to learn more about human beings. This journey helps us explore the complex matrix of technological, social, and ideological factors that influence human food systems and the impact diet has on our social, political, cultural, and health circumstances. Our approach is holistic. That is, we describe human diets and nutrition through the study of all aspects of social and cultural life, as well as examine human diet and nutrition within the context of history and prehistory. This holistic model allows us to fully understand human dietary practices as they fit within the biological, sociocultural, and physical environmental factors that influence food production, distribution, and consumption.

Much of what we will explore throughout this book is contradictory or contrary to our typical ways of thinking and behaving. Consider, for example:

- Although almost every magazine on today's newsstands features articles on how to stay healthy by eating better, over 61% of adults, 13% of children, and 14% of teenagers are overweight. Health problems caused by this epidemic could reverse many of the health gains achieved in the United States in recent decades.
- Many nutritious items considered delicacies in other cultures are not considered edible in North America: horses, dogs,

small birds such as larks and warblers, fish eyes, sea urchins, live, wriggling shrimp, and haggis, which is made of cow's innards such as lung, intestines, pancreas, liver, and head and is seasoned with onions, beef suet, and oatmeal, all cooked together in a sheep's stomach (Farb and Armelagos 1980). In fact, it is possible to make a nutritionally balanced diet from foods that most North Americans would never eat.

- Each year the U.S. government pays farmers to leave portions of their fields fallow and food industries discard millions of tons of safe, nutritious, but cosmetically imperfect foods at the same time millions of people elsewhere go hungry.

People's strategies for meeting their nutrient needs are constantly changing in response to changes in the particular social and physical environments surrounding them. The holistic, or ecological, point of view that is the focus of *The Cultural Feast* will help you to understand the dynamics that influence the widely variable food practices throughout the world. By examining human diets throughout prehistory, history, and contemporary times, we show how dietary practices emerge to meet the nutritional needs of humans living in particular places and times. However, we do not want to suggest that food systems and diets in any one place, at any one time, are optimal. In fact, the wide range of variation in diets around the world and across time is the result of diets needing only to be "good enough" to support human life and reproduction. Food systems and diets that are not "good enough" disappear, or the people who consume them disappear. Those that are "good enough" persist and change in a startling, vast array of ways.

Biocultural Framework for the Study of Diet and Nutrition

Have you ever considered why you eat the foods that you do? Take a moment to jot down what contributes to your eating style. Keep your notes handy and read on.

In this first chapter we discuss a framework within which human dietary patterns can be understood. This *biocultural model,* depicted in Figure 1.1, is like a road map for our discussions of basic concepts regarding the diets and nutrition of human beings. This way of following our discussion draws heavily on conceptual models developed by Gretel Pelto and her collaborators (Jerome, Kandel, and Pelto 1980; Pelto, Goodman, and Dufour 2000) and by other nutritional anthropologists (Sobal, Khan, and Bisogni 1998).

An important feature of this model is the dynamic relationship between its biological, cultural, environmental, and broader socioeconomic components. As you see in Figure 1.1, the arrows point in both directions to illustrate that multiple factors determine our eating behaviors and that our eating practices influence the way we live, think, and relate to the physical environment. Our diet is influenced by the social and physical environment in which we work, the technologies we use to obtain and process food, how we organize ourselves into social groups, and the beliefs and values we have

Figure 1.1
Biocultural Framework for the Study of Diet and Nutrition

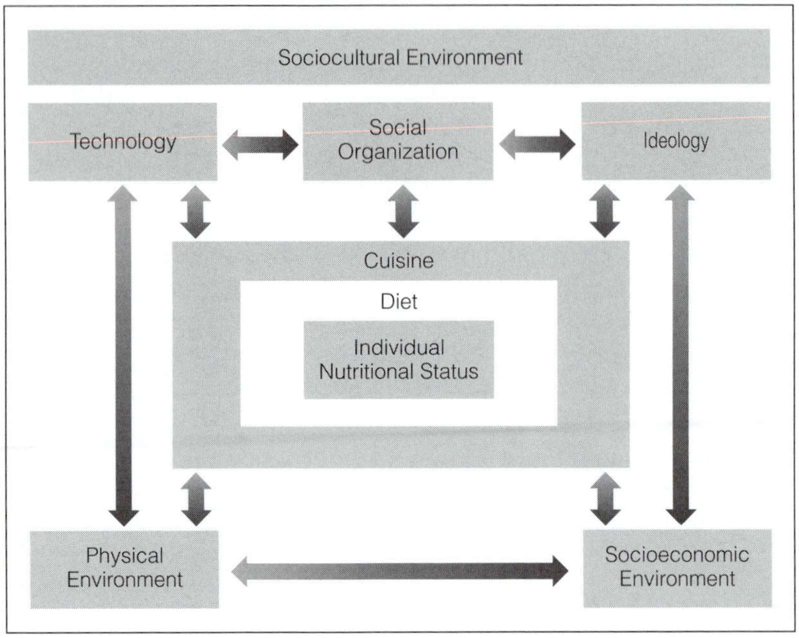

Source: Authors

about food. People's dietary patterns also influence their relations with other people, their ideologies, and the physical environment around them. A classic example of this mutual interaction comes from anthropologist Marvin Harris' analysis of the Hindu prohibition on the consumption of beef. Because of religious and other cultural norms, few Indian people eat beef. However, Harris describes how the prohibition against beef consumption may have developed and become codified into Hindu religion in order to protect cattle, which were more valuable for their other products—milk, dung, and labor—than they were for their meat (Harris 1974). A more detailed discussion of this practice is given in Chapter 8.

Take a look at the notes you made about what factors contribute to your dietary habits. Do any of the factors you listed match those shown in Figure 1.1? In the following sections, each component of the biocultural model will be examined in more detail.

Nutritional Status

At the center of the biocultural model is nutritional status. *Nutritional status* refers to "the health of an individual as it is influenced by the intake and utilization of nutrients" (Smolin and Grosvenor 1997:13). Nutritional status is measured in a number of ways. It can be measured by clinical assessments of health and nutrition, such as diagnosis of the specific diseases that result from inadequate nutrient intake (scurvy, goiter, iron-deficiency anemia), anthropometric measures of height and weight, observing children's growth, assessing body composition (e.g., the amount of body fat); and by laboratory tests to assess nutrient levels in blood, urine, and other body components. We also assess nutritional status indirectly by looking at people's food intake using a set of food intake instruments such as food diaries, food frequency surveys, and 24-hour recalls of food eaten by an individual. These approaches measure the amounts of **nutrients** that individuals and groups take into their bodies. In the biocultural model, we can talk about individual nutritional status or the nutritional status of social groups.

Biological Makeup

The physiology of each plant and animal offers constraints and opportunities for achieving optimal nutritional status. All living things must ingest nutrients in order to live, grow, and reproduce, but the type of chemicals they need to survive and the means by which they obtain and process these vary considerably. Green plants take in carbon from the air, and nitrogen and other elements from the soil, and they use sunlight as an energy source to transform these elements into the organic compounds that form the basis of all life. Animals (and some nongreen plants) must ingest plants and other animals in order to obtain the chemicals they need to live, grow, and reproduce.

As plants and animal species evolve, they may lose their ability to produce certain chemicals that provide the energy, structure, and regulation of body processes. Because the plant or animal must obtain these from the outside environment, they are referred to as **essential nutrients.** Take vitamin C, for example. Most animals produce their own vitamin C by combining chemicals already present in the body. Only humans, most other primates, and a few other animals (e.g., guinea pigs and fruit bats) do not have this ability. To survive they must consume a diet with sufficient quantities of vitamin C. Evolutionists believe the inability to manufacture vitamin C resulted from a **mutation** that occurred millions of years ago. This mutation altered the enzyme that catalyzed the final step in the production of ascorbic acid from glucose. Linus Pauling, a famous biochemist, proposed that this mutation was advantageous because it freed glucose for use as energy. The mutation did not prove lethal because the mutant primate's diet of green leaves and other vegetables provided abundant quantities of the nutrient (a gorilla consumes about 4 to 5 grams of vitamin C a day) and was transmitted selectively to its descendants.

Animals differ in their abilities to obtain essential nutrients from plant and/or animal sources. Some animals, like the koala, which relies chiefly on the leaves and bark of the eucalyptus tree, can obtain nutrients from only a few food sources. Other animals may consume only plants or animal products, forcing them to develop relatively restricted feeding strategies. In contrast, humans and other **omnivores** have the benefit of more flexible physiologies that allow them to draw nutrients from a wide range of plant and animal sources. This versatility poses opportunities and problems. Humans can exploit an exceptionally (perhaps uniquely) wide

nutrients
Chemicals that provide the energy, structure, and regulation of body processes and that are not completely synthesized in the body.

essential nutrients
Nutrients that are indispensable for health and cannot be synthesized by the human body, but must be ingested.

mutation
An alternation in genetic material. If a mutation occurs in a sex cell (sperm or ovum), the change will be passed on to the next generation. Such mutations are rare—only 5 to 100 for every 1 million sex cells. Mutations that alter people's genes after birth (and do not occur in sex cells) have no evolutionary impact.

omnivore
An animal that eats both plants and animals.

neophilia
Interest in trying new things, in this case, trying new foods.

neophobia
Reluctance to try new things, in this case, reluctance to try new foods.

range of environments for food, a feature that has contributed to our species' ability to inhabit much of the earth. Because of the adaptive value of finding new nutrient sources, humans have a natural interest in variety and seek out novel food items. However, this curiosity must be balanced with caution—many potential food sources are toxic or contaminated. Referred to as the "omnivores' paradox," humans' interest in new and varied food sources (**neophilia**) is balanced by their fear of new food items (**neophobia**), especially those that taste different from foods already determined to be safe (Rozin 1988). Unfortunately, our ability to digest a wide range of animal and plant products, combined with a desire for varied and novel foods, may contribute to overeating and other dietary practices that in the long run prove to be unhealthy. In Chapter 2 we discuss the roots of human omnivory in our evolutionary past and their implications for diet and health among living populations.

Human Nutrient Needs

Within a narrow range of variation, humans around the world share similar nutrient requirements and restrictions. No human is able to digest cellulose or avoid toxic reactions caused by eating certain poisonous plants, and all humans require the same 50 or so essential nutrients to stay healthy (see Table 1.1). The common nutrient requirements of humans are the results of the evolution of human beings in specific environments with specific food resources, further influenced by the types of technologies used to obtain food.

Indeed, because of our shared evolutionary past, humans share some patterns of nutrient needs with other members of the primate order—monkeys and apes. However, in some ways, human dietary needs are different from those of other primates and mammals as a result of our specific history of evolution and adaptation. Although there have been some specific adaptations of populations in particular environments, on the whole, humans are physiologically the same animals we were 10,000 years ago before the adoption of agriculture, animal husbandry, and food production technology. This has some interesting implications regarding the impact that diets based on agriculture, animal husbandry, and food technology have on humans' health and nutritional status.

population-level differences
In biological terms, these are differences in the frequency of genetic traits. For example, the prevalence of lactase deficiency is different across populations. In behavioral terms, these are differences in shared expectations, technology, meanings, and values—that is, differences in culture.

Over time, however, some human groups have adapted genetically to the challenges of living in specific environments in ways that make them a bit different from other groups of humans. That is, human groups living over time may evolve specialized adaptations to particular dietary environments. Lactose tolerance offers a classic example of this type of **population-level difference.** While most people in the world are unable to digest lactose—the principal sugar found in milk—after they reach adulthood, some people share a gene that allows them to produce large enough amounts of the enzyme lactase to continue consuming milk products throughout their lifetimes. The frequency with which this gene is found varies remarkably among populations. Almost the entire

Table 1.1   Nutrients That Are Essential for Human Health. These must be ingested as the body does not synthesize them or synthesizes them in small amounts that are not sufficient to maintain health.

Macronutrients for Energy	Essential Amino Acids	Vitamins	Minerals	Essential Fatty Acids
Carbohydrate	Histidine	Vitamin A	*Major minerals (needed in larger amounts)*	Linoleic acid
Fat	Isoleucine	B vitamins:	Sodium	Linolenic acid
Protein	Leucine	Thiamin (vitamin B-1)	Chloride	
	Lysine	Riboflavin (vitamin B-2)	Calcium	
	Methionine	Niacin (nicotinic acid; vitamin B-3)	Potassium	
	Phenylalanine	Biotin	Phosphorus	
	Threophine	Pantothenic acid	Magnesium	
	Tryptophan	Vitamin B-6 (pyridoxine)	Sulfur	
		Folate (folic acid, folacin)	*Trace minerals (needed in very small amounts)*	
		Vitamin B-12 (cobalamine)	Iron	
		Vitamin C (ascorbic acid)	Zinc	
		Vitamin D (calciferol)	Iodine	
		Vitamin E (alpha tocopherol, tocopherol)	Copper	
		Vitamin K (menadione)	Manganese	
			Fluoride	
			Chromium	
			Selenium	
			Molybdenum	
			Other minerals probably essential in VERY small amounts: nickel, silicon, tin, vanadium, cobalt, boron	

population of people living in Sweden, Czechoslovakia, Denmark, and Britain carry the gene for digesting lactose. In northern Italy, the gene frequency is about 49%. In most of Asia, Africa (except among a few groups such as the Tussi, Fulani, and Hilma), and South America it is well below 20%. Virtually no one from Taiwan or Singapore can digest lactose (Allen and Cheer 1996; Durham 1991; Harris 1985).

Studies of the distribution of lactose tolerance suggest that the gene's frequency within a population is related to a reliance on dairying and pastoralism (see Chapter 5) that gives people easy access to milk products. In those societies, individuals with the gene for lactose digestion have had a greater chance of thriving and reproducing, passing on more of their genes to the next generation than those who cannot consume milk products. In other societies, the ability to digest lactose into adulthood has offered no advantage and the gene's frequency has not increased.

Energy metabolism offers another example of population-level differences. In a study of the distribution of diabetes across populations, James Neel (1962, 1982) noticed that adult-onset, non-insulin-dependent diabetes (NIDDM), also called type 2 diabetes, was prevalent among populations that had recently changed from a traditional diet based on hunting and gathering or nonmechanized agriculture to a more industrialized, Westernized diet. The only groups of people who consumed Westernized diets without experiencing high levels of NIDDM were of European descent. He also noticed that NIDDM was most prevalent among peoples whose ancestors had experienced periodic food shortage or expended a good deal of energy to get their food, such as Polynesians whose ancestors had to survive lengthy migrations over the cold ocean with little food. Neel hypothesized that over time these populations developed "thrifty genes," enabling them to do more work per unit of food or to live longer with little food. Obviously, those who carried a "thrifty gene" would have had a greater chance for survival and to pass on this gene to their offspring. Although advantageous under these conditions, the thrifty gene became a liability when a steady source of energy was available. Therefore, Neel also hypothesized that people with thrifty genes have a greater tendency to gain weight and develop NIDDM when they adopt Western diets and sedentary lifestyles. Support for Neel's hypotheses comes from high rates of obesity and diabetes observed in modern Polynesian populations and studies of other human populations that have made similar transitions (c.f. Allen and Cheer 1996; Crews and Gerber 1994; McGarvey 1991).

As another example, the distribution of skin color in the world appears to be related to exposure to sunlight. Those populations that are closer to the equator have darker skin color, and those farther from the equator have lighter skin color. It appears that light skin color is a relatively recent adaptation that occurred for humans over the past 150,000 years when some population groups moved from the tropics to higher latitudes. Because most of the vitamin D humans need is synthesized when sunlight strikes the skin, people at higher latitudes with lower exposure to sunlight will synthesize vitamin D more efficiently if they have lighter skin. People closer to the equator will have high exposure to sunlight and will easily produce sufficient vitamin D. Darker skin color may protect most people from the damaging effects of exposure to solar radiation.

The important point to take from these three examples is that there are population-level differences in the need for, and the ability to access or synthesize, nutrients among groups of humans who have adapted to different ecological and dietary conditions. Even within particular populations there is significant individual variation in the need for specific nutrients. As a result of genetic differences among people and the effects of differences in health, activity levels, gender, and personal dietary experiences, individuals may differ dramatically in the amounts of specific nutrients they need to remain healthy. Obviously, people with higher levels of activity need more energy (calories) than those with lower activity levels. It also appears that for some nutrients, such as vitamin C, people who have higher intakes excrete more of those nutrients and may need higher levels in their diet to maintain health. And women need more of some nutrients, such as iron and calcium, than men. Indeed, when all the potential sources of variation are taken into consideration, it may be that each individual has a virtually unique set of specific nutrient requirements. In practice this is not a particular problem because most diets provide enough of any particular nutrient to satisfy the needs of most people. The Dietary Reference Intakes (DRIs) developed by the Food and Nutrition Board are set at levels that accommodate the range of variation among healthy adults.

In sum, when we think of nutrient requirements we need to remember that each person is unique, all humans are alike in some ways, and some humans are more alike than others. In *The Cultural Feast* our focus is on the common, or *pan-human,* nutritional requirements and the dietary adaptations of particular groups of people with less attention paid to individual variation.

Diet

In the biocultural model, human nutritional requirements and nutritional status are found in the box labeled "diet" (see Figure 1.1). *Diet* refers to the actual foods that individuals or groups consume to meet their nutrient needs. For example, we can describe the diet of the Ju/'hoansi people of the Kalahari Desert as being high in fiber and protein and discuss the specific foods they harvest and the dishes they make from them. Or we can focus on the individual level, examining the nutrient intake of one of the Ju/'hoansi members on a typical day or during a specific period. In this model it is possible to demonstrate that while the available diet (foods and dishes) may provide adequate protein and vitamin C, the foods a single person actually consumes in a given period may not provide adequate protein or vitamin C.

At times we will be using the idea of diet in the collective sense. For example, when we compare the diets of foragers and pastoralists, two groups that exploit distinct food resource niches, diet is used in the collective sense. At other times, we will be concerned with how the foods and dishes in a particular cultural and physical context affect the specific food intake of individuals living in that setting.

Cuisine

Cuisine refers to the foods, food preparation techniques, and taste preferences that are shared by the members of a group of people. The concept of cuisine applies only to groups of people that share a culture. Mexican cuisine, for instance, is in part based on the preparation of maize into specific kinds of dishes, such as tortillas, that are typically understood to be Mexican and which contain flavors that are characteristically Mexican. We can also say that many Americans have a strong preference for sweet foods and that many Mexicans have a preference for spicy foods. (See Figure 1.2.)

Figure 1.2

Traditional Korean cuisine is known for its use of spicy fermented cabbage and grilled meat. Chilies, garlic, ginger root, black pepper, sesame seeds, and scallions are among its many distinctive flavors.

© Robert Brenner/PhotoEdit

Just as nutritional status is placed within diet in our biocultural model, diet is nested within cuisine to demonstrate that a given set of preferred preparation techniques and dishes that characterize a particular culture group has an important impact on the diets of individual members. Consider the dietary impact of a cuisine comprising many high-fat dishes, a strong preference for rich sauces and sweets, and a heavy reliance on deep fat frying. Now compare that diet with a cuisine that features small portions of lean meat and fresh vegetables that are typically baked or grilled with lots of garlic and other spices. It is important to note that while acquired food preferences and tastes that are common among any particular group are powerful influences on diet, people also eat novel food or, at times, simply deviate from the group's preferences. It is also worth noting that a society's cuisine interacts with its members' biological makeup and nutrient requirements. We have already discussed how Polynesians and other populations with a high frequency of "thrifty genes" are likely to become obese when they move to a new society and adopt a cuisine rich in calories and fat.

Cuisines are influenced by the sociocultural and physical environments in which they develop. Both the physical/biological environment and the social/cultural environment provide opportunities and constraints for human food consumption. In fact, it may be more correct to say that people "like what they eat" than to say that they "eat what they like." One of the interesting questions we will address in this book is "Where do cuisines come from?"

The Environment

People live in complex environments that include the physical/biological environment of climate, soils, predators (including microbes and parasites); the social/cultural environment of technology, social organization, and ideology; and the global political and economic environment. Each of these environments provides important opportunities and constraints for human food consumption. Individuals and groups develop strategies that allow them to exploit the environments in which they live in order to meet their nutrient needs. Both individuals and groups develop systems of preferred foods and tastes that reflect the foods physically available and the methods of preparation developed to make those foods edible.

It is tempting to suggest that, in contemporary times, the social/cultural environment has taken on greater importance for humans than the physical/biological environment, but that would miss two important points. The first is that human beings, now and in the past, have always had an impact on their environments, both social and physical. Humans (and all living creatures) change the environment just by existing in it. Humans, with a greater capacity for impact, as a result of using tools (technology), have, throughout their history, had a proportionately greater impact on their environments than other animals. The impact of human technology, which is part of the social environment, may be greater now than in the past, but it has always been characteristic of humans.

The second point is that we are reminded on a daily basis that the physical environment around us continues to provide opportunities and constraints for diet and nutrition. We continually hear reports of droughts, floods, hurricanes, rising ocean levels, and the rise and spread of "new" diseases. The impact of catastrophic events, such as drought and flood, on the ability of humans to produce and benefit from food is real and ongoing. In spite of, and perhaps as a result of, technology and social organization, the physical environment continues to play an important role in shaping human diet and nutrition.

Each of these environmental arenas is interconnected, one affecting the other. However, for the sake of our narrative we discuss each of these in turn, with specific components of the environment constituting the subjects of various chapters of the book.

As we have discussed, the biocultural model also demonstrates aspects of the social and physical environments that influence the basic nutrient needs of humans and the consumption of the foods that satisfy them. The model also suggests a number of potential points of intervention to influence dietary change. The interplay among influences seen in the model is important for understanding our discussions throughout this text.

Physical Environment

The physical environment—the climate, soil and water resources, plant and animal life, and other features of the local landscape—has a significant impact on cuisine and human dietary patterns. Because people can eat only what is available, they develop systems of preferred foods and tastes that reflect both the foods available and the methods of preparation developed to make those foods edible.

The physical environment creates opportunities and constraints on what people eat. Some are obvious: While the arctic and sub-Arctic environments do not allow for the cultivation of crops, many animals are available for consumption. Although the poor soils of many tropical settings, such as the Amazon rainforest, cannot be cultivated continuously over many years, they are suitable for production of root crops, such as manioc, taro, and sweet potatoes, if the soil is allowed to lie fallow for periods in between cultivation. Deserts also provide constraints on food production, but may be relatively rich in wild plant and animal resources.

Humans respond to more subtle aspects of environments as well. Agriculturalists who experience yearly fluctuations in rainfall often cultivate a number of different varieties of the same plant species, including some that yield best in optimal conditions and others that yield better in suboptimal conditions (e.g., Boster 1983). For example, in parts of Central America where droughts are common, farmers plant corn and sorghum in the same field. In years in which there is good rainfall, the corn outgrows the sorghum and produces well. Under drought conditions the corn does poorly, but the sorghum survives and provides a harvest (DeWalt and Alexander 1983). As with other components of the biocultural model, the physical environment and human dietary practices influence each other. All environments in which humans live can be truly understood only within the context of the technologies that humans use to exploit those environments. Even human groups with relatively simple technologies alter the environment, especially the plants and animals in the environment, to meet their needs. In our current world, the activities of a relatively few people can and probably already have changed the global environment.

Sociocultural Environment

Humans live and eat in a world filled with shared understandings. These conventional understandings are manifest in artifacts (tools, houses, clothing), social relationships, values and beliefs, and behavioral patterns. We use the term *culture* to refer to the shared understandings that characterize groups of people and distinguish them from other groups. Culture is dynamic, changing in response to different circumstances and new information.

When talking of culture, we often break the concept down into three subareas: technology, social organization, and ideology. *Technology* refers to the knowledge, practices, and tools a group uses to cope with the physical environment and meet its needs for shelter and subsistence. Technology includes the tools and techniques a society has for producing, processing, and preparing food. Human groups have developed extensive technologies for food getting and processing. The simple tool kits of our Stone Age ancestors contained arrowheads, spear points, and choppers for butchering meat. Later, implements associated with agricultural food production, such as digging sticks, seed drills, plows, and scythes, were developed. In contemporary times, a wide range of tools and "adaptive strategies" are used to produce, preserve, and prepare foods, including foraging, horticulture, intensive agriculture, pastoralism, and industrialized agriculture (Cohen 1974; Kottack 2000). Anthropologists have noted that societies with similar adaptive strategies and technologies share a number of social and cultural characteristics, suggesting that a society's technology interacts with other aspects of its culture. Chapters 5 and 6 describe these technologies, examine the constellation of social and cultural features associated with them, and explore how each of these technological strategies influences the types, amounts, and quality of the foods available to respective groups of people.

Social organization refers to the way a social group organizes its members into families, social strata, communities, and other groupings. Social organization also includes norms that regulate relationships, provide templates for the way work is organized within households, and influence how the day is organized. Local patterns of leadership and political organization; systems of social stratification; ethnic identity; relationships between men and women, adults and children; and the informal ways in which food and other goods are exchanged are also part of a society's social organization. Social organization influences people's diets in many ways including

Figure 1.3
People everywhere cooperate in patterned ways to obtain food.

access to food and the resources needed to obtain and use food (see Figure 1.3). For example, in poor, rural communities, access to land for cultivation and foraging has a major impact on dietary intake. In Chapter 7 we look at how dietary practices and social organization influence each other. Chapter 8 discusses the role of ethnicity and ethnic identity in food use patterns and the ways in which food use serves as a marker of ethnic and class differences.

Ideology refers to the values, preferences, meanings, and beliefs that groups of people share with respect to food. Some of these understandings are connected to systems of religious belief and local medical knowledge. Ideology also refers to symbolic meanings and associated values placed on specific foods. It includes accepted definitions of what constitutes food, the definition of what constitutes meals, and norms regarding the appropriate form and timing of meals. Chapter 8 examines beliefs and values placed on foods by different cultural and ethnic groups and explores the impact of changing ideology and the conflicts over food ideology that are common in our very interconnected world.

Economic and Political Environment

Human dietary patterns are also influenced by the broader economic and political environment. This aspect of the environment refers to the ways humans are organized and stratified within groups and communities and among groups, communities, nations, and regions. It also includes the way in which production, exchange, and consumption of all goods, including foods, are managed. The economic and political environment influences people's access to food and other resources and the capacity to exploit those resources.

agricultural subsidies
Subsidies provided to farmers to grow crops and raise animals. They may be in terms of direct payments to farmers, guaranteed prices for products, or discounted prices for inputs such as seeds, fertilizers, water, or land.

In modern North America, virtually all of the foods of the world are available in large supermarkets for those with the economic means to purchase them. Transnational companies engage in the production and distribution of foods around the globe and bring them to the supermarkets of industrialized countries and urban areas of less developed countries. In rural, poor countries, fewer opportunities exist. Diets depend greatly on access to land for food production or gathering and the equipment needed to exploit the environment.

National and international agricultural and economic policies are especially important. **Agricultural subsidies** affect both producers and consumers. International trade laws influence the kinds of crops that farmers grow, the prices they receive for their products in wholesale markets, and consumer retail costs. Multinational food companies' marketing strategies and distribution networks influence what is available for purchase and the prices at which these products are offered.

Even the way national and international agricultural research organizations approach the development of new seeds, chemicals, tools, and techniques of cultivation may have an impact on who benefits from improved technology and who does not. In Chapter 7 and again in Chapters 9 and 10, we review the ways in which the social and economic environment, national and international policies, and agricultural research and development affect the availability and quality of food.

Food Systems

food system
A chain of interconnected activities that take place in order to get food from the environment into the mouths of people. It includes food production, processing, distribution, marketing, and preparation. It also includes the knowledge and customs that surround food and food consumption.

When using the biocultural model to examine how a society produces, processes, distributes, consumes, and imbues meaning to food, it is convenient to view the components in our model as a **food system**—a set of mutually interacting components. If change occurs in one component, the others must change in order for the system to maintain balance. A good illustration of the interactive and changing nature of food systems comes from Eric Ross' (1987) analysis of the pigs and pork in the Scottish cuisine.

The ancestors of modern pigs lived in forests where they subsisted primarily on acorns. Archaeological evidence shows that the abundance of deciduous forests in Europe made pork an important food for the early domesticators of pigs (Ross 1987). Beginning with the advent of agriculture, many European forests were cut down to plant crops, cut timber, and later, during the industrial revolution, to provide fuel for the smelting of metals. Initially, this process was gradual and the importance of pork in the diet persisted throughout most of Europe. During the Middle Ages, domesticated pigs continued to live in the forests where they were watched by swineherds. The pigs lived primarily on foraged forest foods and human waste products. (Pigs that feed on acorns in the remaining forests of the Iberian Peninsula are still enjoyed as delicacies today.) This system was also reinforced by special laws that protected acorn-growing forests. In the thirteenth century, however, deforestation accelerated, the population grew, and agriculture expanded. By the fifteenth and sixteenth centuries, the raising of pigs was confined mostly to urban areas where pigs foraged in the streets for garbage and ate human waste products. As pigs became more scarce, pork became a highly valued luxury item that only the wealthy and nobility could afford.

In the Scottish Highlands, in part as a result of British colonial domination of Scotland, deforestation was early and rapid. Once introduced, agriculture was confined to the same areas that had been used to raise pigs, creating fierce competition for limited land. As agriculture prevailed and the numbers of pigs declined, the ones that survived gained a reputation for being greedy predators. (Pigs were even accused of eating untended babies.) Again special laws were passed, this time by levying fines on people whose pigs destroyed others' property.

Not surprisingly, as pigs became scarce, pork was no longer featured in the Scottish cuisine and people developed an aversion to its taste. Indeed, after the seventeenth century, some observers described the strong dislike of pork in Scotland as a "deep-seated religious prejudice or taboo" (Ross 1987:26).

With the advent of industrialized pig production in the nineteenth century, pork has reentered the Scottish diet. Though still not an important part of the Scottish cuisine, it is no longer scorned. The rise and fall of pig raising and pork consumption, and the changing value placed on pork as a food in Scottish cuisine, suggests the important interactions between several factors: the physical environment, the technologies used to exploit the environment, the relationships between colonizers and the colonized, and the meanings and values placed on foods. No one factor alone accounts for the presence or absence of pork on the dinner plates of Scots either currently or in the past. However, all have contributed to patterns of food production, distribution, and consumption that influence Scottish diets.

Next Steps

We are now ready to move into discussions of the specific components of the biocultural model and talk about the impact of its various components on the diets and nutrition of people in many different physical and social contexts.

This book is organized into three parts. The first part looks at the evolutionary and historical roots of human dietary practices. In Chapter 2 we examine primate and hominid diets and their implications for diet and health today. In Chapter 3 we place contemporary food systems within a historical context, focusing on the impact of a series of dramatic changes in the ways people interact with the environment to obtain food.

Part II covers modern human populations and how the technology, social organization, and ideology related to food production, distribution, and consumption form a set of interacting phenomena that both influence and are influenced by the foods and diets that people consume. Chapter 4 introduces the concept of culture and describes the influence it has on food practices, including what is considered edible and how groups select certain items to make up their diet. This is followed by a more detailed examination of the major components of culture: technology (Chapters 5 and 6), social organization (Chapter 7), and ideology (Chapter 8).

In Part III we move beyond the concept of culture to explore the need for and processes of change in contemporary nutrition. We will examine some of the important issues in hunger and dietary change and examine issues and various ways to improve human dietary patterns. Chapter 9 describes the world food situation and the numerous factors that contribute to undernourishment, undernutrition, micronutrient malnutrition, and hunger. In Chapter 10 we examine the types of solutions that can be implemented at international, national, and household levels to improve the food supply and make appropriate food available to families and individuals. Chapter 11 focuses on individual dietary behavior, reviewing the role that knowledge and beliefs, attitudes, social norms, self-efficacy, and other factors play in bringing about behavior change. In Chapter 12 and the accompanying highlight, we describe large-scale nutrition intervention approaches that draw on an understanding of a combination of individual, interpersonal factors and broader environmental factors to improve human dietary practices.

Certainly, *The Cultural Feast* isn't the final word on the complex network of food and culture and people. It is simply a beginning. It is our hope this introduction will spark your interest to continue examining the fascinating sociocultural context within which food habits exist. More importantly, we hope this text challenges you to work responsibly and effectively to meet the many food- and nutrition-related challenges of our time.

Diet and Human Evolution

In Chapter 1 we noted the importance of the interactions between the physical and social environments and human physiological needs for nutrients. In this chapter we will draw on the information from our living relatives and extinct species of human-like animals (hominids) to give us some insight into where we came from and whether, as some anthropologists have suggested, "we are what they ate." To do this we need to take a critical look at some of the key concepts of human evolution, such as the notion of adaptation and the idea of "survival of the fittest," and how they may help us understand the kind of creatures that humans are today. We will also examine the way that contemporary understanding of evolution, continuity, and change imposes limitations on how much the study of the diets of nonhuman primates and extinct species of humans and human relatives can tell us about ourselves.

The study of human evolution is both exciting and frustrating. It is exciting because new discoveries demonstrate that rather than being a slow transformation of one species into another, our evolutionary past was filled with an amazing array of extinct relatives—just as we have known to be the case for virtually every other animal. It is frustrating because, unlike a number of other types of animals, we are now the only living species of a once wildly diverse evolutionary group. This means that our closest living relative or relatives—one or more of the large-bodied apes (the orangutan, the gorilla, and the chimpanzee)—are actually very distantly related to us. This, in turn, means that any analogy we make between a living ape or monkey and a human to reconstruct our ancestral diet will be speculation at best. Unfortunately, mining fossils for information on how the creatures we came from behaved is not necessarily any more reliable.

Although we know from the study of living species of animals that individuals of the same species can differ significantly in the foods they eat and the ways in which they acquire food, there is still the temptation to pigeonhole an animal into a category of dietary, locomotory, and social preferences. Not being able to observe extinct organisms walking, moving, consuming, and interacting has not, however, deterred paleontologists from making specific claims about these characteristics in extinct species. Short of throwing up our hands and giving up the possibility of reconstructing our evolutionary past, we should recognize that all of the data available to us at this time has important limitations. Knowing this, we can move toward some understandings and draw some conclusions about what our ancestors ate. We know there is a real evolutionary history of life. Although we will never be able to say with certainty that we have uncovered it, we can try to approach its boundaries. In this chapter we will explore how we might do so.

Exploring the Diets of Extinct Humans Through Paleontology

The study of extinct organisms whose remains become fossilized is called *paleontology.* There are plant paleontologists, or *paleobotanists.* There are also paleontologists who specialize in analyzing extinct animals. *Invertebrate paleontologists* look at the externalized skeletons of organisms such as clams, trilobites, and snails. *Vertebrate paleontologists* look at the internalized skeletons of fish, amphibians, reptiles, and mammals. The group of mammals that includes humans is called **Primates**. Specialists who study these animals are *primate paleontologists.* Primate paleontologists who study human fossils distinguish themselves as *paleoanthropologists.*

Fossils provide the most direct evidence we have of extinct humans. To become a fossil, an animal must first get buried—somehow. We know a lot about some Neanderthal individuals who became extinct only about 27 kya (thousand years ago) because members of this species sometimes buried their dead. We know even more about more recent humans because they consistently buried their dead. Why is it so important to be buried? A grave keeps predators and scavengers from scattering bones around. We know quite a bit about the skulls (the *crania*) and the rest of the skeleton (the **postcranial** skeleton) of a number of early human relatives from the c. 4 mya (million-year-old) species *Australopithecus afarensis,* due to an apparent mass disaster (see Figure 2.1). Perhaps as a result of a flash flood, a group of men, women, and children were quickly buried as they made their way along a dry river bed in what is now Ethiopia. Also, in what is now northern Spain, the fairly complete remains of many individuals of an enigmatic species of extinct human relative, sometimes referred to as *Homo heidelbergensis,* were found at the bottom of an ancient subterranean cave, presumably buried (or possibly sacrificed?) by being dropped down the long airshaft to their grave about 350 kya. Along with these humanlike skeletons are the skeletons of extinct bears and saber-toothed cats that must have fallen by accident into the abyss.

If a body does not get buried immediately, there are lots of potential problems that can prevent it from becoming a fossil. If you are a small land-dwelling animal, your bones might break up and crumble after being exposed to the elements over a few years. If you are a large land animal, your bones might be scattered by scavengers and trampled by herds of migrating animals, and, after being broken up, they might continue to break up and crumble through years of exposure to seasonal elements. If you want to become a fossil, and can't get properly buried, your best bet is to die in or near water, preferably slow-moving water. If your bones end up in a fast-moving stream or river, they are apt to be dashed on the rocks and those bones that do survive can become distributed over a vast area. In a

Primates
The taxonomic group that includes humans, other anthropoids (i.e., Old and New World monkeys and apes), and prosimians.

postcranial
The nonhead (cranial) part of the skeleton.

Figure 2.1

A view of the fossil-rich badlands of the western shore of Lake Turkana, Kenya. Here, Richard ad Meave Leakey are carefully uncovering fossils that were exposed through weathering of the soils.

© Delta Willis/Bruce Coleman Inc.

less active body of water, such as a swamp, lake, or bend in a river, your body would probably sink to the bottom and eventually become covered by sediments. Water also leaches out the *calcium phosphate* that hardens a vertebrate's bones and the *calcium carbonate* that hardens an invertebrate's shell, replacing these minerals with minerals that will eventually turn them into a rock-hard model of what they had once been. Of course, if the deposits containing fossils are never exposed for paleontologists or geologists to find them, the fossils will be hidden forever. Also, if the deposits have been exposed but fully eroded, the fossils they contained will be lost forever.

The larger the animal, the greater the chance that some elements of the cranial or postcranial skeleton will become fossilized and not be destroyed. A number of huge dinosaurs have been found virtually intact. Also, the more recent in age the fossil remains are, the more likely it is that, if exposed, they will not be destroyed by erosion. We also find fossilized remains of tiny animals, such as mice, shrews, and many kinds of **prosimian** (so-called "lower") primates, because even small vertebrates have parts of the cranial and postcranial skeleton that are hard enough to endure.

prosimian
So-called lower primates (e.g., lemurs, lorises, tarsiers).

Teeth

The hardest parts of a vertebrate's skeleton are its teeth. Teeth are covered in a layer of mineralized material called *enamel*. So, even if a mouse's jaw breaks up as it's being dragged along a stream, and its teeth fall out, they might survive. Often, the only remains we have of small, early mammals are their teeth. In part because teeth survive

frugivorous
Principally fruit eating.

folivorous
Principally leaf eating.

proteinivorous
Principally protein eating.

morphology
Form or structure.

Hominidae
The taxonomic group that includes living humans and their closest fossil, non-ape relatives; hominid for short.

Hominoidea
The taxonomic group that includes hominids and fossil and living apes; hominoid for short.

so well in the fossil record, differences in their shapes and configurations are often used to distinguish one species from another and for determining which species may be closely related.

Teeth are also used to infer animals' diets. If you run your tongue over your back, or *cheek,* teeth (on each side, the three *molars* in the far back and the two *premolars,* or bicuspids, in front of them), you will feel that their *occlusal,* or chewing, surfaces are fairly smooth, because the *cusps,* or bumps, are fairly low. In general, low-cusped cheek teeth have been interpreted as indicating a largely fruit-eating (**frugivorous**) and/or leaf-eating (**folivorous**) diet. Tropical fruit bats have very flat molar occlusal surfaces. In contrast, high-cusped cheek teeth, such as are found in some prosimians (e.g., some mouse lemurs and some bushbabies) or insectivores (e.g., shrews, hedgehogs), have been interpreted as indicating an animal that tends to eat insects (is *insectivorous*). The only totally **proteinivorous** primate—the tiny prosimian from islands of Southeast Asia, the tarsier—has very tall, pointed cheek-tooth cusps. It might seem to make sense that an animal's teeth will reflect a certain dietary preference, for it is a commonplace to think that the features of an organism have been shaped by natural selection to a precise situation. However, as we will discuss shortly, it may not be a good idea to be committed to such generalities about an animal being restricted to a particular diet because of tooth shape or **morphology.**

Humans differ from most other primates—the orangutan being the most notable exception—because they develop a thick layer of enamel on their cheek teeth, especially on the molars. It is this thick layer of enamel that in part creates the flatter occlusal surfaces of our cheek teeth. With one outstanding exception—a genus and species called *Ardipithecus ramidus* (White, Suwa, and Asfaw 1994, 1995)—all fossils thought of as **hominid** have thick (or even monstrously thick) molar enamel. What makes *Ardipithecus* even more mysterious is that White and his colleagues have not let anyone else see these fossils. So we do not actually know what these fossils represent: Perhaps an ape? Older fossils, particularly the approximately 16–18 mya Miocene genera (the plural of genus) *Ramapithecus* and *Sivapithecus* from Indo-Pakistan, were once believed to be related to hominids because they had thick molar enamel. General features of hominids are presented in Figure 2.2.

Because evolutionarily minded scientists want to know what these features were good or adapted for, they turned their attention to the reasons Miocene forms and their presumed descendants, the early Plio-Pleistocene hominids, had thick enamel. Their answer: These animals all had eaten hard foods, such as nuts and seeds (perhaps enclosed in tough pods). The reason: In living Old World monkeys and apes, the frugivorous thick-enameled species can masticate these food substances, whereas their folivorous thin-enameled kin cannot (Kay 1981). The ecological distribution of nut-eating mammals also suggested that the earliest hominids may not have been fully terrestrial and/or open woodland- or savanna-dwelling animals.

Figure 2.2

Overview of some features of hominids compared to those of generalized hominoids.

	Locomotion	**Brain**	**Dentition**	**Toolmaking Behavior**
(Modern *Homo sapiens*) 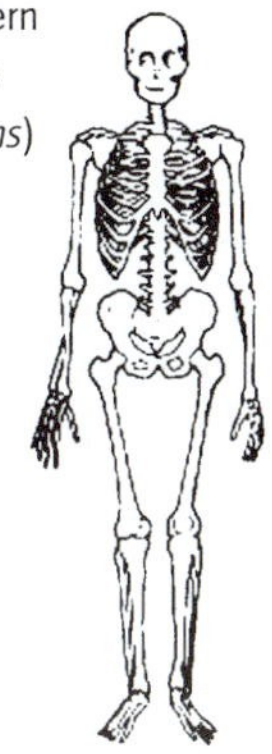	Bipedal: shortened pelvis; body size larger; legs longer; fingers and toes straight, not as long	Greatly increased brain size—highly encephalized	Small incisors: canines further reduced; molar tooth enamel caps thick	Stone tools found after 2.5 mya; increasing trend of cultural dependency apparent in later hominids
(Early hominid) 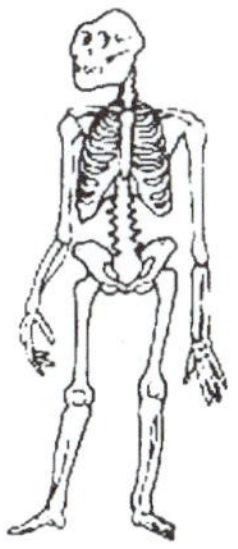	Bipedal: shortened pelvis; some differences from later hominids, showing smaller body size and long arms relative to legs; long curved fingers and toes; probably capable of considerable climbing	Larger than Miocene forms, but still only moderately encephalized	Moderately large front teeth (incisors); canines somewhat reduced; molar tooth enamel caps very thick	In earliest stages unknown; no stone tool use prior to 2.5 mya; probably somewhat more oriented toward tool manufacture and use than chimpanzees — 4 mya
(Miocene, generalized hominoid) 	Quadrupedal: long pelvis; some forms capable of considerable arm swinging, suspensory locomotion	Small compared to hominids, but large compared to other primates; a fair degree of encephalization	Large front teeth (including canines); molar teeth variable depending on species; some have thin enamel caps, others thick enamel caps	Unknown—no stone tools; probably had capabilities similar to chimpanzees — 20 mya

Source: Jurmain, R. et al, 1999 (2nd ed.), *Introduction to Physical Anthropology*, p. 227, Wadsworth.

Skulls and Jaws

Skulls also have a lot to teach us. From a skull, a paleontologist can study features of the face, base, and cranial cavity that provide clues to identifying species and can determine who might be closely related to whom. A feature of the skull that is particularly important to the study of human evolution is the position of the **foramen magnum** (literally, the great hole) at the base of the skull, through which the spinal cord passes. As two-legged, *bipedally erect* animals, we have a foramen magnum that lies pretty much at the center of

foramen magnum
The large hole at the base of the skull through which the spinal cord exits.

sagittal crest
Bony elevation along the midline of the skull that is induced by the chewing muscles' need for more attachment area than the skull itself can provide.

the skull base—which makes sense since our skull is balanced atop a vertical vertebral column. In chimpanzees and gorillas, which walk leaning on the knuckles of their hands, the foramen magnum faces down and back, in alignment with their spines. In totally quadrupedal (four-legged) animals, such as monkeys and prosimians, the foramen magnum faces entirely backward. From muscle scars on the sidewalls of the skull or on the underside of the cheekbones, paleoanthropologists can reconstruct how large the muscles were that move the lower jaw, or *mandible,* in chewing. In general, if a mandible is massive (with the cheek teeth probably also being very large), the chewing muscles (*muscles of mastication*) become so large they produce deep muscle scars on the cheekbone and cause a crest or ridge of bone (a **sagittal crest**) to develop along the midline of the top of the skull. The muscles that attach from the mandible to the side of the skull are the *temporal* muscles. If, as the individual grows, these muscles enlarge so much they cannot be accommodated on the skull, their action (pulling on the tissue—the *periosteum*—that covers bone) causes the deposition of additional bone. This creates bony crests, ridges, or ledges that provide more attachment surfaces for the muscles.

Male gorillas develop huge sagittal crests along the midline of the top of the skull (see Figure 2.3). Given the diet of gorillas— tough, fibrous plants—males, with their large, heavy mandibles, develop huge muscles of mastication, which require larger areas of attachment than the sidewalls of their small braincases can provide. We find sagittal crests and thick cheekbones on skulls of various early hominids, too, and these skulls are associated with very large-boned, and large cheek-toothed, mandibles. Present-day humans do not develop sagittal crests because there is enough space on the skull for the attachment of even the largest chewing muscles. But there are differences between human populations in how much of the available attachment space is occupied depending on how much they use their jaws and teeth. For example, in arctic groups that consume a lot of meat (frozen and/or raw) and use their teeth to soften plant and animal matter for various uses, females typically have muscle scars on their mandibles and skulls that far surpass those seen on the skulls of males from other societies.

The Postcranial Skeleton

Small ankle and wrist bones are compact and, therefore, also end up in fossil collections. The different shapes of ankle and wrist bones can provide possible clues to how an animal got around. In primates, these bones suggest whether the animal lived exclusively in trees or on the ground, or spent time in both environments. As with teeth or other parts of the skeleton, there has been a tendency among functional morphologists to think that a bone or bones of specific shapes were "selected" for that morphology in order for the animal to become adapted to a particular situation. This assumption has been at the base of debates on the evolution of human **bipedalism** (walking on two legs). We humans may use our skeletons in

bipedalism
Locomoting or getting around on two legs.

certain ways to get about, or **locomote,** but that doesn't mean ear-lier hominids with similar morphologies necessarily locomoted exclusively or even in the same way we do.

Bones of the hand, arm, and shoulder girdle and of the foot, leg, and pelvic girdle would seem to be able to provide specific informa-tion on how an animal might have gotten around and, conse-quently, how it might have exploited its environment in search of food. In quadrupedal mammals, such as horses and mice, the fore-limbs and hindlimbs both lie underneath the animal's body. The forelimb is tethered pretty securely directly under the downwardly pointing *scapula* (shoulder blade), although not as rigidly as the hindlimb. Back and forth movement is easy to do, but rotation is not. The shoulder joints of all apes are like ours in being widely sep-arated, outwardly facing, very mobile, and capable of a great range of rotation. Only in the orangutan is the hip joint as loose as its shoulder joint. The configuration of our shoulder joint allows for some degree of **brachiation** ("swinging through the branches") with the body suspended below the branches.

Living humans and gorillas are primarily land-dwelling, or *ter-restrial,* animals, though, and this is supposed to be reflected in both having longer thumbs and shorter and straighter finger and toe bones than the tree-dwelling, or *arboreal,* **apes,** the orangutans, gib-bons, and siamangs. Although we might think of chimpanzees as being fairly terrestrial, their thumbs are fairly short and their finger and toe bones are long and curved. Various early fossil hominids, including "Lucy" (*Australopithecus afarensis*), had long, curved finger and toe bones, which has been taken as evidence of at least some amount of tree dwelling in their behavioral repertoire.

Most primates are distinguished from other mammals in having a thumb that can move away from the other digits—fingers—of the hand, but which can also be used with other digits of the hand for grasping. The primate thumb is thus described as *divergent* and *opposable*. A grasping hand allows young primates to hold on to the fur of the mother and use branches of different sizes for locomotion and the search for and acquisition of food.

Humans are the only living primates whose feet are not like their hands (i.e., they are not "four handed"). Most primates have an opposable first digit on their feet as well as on their hands. The divergent big toe of apes and lemurs, for instance, allows them to use their feet as they do their hands. The human big toe, however, is permanently aligned with the other toes of the foot. For decades it was believed that fossil hominids had feet like ours. Most recently, toe bones of a new (but unnamed) species of *Australopithecus* or even of a new genus were found which indicate that the big toe was opposable to some extent (Clarke and Tobias 1995). Comparisons of these toe bones with the previously known partial foot of a sup-posed specimen of early *Homo* (*Homo habilis*) showed that the latter hominid also had a divergent big toe.

Although humans and the large-bodied apes differ from other primates in having broader upper parts (*ilia*) of their pelvic bones, humans have by far the broadest (from front to back), shortest, and

locomotion
The way in which an animal gets itself about.

brachiation
Locomotion by means of arm swing-ing with the body extending below the support.

apes
Tailless anthropoid primates, includ-ing gibbons and siamangs (small-bodied apes) and orangutans, chimpanzees, and gorillas (large-bodied apes).

most posteriorly expanded ilia. Humans are also distinguished from the large-bodied apes, and mammals in general, in having a knee joint that is not symmetrical on the inner and outer sides. Our femur is oriented down and in under our bodies, with the result that our knees lie under our bodies. The angle on the inner side of our knee joint is greater (more obtuse) than the angle on the outer side; this produces what is called the *carrying angle*. Anatomists have cited these features of the *ilium* (the singular of ilia) and femur as reflective of our being bipedal animals. The ilia of early hominids are in general similar to ours, but they are more expanded anteriorly and, curiously, they flare out considerably (like a platter). The knee joint of early hominids also had a more severe carrying angle than ours. Putting all of the postcranial features together, some paleoanthropologists have reconstructed a much more arboreal lifestyle for various early hominids than the habitually bipedal, terrestrial lifestyle that had been thought to characterize these extinct hominids.

What Is Adaptation?

It is generally believed that the physical and behavioral attributes of an organism represent adaptations that place it in perfect harmony with its environmental circumstances. This notion is a consequence of a picture of evolution in which organisms are constantly and gradually being fine-tuned to ever-changing conditions around them. It also derives from the view that natural and other kinds of selection act to choose the most beneficial or selectively advantageous trait from a seemingly limitless realm of variations. Although these ideas are familiar to everyone interested in evolution, they are not necessarily accurate. There is a difference between "fine-tuning," via selection and adaptation, the features that individuals of a species have, and the emergence of those features in the first place. For example, it may make some sense to inquire whether it is more advantageous—more adaptive—for a peacock to have gaudier plumage or for a buck to have larger and more branching antlers, but it doesn't make sense to ask whether it is more advantageous to have feathers than to have antlers. Unfortunately, Darwin confused adaptation (the "fine-tuning") with evolution (the emergence of novelty), and this is still done today. As the "father" of fruit-fly genetics, Thomas Hunt Morgan, suggested, it would have been more appropriate for Darwin to have titled his most famous work *On the Origin of Adaptation by Means of Natural Selection* than *On the Origin of Species by Means of Natural Selection* (see review in Schwartz 1999).

Another consequence of Darwin's portrayal of natural selection is that it is a force that is constantly choosing variations believed to be the most advantageous, thereby creating the "fittest" individuals. One may be able to construct laboratory or controlled experiments in which the features that the designer considers the most advantageous traits are selected over those considered to be the least advantageous, but this may not be how things work in nature. There is, however, a more realistic way to look at selection, that is, to focus on the elimination of the unfit (say, by death) rather than the selection of the fittest. The difference between these two views is important: Darwinian selection implies a constant culling and improvement of the cream of the crop, whereas the idea based on elimination of the unfit emphasizes the survival of the just "OK." There is something appealing about the notion that species evolve to be only as good as is necessary, especially because it better characterizes the biological world around us, including our own species.

Although we may see an animal eating certain foods and behaving, or locomoting in certain ways, this does not mean that these are the only foods and activities the animal can embrace. Take the case of the chimpanzee (see review in Strier 2000). In the 1960s, at the recommendation of Louis Leakey, Jane Goodall went to Gombe, in Africa, to study the tropical rainforest chimpanzees as a way to learn about early hominid behavior. These chimpanzees were active both on the ground and in trees, bearing their weight on the second-fifth knuckles of each hand to get about on the ground and along the sturdier branches and also brachiating as necessary. When first studied, their diet appeared to consist of plants and fruits. Subsequently, however, other aspects of chimpanzee dietary and locomotory behavior came to light.

Goodall herself later found that some of the chimpanzees in the Gombe area complemented their "folivorous" and "frugivorous" diets with protein by eating termites. To the surprise of primatologists and paleoanthropologists worldwide, these "vegetarian" apes were actively consuming "meat." More surprises were to follow with studies of chimpanzee groups that lived in the fringe zone between the forest and the grassy savanna. Chimpanzees were observed spending more time on the ground (which makes some sense in a forest-savanna fringe zone), assuming more bipedal poses, and sometimes engaging in coordinated hunting of other animals, including baboons and other monkeys. Probably because field workers were now clued in to looking for it, hunting soon became a more frequently observed behavior of other chimpanzee groups, as well. More recently, a group of forest-fringe dwelling chimpanzees was found to include figs as a major part of their diet. Elsewhere, in an area where lava flows prevented water from soaking into the ground and kept vegetative growth scattered, chimpanzees were seen digging out from crevices in the rocks huge rootstocks whose seeds had managed to sprout and take hold. Clearly, chimpanzees, like most organisms, are opportunists, trying to survive in the situations in which they find themselves. If organisms were perfectly adapted to a particular circumstance, this would not be the case. Also important to realize is that no chimpanzee changed its dental or skeletal morphology as a result of this change in its dietary or locomotory lifestyle. Essentially, in tooth and bone, a common chimpanzee from the depths of the African tropical rainforest looks just like a common chimpanzee from the forest fringe.

Another example of opportunism comes from studies of lemurs, which constitute a number of prosimian species that inhabit

Madagascar, a large island that lies off the east coast of Africa in the Indian Ocean. Of the many species of lemur recognized by primatologists is *Eulemur mongoz,* which is commonly referred to as the mongoose lemur (see Curtis 1997; Tattersall and Sussman 1975). In addition to Madagascar, the mongoose lemur is found on the small Comoro islands, Moheli and Anjouan, having probably been imported there some centuries ago by travelers. The mongoose lemur is the only nocturnal Malagasy lemur, although the degree of nocturnality differs for some groups depending on the season, and between groups even living within 100 kilometers of each other. The Moheli mongoose lemur is also nocturnal, but on the island of Anjouan, this prosimian is nocturnal only in the warm, seasonal lowlands, where it typically forms social groups of a few males, a few females, and some juveniles. In the other locales, it typically lives in pair-bonded "family" groups. In the cool, humid central highlands of Anjouan it is diurnal.

Like many lemur species, the mongoose lemur is basically arboreal, getting around quadrupedally through the canopy by running, walking, and leaping. The mongoose lemur is also opportunistically folivorous and frugivorous depending on the season and the different trees in its territory. On Madagascar, fruit is the main foodstuff for some groups of *Eulemur mongoz,* with nectar being supplemental in the wet season and leaves in the dry season. Other Malagasy mongoose lemurs turn almost entirely to consuming nectar and pollen during the dry season. While sitting on a branch, the animal brings the fruit or leaves to its mouth with its hands. When getting nectar and pollen, the animal often hangs upside down by two feet or even one foot and draws the flower or petal toward its mouth with its hands. As with the chimpanzees, there are no detectable differences in dental, cranial, or skeletal morphology between specimens of mongoose lemur.

These examples provide evidence that aspects of an animal's dietary and locomotory behavior cannot necessarily be reconstructed from features of its teeth and skeletal anatomy. Clearly, the relatively low-cusped molars of a common chimpanzee are perfectly capable of masticating leaves, termites, or the flesh of a monkey. Extrapolation from molar morphology to diet would at best have provided a small clue to one element of its potential and real dietary repertoire. Similarly, one could have predicted from its skeleton that the mongoose lemur would have had a hindlimb-dominated form of quadrupedalism that would allow it to jump and leap, but one would never have concluded that this animal hangs upside down to harvest nectar and pollen. Dental morphology would also fail to predict the presence of pollen in the lemur's diet.

Humans are also opportunistic, although in a slightly different way. In contrast to being more or less uniform in dental and skeletal morphology, it has long been known that, on average, various human groups—such as Polynesians, Australo-Melanesians, Far Easterners, Australoids, Europeans, and sub-Saharan Africans—can be distinguished from other human groups on the basis of cranial shapes as well as by dental and postcranial characteristics (see review and references to Krogman and Iscan, Stewart, and Turner, in Schwartz 1995; also Howells 1989). Yet we know that a human transplanted from one part of the world to another can survive, even if the dietary constituents of the two areas are very different. There may be temporary difficulties in digesting a diet with more meat or plant foodstuffs than one is accustomed to, but, eventually, one gets used to the new dietary emphases. This is not necessarily the case with genetically based conditions, such as *lactose intolerance* or *phenylketonuria* (discussed later), where novel foodstuffs may continually create discomfort of one kind or another or even severe disability. But, on the level of chewing and ingesting one kind of diet versus another, any human is just like any chimpanzee or lemur: If it's a matter of their survival, they can do it.

It looks like many of the primates, including humans, are opportunistic omnivores. That is, they vary their diets depending on what foods are most available in the place in which they live, without changing the "hardware" of teeth, muscles, crania, and guts. Looking at the "hardware" alone may not give us much insight into the variety of diets for many existing and extinct primates.

Using Chemistry to Infer the Diets of Extinct Hominids

Within the past 10 years, techniques for analyzing the chemistry of prehistoric bone have become sufficiently sophisticated and reliable that it is possible to infer some things about our extinct relatives' diets. In most cases, teeth are perfect for these tests because the hard, outer covering of enamel protects the inside of the tooth from being contaminated. A small drill hole from which to extract a sample from within the tooth is all that is necessary. A postcranial bone can also be used. The most commonly used tests are *carbon* and *nitrogen stable isotope analyses. Isotopes* are chemical elements (such as carbon and nitrogen) that have the same atomic number but different atomic weights.

Carbon stable isotope analysis has focused on the ratio of the amount of the carbon 13 (^{13}C) isotope compared with the carbon 12 (^{12}C) isotope ($^{13}C/^{12}C$ ratios, also referred to as $\delta^{13}C$ values) in **collagen,** which is the predominant *protein* found in bones and tooth pulps, as well as on $\delta^{13}C$ values in tooth enamel (Lee-Thorpe and van der Merwe 1993; Sponheimer and Lee-Thorp 1999). During **photosynthesis,** plants (which are at the base of the food chain) discriminate against taking up ^{13}C, which is the heavier of the two isotopes. That is, they prefer to take up ^{12}C from the environment. Plants discriminate more against ^{13}C (and thus have proportionately higher levels of ^{12}C) when they follow what's identified as the C_3 pathway of photosynthesis than when they follow the C_4 pathway. That is, C_3 plants have greater $\delta^{13}C$ values (sometimes twice as high) than C_4 plants. Consequently, classes or types of plants can be distinguished by their differences in $\delta^{13}C$ values. In warm climates, which would have characterized Africa during the Plio-Pleistocene,

collagen
The predominant protein in bones and tooth pulps.

photosynthesis
The chemical process green plants use to convert carbon dioxide and water, in the presence of sunlight, into organic compounds.

grasses follow a C_4 pathway while other plants (trees, shrubs, forbs, corms, and tubers) follow a C_3 pathway. When plants are consumed, their $\delta^{13}C$ values become incorporated into the consumer's bone collagen (protein) and tooth enamel.

Because bone is constantly rebuilding itself throughout an individual's life, $\delta^{13}C$ values will reflect the plants consumed until the individual dies. Teeth, however, do not represent an ongoing tape recording of an individual's life history because, once they erupt into the jaws, they have essentially completed their course of development. This applies especially to the crown of the tooth, whose size and shape are set once its enamel covering is deposited. Thus, a $\delta^{13}C$ value obtained from the enamel of a tooth will reflect diet only during the period of enamel deposition of that tooth. Because an individual's teeth develop and erupt at different times over a period of some years, it is possible that $\delta^{13}C$ values could vary from tooth to tooth as a reflection of shifts in diet corresponding to life changes from childhood to young adulthood, when the last teeth in the jaws (in hominids, most likely the third molars, or "wisdom teeth") erupt. After this event, teeth would not be able to provide evidence of an individual's subsequent dietary habits, but it is probably likely that the components of the individual's diet would not change significantly after the first (deciduous, or "milk") set of teeth had erupted into the jaws and the individual could chew with some efficacy.

In addition to $\delta^{13}C$ analysis, study of the levels of a isotope of nitrogen—$\delta^{15}N$—in bone collagen can also provide dietary clues. Whereas $\delta^{13}C$ analysis identifies the type and relative amount of plant foods that make it into an organism's body either directly or through consumption of herbivores, $\delta^{15}N$ analysis attempts to identify the relative amount of flesh an organism consumed. Nitrogen stable isotope values reflect the **trophic level** of an organism in the food chain. The food chains (i.e., number of organisms) in aquatic systems are basically longer than in terrestrial ones (the latter often consisting only of plant-herbivore-carnivore). Thus, a fish or a carnivorous aquatic bird or mammal will be expected to have higher $\delta^{15}N$ levels than an herbivore, and if one of these organisms is consumed by a hominid, its $\delta^{15}N$ levels will be reflected in the hominid's bones.

Other bits of evidence of a hominid's diet can be reflected through the study of the remains of other animals, but, as you can imagine, this enterprise is not free of problems. When you are dealing with early hominid sites, you typically find either the hominid remains or the animal remains at a site. Given the possibility—and these days the evidence is increasingly pointing in this direction—that more than one hominid species roamed the landscape at the same time, you cannot be certain if only one or if all contemporaneous hominids exploited the same food resources. The same difficulty arises when analyzing stone tools. First, just because you find only stone tools in the archaeological record, this doesn't mean that these were the only materials hominids used to fashion tools. Stone tools do not decompose, as bone and wood tools might. Second, it is more common to find stone tools at butchering sites than it is to

trophic level
An animal or a plant's position in the food chain.

find them associated with any given hominid, especially the earlier species. Consequently, you might be able to reconstruct how the animal was butchered and dismembered, but you would be hard put to state without reservation that you know which hominid or hominids made the tools that butchered, or even killed, the animals. Recently, microscopic and chemical analyses of residues and tissues preserved on the edges of stone tools have been brought to bear on identifying the animals on whom the tools were used. Again, however, there is the problem of associating the tools with a particular hominid. The safest associations are very late in human evolutionary history, in the last 100,000 years, and most often in places in which we are trying to distinguish Neanderthals from modern humans, *Homo sapiens.*

Our Place in Nature

You would think by now—the twenty-first century—that physical anthropologists and paleoanthropologists would have sorted out the basic pattern of human evolution, with new fossil discoveries neatly fitting into this scheme. This, unfortunately, has hardly been the case, largely because of the odd history of looking at humans as the unique end point of the evolution of our group. Even as recently as the first edition of this textbook (1984), human evolution was portrayed as a simple linear transformation of one species into another, culminating in us—*Homo sapiens.* In this context, it was a simple matter to talk about the diet of "the ancestor of us all" or the diet of a particular species. It was also easy to turn to other primates, such as apes, and try to transfer their dietary behaviors and proclivities onto our earlier ancestors. As we saw earlier, however, we know better now. We can't do that. Similarly, we can no longer think of human evolution as a stream of change flowing from an "apelike" ancestor to ourselves. Recent discoveries and reanalyses of the known human fossil record have made it clear that our evolutionary past was filled with many different species. This is part of the exciting aspect of human evolutionary studies. The frustrating part is that it is becoming more and more difficult to claim with certainty which species are most closely related to which other species. Rather than being a straight line or linear—like a tree with a single trunk—human evolution is a "bush," with lots of twigs and branches. How they all fit together will be the work that lies ahead of us.

A Brief Who's Who of the Early Hominids

Although a great deal of debate still surrounds the discussion of early hominid species, at least 18 hominid species are currently named in the literature, some of which are illustrated in Figures 2.3 and 2.4. These are:

- *Sahelanthropus tchadensis* (6–7 mya), West Africa
- *Orrorin tugenensis* (c. 6 mya), East Africa

- *Ardipithecus ramidus* (c. 4.4–5.8 mya), East Africa
- *Australopithecus anamensis* (c. 4.1–4.0 mya), East Africa
- *Australopithecus afarensis* (c. 3.7–3.4 mya), East Africa
- *Kenyanthropus platyops* (c. 3.5 mya), East Africa
- *Australopithecus bahrelghazali* (c. 3.5–3.0 mya), West Africa
- *Australopithecus africanus* (c. 3.6–1.5 mya), South Africa
- *Australopithecus garhi* (c. 2.5 mya), East Africa
- *Paranthropus aethiopicus* (c. 2.5–1.0 mya), East Africa
- *Paranthropus robustus* (c. 1.8–1.0 mya), South Africa
- *Paranthropus boisei* (c. 1.8–1.3 mya), East Africa
- *Homo habilis* (c. 1.8–1.2 mya), East Africa
- *Homo rudolfensis* (c. 1.5 mya), East Africa
- *Homo ergaster* (c. 1.8–1.4 mya), East Africa
- *Homo erectus* (c. 1.7 mya–36 kya, if you include virtually all the Asian hominid specimens)
- *Homo antecessor* (c. 780 kya), Europe
- *Homo heidelbergensis* (c. 600–300 kya), Europe and Africa
- *Homo neanderthalensis* (c. 300–27 kya), Europe
- *Homo sapiens* (c. 100 kya–present), Almost global

Clearly, even if we stick only to this list, there was a lot more going on in human evolution than the traditional unilinear model of one species being transformed into another allowed for. Some of these species, such as *Ardipithecus ramidus* and *Australopithecus bahrelghazali*, which these many years later have not been made publicly available, may not stand up to future scrutiny. Other species, such as *Homo habilis* and *H. erectus*, are a hodgepodge of specimens of quite different morphologies that were placed into one or the other species when it was more universally believed that only one species of hominid could have existed at any particular time. Even species that have more recently been accepted in the literature, such as *H. ergaster* and *H. heidelbergensis*, appear to be unnatural associations of specimens that potentially represent different species. And, to bring this discussion to the near present, it seems that various fossils placed in our own species represent different species. Who knows how many species are currently masquerading as *habilis, erectus, ergaster, heidelbergensis,* and even *sapiens*? Indeed, with the recent discoveries of *Sahelanthropus, Orrorin,* and *Kenyanthropus,* who knows how many different genera and species of hominid really existed in the past?

A consequence of realizing that our fossil record represents one of great diversity of species is that it becomes increasingly difficult to be certain who is related to whom. In addition, with many different species coexisting in the same regions, it also becomes more difficult to be certain which species had been the consumer or hunter of the remains of animals and plants found at archaeological sites.

Yet, coming to recognize that there had been numbers of different species in our evolutionary past is a great step forward. It makes our appreciation of what can and did happen in human evolution

Figure 2.3

Male gorilla (upper left), South African Paranthropus robustus *(middle top), East African* Paranthropus boisei *(upper right and lower left), South African* Homo *from Swartkrans (middle bottom, and South African* Australopithecus africanus *from Sterkfontein (lower right). Note the pronounced sagittal crest along the midline of the skull in the gorilla, and the preserved bases of sagittal crests in the skulls of* P. robustus *and* P. boisei. *Also note the huge cheek teeth and the very tiny six anterior teeth in the lower jaw of* P. boisei *(lower left); the proportions would be the other way around in* A. africanus.

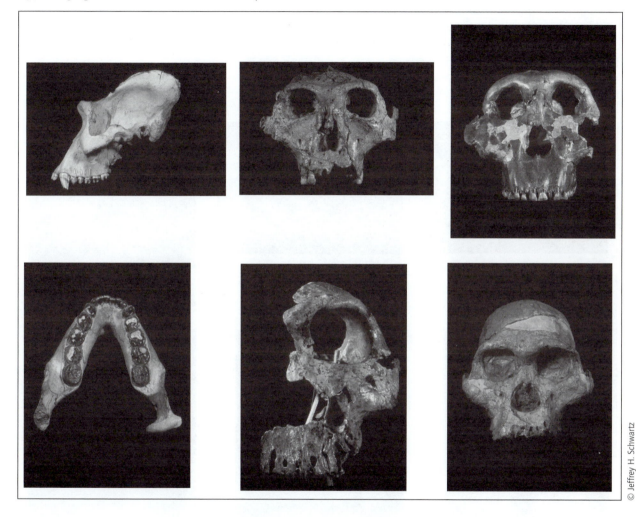

that more real. For instance, rather than thinking, as Darwin did, that a knuckle-walking apelike ancestor became gradually converted into a bipedal human, we now know that different species of early hominids not only looked different from one another in cranial and dental morphology, many of them also differed from one another in the relative proportions of their limbs, in the length and curvature of finger and toe bones, and even in the degree to which the big toe was divergent from the other toes. Perhaps all of these hominids could have walked upright on two legs, but they were not all committed to bipedalism in the same way we are. In reality,

Figure 2.4

A sampling of skulls of the genus Homo. *The one from the site of Sangiran in Java (upper left) is definitely* Homo erectus. *The skull from Zhoukoudian, China (upper middle) has traditionally been included in* H. erectus *but may be a different species. As with* H. erectus, *a variety of differing skulls have been placed in the species* H. heidelbergensis, *including the Kabwe skull from Africa (upper right) and the Petralona skull from Greece (lower right). Specimens attributed to* H. neanderthalensis *(such as the skull from Gibraltar), (lower left) are, however, much more similar in their peculiar cranial features. Fossil and living* H. sapiens, *as well, are distinct in their own right, as can be seen in the skull from Predmostí in eastern Europe (lower middle). When looking at these skulls, pay attention to differences in, for instance, the shape of the brow ridge, forehead, cheekbone, nasal aperture, and lower face.*

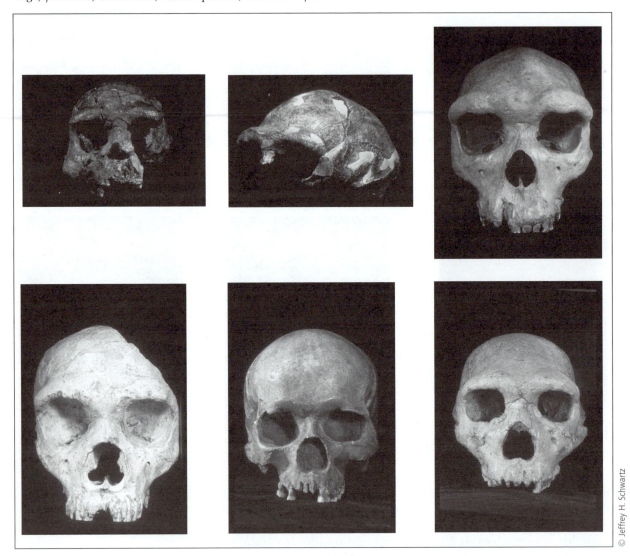

it is unlikely that any early hominid actually strode bipedally exactly as we do.

Given the diversity of early hominid species and the different packages of primitive and derived dental and postcranial configurations each species displays, it is obvious that each species would also have been able to exploit its surroundings in different ways. For example, the funnel-shaped chest, long arms, and long and curved finger and toe bones of *A. afarensis* suggest the possibility of arboreality, as does the somewhat upwardly oriented scapular head known for other australopiths. In addition, if the reconstructed range of habitats available to early hominids is correct (e.g., lake and river margins, woodland, bushland, and savanna) (Potts 1998), it is likely that the same species of hominid would probably have been able to travel on or use a variety of substrates, and thus would potentially have had access to a diversity of food resources. Consequently, it is impossible to conceive of a specific "ancestral hominid diet," consisting of the same food sources in the same frequencies and having been collected in a similar way. But we can consider the available clues.

What Did Early Hominids Eat?

Traditionally there have been two basic ideas about early hominid diets: (1) the robust australopiths were the most herbivorous, and (2) compared even to the gracile australopiths, the emergence of *Homo* was marked by a shift to a greater consumption of protein. Regardless of the scenario, australopiths have generally been seen as broad-based herbivores that opportunistically scavenged meat from available carcasses. How does the evidence stack up? Various analyses have been brought to bear on this question.

Lee-Thorp and van der Merwe (1993) analyzed carbon stable isotope levels in the teeth of hominid fossils from the South African site Swartkrans (1.8–1.0 mya) that had been identified as *Paranthropus robustus* and found that, although C_3-based plants were the predominant class of plant, there was also a notable presence of C_4 grasses. This finding does not suggest a specialized diet. In a follow-up study on carbon isotope values in Swartkrans specimens that had been identified as belonging to *Homo,* they found no significant difference between these specimens and specimens of *Paranthropus* (Lee-Thorp, Thackeray, and van der Merwe 2000). It would thus seem that these *Homo,* too, had an unspecialized C_3-C_4 plant diet. Teeth attributed to *Australopithecus africanus* from Makapansgat, South Africa (c. 3 mya), a site far north of Swartkrans, had average $\delta^{13}C$ values (–8.2%) that were similar to those of *P. robustus* (–8.5%); the range, though, was greater (5.7% for *A. africanus,* 3.2% for *P. robustus*) (Sponheimer and Lee-Thorp 1999). Although it is possible that these hominids ate C_4 grasses and sedges, it is at least equally likely that they consumed grazing animals (whose carcasses they probably scavenged).

Morphological and *morphometric* (measurement) comparisons of the teeth and jaws of *Australopithecus afarensis*, *A. africanus,* and *A. anamensis* with those of living and Miocene (10–18 mya) apes suggest the following for the hominids (Teaford and Ungar 2000): The relatively larger surface area of their molar tooth series would have favored crushing foods. Their relatively smaller incisors would have prevented them from eating foods (such as husked fruits) that require incisal preparation. Their thicker mandibular bodies (*corpora*) would have withstood high occlusal loads. In addition, their thick molar enamel would have resisted cracking and abrasion. Putting these attributes together, Teaford and Ungar concluded that these early hominids (spanning c. 4.2–2.5 mya) would have had difficulty dealing with tough foods (such as tough fruits, leaves, and meat), but would have been able to deal with abrasive as well as nonabrasive, hard as well as brittle, and soft foods. If chewy "meat" is excluded from their dietary repertoires, then the $\delta^{13}C$ values determined for specimens of Makapansgat *A. africanus* and Swartkrans *Paranthropus robustus* and *Homo* indicate that these hominids ate grasses and sedges directly, or insects, or both. Because stone tools older than 2.5 mya are not yet known, it might have been difficult for *A. africanus* to detach manageable pieces of meat from available large-mammal carcasses, but this would not have been a problem later on, when both *P. robustus* and *Homo* probably used the stone tools found at contemporaneous sites. There is, however, some evidence for consumption of large mammals as well as of invertebrates.

Analyses of wear patterns on the abraded ends of bone and horn-core (what is preserved of the horn of an animal) tools from Swartkrans, as well as from other South African sites (Sterkfontein and Drimolen), indicate that while some tools were probably used for digging up tubers (Brain and Shipman 1993), others were used for digging into termite mounds (Backwell and d'Errico 2001). The latter might seem surprising, but African termites do eat grass, which is a C_4 plant. As such, these insects could be considered "bite-sized" grazers. We do not, of course, know which hominid used the tools. However, given the carbon isotope data, it would seem that *Paranthropus* and *Homo* at Swartkrans and Drimolen, as well as *Australopithecus* and *Homo* at Sterkfontein, all ate tubers and termites.

In the ancient lake region of the site of Bouri, Ethiopia, in addition to the fossil remains of *Australopithecus garhi* (Asfaw et al. 1999), stone tools and some number of apparently butchered large-mammal bones were also found (de Heinzelin et al. 1999). At 2.5 mya, this is the earliest evidence of stone-tool production and the oldest indication of hominid modification of animal bones (two other sites in East Africa—Koobi Fora, Kenya, and Olduvai Gorge, Tanzania—provide evidence that is more than half a million years more recent). Cut marks on a mandible and tibia of an antelope and a femur of a primitive species of horse probably reflect dismembering and defleshing, while percussion pits (no doubt from a hammerstone) indicate (as at Koobi Fora and Olduvai Gorge) that bone was fractured for its marrow (a concentrated source of calories). Because, as published, *A. garhi* compares well in dental and facial features to the australopiths that Teaford and Ungar (2000) analyzed, it appears that not only *A. garhi*, but also many other species of early hominid, consumed the meat of large mammals (but with what frequency is not known).

These discoveries and analyses in part support Milton's (1999) argument that, even though early humans may have had the gut anatomy and digestive *kinetics* (motions or activities) of herbivores, they would have supplemented their diets with meat-derived proteins. Although Milton suggested that such a dietary shift occurred with the emergence of early *Homo,* a tendency in this direction seems to have been more ubiquitous among early hominids. Thus, although Aiello and Wheeler (1995) argued that increased consumption of animal products was essential to the evolution of the large hominid brain, these sources of protein were clearly part of some, if not all, early hominid dietary regimes millions of years before brain size increased significantly.

What Can We Say About the Diets of Fossil *Homo*?

Although australopiths and other early hominids may have been distributed over great expanses of the African continent, they were still African-based hominids. It seems, however, that just under 2 million years ago at least one species of *Homo* migrated out of Africa into Eur-Asia. The discovery of a unique collection of hominid crania and jawbones from the Central European site Dmanisi (c. 1.8 mya), in what is now the Republic of Georgia, certainly attests to early human migration from Africa. The fairly complete skeleton of a 9-year-old boy from the site of Nariokotome (c. 1.5 mya), Kenya (Smith, in Walker and Leakey 1993) presents a body form like ours that would have been committed anatomically to striding locomotion: good for steadily, if not slowly, traveling long distances at a time (Ruff and Walker, in Walker and Leakey 1993). The long, lanky body of this Nariokotome youth, possibly as tall as 6' 1", suggests a population, like the Turkana or Masai in modern-day Africa, that was adapted to desertic conditions (Ruff and Walker, in Walker and Leakey 1993). In turn, this suggests that this kind of hominid was not restricted to woodlands and forest fringes but, rather, was capable of traveling across open, and hot, expanses of savanna. In the broader picture, these anatomical and physiological elements, conducive to long-distance foot travel, were essential for hominids being able to leave Africa, perhaps as they followed migrating herds or sought more distant areas in search of food resources.

A few years prior to their discovery of the Nariokotome youth, Walker and his colleagues found another, but much less complete, skeleton from the same time period (c. 1.5 mya). The odd thing about this specimen was that the walls of its bones were thickened and the bone itself looked coarsely woven, which suggested a disease or disorder (Walker, Zimmerman, and Leakey 1982). A microscopically analyzed thin-section of one of the bones revealed pathologically formed bone, as happens when an individual consumes too much vitamin A (Schwartz 1995). One food source that is high in vitamin A is liver. Could this individual, as Walker et al. suggested, have overdosed on vitamin A by eating too much liver? In various human societies, such as among the arctic Inuits, leaders and others deemed important have first access to the best parts of a kill and often choose the liver. Perhaps this ancient hominid had also enjoyed a similarly special status in its group, which came, unfortunately, with certain physiological consequences. There is, however, another possible explanation for *hypervitaminosis A* (too much vitamin A). This individual could have habitually over-indulged on honeybee eggs, pupae, and larvae, which are concentrated sources of vitamin A (Skinner 1991). Although not what one thinks of as "meat" eating, nonetheless the consumption of honeybee broods does qualify as carnivory.

As this East African specimen further suggests, meat eating was most likely part of the dietary repertoire of early hominids, including early species of *Homo*. But how was the meat procured? Curiously, the stone tool (*lithic*) record remains unchanged in Africa until about 1.5 mya, when small, crudely made Olduwan tools begin to be replaced by large, bifacially worked handaxes characteristic of the tool industry archaeologists identify as the Acheulean. But outside of Africa, simpler lithic technology persisted, as is evidenced by crude chopping tools found at a site in Pakistan that is at least 1.6 mya and crude "chopper-chopping" tools found at Zhoukoudian and other sites in China at more than .5 mya. With the Central European Dmanisi hominid fossils being dated to a c. 1.8 mya site and various Javanese *Homo erectus* specimens to more than 1.7 mya, evidence of migration out of Africa clearly predates the 1.5 mya appearance in Africa of a new type of stone-tool technology. Consequently, we must abandon the long-held notion that hominids left Africa as a result of technological innovations and realize that, for at least 1 million years in Africa and longer when one includes Asia, a multiplicity of hominid species went about their business of surviving—acquiring foods of whatever kind, defending themselves—using essentially the same stone (and probably also bone) tool kit.

At about 400 kya, we begin to see evidence of major changes in hominid activity patterns. At the site of Terra Amata in southern France, as well as at the slightly younger site of Bilzingsleben in Germany, hominids apparently built large oval huts from saplings pulled together at the top and also dug shallow fire pits or hearths, which suggests that fire had been domesticated. In addition to providing warmth, these fires might have been used for cooking food, including meat from animals that had been brought down by spears. An amazing set of fully preserved wooden spears, measuring 6 feet in length, were unearthed at the c. 400 kya site of Schöningen, Germany (Thieme 1997). These spears were purposefully carved to be top-heavy, which means they could be thrown great distances with accuracy, probably to bring down large mammals.

Although representatives of the Neanderthal group date back at least to 500 kya, it is with the Neanderthals themselves that we have the most information about diet (see review in Tattersall and Schwartz, in press). In this regard, we might turn to tooth wear, which, uniquely among fossil hominids, is quite profound on Neanderthal anterior teeth. In fact, notable wearing down of the anterior teeth is seen even in juvenile Neanderthals. Typically, the wear that accrues creates a bevel to the outer surfaces of the anterior teeth, particularly the upper incisors. The only modern analog we have of this kind of peculiar tooth wear comes from the Inuits, who use their front teeth not in the consumption of food, but in the preparation of hides. Otherwise, gross cheek-tooth wear was apparently no different in Neanderthals than in Upper Paleolithic *Homo sapiens*. Neanderthals were also similar to arctic-dwelling *H. sapiens* in that they rarely developed dental caries (cf. Schwartz, Brauer, and Gordon 1995; Schwartz and Tattersall 1999; Tattersall and Schwartz, in press). Reminiscent of recent arctic populations, some Neanderthal specimens show significant amounts of enamel chipping (ibid.). The rarity of dental caries probably indicates low levels of fruit-derived fructose in Neanderthals' diet, while enamel chipping reflects a considerable amount of grit in the food they consumed (e.g., plant rootstocks). But the food one consumes, as well as the substances adherent to it, also leave minute bits of information in the form of striations or striae—scratches, but also grooves and pits—that can be used to reconstruct diet.

Microscopic analysis of tooth-wear striations demonstrates that the teeth of nonagricultural, relatively carnivorous modern humans (e.g.,the Inuit) bear fewer striae than more omnivorous modern humans (e.g., the San) (Lalueza, Perez-Perez, and Turbon 1996). The frequency of striae on Neanderthal teeth suggests that they had a meat-heavy diet. $\delta^{13}C$ values in the few Neanderthal specimens so far studied are compatible with the notion that Neanderthals consumed open-field, grazing herbivores (which they procured by either hunting or scavenging). This former inference can also be made with some confidence from the study of stone tools, which, if deposited in the right circumstances (e.g., not too much surrounding acidity), sometimes preserve microscopic elements of the plants and/or animals on which they had been used. A recent investigation of residues on the working edges of Middle Paleolithic, Mousterian tools from two Neanderthal sites in present-day Crimea found evidence not only of mammal hairs, but also, and quite unexpectedly, of feathers from waterfowl (Hardy and Kay, in Tattersall and Schwartz, in press). All of this is not to say, however, that there wasn't any plant food in Neanderthals' diet. At Kebara Cave in the Near East (c. 60 kya), where a partial Neanderthal skeleton was found, remains of lentils as well as of specifically unidentifiable medium- and large-sized legumes were recovered (Lev and Kislev, cited in Bar-Yosef 1995).

A recent study of $\delta^{15}N$ values in collagen taken from Neanderthal bones from various inland European sites suggests that the exploitation of aquatic resources (mammals and birds that prey on smaller aquatic animals, including fish and mollusks, or these aquatic animals themselves) was notably lower in the Neanderthals than in the Upper Paleolithic *H. sapiens* included in the sample (Richards et al. 2001). Based on zooarchaeological evidence, Neanderthals who lived near the Mediterranean did consume shellfish to some extent.

While it is relatively safe to say that dietary information on Neanderthals is, indeed, based on Neanderthals, it is becoming clear that not all fossils that have traditionally been thought to represent *Homo sapiens* are, in fact, *H. sapiens* (Schwartz and Tattersall 2000). In the study cited above that sought to compare tooth-wear striations and inferred diets of Neanderthals with those of *H. sapiens*

(Lalueza et al. 1996), the "*H. sapiens*" sample was mixed and included true fossil *H. sapiens* (e.g., Cro-Magnon 4, Abri Pataud, Qafzeh 9) as well as a specimen that is not. Interestingly, all of the supposed *H. sapiens* had a higher incidence of tooth-wear striations than the Neanderthals, suggesting they consumed a diet with more vegetal matter in it.

Of further interest, regardless of the actual number of hominid species present in the Middle and Upper Paleolithic, is the general observation that, throughout the Paleolithic, all hominids survived by hunting and gathering wild resources (Klein 1999). It is only at about 9–11 kya that we see a major change, which is primarily restricted to the Near East, with the appearance of settlements and the domestication of plants, animals, or both (Bar-Yosef and Meadows 1995). In the Near East, at least, at this time, *H. sapiens* does appear to have been the hominid in question (but, in South Africa, another species existed up until c. 12 kya [Schwartz and Tattersall, in press]). In Chapter 3 we will discuss the shift from hunting and gathering to food production (the Neolithic revolution) in more detail.

A number of authors have suggested that modern humans have a twenty-first-century diet and a Paleolithic body. The examination of the record of human evolution that we have available now, and what we know of other primates, suggests that the likelihood that there is a single, limited "Paleolithic" physiology is incorrect. Paleolithic humans, like contemporary humans, exploited many different types of environments and diets, and probably developed different ways of adapting to those circumstances. This does not mean that new ways of producing and processing food have not had an impact on human nutritional health. In Chapters 3 and 5, we will examine this issue again in light of information on the impact of the Neolithic revolution on food availability and what we can know of diet change from examining the diets of contemporary foragers.

Summary

- As a result of a number of factors, it is difficult to reconstruct the diets of extinct animals, including hominids.
- Even with existing evidence, the job of reconstructing hominid diets is made more complicated because there is still disagreement about how many species of hominid there were at any one time, and it is not always possible to associate hominids directly with tools or the bones of butchered animals. One has to put together the pieces of a jigsaw puzzle, but it is a puzzle that will probably always have a number of pieces missing. At best, one should investigate as many areas as possible and realize that the information gathered may illuminate only a single aspect of an array of possibilities.
- It has been an accident of intellectual history that humans have thought of their own evolution as being essentially the transformation of one species gradually into another, which is unlike the way we think of the evolution of all other contemporary animals.

Like the evolutionary histories of other animals, the evolutionary history of humans was filled with a diversity of species. At least 18 species of hominid are currently recognized in the literature. Of course, as we reviewed, there are many more species that await more widespread recognition—which will happen. There is not, then, a single unbroken line from early hominids to contemporary humans. This complicates our ability to determine how the diets of hominids might have influenced modern human physiology.

- As suggested by both the fossil record and evidence from the biology of development, the origin of species, if recognized by the emergence of novel features, occurs rapidly, not gradually. Adaptation and natural selection relate to the survival, not the origin, of species.

- However, we can speculate with some confidence that, similar to primates in general, hominids, as they still are, were probably very opportunistic consumers and exploiters of their immediate circumstances. Fossil humans, like contemporary primates and modern humans, appear to have consumed a large variety of foods, and changed diet in response to what was available, while not necessarily changing any of the anatomical hardware. One should not confuse an animal's capacity for opportunism with evolutionary processes.

- Just as it is clear that we cannot think of there having been only one hominid species existing at any one time period, we cannot think of there being either a generic "hominid diet" or specific "hominid diets."

- Clues to what hominids might have eaten have been sought by studying
  - Carbon and nitrogen stable isotopes in hominid bones and teeth
  - Tooth enamel thickness
  - Tooth morphology
  - Tooth wear
  - Dietary behavior of other primates
  - Animal bones and shells from archaeological sites

- Clues to how hominids might have been able to exploit their environment in search of food come from studying
  - Hominid cranial and postcranial skeletal anatomy
  - Locomotory and positional behavior of other primates

Web Sites on Human Evolution   A user-friendly Web site was put together by the Institute of Human Origins: www.becominghuman. org. From there, one can access a huge array of different sites providing less or more detailed information.

HIGHLIGHT Lactose Intolerance

Probably the best-known genetic adaptation to dietary factors is the variation in the ability to digest milk among different human populations. In response to the multitude of milk-mustachioed celebrities asking us, "got milk?" many reply, "not milk!"

Mammalian milk contains the sugar *lactose,* which the body breaks down with the enzyme *lactase,* to the simple sugars *glucose* and *galactase.* Only then can it be absorbed into the bloodstream. If people are lactase deficient, the lactose is not spliced into its simple, absorbable sugars and can stay in the intestine, wreaking havoc by fermenting and causing an array of uncomfortable symptoms. This fermentation produces carbon dioxide, acids, methane, and hydrogen. The gas production can lead to abdominal distension, flatulence, diarrhea, intestinal pain, and sometimes cramping, usually occurring 30 minutes to 2 hours after milk intake (Quak 1994). Not all people who are lactase deficient have symptoms; those who do are considered to be *lactose intolerant.*

In human populations, virtually all infants and young children can digest the lactose in milk, an obvious necessity for survival, given that human milk contains 7% lactose (Filer and Reynolds 1997). However, as most children grow older they produce less and less lactase, and between the ages of 3 and 7, most are lactase deficient. However, in a few populations, lactase activity is present in adulthood. These include mainly those of northern European ancestry, along with those descended from people in a few pockets of the Mediterranean, as well as parts of Central Africa.

For years, scientists were unaware of the prevalence of lactase deficiency throughout the world. They were perplexed by the strong preference for milk in some societies and the equally strong aversion in others. Because milk is a relatively inexpensive source of calories, protein, calcium, and other essential nutrients, as well as being a versatile ingredient, it was assumed that everyone would want to drink it.

When relief agencies in the United States, Canada, and other nations donated shipments of powdered milk to developing countries, they were surprised by the response. The people of Colombia and Guatemala used it as whitewash, the Indonesians took it as a laxative, the Kanuri of West Africa believed it was a food of evil spirits, and many groups simply threw it away (Farb and Armelagos 1980). Where the milk was consumed, it often produced uncomfortable symptoms in the very people the relief agencies had intended to help (Nelson and Jurmain 1979).

Some scientists attempted to explain this widespread aversion to milk as the result of some societies' reluctance to take up the strange practice of keeping animals for the milk they produce for their young and drinking it. As Simoons (1983), a pioneer in the study of lactase deficiency, points out:

That explanation did and still does make sense, for, when asked, nonmilking persons would point out that manipulating the udder of an animal was indeed a strange procedure. They also noticed that the end product of milking was a white animal secretion, a substance that when consumed, could make a person ill. It was suspected that such an illness was psychosomatic in origin, an understandably strong reaction to an alien and somewhat revolting food. (pp. 211–12)

Prevalence of Lactose Intolerance

The view that milk aversion is due to beliefs alone was dispelled in the late 1960s, however. At that time, Simoons (1970, 1971) began assembling data on the history and geographic distribution of dairying and milk consumption throughout Asia and Africa. These findings were then compared to the distribution of high and low prevalence of lactose intolerance among the world's people. Societies that have traditionally practiced dairying and milk consumption, mainly Northern Europeans and some herding groups in Africa and the Middle East, were found to have a low prevalence of lactase deficiency, whereas those whose ancestors did not rely on domestic animals and who avoided milk had high frequencies of deficiency (see Figure H2.1).

The incidence of lactose intolerance in various population groups is shown in Table H2.1. Almost 90% of Asian Americans and

Figure H2.1

Shaded areas indicate populations in which older children and adults have difficulty digesting lactose.

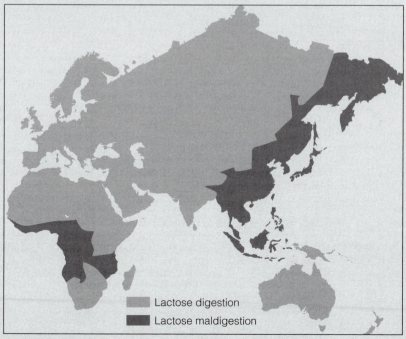

Lactose digestion
Lactose maldigestion

Source: Authors

Table H2.1 Estimated Incidence of Lactose Maldigestion Among Older Children and Adults in Different Population Groups

Incidence of Lactose Maldigestion	
Asian Americans	90%
Africans	70
African Americans	70
Asians	65 or more
American Indians	62 or more
Mexican Americans	53 or more
U.S. adults (overall)	25
Northern Europeans	20
American Caucasians	15

Source: J. E. Brown, *Nutrition Now,* 3rd ed. (Belmont, CA: West/Wadsworth; 2002), pp. 7–11.

autosomal
Any chromosome other than a sex chromosome.

70% of Africans and their descendants living in the United States have difficulty digesting lactose compared to only 15% of Anglo or white North Americans.

Most research suggests a person's ability to digest milk is under genetic control, inherited as an **autosomal** dominant trait (Flatz 1987). But why have some groups inherited a genetic trait that allows them to digest milk as adults, making them different from most of humanity and land animals? One of the best-supported explanations points to the advantages derived from milk consumption. According to this cultural-historical hypothesis, all early foraging populations were lactase deficient. About 10,000 years ago, humans began to domesticate animals, raising them for meat and hides. Domesticated animals offered another resource: milk. However, to take advantage of this new food, people had to be able to produce lactase throughout adulthood (Johnston 1982). At some unknown point in time, a mutation occurred that enabled humans to produce the enzyme lactase and digest milk as adults.

In societies that made use of domesticated animals, individuals with this trait for tolerating lactose had an advantage in being better nourished than those who could not make use of this new resource. As dairying practices developed in some societies, lactose-tolerant individuals survived in greater numbers and gave birth to more offspring who also inherited the same advantage. Through natural selection, the frequency of the genetic trait for digesting lactose increased in dairying societies until the majority of people were able to digest milk throughout adulthood (Farb and Armelagos 1980; McCracken 1971).

Although this hypothesis has received the most widespread support, other explanations for why some people have lactase activity into adulthood have been put forward. In a slight variation of the cultural-historical hypothesis described above, Cook (1978) has suggested that the ability to digest lactose originally evolved among desert nomads in the Arabian peninsula and was selected for because the trait would have helped people absorb the water and electrolytes in milk, an important advantage in a hot, dry climate.

It is important to remember that lactose tolerance is not always predictable. In the Tokelauans of the Pacific, for example, there is a wide range in ability to handle lactose. Those who have some European ancestors, determined by constructing detailed genealogies, are more likely to be lactose absorbers than those who are not (Cheer, Allen, and Huntsman 2000).

Should Milk Be Promoted So Widely?

Between 30 and 50 million Americans are lactose intolerant. The condition occurs in 15% of European Americans, 53% of Mexican Americans, 62% of Native Americans, 80% of African Americans, and 90% of Asian Americans (National Institutes of Health 1994). Given that so many Americans have difficulty digesting milk, it is interesting that milk is seen as a staple food in America and included as a key food in all government dietary recommendations, as

BOX H2.1

Some Definitions

Primary lactase deficiency is the most common type and the one discussed in detail in this highlight. It occurs as a normal physiological process in most mammals in which the production of lactase in the small intestine is reduced between the ages of 3 and 7 years.

Secondary lactase deficiency is usually a temporary condition whereby low levels of lactase occur as a result of an underlying disease that affects the gastrointestinal tract. Even having stomach flu can temporarily decrease the amount of lactase made in the intestine.

Congenital lactase deficiency is a very rare genetic abnormality in which the enzyme lactase is very low or absent at birth.

Milk allergy   Some individuals have trouble digesting milk not because they lack a digestive enzyme, but because they are allergic to milk protein. Their immune system responds to the proteins in dairy products with symptoms including rashes, hives, wheezing, redness, stuffy nose, and runny eyes.

well as being served in the School Breakfast and Lunch Programs and distributed in the Special Supplemental Nutrition Program for Women, Infants and Children Program (WIC), both programs of the U.S. Department of Agriculture.

Some are challenging the long-held belief in the United States that milk is the ultimate beverage. According to Milton Mills of the National Medical Association, an organization of 20,000 African American physicians, the U.S. Dietary Guidelines, although it may be unintentional, are a form of institutionalized racism (Mills 1999). The Congressional Black Congress wrote to President Clinton in 1999 stating that the U.S. Dietary Guidelines "demonstrated a consistent racial bias" (Carroll 2000). The Physicians Committee for Responsible Medicine, a Washington, D.C.-based group that promotes preventive nutrition, alleges racial bias in the formulation of the Dietary Guidelines and the associated Food Guide Pyramid. They charge that American minorities are badly served by the Dietary Guidelines that take little notice of their particular needs and hold the position that the Dietary Guidelines should make dairy products optional (Muwakkil 2000).

But dairy products have much to recommend them. They are one of the most concentrated food sources of calcium, as well as one of the best absorbed. Dairy products provide 73% of the calcium in the U.S. food supply (Gerrior and Bente 1997). When dairy products are eliminated, calcium intake is often compromised. Because of the increased importance of calcium and its relationship to the prevention of hypertension, osteoporosis, periodontal (gum) disease, weight loss, and possibly some types of cancer, maintaining calcium intake is an important goal. The recommendations for calcium intake have continued to rise over the years, with 1000 mg a day recommended for adults 19 to 50 years of age and 1200 mg a day recommended for those over age 50 (see Table H2.2). As Table H2.3

Table H2.2   Calcium Recommendations

Age Group	Daily Recommended Calcium Intake (mg)
0–6 months	210
6–12 months	270
1–3 years	500
4– 8 years	800
9–18 years	1,300
19–50 years	1,000
51–70 years	1,200
Pregnant and nursing women	1,300

Source: Calcium Recommendations from the Institute of Medicine, 1997, from National Digestive Diseases Information Clearinghouse.

shows, it is difficult for most people to get adequate calcium without some dairy products. In addition to calcium, milk is a good source of protein, riboflavin, and vitamin D and has a mild flavor that can be used in a variety of dishes.

A recurring theme throughout this book is that many things are not as simple as they seem. Lactose digestion falls in that category. For most people, lactose digestion is not an all-or-none phenomenon. Recent evidence shows that people with low levels of lactase may be able to drink their milk and tolerate it too. Most people with laboratory-confirmed low levels of the enzyme lactase could drink

Table H2.3   Calcium and Lactose Contents of Selected Foods

Food	Calcium Content (milligrams)	Lactose Content (grams)
Vegetables		
1 cup cooked broccoli	94–177	0
1 cup cooked bok choy	150	0
1 cup cooked collard greens	148–357	0
1 cup cooked kale greens	94–179	
1 cup cooked turnip greens	194–249	0
Dairy products		
1 cup milk (whole, low fat, skim, or buttermilk)	176	6–7
1 ounce processed cheese	159–219	2–3
4 ounces sour cream	134	4–5
1 cup plain yogurt	274–415	12–13
1 cup ice cream/ice milk	176	6–7
Fish/seafood		
1 cup oysters	225	0
3 ounces canned salmon with bones	167	0
3 ounces sardines	371	0

Source: National Digestive Diseases Information Clearinghouse/National Institutes of Health.

one serving of milk with a meal or two servings of milk divided between meals throughout the day without experiencing symptoms (Suarez and Savaiano 1995; Suarez et al. 1997). In fact, only a small fraction of the lactose maldigesters have a complete intolerance. One reason is that many still make lactase in reduced amounts, which allows at least some dairy products to be digested. In addition, there are bacteria in the gut that digest a portion of the lactose, even when no lactase is present. And the more consistently a person eats dairy products, the larger the colony of lactose-digesting bacteria in the gut.

 **H2.2**

How Lactose Intolerance Is Determined

If a person consistently suffers bloating or gas after drinking a small amount of milk, a diagnosis of lactose intolerance can be made without a laboratory work-up. But in cases where it is uncertain, a diagnostic test may be in order. One test involves drinking a large dose of the milk sugar *lactose* on an empty stomach and then measuring the rise in blood sugar. A more sensitive test measures the hydrogen level in the breath before and after a lactose dose. If the person can't digest the lactose, the hydrogen concentration in the breath will increase.

Researchers at Meharry Medical College gave gradually increasing amounts of lactose to a group of African Americans with low lactase levels (Johnson et al. 1993). Each one started out with a cup of milk that had just 5 grams of lactose per cup, compared to the usual 12 grams. If he or she didn't suffer any unpleasant symptoms for 2 to four days, the lactose content was increased by 1 gram for another few days. After 6 to 12 weeks, 75% of the people could drink at least 1 cup of milk without feeling any symptoms.

This reinforces the experience of those who have participated in supplementary feeding programs for children who have restricted lactase activity. In Ethiopia, for example, youngsters who had unpleasant symptoms during the first week of drinking a daily cup of milk tolerated milk with no problems after that point. In Colombia, children who had symptoms when milk was first introduced at their school could handle it with no symptoms 6 months later (Tufts University 1994).

Research shows that sometimes a person may think he is lactose intolerant, not because of an inability to digest lactose, but because of a belief that milk products disagree with him. One study checked for lactose maldigestion in more than 150 African Americans who said that they could not drink a glass of milk without discomfort. Lab tests found that only 58% of them were actually lactose maldigesters. The researchers then gave the subjects both regular and low-lactose milk and didn't tell them which was which. One in three reported symptoms no matter which beverage they drank. For some, the idea of milk made them sick, rather than the actual lactose level (Johnson et al. 1993).

Although adult lactase persistence is the result of genetics, it does not entirely predict the variation in milk drinking in human populations. The great deal of variation seen in milk drinking practices reflects a more complex biocultural interaction than would be implied by a single genetic trait.

Nutritional Implications

Health care workers should be aware that an estimated 70% of the world's population has reduced lactase levels after early childhood. Only 30% of humans retain a lifelong ability to easily digest lactose, including northern European and some Mediterranean and central African groups. This means a high percentage of people from Asian, African, Native American, and Hispanic ancestry may not handle lactose-containing products easily. The range of tolerance level is wide: Some might note symptoms after drinking a cup of coffee with a few teaspoons of milk, and others might not be symptomatic unless they drink considerably more milk. See Box H2.2 for a brief description of some tests used to measure lactose sensitivity. In addition to these physiological considerations, there may be cultural norms surrounding eating dairy products.

There are a number of creative solutions for helping people meet their calcium needs with this challenge involved. One is to encourage eating lots of nondairy calcium-containing food as shown in Table H2.3. Broccoli, kale greens, molasses, almonds, and sardines with bones are all good sources of this important mineral.

Calcium supplementation is another possibility, but there is a wide range in bioavailability of various calcium supplements. The body better utilizes calcium supplements that are coupled with vitamins D and K. Taking the pill with meals also helps with absorption of the mineral. And it helps to spread intake of calcium throughout the day, because the more you take in at one time, the less you absorb. It is recommended that no more than 500 mg of calcium is taken in one dose.

Foods fortified with calcium are another option. In 1999 more than 100 new calcium-fortified foods hit the market. These foods could give the consumer a false sense of security, however, because the calcium isn't always well absorbed from these foods. For example, though some soymilk is calcium fortified, the calcium is not as well absorbed as that in cow's milk. The amount of calcium absorbed from soymilk is typically 25% less than the amount absorbed from cow's milk (Heaney 2000). Components in soymilk are thought to partially block calcium from being absorbed.

As mentioned earlier, most people with low lactase levels can increase their tolerance to milk products by gradually increasing them in the diet, thus increasing lactose-digesting bacteria in the gut. Drinking milk with meals or snacks seems to help people tolerate it better.

Some dairy products are better tolerated than others. Aged or hard cheeses have much of the lactose removed when the lactose-containing whey is removed. In mature ripened cheese, lactose totally disappears within 3 to 4 weeks (McBean and Miller 1998). Fermented dairy products have been used for centuries in some lactase-deficient populations. Yogurt is popular among people of the Middle East. The use of fermented milk products explains why some South African tribes and other groups are cattle herders and milk consumers despite a high prevalence of lactase nonpersistence. Fermentation lowers the lactose level of milk considerably, while leaving many other nutrients unchanged. Yogurts with active cultures provide their own lactose-digesting enzymes and continue to digest lactose even further once in the warm environment of the gastrointestinal tract. The bacteria are inactive or dormant at refrigerated temperatures, so it is best to let the yogurt sit at room temperature about a half hour before eating it to get the lactose-digesting effect. Other fermented products that may be well tolerated include buttermilk, sour cream, and acidophilus milk. A variety of dairy products that are more easily tolerated by those with low lactose levels are pictured in Figure H2.2.

Lactose-free and reduced-lactose milks are now on the market as well. They generally have about one-third of the lactose of regular milk but are almost twice as expensive. The WIC program offers these as a part of their food package to people who cannot tolerate milk. Enzyme drops can be added to milk that is then allowed to stand for 24 hours. These milks are sweeter than regular milk because the lactose is broken into sweeter forms of sugar. Enzyme tablets are also available to be taken by lactose-intolerant people before eating dairy products. They are costly but may be useful in some cases.

The "Other" Milks

Cow's milk is not the only milk. There are many nondairy milk products that are popular in various cultures. Though they have quite different nutrient composition than cow's milk, they are nutritious in their own right. For millennia, the Japanese and Chinese have made milk from soybeans. Recently, soymilk has become popular in the Western world and is available in an array of flavors and convenient cartons. Small machines designed for making fresh soymilk in the home are now available. They grind the beans and filter and brew five cups of fresh soymilk in less than 20 minutes. Soy ice cream ("Ice Bean") and yogurts (soygurt) are available as well.

The American Indians of the Southeast made milk from ground hickory nuts. In the Middle East, sesame seeds are used to make a milk product. Perhaps one of the most delicious nondairy milks is almond milk. This can be made by liquefying 1 cup of raw almonds, 1 quart of water, and 1 tablespoon each of honey and oil in a blender. The milk is then strained and is suitable for the many uses to which other milk is put. Similar milk can be made from cashews.

Figure H2.2

Dairy products generally well tolerated by people with lactose maldigestion.

© Scott Goodwin Photography

CHAPTER 3

Food in Historical Perspective: Dietary Revolutions

In the early twenty-first century, most of us take food production for granted. We can cruise up to a fast-food take-out window, stick our hands out the car window, and have an entire meal handed to us in minutes, without giving a thought to the complex history and systems that produced the food. Only about 1.5% of Americans are full-time farmers. While some suburbanites and city dwellers have garden plots, most people are completely removed from food production and few are aware of the impact agriculture has had on the development of human society. As we shall see in this chapter, the domestication of plants and animals is one of the major cultural revolutions humans have undergone in order to adapt to their physical and social environments. The agricultural revolution of the Neolithic era laid the groundwork for the subsequent industrial and scientific revolutions. These revolutions have increased food production, fostered a global network of food distribution, and reshaped many aspects of modern life.

The Agricultural Revolution of the Neolithic Era

About 12,000 years ago, a dramatic change in the way humans acquired their food began to unfold. In the Middle East, sub-Saharan Africa, Southeast Asia, China, and Central and South America, people gradually shifted from food collection (foraging) to food production. Although the transition took thousands of years (it was more of an evolutionary than a revolutionary process), we consider this a major cultural revolution because of its important long-term impact on how humans interact with their environment.

Before the advent of agriculture most people lived in small groups and moved from place to place in search of food. In general, populations were small. Without substantial food production, the diversity and density of plant and animal resources were such that human groups needed to exploit a relatively large territory. (It should be noted, however, that in a few areas of the New and Old Worlds, foraging groups living on riverbanks and in coastal areas with access to ample shellfish and plant foods were able to develop sedentary lifestyles.)

Nomadism, or moving from place to place daily, weekly, or seasonally, allowed human groups to include a wide range of foods in their diets without depleting any one resource. As people began to domesticate plants and animals, and actively manage the production of their food supply, the food supply became more stable and abundant (although, as we will see, less diverse), allowing humans to settle into villages with an accompanying rise in population density and an overall increase in the total human population. The establishment of large sedentary villages, made possible by agriculture, had significant effects on many other aspects of human life and provides the foundation for the development of civilization.

The adoption of food production—that is, the growing of plants (agriculture) and the management of domesticated animals (animal husbandry)—resulted in what has been called the *agricultural revolution.* The crucial feature of the agricultural revolution was domestication, or control over plant and animal reproduction. Domestication involved the genetic transformation of wild species into domesticated species that flourished under human manipulation through selective breeding. By breeding only those species with desirable characteristics (flavor, yield, ease of cultivation, or, in the case of animals, docility), people were able to mold their crops and animal stock to meet their needs and preferences (see Figure 3.1).

Many questions about how, why, and where plants and animals were domesticated remain unanswered. We now know that contemporary foragers (see Chapter 5) manage the plants and animals in the environments in which they live, though not to the extent farmers and herders do. Agriculture probably started simply enough as people noticed that plants spring up from seeds tossed on the ground. Young animals that were captured alive may have been brought to the home base and became tame, as they grew dependent on people. It is thought that women were responsible for much of the development of agriculture, as they probably did much of the gathering of plants and capturing of small animals and were probably more attuned to the plants in the environment. As we can see from contemporary foragers, women tend to stay closer to the home base than men. Women were in a position to observe the growth of plants from seeds and care for captured animals.

We know that people switched very slowly from harvesting wild species to planting selected varieties. At first, the cultivated varieties served only as supplements to the wild plants and animals they consumed. Through time, people grew increasingly dependent on cultivated plants and animals until agriculture produced the vast majority of foods eaten (Kottak 2000). As shown by evidence from the Tehuacan Valley of Mexico and settlements in pre-Columbian Kentucky (described in the following two sections), the change toward dependence on agriculture was not always swift and, at least in the short term, not always healthful.

Development of Agriculture in the Tehuacan Valley

To illustrate how people shifted from foraging to food production, we will describe the agricultural revolution that occurred in the

Figure 3.1

The wild grasses from which corn originated (A) are far less productive than the corn domesticated 5,500 years ago (B) or modern varieties (C).

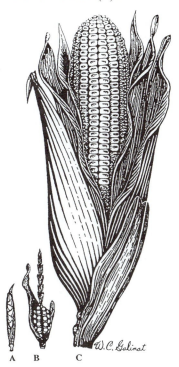

A   B   C

Bering Strait

The 56-mile-wide portion of the North Pacific Ocean that separates Asia from North America (Alaska). At certain times in its geologic past it has been dry land.

riverine

Located or situated on the banks of a river.

mastodon

A primitive elephantlike animal, now extinct.

tapir

A large mammal having short, stout limbs and a long snout.

agave (or century plant)

A plant used for food, fiber, and ornament. Some parts of the plant may be eaten or roasted. The sap from the plant can be fermented to make a mildly alcoholic drink called *pulque*. It is named century plant due to the mistaken notion that the plant blooms only once in 100 years. Actually the plant blooms just once, at the end of its life, which can range from 5 to 35 years.

central highlands of Mexico. We have selected this region because extensive archaeological research has been carried out there, giving us a relatively good understanding of the domestication process.

Archaeological data from human occupation in the New World begins perhaps as early as 25,000 years ago when it is thought that people had the opportunity to cross the **Bering Strait.** Certainly by about 13,000 B.P. (before present) there is widespread evidence of humans in what are now the Americas. Most likely these hunters and gatherers followed large game animals into the new lands, completely unaware that they had crossed into a new continent. As these early bands spread throughout North and South America, they adapted to the diversity of specific environments they encountered. Along coastal and **riverine** areas, fish, shellfish, and other aquatic resources were foraged. Throughout much of the Americas, **mastodons,** mammoths, big-horn bison, camels, **tapirs,** and other large mammals lived in great abundance and supported big-game hunting societies. As these animals became extinct, about 11,000 B.P., perhaps as a result of overhunting or the human impact on the environment as a result of burning, seed collection and smaller game became more important in the diets of the first Americans. In several regions of Central and South America, groups began to domesticate plants and animals. Eventually, agriculture spread to most societies in the Americas.

In Central America, in what is now Mexico, the transition from foraging to domestication of plants and animals began between 7000 B.P. and 10,000 B.P. Caves in the Tehuacan Valley in Central Mexico were inhabited for thousands of years and offer us a rare view of the dietary changes accompanying the shift from foraging to agriculture. Because of the extreme dryness in the caves, remains of many foods and human feces were preserved. The dried feces, called *coprolites,* contained parts of a number of plants, including *setaria* (a kind of millet), cactus, **agave,** squash, gourds, and chili peppers, as well as pieces of charred snail shells, bones from mice, lizards, snakes, and birds, feathers, and fragments of shells. At another Mexican site, coprolites contained bits of uncooked grasshoppers and the shells of snails that had been eaten raw. Analysis of these archaeological finds suggests that the people living in the Tehuacan Valley 10,000 B.P. obtained all of their food from hunting and gathering. Their diet contained a mix of animals and foraged plants, including the wild ancestors of domesticated squash, avocados, corn, and beans.

The first evidence for cultivated plants in this region comes about 7000 years B.P. At this time we find cultivated varieties of squash, corn, bottle gourds, a variety of beans, a fruit called *zapote,* domesticated chilies, avocados, setaria, and amaranth. Diets still contained a large proportion of wild plants, but meat consumption had dropped considerably.

Over the next 5,000 years, the Tehuacan inhabitants relied more and more heavily on cultivated crops, until about the time of Christ, when domesticated plants made up almost their entire diet. Wild plants and animals contributed only a small portion to their nutrient intake. Over time the Mexican diet became even more narrowly focused, with corn, beans, and squash serving as its core (DeWalt 1983).

The rich archaeological record of people living in one area over thousands of years demonstrates the slow transition with which they changed from a foraging to an agricultural existence. The gradual way in which the production of food developed may have been due, in part, to the nutritional risks inherent in an agricultural way of life. Diets of hunters and gatherers include a wide variety of plants and animals and, therefore, tend to be nutritionally well-balanced. Agriculturists, on the other hand, typically rely on a limited number of cultivated crops. If these crops do not contain a balance of nutrients necessary for survival, as is often the case, wild foods must be used as supplements.

In Mexico, full dependence on agriculture had to wait until a group of foods were domesticated that could sustain human populations as adequately as the more traditional diet obtained through foraging. Not until corn, beans, and squash were combined did agriculture adequately meet the protein, energy, and vitamin needs of humans.

Corn does not provide sufficient protein to sustain life. As a staple, it is deficient in lysine and tryptophan, two amino acids that must be present to make up the complete protein essential in human diets. However, when corn is combined with beans (a good source of lysine and tryptophan), together they provide a high-quality protein mixture capable of supporting human populations. Squash seeds also make a good protein supplement to a corn diet. See Focus 3.1 for more information about protein complementation.

Corn and beans were not domesticated at the same time and place in all parts of the New World. Archaeological evidence compiled by Kent Flannery (1971) reveals that wild deer, rabbits, waterfowl, and insects occupied an important dietary role in parts of Central America even after the cultivation of corn appeared. Not until cultivated beans and corn diffused into the same areas did hunting give way to a complete agricultural way of life (Kaplan 1971).

Methods used to prepare corn also may have influenced the rate at which people became fully dependent on agriculture. In this instance, niacin (a B vitamin) proved to be the limiting factor. Corn contains a fair amount of this essential nutrient, but 97–98% of it is in the form of niacytin (Kodicek and Wilson 1959), which is biologically unavailable to the human body. Niacin is an important substance in human energy utilization, and deficient intake of niacin leads to the condition known as *pellagra*. Pellagra, a disease characterized by the "four D's"—diarrhea, dementia, dermatitis, and death—was common among corn-dependent people of the southern United States, the Mediterranean, and Africa until the early twentieth century when the cause of the problem was discovered and niacin supplementation became common. The word "pellagra" dates from eighteenth-century Italy. Crop failures in 1721 forced peasants to rely almost exclusively on corn. Many developed a sickness they attributed to eating corn. Pellagra meant literally "rough skin" (Fussell

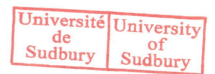

1999). While common in some areas, pellagra has not been a problem in Mexico and Central America (May and McClellan 1972).

Central Americans are protected from this debilitating disease by their traditional methods of preparing corn for use as tortillas. Throughout Mesoamerica, corn is processed by heating the grain in a solution that contains an **alkali.** In Central Mexico, lime is added to the cooking water, while in some other areas, ashes are used. The corn-lime-water mixture is cooked, and, after heating, the product (*nixtamal*) is allowed to stand, usually overnight. The liquid is poured off and the *nixtamal* is washed and made into a dough (*masa*). Formerly, the grinding was done in a stone **quern** (*metate*) with a grindstone called a *mano. (*Today the grinding is usually done at a commercial mill, where the customer is charged for the amount of *nixtamal* that has been ground.) The *masa* is patted or pressed into thin cakes and baked on a clay or iron griddle into tortillas.

Soaking corn in a heated alkaline solution for a sufficient length of time releases all of the bound niacin. Kodicek and Wilson (1959) found that 100% of the niacin in corn tortillas became biologically

alkali
A compound capable of neutralizing acids.

quern
A small hand mill used for grinding grain.

FOCUS 3.1   A Protein Primer

Amino acids are the building blocks of protein. Human tissue contains 22 different amino acids. Of the 22 amino acids, 13 can be made by the body; the other 9 must be obtained from foods we eat and are called *essential amino acids* (EAAs). Our bodies must have all 9 EAAs for maximum protein use.

Protein is found in a variety of foods: meat, fish, dairy products, eggs, beans, grains, nuts, and vegetables. But all proteins are not created equal. Animal foods contain all 9 EAAs and are easily utilized by the body. Most plant foods, however, contain limited amounts of one or two amino acids. For this reason, single-item diets, such as those made up almost solely of corn or yams, can lead to protein deficiency.

But if a diet contains several different plant foods, protein deficiency does not occur. The reason for this is that some plant foods have generous amounts of amino acids that others are lacking. If plant foods are combined, the strengths of one can complement the weaknesses of another and together they make a high-quality protein. This is called *protein complementation*. As long as the protein from plant sources is reasonably varied and there are enough calories, plant sources of protein can provide adequate protein. In addition to plant foods complementing one another, the body has a reserve of amino acids that can be used to complement dietary proteins. The reserve comes from three sources: enzymes secreted into the intestine to digest proteins, intestinal cells sloughed off into the intestine, and a pool of free amino acids, in the intracellular spaces of the skeletal muscle (Messina 1996).

Plant food can be divided into three broad groups based on EEAs' strengths and weaknesses. The groups are (1) whole grains (wheat, rye, barley, rice, corn, etc.), (2) legumes (beans, peas), nuts, and seeds, and (3) vegetables. Vegetables and the legume group generally compensate for the EEAs underrepresented in the grain group. Even within groups, the proteins often complement each other to some extent, because all foods have a slightly different collection of amino acids. For example, legumes complement the protein of nuts and seeds. Dairy products, eggs, and meats can improve the protein efficiency of any of the groups.

Interestingly, before scientists discovered the need for essential amino acids, complementary protein combinations evolved spontaneously as the basis of many cuisines. Examples include Chinese soy products and rice; African sorghum or millet and cowpeas; lentil curry and rice from India; Italian pasta and beans (*pasta e fagioli*); and soup beans and corn bread from the southern United States.

available after the treatment procedures outlined above had been followed.

Lime treatment of corn also significantly increases its calcium content. Untreated ground corn contains about 25 milligrams of calcium per 100 grams while an equal amount of tortillas has about 140 milligrams of calcium. This calcium undoubtedly comes from the lime used in making the *nixtamal*.

Cooking corn in an alkaline solution also affects its amino acid pattern (the ratio of essential amino acids). The overall protein content is lowered, and the amounts of specific amino acids changed, by the lime treatment. For example, 21.0% of the leucine, 18.7% of the arginine, 12.5% of the cystine, and 11.7% of the histidine are lost, in contrast to only 5.3% of the lysine. The proportionally large loss of leucine in corn after lime treatment changes the ratio of isoleucine to leucine, making it more usable by the human body (Bressani and Scrimshaw 1958).

Looking at **ethnographic** data, Katz, Heidger, and Valleroy (1975) found that most Central American groups that rely on corn as a dietary staple use some kind of alkali treatment process in preparing corn for human consumption. In cultural groups that had no alkali treatment, corn made up a minor part of the diet and was heavily supplemented with either game and/or other vegetable foods that would supply the deficient nutrients.

It seems reasonable to hypothesize, then, that reliance on corn as a major staple may have been delayed in the Tehuacan Valley until an alkali treatment process was developed.

Archaeological evidence needed to test this hypothesis is difficult to find. Tools used to grind corn, *metates,* date back to 5000 B.P., before corn became the major dietary staple. However, this does not tell us whether or not lime or ashes were used in processing the corn. By 3000 B.P., tortillas were being baked on clay griddles called *comales*. But clear evidence of lime-soaking pots does not appear until about 2100 B.P. (Katz et al. 1975).

The archaeological record of the shift from hunting and gathering to agriculture in the Tehuacan Valley illustrates two points. First, the transition took thousands of years. Even after people learned to cultivate crops, they continued to forage for wild plants and animals. Second, full dependence on agriculture may have been delayed until the food system could be adjusted to provide adequate nutrition. In Mexico, the combination of crops and the development of food-processing methods may have been necessary before people could give up a foraging way of life.

ethnography
The descriptive study of contemporary cultures.

Nutritional Consequences of the Agricultural Revolution: A Comparison of Foragers and Agriculturists

Most human societies have made the shift from foraging to domestication. Only a few groups, now living in areas of the world that are not suitable for agriculture, still live as our ancestors once did. (The contemporary hunters and gatherers will be discussed in more detail in Chapter 5.)

The agricultural revolution had a profound impact on human dietary practices and nutritional status. As we have already seen, hunters and gatherers generally consume a diet comprising a wide variety of small animals, insects, tubers, seeds, berries, nuts, fruits, and other vegetables. Besides ensuring an adequate supply of essential nutrients, this diversity helps guarantee survival during climatic disturbances that affect the availability of any particular type of food. Times of food scarcity certainly exist, but famine is a relatively infrequent occurrence (and chronic malnutrition, even rarer) among hunters and gatherers.

With the advent of agriculture, the picture changes. Dependence on a small number of cultivated crops or domesticated animals increases the risk of widespread famine. A less diversified diet makes it far harder to achieve an adequate balance of essential nutrients, especially protein and certain vitamins. Vitamin deficiency diseases are especially problematic in grain-dependent communities.

Claire Cassidy (1980) has attempted to assess the nutritional impact of the introduction of agriculture on pre-Columbian Native Americans by examining the skeletal remains of two precontact villages in Kentucky. Foragers inhabited the older village, Indian Knoll, about 5,000 ya. The more recent site, Hardin Village, was inhabited by agriculturalists about 1,000 B.P.

Remains of food refuse at each site suggest very different diets. At Indian Knoll, large quantities of river mussels and snails were consumed. Other meat sources included deer, small mammals, wild turkeys, box turtles, fish, and occasionally dogs. The archaeological remains of wild plants were incomplete, but similar sites contain hickory nuts, walnuts, acorns, elderberries, persimmons, sunflower seeds, and other wild berries (Cassidy 1980). The agriculturalists at Hardin Village relied primarily on cultivated corn, beans, and squash. These were supplemented with deer, eel, small mammals, wild turkeys, box turtles, and wild plants.

Cassidy examined 296 skeletons from Hardin Village and 285 skeletons from Indian Knoll. She summarizes the data on health derived from careful analysis of the bones and teeth as follows:

1. *Life expectancies for both sexes at all ages were lower at Hardin Village than at Indian Knoll.*
2. *Infant mortality was higher at Hardin Village.*
3. *Iron-deficiency anemia of sufficient duration to cause bone changes was absent at Indian Knoll, but present at Hardin Village, where 50% of cases occurred in children under age 5.*
4. *Growth arrest episodes at Indian Knoll were periodic and more often of short duration and were possibly due to food shortages in late winter. Those at Hardin Village occurred randomly and were more often of long duration, probably indicative of disease as a causative agent.*

5. *More children suffered infections at Hardin Village than at Indian Knoll.*

6. *The syndrome of periosteal inflammation [a swelling of the outermost layer of the bone] was more common at Hardin Village than at Indian Knoll.*

7. *Tooth decay was rampant at Hardin Village and led to early abscessing and tooth loss. Decay was unusual at Indian Knoll and abscessing occurred later in life because of severe wear to the teeth. The differences in tooth wear rate and caries rate are very likely attributable to dietary differences between the two groups.* (Cassidy 1980:138)

Cassidy concludes, therefore, that the agricultural Hardin Villagers were less healthy than the Indian Knollers, who lived by hunting and gathering. In Cassidy's opinion, most of the health conditions were related to dietary factors, especially the lack of animal protein in the agriculturists' diet. Goodman and Armelagos (2000) draw similar conclusions from the remains at Dickson's Mounds in Illinois.

Despite a higher incidence of malnutrition and disease in the agricultural population, domestication of plants and animals was associated with population growth. Hardin Village, like millions of agricultural communities, increased significantly, growing from 100 to 300 people over a 150-year period. Thus, although overall health was poorer, food production allowed a much larger population to live together than the previous way of life could sustain (Cassidy 1980).

Social and Political Consequences of the Agricultural Revolution

Once well established, the domestication of plants and animals initiated a stream of changes in social and cultural organization. As we have just seen, farming and animal husbandry provided sufficient food for population growth and the establishment of large, sedentary villages. With this development came the opportunity for increased social interaction (see Figure 3.2). Agriculture also produced surplus food that fostered trade with people in other ecological zones, thereby gaining access to previously unknown commodities. These surpluses also meant that, for the first time in history, not everyone had to be involved in food-getting activities. People began to master ceramics, bronze casting, brick making, masonry, and other skills. Some people also succeeded in trade and the acquisition of surplus wealth, giving rise in many places to an affluent and powerful merchant class.

As population density increased, competition for limited water, land, and other basic resources occurred. The tasks of organizing production, maintaining order in an increasingly complex society, and defending the community were gradually taken over by political/

Figure 3.2
Food surpluses allowed Native American Indians to settle in permanent villages and develop relatively complex social structures.

© 1992 Stock Montage

religious/military hierarchies that led, in turn, to the rise of states and empires. This type of social structure was quite different from the way of life of foragers, who had no specialized laborers and held all possessions in common. Foragers had no class system and, of course, no organized government. But as the agricultural revolution unfolded, people became divided for the first time into rich and poor, rulers and ruled, priests, warriors, and artisans.

The social stratification that occurred with the development of settled agriculture changed humans' patterns from homogeneous ones, where everyone who lived in the same area ate similar foods, to heterogeneous diets, where food patterns of a region varied considerably among people of different rank or status. In ancient Greece, for instance, aristocrats ate more meat and a wider variety of foods than the poor, who ate little more than coarse bread. As a result, skeletons of ancient Greek aristocrats were found to be, on average, 3 inches taller than those of the common folk (Gordon 1983).

By the end of the fifteenth century, agriculture had brought cities and complex political systems to most parts of the world. Trade networks linked many parts of the world and the age of exploration, which would greatly expand these networks, was about to begin. As we shall see in the next section, food played an important role in the political developments of this era.

The Search for Spices

"One teaspoon cumin, one-fourth teaspoon ground coriander, and five turns of the pepper mill . . ." As twenty-first-century cooks measure seasonings into a soup pot, they rarely give thought to the impact these spices have had on the course of history. Though spices are a pleasing addition to some dishes, they hardly seem to be a matter worthy of global concern. But in the fifteenth century, spices were so highly valued in Europe that the quest for them stimulated the exploration of new lands and brought nations into keen competition.

Why this passion for spices? There are several reasons. First, spices preserved foods and disguised the flavor of food partially spoiled (remember, these people lacked refrigerators). But spicing food was more than a mere cover-up; it became a "mania" (Redon 1998). The culinary fashion of the late medieval period, as shown in recipes from this time, called for a great variety of spices used in precise combinations, even detailing the proper moment to add the spices. Moreover, some spices were believed to have curative properties. Spices were also used to make incense for religious ceremonies and believed to improve sexual functioning.

Many of the most desirable spices could not be produced in Europe. Europeans first discovered cinnamon, pepper, ginger, nutmeg, mace, and other delightful spices during their invasions into western Asia during the Crusades. But the origins of these commodities were places far to the east (China, India, and the East Indies), making them extremely expensive—so expensive that pepper was often called "black gold."

Marco Polo's late-thirteenth-century trek and subsequent book helped stimulate the demand for these exotic flavorings. Spice traders could not use Polo's overland route, however, because of the many geographic barriers and growing strength of the unfriendly (to Europeans, that is) Ottoman Turks. For this reason, Muslim traders dominated the spice trade by bringing the goods across the Indian Ocean to as far away as the Mediterranean. From there, Venetian merchants had a monopoly for distribution in Europe.

By the late fifteenth century, some European monarchs were willing to finance sailing expeditions from western Europe to Asia in hopes of cutting out the costly middlemen.

Systematic probing of the west African coast under the guidance of the Portuguese prince, Henry the Navigator, resulted in cargoes of pepper, as well as the valuable knowledge that it was possible, contrary to the speculations of ancient geographers, for humans to survive in the region near the equator. In 1488 Bartholomew Dias navigated his way around the Cape of Good Hope (the southern tip of Africa), and in 1498 Vasco da Gama followed up on this by sailing all the way to India. The Indian ruler of Calcutta refused to exchange spices for the glass beads and woolen cloth da Gama had brought to barter, but he did accept gold and silver. Da Gama's disappointment in having to pay cash for spices was alleviated by the high price he got for them when he returned to Lisbon.

With a direct route established, the Portuguese were in a position to displace the Turks and Italians in the lucrative spice trade. But, within 25 years, other European nations were as dissatisfied with the Portuguese domination of the trade as they had been earlier with the Turks and Italians. Gradually the Dutch, and then others, began to make inroads into the spice trade, setting the stage for increased international competition and the European domination of ever-larger portions of Asia.

Attempting a different route to East Asia was, of course, Christopher Columbus. King Ferdinand and Queen Isabella of Spain financed Columbus' trip westward in hopes that a direct route to the "Spice Islands" would be discovered and be a boon to the royal treasury. In 1492 Columbus reached the Americas (though he maintained until his dying day that he had not discovered a new continent but islands off the coast of Asia); he returned home just about empty-handed. Undaunted, he made three additional trips to the "Indies" only to bring back pineapples that wouldn't grow in European soils and a few other items that were not particularly impressive to the voyage's financers. Although Columbus did not bring the variety and volume of spices the Queen had hoped for, he did change the course of history by opening the Western Hemisphere to European domination.

So, the next time you eat a well-seasoned dish, take a moment to consider the political, geographic, and economic changes the pursuit of spices spurred hundreds of years ago.

The Exchange of Food Between the Old and New Worlds

Although the New World produced few of the spices that helped spur the overseas voyages of exploration, it did contribute many new foods to the Old World diet. For various reasons, the European settlers rejected many of the foods the American Indians enjoyed. In some cases, these foods simply weren't exportable or wouldn't grow well in alien soil. In others, they conflicted with Europeans' cultural sensibilities. Spanish *conquistadors* (sixteenth-century Spanish conquerors in the Americas) rejected Indians' crawling delicacies—ants, spiders, and white worms, newts and iguanas—although Columbus reported that some of his men did find iguana to be tasty (Tannahill 1988). The Indians seem to have been more accepting of European food practices. We have accounts of them heartily eating European prepared meats, vegetables, and even ship's biscuit (a hard, tasteless affair). A group of New England Indians liked everything a ship's captain offered them except the mustard "whereat they made many a sowre face" (Axtell 2001:39).

While not every food item was embraced in the cultural exchange, the story of European settlers colonizing the Americas does illustrate how the diets of different cultures are affected by contact with each other. The exchange of foodstuffs between the colonists and the natives altered dietary patterns radically.

In the early 1600s, English colonists settled the eastern coast of the present-day United States, bringing with them the gastronomic tastes of the British Isles. The colonists had come to a land of plenty. Their new land was well stocked with fish and game, a variety of crops were cultivated by the natives, and they could gather wild mushrooms, cherries, nuts, and berries. Why, then, did some of the new arrivals starve?

First, the colonists did not have much experience making their way in the wilderness. In the first permanent English colony, Jamestown, Virginia, most of the colonists were hoping to get rich quick via precious metals (of which, it turned out, there were none) and were poorly prepared for the hard work and endurance required of them in the New World. The Pilgrims and Puritans who settled New England were mostly middle-class merchants, tradespeople, artisans, or landowners who managed properties worked by hired laborers. They were largely unprepared to reap the bounty of their new homes. Poor timing (the Pilgrims arrived in the month of December) accounted for many of their initial difficulties in obtaining food.

The colonists' survival also was jeopardized by their reluctance to accept new foods available in America. The settlers relied on shipments sent from England rather than collecting fish and other foods that were in abundant local supply. When shipments failed to bring sufficient quantities of food, the colonists began to adopt some of the staples in the native Indian diet. In Jamestown this was done reluctantly. The Englishmen there considered the Indians' corn to be an inferior type of wheat and called it "savage trash." Once faced with eating the alien grain or starving, however, they ate corn (Fussell 1999). As you might expect, people are much more likely to accept new foods if they are similar in some way to the foods they already eat. For this reason, two important Indian vegetables—beans and squash—made their way into the diets of the colonists. The Indian bean looked much like the European broad bean and was accepted readily. Squash was adopted too, for though there were no true squashes in Europe, there were edible plants that resembled them.

A well-known example of the use of native and European foods by colonists is that of Thanksgiving Day, which originated in the festivities held by the Plymouth Colony in the autumn of 1621. On this historic day, about 90 Native Americans shared a bountiful meal with the Pilgrims, who were grateful for the end of a difficult year, a successful harvest, and the Indians' help in learning how to secure food in the new land. The menu that day surely featured the American fowl, the turkey, and American vegetables: corn, squash, and pumpkin. But the pumpkin was stewed in cinnamon of Asian origin. And the table likely also offered Old World carrots, onions, and cabbage (Toops 2000).

Although the colonists grew more open-minded about indigenous foods, they were also working hard to import and develop their beloved European foods: apples, apricots, pears, and a variety of vegetables and beans. The Native Americans were quite receptive to these unfamiliar foods. In a short time after their introduction, apple, peach, apricot, and pear trees were grown by many Native Americans. Vegetables from the Old World were assimilated smoothly into the native bill of fare, partly because they looked similar to Indian foods. Lettuce and cabbage resembled the greens so prevalent in the Native American diet. Lentils were accepted as another kind of bean, while onion, turnips, and beets looked somewhat like indigenous tubers.

The introduction of domestic animals by the Europeans was also accepted and added to the Indians' predominantly vegetable-based diet. When cows came to Florida about 1550, for the first time Native Americans had access to milk and beef. Sheep also became popular among some tribes. But it was the pig that was most readily adopted into the native food system. It has been claimed that many Indians developed "an unhallowed passion for pork" (Tannahill 1988:223).

The most significant American crops introduced to the Europeans were the potato (which originated in South America) and corn. The British use the word "corn" as a generic term for grain. When corn was imported to England, it was referred to as *maize*, or Indian corn. The prolific potato would do much to help feed the rapidly growing British population in the eighteenth century and beyond. And corn, while an important human food in many places, became particularly important as animal feed for the burgeoning farm animal population. By directly or indirectly serving to feed the masses, corn and potatoes helped lay the foundation for the industrial revolution.

Also, the movement of foods across Asia and Europe and then across the Atlantic Ocean (in both directions) reminds us that the process of globalization with respect to food is not a new one. The process clearly started as domesticated plants and animals moved from group to group early in the agricultural revolution. It accelerated as more contact among the continents took place. Humans have had a globalized food system for a long time.

The Industrial Revolution

The *industrial revolution,* the use of innovative technologies and reorganization of the workforce to generate mass production of commodities, was only possible as a result of agricultural change. Once launched, the industrial revolution brought, and continues to bring, dramatic changes in the ways people earn their livelihood, the conditions of their material existence, and the patterns of their daily lives. Not since the Neolithic agricultural revolution had people so drastically changed how they related to the world around them. The agricultural revolution brought farms and villages into existence; the industrial revolution gave birth to factories that mass-produced a seemingly endless array of consumer goods and densely populated industrialized cities. These changes also strongly affected dietary patterns in the new urban centers.

In the late eighteenth century, England began to experience rapid acceleration in manufacturing. This development is usually considered the beginning of the industrial revolution. During the nineteenth century, other European countries and the United States began to move rapidly toward industrialized economies. Throughout history, new discoveries had always stimulated other discoveries and developments, but this process, for the most part, had been slow. What is so striking about the industrial age was the speed of change: The rapid technological advances set off a landslide of innovation, some technological, others social and political. But the industrial revolution could not have been launched without changes in food production that made it possible to feed a large urban population.

Agricultural Change in England, Seventeenth and Eighteenth Centuries

In Britain, the industrial revolution was closely related to the changes made in farming and livestock breeding practices beginning as early as 1500, but taking hold during the eighteenth century. These changes came about when large landowners sought to make farming more efficient and productive. Throughout the Middle Ages, various arrangements of commonly owned lands, open fields, and semicollective methods of farming prevailed. Although plots of land were owned individually, a single farmer's land was in small strips scattered across the village lands. In such a setting, land was worked and decisions were made communally. Everyone's animals had grazing access to uncultivated areas—woodlots and land left fallow to replenish itself. Although equitable, the system was less than efficient and hindered the introduction of innovations. For example, poor farmers saw little need for machines that could put people out of work.

The picture began to change as early as the sixteenth century when some large landholders advocated enclosures—consolidating and fencing (usually by hedges) their holdings. Enclosures were originally used to increase the profitabililty of raising sheep. By the eighteenth century, enclosure was often done to increase the efficiency of producing crops.

Once a large landholder had his property "enclosed," he was free to strive to maximize profits on his holdings. He was likely to plant new fodder crops like turnips and clover, attempt the improved breeding of livestock, use new methods of crop rotation, and introduce new machinery. Such innovation was encouraged by eighteenth-century reformers Charles "Turnip" Townshend (1675–1738), who gained his nickname for his untiring advocacy of a certain crop, and Jethro Tull (1674–1741), inventor of horse-drawn farm machinery. The "enclosure movement" not only led to vast increases in agricultural production, it also displaced a great number of agricultural laborers. As early as 1516, Thomas More, in *Utopia*, lamented the disruption of rural life due to enclosures to raise sheep: "[These animals] that were wont to be so meek and tame . . . now . . . eat up and swallow down the very men themselves." As time went on, the number of displaced agricultural workers increased. These people often sought work in the rising manufacturing cities. This influx of people provided much of the labor necessary for the industrial revolution to gain momentum.

Food and the Industrial Revolution

The pace of industrialization and urbanization varied from country to country, but by the late nineteenth century most of western Europe and the United States saw tremendous productivity in agriculture and growth of the industrial sector with accompanying urbanization. The shifts in livelihoods and locations were dramatic. In 1500 in England, an estimated 80% of the population worked in agriculture, with most farmers producing only enough to support their household. By 1850 only 20% of the English workforce was employed in agriculture, largely to produce food for the marketplace. In the United States, at the time of the first census in 1790, only 5% of the population lived in cities. By 1920 the percentage had increased to 51% and by the year 2000, 80% were living in metropolitan areas.

In the initial decades of the industrial revolution, urban working and environmental conditions were especially deplorable. Bad housing and poor sanitation were combined with dehumanizing working conditions. After a long, hard day in a noisy, dark, and stifling factory, the working class went home to overcrowded slums. Cooking facilities were limited. The water supply was often contaminated by industrial pollutants and waste materials from humans and animals.

Hundreds of thousands of people died from disease and malnutrition. As students of nutrition know, the two go hand in hand, as an inadequate diet lowers the body's resistance to infection—and an inadequate diet it was. A typical meal for the British working class consisted of tea and boiled potatoes, occasionally supplemented with bits of bacon. Bread, cheese, and porridge were also common foods. A survey taken in London during the 1840s showed that in one working-class district, bread was the only solid food given to children in 17 out of 21 meals in a week (Tannahill 1988). As you might expect, vitamin deficiencies were widespread. The health effects of urbanization can be illustrated by data from the British military. In 1840, while 60% of rural recruits were deemed fit for service, an astonishing 90% of recruits from urban areas failed to make the grade.

Industrialization influenced not only what people ate but also when and where they ate it. Agrarian Europeans had traditionally eaten two meals a day, a large meal at noon after a strenuous morning in the fields and a lighter meal at the end of the workday in the late afternoon. Industrialization introduced a new rhythm to the workday. Because people now worked 10 to 12 hours a day, they consumed a breakfast large enough to provide enough energy for the first half of the day. Lunch, previously food served during a 2- to 3-hour break in the middle of the day, was shorter, less caloric, and eaten at the factory so that the worker could get back to the machines as soon as possible. After a long day at the factory, workers went home and consumed a third meal.

While the working class struggled in the first part of the nineteenth century, the middle class of industrial England increased in number and influence. They included factory owners, bankers, merchants, shippers, lawyers, and clergymen. Members of this more affluent class had access to a much wider variety of foods and a better quality diet—more meat, butter, cheese, a finer bread, and occasional fruits and vegetables.

The Emergence of National Cuisines

As people migrated into cities during the late eighteenth century, regional cuisines developed based on popular cooking styles of the major urban centers. Mass production of foods and advertising elevated some of these urban food styles into national cuisines. For example, the pasta of Milan, Turin, and Rome expanded into an Italian cuisine, while Parisian bread, butter-based cooking, and red wine became a national French cuisine. English fish-and-chips emerged, as did American hot dogs, canned beans, and white bread. By the end of the nineteenth century, most major countries had their own national cuisines, foods that immediately struck visitors as characteristic. The association of certain dishes with nationalities reflected the emergence of the nation-state in the political life of the Western world in the nineteenth century (Gordon 1983).

Transportation, Refrigeration, and Canning

By 1850 it was becoming clear that the ever-increasing number of urban laborers needed to be supplied with more abundant, cheaper food. If not, they would be unable to continue to provide the labor needed for the new industrialized economy. This situation, first recognized in England, was soon noted in other European countries and the United States as well. While the dietary problems the workers faced were due, in part, to the new industrial society, the solutions, too, lay in the new economic conditions. Technological advancements in transportation, refrigeration, and food processing—all capable of being mass produced and widely disseminated—improved the diet of the working class considerably by making food cheaper, expanding the variety of foods available, and keeping foods fresh longer.

Transportation

Before the industrial revolution, most people in the United States and Europe had access to locally grown fresh foods. This changed, however, as cities expanded and it became more difficult to provide food to urban dwellers, who now lived miles from where their food was produced.

Land transport was limited because only small amounts of food and other products could be hauled by oxen or horses over narrow, bumpy roads. The construction of canals greatly improved the situation in some areas, and food prices declined. In the United States, flour, which had sold for $16 a barrel in coastal cities, dropped as low as $4 when the Erie Canal was opened in 1825. Not only did the building of canals promote the exchange of goods, it also encouraged agriculture by giving farmers access to urban markets.

Figure 3.3

Before refrigeration, everyone in a community helped prepare meat. This hog-butchering tradition continues today in French-speaking Cajun communities in Louisiana.

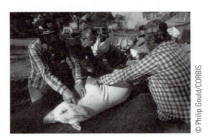

© Philip Gould/CORBIS

With the development of the steamboat and railroads in the early nineteenth century, transportation and food became even cheaper. The railroad not only improved people's diet by increasing the number of foods available, but it improved the quality of the foods as well. Foods arrived at their destinations fresher and thus more nutritious. Previously, meat had been transported on the hoof, often traveling long distances to reach the slaughterhouse. The long hike strengthened the cows' muscles and produced stringy, tough meat. Cattle transported by train were not only tastier, but also cheaper, because they lost less weight on the journey from pasture to slaughterhouse. Pork improved too. Before the advent of the locomotive, pigs had been bred partly for their ability to walk to market. Now that this was no longer necessary, breeders could focus on developing stock that produced rich meat.

Refrigeration

One of the early solutions to keep food fresh in hot weather was the icehouse. Ice was cut from a nearby frozen pond and stored in the icehouse, often lasting through much of the summer. Ice cutting was a difficult and relatively expensive process. An icebox was patented in 1803 but was little used until an efficient horse-drawn ice cutter came into use in the 1820s.

In the 1830s, machines that made ice were patented and soon ice became ever more accessible. Many houses built in the United States in the mid 1800s and the early 1900s featured a small porch built outside so that the iceman could deposit the machine-made ice into an icebox. A household would leave a note indicating the amount needed, and the iceman would slide a block of the specified size onto a thick rubber apron and deposit it into the icebox. A Boston family buying ice in this fashion in 1855 paid $2 a month for the delivery of 15 pounds of ice a day (Root and de Rochemont 1976). Although popular, this method of cooling food was not universal. In the 1920s, an ice industry study determined that about half of urban households had the appliance; surely, far fewer rural homes were equipped (Strasser 1982). Still, even homes without iceboxes benefited from refrigeration because transportation facilities, warehouses, and retail outlets used cooling techniques that affected the quality and choices available in the marketplace. Mechanical refrigeration was patented as early as 1834; however, it was more than a century before the relatively cheap mass production and electrification made the home refrigerator commonplace.

The combination of improved refrigeration and faster transportation greatly expanded available foodstuffs around the world. Beef preserved by refrigeration was shipped from the United States to Great Britain; bananas from the West Indies could now be shipped to Europe; and other fruits, vegetables, dairy products, and eggs were distributed throughout the United States and Europe. Vitamin deficiencies and incidences of food poisoning decreased dramatically as a result of this expanded and well-preserved food supply. As we shall see in Chapter 6, however, a food supply produced far from home is not without its problems.

Canning

Although salting, fermentation, and other methods have been used to process food for many thousands of years, modern food processing began in 1809 when Nicholas Appert invented vacuum-packed, airtight glass bottles for foods. For his invention, Appert won a prize (2000 francs) from Napoleon, who needed to supply food for his armies. Napoleon kept this process a French monopoly by treating it as a military secret until others discovered the same principle.

The next step, using tin cans in place of breakable glass bottles, started in England. By 1818 a canning factory was turning out corned and boiled beef, veal, carrots, and vegetable soup. Though canned food was cheap and convenient, it did not get high marks for taste. Canned meat, for instance, was described as coarse, stringy, fatty, and generally unappetizing. Safety posed another problem. During the initial stages of the canning industry, sterilization processes were poorly understood and the larger cans of meat were often contaminated. Later, standardization of temperatures and processing times made canned food a safer product. By the end of the nineteenth century, canned foods were providing industrialized populations with a diversity of fruits, vegetables, and meats not previously obtainable. As with refrigeration and improved transportation systems, advances in the food-processing industry helped offset some of the negative impact of the industrial revolution on laborers' diets. A final innovation, the mass production of canning jars in the second half of the nineteenth century, allowed for the economical home canning of foods.

Unforeseen Drawbacks of Food Processing

In some cases, technological developments in food processing have been a mixed blessing. Consider condensed milk. When used during the American Civil War, canned condensed milk provided a safe supply of nutrients that soldiers could carry with them. Later, however, some manufacturers began making cheaper brands from skimmed rather than whole milk. The resulting product lacked the fat-soluble vitamins A and D. However, because it was cheaper, consumption of the skimmed product rapidly increased, especially among infants and children. When it was realized that many infants and children raised on this milk developed rickets (though, because the notion of vitamins was unknown at the time, the correlation between skimmed milk and rickets was a mystery), British legislation ruled that condensed skim milk must carry a label stating it was skimmed and not suitable for feeding infants and children. Unfortunately, the mandatory labeling did not have the desired effect. A report published 17 years later revealed that many poorer families continued to use it because of its low cost, and many consumers believed that the labels' warning—"Machine Skimmed"—meant that it was a technologically superior, not a nutritionally inferior, product.

The story of white bread illustrates a common problem encountered with food processing: the removal of nutrients. In 1840 a new method of milling flour was invented, using iron rollers to process grain more rapidly and give flour a more consistent quality. The old mills removed the bran, but not the germ, from the wheat kernel. Although some nutrients and fiber were lost when the bran was removed, much remained in the germ, making flour a relatively nutritious food. In contrast, the new mills removed both germ and bran. This delighted the millers, bakers, and shopkeepers because the white flour did not become rancid as quickly as wheat flour containing the oil of the germ. Whiter flour also appealed to some consumers who liked the lighter baked products it yielded.

Because a somewhat varied diet was available in industrialized countries, the removal of B vitamins from flour did not create an immediately evident nutrition problem. The missing vitamins were often supplied by other foods. But in Southeast Asia and other places where a more limited variety of foods were available, the introduction of a new milling process had a swift and corrosive effect on the health of the people. There, the new mills were used to process rice, the dietary staple. As with flour, the new process removed most of the essential B vitamins. As people switched to polished rice, *beriberi*—the vitamin B-1 deficiency disease that affects the nerves, heart, and digestive tract—swept through the population. But at the time, the cause of the beriberi epidemic was unknown. (See The Discovery of Vitamins, p. 68, for how nutritional science has dealt with these problems.)

While technology advanced in leaps and bounds in Europe and North America, industrialization moved at a snail's pace in other countries. In order to continue to expand industrialization, Europe and America tried to find overseas markets for their products. The nineteenth-century pursuit for colonies was, in part, a search for overseas markets. In order for Western nations to maintain their own markets, industrial development in overseas countries was retarded, in part by denying colonial subjects the technical and managerial skills necessary for industrialization. As a result, large sections of Africa and Asia remained hundreds of years behind Western countries in terms of economic growth.

As "progress" swept across the United States, Europe, and later Japan, other nations lagged increasingly behind. By the close of the nineteenth century, the social, economic, and dietary distinctions between "developed" and "underdeveloped" countries were undeniably clear. (The relationship between the developed and underdeveloped worlds, and its consequences for diet and disease in the modern world, is discussed in greater detail in Chapters 9 and 10.)

The Scientific Revolution

The seventeenth and eighteenth centuries brought with them yet another important cultural revolution that would ultimately have a major impact on agriculture and industry and on dietary practices.

In the seventeenth century, "natural philosophers" such as Kepler, Galileo, Newton, and Bacon rejected the notion that nature was mysterious and capricious. They believed the workings of the world were governed by "natural laws" that are intelligible to humans. And if nature can be understood, its actions are predictable. Natural science, based on the combination of experiment, mathematics, and reason (exemplified by Newton's universal law of gravitation), fostered a sense that humans would one day control nature. As Francis Bacon confidently extolled, "Man is the architect of his future." It would not be until the nineteenth century that science would take a leading role in technology and medicine, and even now Bacon's dictum may seem presumptuous. But it is important to note that the scientific revolution ultimately led to our current level of knowledge about human nutrition and enabled us to exert an unprecedented control over food supply, health, and physical well-being.

Adulteration of Food

The nineteenth century brought significant advances in technology and medicine. One of the first substantial impacts of science on food came when chemists brought popular foods into their laboratories and found, much to the manufacturers' chagrin, that many foods contained questionable ingredients. As early as 1820, a British scientist, Frederick Accum, published *A Treatise on Adulteration of Food and Culinary Poisons*. This book revealed that commercial breads were laden with alum, poisonous salts of copper and lead gave candies their bright colors, and cheeses were laced with red lead to give them an appealing orange color. Accum's work created a public stir, but not enough to drive the Parliament to action. But 30 years later, when a newly created "Analytical and Sanitation Commission" issued similar findings, the Parliament responded by passing the first British Food and Drug Act (Tannahill 1988).

In the United States, the government was even slower to follow up on scientific findings. When Dr. Harvey Wiley, chief of the Bureau of Chemistry of the United States Department of Agriculture (USDA), conducted experiments in the early 1900s demonstrating that some food additives were dangerous to health, food manufacturers attempted to have him removed from office. Although they were not successful, Wiley later resigned in frustration.

Wiley's reports did not make an immediate impression on the public or on federal legislation. Some consumers had a difficult time believing that their favorite foods were dangerous when labels and advertisements promised otherwise. But various "muckraking" writers continued to advance the message that the food supply was cause for concern. The dramatization needed to alert the public to the danger of adulteration came in 1906 with the publication of Upton Sinclair's book, *The Jungle*. Sinclair's book described in graphic detail Chicago's meat-packing plants as filthy, rat-infested buildings where spoiled meat was chemically treated and handled by tubercular

Figure 3.4

The Federal Meat Inspection Act of 1906 required that all interstate beef be inspected. USDA inspectors look for signs of spoilage and contamination with E. coli.

Courtesy U.S. Department of Agriculture

workers. The book, read widely throughout America, created a public protest. Under the spur of aroused public opinion, the U.S. Congress passed two measures to protect consumers. One arranged for regular meat inspection services; the other established the Pure Food and Drug Act of 1906, designed to prevent the "manufacture, sale, or transportation of adulterated, misbranded, poisonous or deleterious foods, drugs, medicines and liquors" (Robinson and Lawler 1977:274). (See Figure 3.4.) It should be noted that, despite great strides in ensuring safe foods, the meat-packing industry is still far from perfect. The dangers of contracting *Salmonella* poisoning from poultry and several deadly outbreaks of *E. coli* poisoning from ground beef show the need for further efforts toward food safety. As a result of a 1998 listeriosis outbreak caused by tainted hot dogs and lunch meat that killed at least 15 people and caused several miscarriages, the Sara Lee company agreed to give $3 million to finance university research projects on food safety (Zimmerman 2001). A best-selling book in 2001, *Fast Food Nation,* has done much to advance public awareness of the potential dangers in our food supply. The book has been likened to Upton Sinclair's muckraking classic of almost a century before.

Food Preservation

Food preservation also owes much to scientific discoveries. Although Appert's 1809 innovation in canning was significant, Appert did not have a clear idea as to why his method worked. And much of what was canned in the early nineteenth century was of questionable quality and safety. Not until the 1860s, when Louis Pasteur laid the groundwork of the science of microbiology, could food safety be better accomplished. Pasteur discovered that it was microorganisms that caused the spoiling of wine, beer, and milk. And he found that by heating and re-cooling the liquids he could preserve freshness—a process we know as **pasteurization.** From this scientific breakthrough came the ability to identify dangerous microbes in many food products and then devise methods to prevent contamination or destroy the organisms. The great variety of foods, found in an array of styles—frozen, canned, dried, etc.—on store shelves, are reasonably safe because of the work of Pasteur and his successors.

pasteurization
Partial sterilization of a substance achieved by holding it at a high temperature for a short time to destroy objectionable microorganisms without significantly changing the chemical makeup of the product.

The Discovery of Vitamins

It has long been known that relatively small amounts of certain foods had positive effects on health. For example, since the mid-eighteenth century, people have known that seamen could avoid getting *scurvy* (a debilitating and often deadly diseased caused by a vitamin C deficiency) by having access to citrus products. Most European navies gave sailors a daily ration of lemon juice. The British sometimes used the cheaper lime juice, giving rise to the nickname "Limey" to refer to British sailors (Tannahill 1988). But even then it was not clear what made citrus so beneficial.

By the mid-nineteenth century, through the work of European chemists, the foundation of modern nutrition science was laid when it became possible to classify foods as carbohydrate, protein, or fat based on their percentages of carbon, hydrogen, and nitrogen and when it was recognized that all three classes of foods were needed in the human diet. Soon the importance of minerals also became apparent. Still, the understanding of a fundamental component of a healthful diet—vitamins—was missing. The story of their discovery is an interesting one.

As noted earlier (p. 66), a high incidence of beriberi was reported in Southeast Asia in the late nineteenth century. The colonial authorities in the Dutch East Indies (Indonesia) became so concerned about the number of deaths that they dispatched a research team to determine what could be done. The findings of this investigation led to one of the greatest scientific discoveries in the history of nutrition: the identification of vitamins.

For two years, the Dutch researchers concentrated on the possibility that beriberi was an infectious disease. The year was 1886, only a short time after Pasteur and Koch had shown that tuberculosis, anthrax, and other diseases were caused by microbes. These discoveries had astounded the world, and it was only natural that the researchers should think in terms of microbes and infection. Scientific workers often deplore the limited amount of funds available for research, but for Dr. Christian Eiijkman, it turned out to be a blessing in disguise. The hospital at which he was studying had such limited funds that the experimental hens had to be fed with leftovers from the ward kitchen. Eiijkman observed that the hens fed on a diet of polished rice developed an inability to walk and exhibited other symptoms similar to beriberi. At first he thought the "beriberi germ had infected them," but then he noticed that the hens recovered promptly when fed rice bran. This information was soon applied to humans, and hundreds of patients who had dragged themselves into the Buitenzorg beriberi hospital on swollen legs were able to walk out only a short time later, fully recovered.

It was not until 1901 that Eiijkman identified the importance of the rice germ within the bran, and even then he had not isolated what it was in the germ that had the healing effect. But his work led others to seek a new category of food components so essential to health and survival (see Figure 3.5). (His work also netted him a Nobel Prize.)

In 1913 American biochemists isolated what would become known as vitamin A. Over the next few decades, biochemists isolated and chemically identified the various vitamins most of us are so conscious of today. Once scientists synthesized these food components, they could be added to milk, breakfast cereals, and breads in hopes of diminishing vitamin deficiencies. The first half of the twentieth century is referred to as the "golden age of nutrition," when vitamins were isolated and linked to deficiency diseases. Understanding the properties of vitamins in food has paved the way for scientists today as they continue to fine-tune the understanding of the health-promoting compounds found in food.

Figure 3.5

Whole-wheat kernel. The whole-wheat kernel has three parts: the endosperm, the bran, and the germ. Most of the vitamins, minerals, and fiber are found in the bran and germ. Because whole-wheat flour is ground from the whole-wheat kernel, it is rich in vitamins and minerals. New milling techniques, developed in 1840, removed both the bran and germ from the whole-wheat kernel, leading to widespread vitamin deficiencies in Southeast Asia.

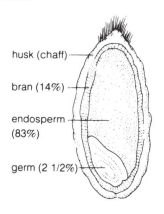

husk (chaff)

bran (14%)

endosperm (83%)

germ (2 1/2%)

Current research suggests that *phytochemicals,* biologically active compounds found in plants, may help slow the aging process and reduce the risk of many diseases, including cancer, heart disease, stroke, high blood pressure, cataracts, osteoporosis, birth defects, and urinary tract infections. Phytochemicals do not fit the classic definition of a nutrient. That is, their absence is not related to an acute deficiency disease and they provide no calories. They are, however, increasingly being viewed collectively as key players in maintaining optimal health. According to the authors of a comprehensive review of phytochemicals, the recognition of the importance of the literally thousands of phytochemicals in foods has opened an exciting era in nutrition research that has been termed the "second golden age of nutrition" (Messina 1996).

Complicating Factors Associated with Modern Food Technology

Throughout the twentieth century, technological know-how and scientific knowledge have increased steadily in all industrialized nations, and this has had tremendous impact on the amounts and types of food available to people in industrialized countries. Chemical fertilizers and pesticides mass-produced in factories and applied by farm machinery likewise mass-produced in factories have created farm yields far beyond what Jethro Tull and "Turnip Townshend" could ever have imagined. Transportation and communication technologies based on scientific knowledge have fostered what could be construed as a new revolution—globalization. *Globalization* refers to wide distribution of these food products around the world, often after they have been modified in some way for purposes of preservation. Thus, the ongoing and intertwining agricultural, industrial, and scientific revolutions have created a global system that makes possible the most abundant, reliable food supply ever known to humankind.

Unfortunately, this bounty is not without drawbacks. Some of the chemicals that facilitate such fantastic yields may be harmful to the humans who ingest the final product, as well as to birds, fish, and other animals whose habitats are affected by the chemicals. While farm machinery is efficient at planting and harvesting, it is also the cause of devastating erosion of topsoil resources. Questions have also been raised about new methods to increase milk production through use of synthesized growth hormones and the use of radiation to increase the shelf life of fruits and vegetables. Genetic modification of foods—insertion of genes from one plant into another plant in an effort to increase yield, resistance, and nutrition—has drawn fire because of the potential health and environmental risks, including these new food products causing allergic reactions in some people. (See Highlight 6 for a more detailed discussion of genetically modified foods.)

These concerns extend beyond health and environmental issues. Questions have been raised about the wisdom of farmers increasingly focusing on specialized commodities for a world market rather than a variety of crops for local consumption. Such market focus brings the producers increasingly under the influence of the multinational corporations that dominate world markets. It is feared that globalization will result in the worldwide distribution of wealth that will grow increasingly lopsided in favor of the holders of technological and scientific power. As their economic power grows, it is likely their political power will, too.

Health problems have also been associated with the abundance and convenience of the food supply in industrialized countries. High intakes of refined carbohydrates, salt, cholesterol, and saturated fats have become common, while dietary fiber consumption has dropped. Little physical exertion is required to obtain the abundant supply of calorically dense foods. It is not surprising, then, that obesity, cardiovascular disease, and diabetes are major killers among humans who evolved for millions of years as hunters and gatherers.

Modern-Day Adaptations

In the last 40 years, science and technology have produced yet another potential environmental problem: a food supply containing a large menu of chemical compounds never before ingested by humans. We know little of these chemicals' effects on liver function, nerve tissues, the immunological system, fetal development, or biochemical genetics (Gaulin and Konner 1977).

The questions we must ask ourselves today are "How are we adapting to these new environmental stresses, and what will these adaptations bring for life tomorrow?" (See Focus 3.2.) There is no doubt that we are responding to dietary changes on all three levels: genetically, physiologically, and behaviorally. Genetically, this is evident in the different rates of fertility, morbidity, and mortality found among people in industrialized and nonindustrialized nations. We cannot predict the direction or rate of genetic changes brought about by natural selection, but we do know that humankind is still undergoing biological evolution (Vander 1981).

Physiologically, the picture is much clearer. "States of health and disease are expressions of the success or failure experienced by the organism in its efforts to respond adaptively to environmental challenges" (Dubos 1980). The fact that industrialized societies have achieved record lows in infant mortality rates and enjoy a long life expectancy attests to our physiological adaptability, while recent warnings about the sustainability of agricultural methods, the potential carcinogenic effects of food additives, and the ever-growing number of occupational hazards remind us that our physiological adaptability also has limits—limits that are being exceeded by our modern, industrialized way of life.

Behaviorally, we are trying numerous adaptive strategies. In the United States, for example, many chemicals (cyclamates, DDT, the dye Red No. 2, the soil fumigant EDB, and dioxin) are restricted by government agencies. Also, an increasingly large number of us are

FOCUS 3.2   Obesity in Modern Industrialized Societies: Have We Outgrown Our Genes?

If genetic adjustments take hundreds of generations to unfold, how do they keep pace with an environment that changes as rapidly as that found in modern industrialized societies? Unfortunately, they don't. Many dietary practices may simply be too new for our bodies to have adapted genetically. This does not mean that all dietary changes are necessarily bad just because they are new. After all, as we have already seen, our bodies, like those of other primates and our ancestors, are amazingly adaptable and respond both physiologically and behaviorally to many environmental stresses.

Nevertheless, there is evidence that our bodies, prepared by several million years of natural selection that operated when humans obtained their food as foragers, are not designed for many aspects of the modern industrial lifestyle. In stark contrast to the forager mode, the industrial lifestyle provides a continuous supply of calories, little motivation to exercise, a relative lack of fiber, and a preponderance of refined carbohydrates and saturated fats (Gaulin and Konner 1977). This lifestyle, in turn, has been associated with an increased prevalence of obesity, diabetes, hypertension, and cardiovascular disease.

Perhaps the best illustrations of modernization's effects on our health status come from societies that have emerged rapidly from a traditional to a more industrialized way of life and diet. For example, among many North American Indian tribes, obesity and diabetes were extremely rare prior to 1940. But as these peoples shifted from an agricultural to a more industrial way of life and replaced their traditional diet with one abundant in refined carbohydrates and calories, obesity and diabetes became epidemic. While rates are still low in a few North American tribes, obesity and diabetes are now exceedingly common in many groups.

There is strong evidence that obesity has been increasing in epidemic proportions for the past 30 to 40 years. The most extreme rates are found among the Pima of Arizona and the Cherokee of North Carolina, with over 60% of all adults and 80% of those 60 years or older classified as overweight (Caballero et al. 1998; Story et al. 1999). Other populations (Polynesians, Micronesians, and the Aborigines of Australia) also have experienced dramatic rises in obesity, cardiovascular disease, hypertension, and diabetes as they have entered the modern era. (See Highlight 4 for more information on body image, obesity, and eating disorders.)

---

making changes in our personal consumption patterns, switching to organically grown vegetables or passing up most of the highly processed convenience foods that now stock supermarket shelves.

Although it is difficult to predict the direction in which these changes will take us, we do know that culture offers us the most rapid and flexible way to adapt to an ever-changing environment. In the next chapter we will explore the concept of culture and its potential for helping us deal with the nutritional challenges facing us today.

Summary

- Our diets have changed dramatically through history. We have gone from foraging wild foods in the wilderness to shopping in supermarkets for such items as frozen dinners and produce grown thousands of miles away. The chain of events that occurred between these two dramatically different ways of securing food can be divided into three major revolutions: agricultural, industrial, and scientific.
- About 12,000 years ago, a change in the way humans got their food began to unfold as people gradually shifted from food col-

lection to food production. Nutritional status was affected as people switched from eating a wide variety of foraged foods to a small number of domesticated plants and animals. The additional life expectancy and increase in population that are associated with agriculture are the basis of modern civilization.

- During the fifteenth century, the search for spices led Eruopeans to explore new lands. This led to exchange of foods between regions and to diversified diets and to the globalization of food.

- During the eighteenth and nineteenth centuries, the industrial revolution gave birth to factories and industrialized cities. The nutritional status changed for the worse initially as people moved from farms into overcrowded cities.

- Technological advancements in transportation, refrigeration, and food processing gradually improved diets by making food cheaper, expanding the variety of foods available, and keeping foods fresh longer.

- The scientific revolution of the nineteenth and twentieth centuries has led to our current level of knowledge about human nutrition and enables us to exert an unprecedented control over our food supply, health, and physical well-being. Scientific developments in food processing have had both good and bad effects on the food supply. While the discovery of pasteurization increased the safety of food, the milling of grains led to widespread vitamin deficiencies in some parts of the world. The discovery of vitamins paved the way to a new understanding of food and its effects on health.

HIGHLIGHT

Vegetarian Diets: Then and Now

Human populations have sometimes been obliged to eat plant-based diets due to poverty or the scarcity of animal foods. The term "vegetarianism" describes the practice of choosing to abstain from flesh foods because of religious, spiritual, ethical, health, or environmental reasons. Once considered an eccentric practice, vegetarianism is now quite popular and the subject of much scientific research. And while many government publications do not specifically suggest a *vegetarian* diet, they do recommend that the majority of the diet come from plant foods, with smaller amounts of animal foods such as meat and dairy products included.

To gain a better appreciation of this increasingly popular way of eating, we'll consider the history of vegetarianism, why people choose this way of eating today, and the various types of vegetarians. We'll wrap up with a look at the nutritional considerations of vegetarian diets.

A Brief History of Vegetarianism

The most significant examples of a religious basis for vegetarian diets are found in religions of the Asian culture. Hinduism, though not requiring a strictly vegetarian diet, promoted a strong tradition of vegetarianism for more than two millennia. Vegetarianism is more widespread in Buddhism and Jainism, which promote nonviolent treatment of all beings and so prohibit the killing of animals for food (Whorton 2000).

Religion has been a component of the history of vegetarianism in the West, but philosophical and scientific influences have been stronger contributors. Vegetarian doctrine in the West can be traced to Pythagoras, the Greek natural philosopher of the sixth century B.C. (known for the geometric theorem). He believed that human spirits were reborn in other creatures and so animals were as worthy of moral treatment as humans were. Killing an animal was seen as murder, eating it was seen as cannibalism (Whorton 2000). In the fourth century B.C., Plato wrote in *The Republic* that a vegetarian diet was best suited for the ideal society because plant foods were better for health and required less land to produce than animal foods.

In later antiquity, during the first two centuries of the Christian era, Ovid and Plutarch both described slaughtering animals for food as cruel treatment of innocent animals. Plutarch's essay "Of Eating of Flesh" added health arguments, stating that flesh foods caused "grievous oppressions and qualmy indigestions" and brought "sickness and heaviness on the body" (Plutarch 1998:14). Porphyry, of the third century, promoted a flesh-free diet for spiritual purification (Porphyry 1965).

With the Fall of Rome and the spread of Christianity across Europe, vegetarian ideas became less pronounced. Christian thinkers such as Saint Augustine of Hippo (354–430) and Saint Thomas Aquinas (1225–1274) proposed that only people had free will, rationality, and immortal souls and that animals were placed on the earth for the use of humans. Using Genesis 1:28 as a guide, they saw meat eating as the right of humans: "God blessed them and said to them 'Be fruitful and increase, fill the earth and subdue it, have dominion over the fish in the sea, the birds of the air and every living thing that moves on the earth.'" Vegetarianism was practiced during these times in some Christian abbeys, however.

Both medical and moral arguments were central to vegetarian philosophies throughout the 1700s. Thomas Tyron wrote the most broad-based argument to date for a vegetable diet in *The Way to Health, Long Life and Happiness* (1683). He quoted Genesis 1:29 to support vegetarian diets: "I give you every seed-bearing plant of the whole earth and every tree that has fruit with seed in it. They will be yours for food." Tyron also raised the issue of health to a new level, noting "Nothing so soon turns to putrification as meat" (Tyron 1683:376).

A vegetarian "renaissance" unfolded in the nineteenth century. This new view was woven into the American humanitarian reform movements. It was not uncommon for vegetarians to be involved in other issues of social change such as antislavery, women's rights, and utopian socialism. Sylvester Graham (as in the cracker) was a Presbyterian minister, turned temperance lecturer, who promoted an all-inclusive program of physical and moral reform that included vegetarian practices (Whorton 2000). Later the term vegetarian was coined, originating from the Latin word *vegetus*, which means "vigorous" or "active." The term has led many to believe that vegetarians eat only vegetables, an inaccurate notion.

The formation of the Bible Christian Church, founded by William Cowherd in 1809 in England, is said to be the start of the organized modern vegetarian movement. Followers believed that a vegetarian diet kept one in better health so as to be better able to serve God. They formed the Vegetarian Society in 1847. Cowherd gave lectures and distributed pamphlets that taught that eating plant-based diets would lead to universal brotherhood, increased happiness and a more civilized society. This group still exists and is now called the Vegetarian Society of Great Britain.

In some people's minds, Darwin's theory narrowed the distance between humans and animals and so disrupted the religious and philosophical defense for humans to eat meat. The vegetarian movement continued to build in the twentieth century with proponents such as Seventh-Day Adventists, George Bernard Shaw, and Mohandas Gandhi. John Harvey Kellogg (the man we have to thank for cornflakes) headed the Seventh-Day Adventist Battle Creek Sanitarium where he invented nuttose, the first meat analog made from peanuts and flour. Visitors to his sanitarium included Admiral Richard Byrd, John D. Rockefeller, J. C. Penney, Mrs. Knox of gelatin fame, Alfred duPont, Montgomery Ward, and Thomas Edison (Carson 1957).

Upton Sinclair wrote *The Jungle* in 1905, featuring graphic descriptions of the unsanitary conditions in slaughterhouses, which brought more people into the vegetarian movement. A popular restaurant in New York City at the turn of the century was the Physical Culture and Strength Restaurant, which featured vegetarian food (Messina 1996).

By the mid-twentieth century, with the discovery of vitamins and government-sponsored food guides, meat-based diets were promoted as the most healthful. All food guides developed at the time encouraged generous helpings of both meat and dairy products. Nonetheless, in 1942 the Gallup Poll showed that 2% of the U.S. population, or 2.5 million Americans, were vegetarian. In 1944 the term *vegan* was coined for people who ate no meat, dairy products, or eggs, and the Vegan Society was formed in Great Britain the same year (Messina 1994).

During the social upheavals of the 1960s and 1970s, a variety of influences converged that had major effects on the course of vegetarianism. These included an increased understanding of the relationship between diet and health, an interest in Eastern philosophy and religion, anti-war sentiment, and the environmental and animal rights movements. Out of this stew of social change, the modern era of vegetarianism grew.

There were two books published during this era that had major impact on increasing the popularity of the practice of vegetarianism. The 1971 publication of Frances Moore Lappe's book, *Diet for a Small Planet,* linked diets to global concerns and focused attention on the negative effects of meat production on the planet. The discovery of the meat, dairy, and egg industries' harsh treatment of animals described in Australian philosopher Pete Singer's 1975 book, *Animal Liberation,* galvanized the animal ethics movement.

The number of vegetarians in the United States doubled between 1985 and 1992. A 1992 Gallup Poll showed that 12 million Americans identified themselves as vegetarian. Nearly half had been eating this way for 10 years, and a quarter had been vegetarian for more than 20 years (Messina and Messina 1996). As throughout history, vegetarianism in this age is often part of an overall worldview.

Why People Adopt Vegetarian Diets Today

Knowing why people choose a particular type of diet is useful to health care practitioners in providing sensitive and culturally appropriate care. Vegetarians are somewhat varied in outlook and background—and those who call themselves *vegetarians* have a wide range of diets. One study found that vegetarians are more knowledgeable about nutrition than meat eaters. Another showed that vegetarians had read on average six books on the subject of nutrition, compared to two books among nonvegetarians. Vegetarians tend to have a strong internal locus of control when it comes to health, agreeing with the statement "How healthy you are mostly depends on how you look after yourself" more often than meat eaters do (Messina 1996).

The three main reasons for choosing a vegetarian diet today are health, animal ethics, and environmental issues. Other reasons include world hunger, religion, and aesthetics. In the following sections, we include the words of real-life vegetarians gathered in surveys of Western vegetarians and collated in the book *The New Vegetarians* (Amato and Partridge 1989). This will hopefully expand your understanding of this diverse and growing group and be of help if you find yourself working with people who practice vegetarian diets.

Health

Health is one of the most common reasons people give for following a vegetarian regime. In one survey, 46% of vegetarians interviewed gave health as their main motivation (Kim et al. 1999).

People eating vegetarian and semivegetarian diets have lower rates of the chronic diseases that typically plague Western countries including heart disease, obesity, hypertension, diabetes, cancer, osteoporosis, and gallstones (White and Frank 1994). Observations of people living in Western countries and developing countries support this conclusion. Migration studies also reinforce the overall health benefits of plant-centered diets. When people move from countries with semivegetarian diets to countries with animal-based diets, their incidence of the aforementioned diseases increases. Also, when people in developing countries become more affluent and start eating more animal foods, their rates of chronic disease rise as well.

Vegetarians tend to be more likely to have a cluster of healthy behaviors, such as regular physical activity and abstention from smoking and drinking alcohol, than meat eaters do. These behaviors may be, in part, responsible for the lowered disease rates of vegetarians, but diet is clearly an important contributing factor.

In general, vegetarians eat slightly less overall fat than omnivores, but considerably less saturated fat and cholesterol. Vegetarians consume 50–100% more fiber than nonvegetarians, helping to explain their reduced risk of colon cancer. In addition, vegetarians take in more of the three primary vitamin antioxidants, beta-carotene and vitamins C and E, as well as more phytochemicals, which appear to be protective against chronic disease. According to the authors of a comprehensive book on phytochemicals, "Vegetables and fruits contain the anticarcinogenic cocktail to which we are adapted. We abandon it at our peril" (Steinmetz and Potter 1990:427).

But it should be noted that just because someone is vegetarian doesn't guarantee that he or she eat healthfully. Some vegetarians just leave out varying amounts of animal foods and load up on foods with little nutritional value, not taking care to eat a balance of foods. Like all healthful diets, vegetarian diets need to follow the fundamental rules of nutritional science. And because meat is an excellent source of many vitamins, minerals, and high-quality protein, those who avoid it may not meet their nutritional requirements unless they are careful to eat a variety of foods. In places where food is limited in amount or variety, consumption of some meat can greatly improve nutritional status, particularly in children.

Animal Ethics

For some people, the suffering of animals is the central or even the only issue for adopting a vegetarian eating style. The same techniques that make for an efficient and inexpensive meat and dairy supply are challenged by the animal ethics movement. According to the book *Deep Vegetarianism,*

> *The special wrongfulness of factory farming lies in its typically brutal procedures which include lifelong confinement in artificial, barren environments; debeaking to prevent stress-induced cannibalism; separation of veal calves and piglets from their mothers in early infancy; crowded cages; social isolation; suppression of grooming and other species-specific behaviors; and cruel transportation, holding and slaughtering methods. (Fox 1999:77)*

People who follow vegetarian practices for these reasons are often passionate in their beliefs. According to one vegetarian: " I felt in my heart, mind and soul that eating an animal was totally insane, inhumane, and I knew it was wrong. My conscience told me so" (Amato and Partridge 1989:35*).*

Environmental Issues

Some people come to vegetarian eating practices via the environmental movement. For these people, the main reason to eat in this particular manner is to minimize the destructive effects of livestock production on soil, water, forests, and wildlife. As one man put it: "My concern for the environment is what keeps me going. I'm saying 'no' to the fast-food hamburger chains' destruction of rainforests for grazing land. I'm saying 'no' to the overgrazing of the continent of Africa and resulting starvation of millions and 'no' to human aggression to all living creatures" (Amato and Partridge 1989:49).

The list of concerns that environmental vegetarians have is lengthy. They argue that livestock and agribusiness farming places a strain on natural resources and quality of the environment. One-half of the water used in the United States is used in livestock production. The 2 billion tons of manure produced annually by farm animals contaminates soil and waterways. And from Mexico to Brazil the number-one factor in the elimination of Latin America's tropical rainforests is cattle grazing (Fox 1999).

World Hunger

Some people eat no meat because they see meat consumption in the West as being, in part, responsible for world hunger. They point out that a larger number of people could be supported by using a given area of land to grow crops directly for human consumption than by using the land to grow crops to feed animals. In fact, 240 million tons of grain are fed to livestock in the United States each year, enough to feed approximately 800 million people a vegetarian diet (Pimentel 1997). One vegetarian motivated by this argument described his rationale like this: "I was very concerned with the suffering of others on a world-wide level and felt that my eating of meat contributed to the problems so I had to stop. Feeding the food that people can eat, such as grains and beans, to animals for slaughter is crazy when there are so many people starving" (Amato and Partridge 1989:47). On the other side of the issue, however, are experts who argue that livestock product may even increase the yield of high-grade protein and by-products extracted from renewable resources. (See Chapter 9 for a more detailed discussion of livestock production and world hunger.)

As we will discuss in Chapters 9 and 10, world hunger is an enormously complex problem with many contributing factors that keep food from being equitably distributed. A shift in consumption practices wouldn't necessarily put the foods in the stomachs of those who are hungry. We must therefore be on guard in seeking a one-dimensional explanation of the problem, although it is still viable to identify how meat production contributes to world hunger.

Religion

Religious beliefs or a desire to deepen one's spiritual practice is a motivating force for many practicing vegetarians. Religions that traditionally recommend vegetarian diets include Buddhism, Hinduism (for members of some castes), Jainism, the International Society for Krishna Consciousness, Seventh-Day Adventism, and Mormonism. According to one Mormon: "I first gave up meat because our religion asks us to eat meat only if we need to sustain ourselves. In the Doctrines and the Covenants—which the Mormons consider to be a revelation from Jesus Christ through Joseph Smith—it says that it is pleasing to the Lord to eat meat only in times of cold winter or famine. I interpret that to mean we should eat meat only to sustain ourselves when we can't get other food" (Amato and Partridge 1989:43).

Some people belong to a religion that doesn't formally recommend vegetarianism, but they may follow a meatless diet because it is consistent with their own personal understanding of the teachings. A book titled *What Would Jesus Eat?* (Colbert 2002) recommends that Christians follow a Middle Eastern diet, considered to be one of the most healthful cuisines in the world.

Aesthetics

Another group of meatless eaters give up flesh purely for aesthetic reasons. They find it unappealing or even repulsive. They have often felt this way since childhood, finding the texture and flavor of meat unappetizing. One man said: "I've never been drawn to meat. I have never looked up and seen a rabbit or any other animal and salivated and thought that was what I wanted to eat. It has always been fruit that I found to be appealing, even as a child" (Amato and Partridge 1989:45).

On the positive side, some people like the taste of vegetarian food and are drawn to the practice for that reason. As one survey respondent described, "Part of what sold me is how delicious vegetarian cuisine actually is" (Amato and Partridge 1989:45).

Types of Vegetarians

There is a wide range in the diets of those who call themselves vegetarians. The list below describes the general categories of vegetarians. These forms of vegetarianism are not necessarily observed without exception, but for the most part are upheld. For example, in a study of female physicians who identified themselves as vegetarian, more than half reported eating animal flesh, mainly poultry and fish, at least one time in the month preceding the survey (White, Seymour, and Frank 1999). People also come up with clever ways of describing their personal dietary habits. One author knows a man who terms himself a "pancake-o-vegan" because he eats pancakes made with milk and eggs almost every Sunday morning, but abstains from milk and eggs the rest of the week. Some people call themselves vegetarians, but do eat meat upon occasion.

Lacto-ovo vegetarians eat dairy products (lacto) and eggs (ovo), but no meat. This is the most common type of vegetarian.

Lacto-vegetarians eat dairy products, but no meat or eggs.

Ovo-vegetarians eat eggs, but no dairy products or meat.

Vegans (pronounced vee-gans) eat no dairy products, eggs, or meat and generally avoid honey and leather products too.

Natural hygienists eat plant foods combined in particular ways and fast periodically.

Fruitarians eat fruit, nuts, seeds, and some vegetables.

Raw foodists eat only uncooked nonmeat products.

Semivegetarians include small amounts of meat in their diets, but predominantly eat foods from plant sources.

Nutritional Considerations of Vegetarian Diets

Vegetarian diets have the potential to be very healthful. But because meat is an excellent source of high-quality protein and a number of vitamins and minerals, its exclusion in the diet requires that special effort be made to ensure that nutrient needs are met. In general, the more inclusive the diet, the easier it is to meet nutritional needs. And the more foods excluded from the diet, the more careful the meal planning needs to be (see Figure H3.1). Special care needs to be taken particularly when feeding children vegetarian diets. In situa-

Table H3.1 For Vegetarians: Where the Necessary Nutrients Are

Nutrient	Vegetarian Sources (lacto-ovo)
Protein	Legumes, soy foods, nuts, seeds, whole grains, potatoes, corn (eggs and dairy foods).
Iron	Leafy greens (spinach, mustard, chard), dried fruits (apricots, raisins, prunes), whole grains, oatmeal, beans, soy foods, lentils, blackstrap molasses, pulses (peas, beans, lentils), fortified breads and cereals.
Calcium	Tofu (if processed with calcium), leafy greens (except spinach, beet, rhubarb, chard), broccoli, almonds, figs, blackstrap molasses, sea vegetables, fortified foods (cereal, soymilk, juice), (dairy foods).
Vitamin D	Fortified foods (soymilk, cereal) and (fortified milk). Plus some unprotected sun exposure or rely on a multivitamin.
Zinc	Whole grains, lentils, branflakes, wheat germ, soy foods, sesame and sunflower seeds.
Vitamin B-12	Fortified foods (soymilk, cereal), nutritional yeast (check label), (eggs and dairy foods). Or rely on a multivitamin.

Source: Based on chart from *Environmental Nutrition Newsletter 2000* (New York: Environmental Nutrition, Inc.).

tions where food is limited in terms of quantity or variety, small amounts of meat can boost the nutritional status of children.

Lacto-ovo vegetarian diets that include a variety of plant foods and nonmeat animal foods can easily meet nutritional needs. Zinc and iron are the main nutrients that may be in short supply with this type of diet. *Ovo-vegetarians* need to also consider calcium needs (see Highlight 2 on lactose intolerance for a discussion of meeting calcium needs without milk products). Vegans, who shun all animal foods, need to watch for the nutrients listed for the other type of vegetarian diets above as well as be aware of vitamin B-12 intake. Vitamin B-12 represents a special challenge for vegans. It is mainly found in animal foods, and that which is found in products such as tofu and tempeh is an inactive analog. Long-term deficiencies of this nutrient can lead to nerve damage. See Table H3.1 for sources of vitamin B-12. Although strict protein-combining is not necessary for vegetarians, eating different sources of protein throughout the day—like rice and beans—ensures that essential amino acids needs are met to form complementary proteins.

The *natural hygienist* diet is hard to assess for nutritional adequacy because there is such a wide range of practices. When cooked grains and beans are used, the diet may be nutritionally adequate. The emphasis on raw foods would make the diet problematic for children. *Fruitarians* eat diets based on fruit. Although fruits are high in vitamins, minerals, and phytochemicals, they are devoid of protein and low in important nutrients such as calcium. This diet is generally not recommended, and it is *never* recommended for children. Adherents to *raw-foods* diets believe that this pattern closely resembles the natural eating pattern of humans. While there is merit in eating raw foods to maintain nutrient levels in some foods, cooking improves the digestibility of some other foods. The premise for a raw-foods diet is not well supported, but it can be sufficient if sprouted grains and beans are included. Diets that are 100% raw take considerable planning to be adequate. This diet is not recommended for children either (Messina 1994:13).

Figure H3.1
A comparison of the USDA Food Guide Pyramid and Vegetarian Food Guide Pyramid.

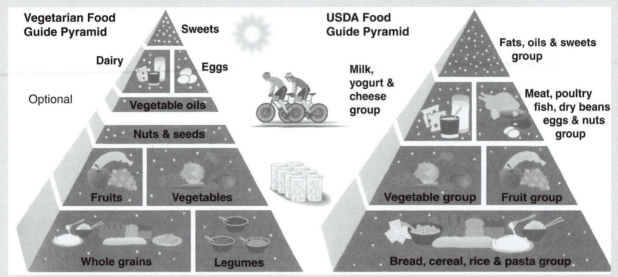

Note: A reliable source of vitaman B12 should be included if no diary or eggs are consumed.

Source: Haddad, et al. Vegetarian food guide pyramid: A conceptual framework. *American Jouranl of Clinical Nutrution,* 70 (1999) 615S–619S. United States Department of Agriculture, 1992.

PART II

Food and Culture

CHAPTER 4

chutney
A condiment that is made with fruits and sometimes added raisins, nuts, onions, and seasoned with spices.

Eating Is a Cultural Affair

If you have traveled to other countries, you probably have noticed that foods considered a delicacy in one society may provoke amusement or disgust in another. For instance, as a North American, you may not share the Iranian's taste for **chutney,** made from fruits allowed to rot in a spicy sauce. When encountering puppy hams and suckling pigs featured in Chinese cuisine, you may fail to appreciate the careful breeding that has gone into making these delicacies. It may surprise you to learn that natives from the Gilbert Islands are equally startled at the idea of eating eggs, and some Brazilian Indians react similarly to milk. To them, the milk of other animals is an excretory secretion as distasteful as urine or feces.

Differences in what is considered edible are well illustrated by various societies' reactions to *entomophagy*—insect eating. Europeans arriving in the New World were shocked by the sight of Indians eating large, fat spiders and white worms. This aversion to insect eating is not widespread; from ancient times to the present, insects have been a valuable nutritional resource, as well as a taste treat, for most of humankind. Insects like those shown in Figure 4.1 can be a valuable source of protein, calcium, and a wide variety of vitamins and other minerals. Most societies that practice entomophagy prefer larger bugs cooked (beetles are shelled, fried, and mixed with sugar to make a sauce; grasshoppers are roasted with soya sauce), but smaller varieties may be eaten raw. In some places, lice and fleas are eaten immediately while delousing friends or relatives. Although most peoples are quite selective about the types of insects they eat, some societies consume just about anything they catch. The complete absence of insects in European and North American diets is unique. (See Focus 4.1.) Europeans and North Americans are not alone, however, in their decision not to use many potentially nutritious resources in the environment. Although *Homo sapiens* as a species has eaten just about everything physiologically possible—including a wide range of poisonous plants, clay, blood, urine, raw as well as rotted meat, and dried cowhide—no group has used all available nutritious substances for food. Everywhere people develop food prohibitions and preferences. As a result, all societies overlook valuable resources, fail to develop technologies to raise nutritious plants and animals, and reject items for ideological reasons. The Turkanoans live in one of the most diverse ecosystems in the world—the Amazon rainforest—but their diet relies heavily on cassava and fish, with 80% of their calories coming from cassava bread (Dufour 2000b). Even foragers living in harsh environments exercise discrimination in choosing what they will eat. The !Kung Bushmen,

Figure 4.1
Insects are a valuable source of nutrients.

FOCUS 4.1   Entomophagy

Insects far outnumber any other animal on earth, in both the number of species (over 800,000) and the number of individuals. Many groups capitalize on this fact and make insects a nutritious part of their diets. In parts of India, for example, insects are eaten with rice, in Burma they are used in stews, and in Egypt grasshoppers are skewered on sticks, roasted over fires, and then dipped in butter for a tasty snack. Some Peruvians use beetles as flavoring, and in Thailand over 50 insect species are consumed; they are roasted, boiled, sautéed, or pounded and added to pastes. The Kiriwinians of Melanesia gather yellow leaf ants and eat them raw, while their Chimbu neighbors in New Guinea stuff caterpillars into bamboo sticks to roast over an open fire (Meyer-Rochow 1973).

As a food source, insects offer many benefits. They are rich in calories and protein, carbohydrate, water, vitamins, and minerals. In terms of fat content, insects are a lot like fish and poultry, with between 10% and 30% of their weight made up of fat. Most edible species are cholesterol free and have a high ratio of polyunsaturated to saturated fats. They also contain relatively large proportions of linoleic acid and linolenic acid, which are both essential fatty acids, and the arachidonic acid precursors for the hormone needed to regulate metabolism in many tissues (DeFoliart 1991; Jonjuapsong 1996). By eating an average of 150 ants per day, the !Kung Bushmen, foragers in Africa's Kalahari Desert, enhance their diet with 60 g of protein, 118 mg of calcium, 48 mg of iron, 142 mg of copper, and 220 mg of phosphorous, as well as vitamins B-1 (thiamin) and B-2 (riboflavin) (Hetzel 1978).

People also eat insects for enjoyment. Sugar ants are a tasty delicacy to Australian Bushmen; the ants' abdomens are swollen with nectar, providing sweetness followed by the sharp astringent taste of formic acid. Termites and roasted grasshoppers are also considered delicacies in some societies.

Entomography offers important ecological advantages, too. Insects are readily available almost everywhere and do not require large rangelands to produce. Compared to beef cattle and sheep, they are very efficient in converting energy into high-protein foods. The Efficiency of Conversion of Ingested Food (ECI) rate is only 5.3 for sheep and 10 for beef cattle. These rates indicate that only 5.3% and 10%, respectively, of these animals' diets are converted into meat. Insects are much more efficient: The ECI of the pale western cutworm is 16, whereas the German cockroach converts 44% of its diet into edible meat (Dunkel 1998).

Many societies also use insects for medicinal purposes. The people of Alagoas, Brazil, use a variety of insect parts and products to prevent and treat disease, including the whole body, nest or cocoon, honey, wax, gut, abdomen, exoskeleton, visceral mass, secretion, **oil scutellum nest,** and legs. For example, Alagoans drink tea made from cockroaches to treat colic, and balding men use a paste made of houseflies to stimulate hair growth (Geraldo, Marques, and Costa-Neto 1997). In Niger, grasshoppers, ants, crickets, bees, butterflies, termites, and houseflies are used routinely to treat a variety of illnesses, enhance energy, increase crop production, and ward off evil spirits (Fasoranti 1997). The Chinese use ants to increase blood circulation and metabolism, bolster the immune system, and treat rheumatoid arthritis (Chen and Ahre 1994). Many promising anticancer drugs and other active pharmaceutical products have been isolated from butterflies, beetles, and other insects (Costa-Neto and Nogueira de Melo 1998).

Because entomography has so many advantages, the almost total absence of insects from Euro-American diets is being challenged. Some colleges and universities now offer courses designed to increase the acceptance of insects as animal and human food (Dunkel 1996). The Audubon Institute and various museums throughout Canada and the United States host insect awareness feasts, festivals, and fairs where people can sample food insects, acquire recipes, and learn about the benefits of entomography. Food insect cookbooks, commercially produced food insect products, and a scholarly publication—*The Food Insects Newsletter*—are now available. Although it is unlikely that insects will become a regular part of our diets, the steady increase in insect consumption among young people during the 1990s suggests that the Western taboo on insect eating may be changing (Dunkel 1997).

For those interested in trying food insects, a final word of caution is in order: As with wild mushrooms, not all insects are safe to eat. Certain mealworms can create dangerous reactions in people who are allergic to shellfish, and monarch butterflies and some beetles contain carcinogens and other poisonous substances. For these reasons, it is best to eat insects known to be edible, avoid raw or live bugs, and follow kitchen-tested recipes (Dunkel 1996).

oil scutellum nest
A massive structure, shieldlike in shape, found inside the nest of stingless bees.

for example, derive 90% of their vegetable diet from only 23 of the 85 available edible species. Of the 223 local species of animals known to the Bushmen, 54 species are classified as edible and of these only 17 species are hunted on a regular basis (Lee 2000).

Culture has a pervasive influence on human diet; it affects what we eat, when we eat, how we prepare our food, and with whom we share it. This chapter examines the concept of culture and the importance of a cultural perspective for understanding human dietary practices. In our exploration of the cultural context of dietary habits, we draw on examples from a wide variety of societies, past and present. In some examples, the customs and beliefs we describe are no longer practiced in their traditional forms. In a few instances, the practices or the societies themselves no longer exist. The ethnographic examples given in this book, therefore, should not be taken as the final word on how these people live today. Whenever possible, we attempt to alert the reader to this by noting briefly how food practices and other cultural traits have changed. However, because we do not always know the extent or nature of change experienced by some societies, we simply refer to their customs as "traditional" or note the period when the observations were made, either as part of the text or as reference citations at the end of the passage.

Some Definitions

When discussing culture, it is important to distinguish between the term "culture," used in the general sense to refer to humans' capacity to create and share meaning, values, and patterned ways of thinking, and the term "a culture," used to describe the set of meanings, knowledge, values, and norms shared by a specific group of people. Culture, without the modifier "a," is used to describe the potential we have to share patterned ways of thinking, feeling, and behaving that arise from social interaction with other people. Culture contrasts with patterned responses that are biologically determined or learned directly from interacting with the natural environment. "A culture" refers to the set of norms, beliefs, values, and other conventional understandings shared by members of a *specific identifiable social group,* such as an ethnic group, class, professional organization, corporation, or discipline (e.g., public health, nutrition, and medicine). The unique combination of cultural elements or traits (e.g., information, meanings, taste preferences, and symbols) transmitted to the group's members comes to characterize it and distinguish it from other social groups. When talking about small groups that make up a larger social unit, some people use the term "subculture" to describe the rules, values, information, and meanings that its members share. If you have ever changed careers or started a new type of job, you may have encountered the frustration and challenges associated with learning a new "organizational" culture (Schein 1992). The transition from high school to university life also requires learning a new culture or subculture.

Culture as a Mechanism for Responding to the Environment

As mentioned in Chapter 1, people live in complex environments that include the physical, social, cultural, and broader geopolitical environment. Culture influences how people respond to the opportunities and constraints these environments pose. Culture influences how a society produces, prepares, and distributes the food needed to sustain growth and bodily functions, enabling it to meet the nutrients needed for human life. In many cases, the cultural solutions that groups devise are quite successful. When people learned to use ice to keep foods cold, the number of food-borne illnesses dropped significantly. Some cultural solutions, however, prove to be maladaptive. The use of **fluorocarbons** for refrigeration, for instance, has contributed to the depletion of the ozone layer and an increased rate of skin cancer. Of course, social groups that fail to develop successful strategies soon die out. The Shakers, a religious sect that prohibited all sexual relations between its members, has not been able to maintain enough members to transmit its culture to a new generation.

As a means of responding to the environment, culture can be divided into three major sections: **technology, social organization,** and **ideology.** Although each of these environmental arenas is interconnected, one affecting the other, it is also useful to discuss each of them separately, noticing how they influence and are influenced by human dietary patterns.

Technology refers to the knowledge, practices, and tools a group has developed to help its members respond to the physical environment and meet their basic biological needs for subsistence, shelter, and protection from natural enemies. A society's technology allows its members to modify the environment and develop other ways to cope with environmental challenges. Technology includes the norms, values, and information a social group shares for producing food and other goods, distributing resources, constructing houses, making clothes and tools, and transporting people and goods. Food processing is another aspect of technology that has an important impact on dietary intake.

Social organization refers to the norms, beliefs, values, and other conventional understandings a group has developed to regulate social life. Social organization deals with the ways humans organize themselves to achieve common goals, including marriage, reproduction, familial relationships, friendships, political alliances, and law and order. In addition to a society's rules about how the social group *should be* structured, social organization refers to how it *actually is* structured. We may speak of a culturally shared definition of the "American nuclear family" as two parents and their children, and at the same time describe the increasing proportion of single-parent households in the United States.

Ideology, the third adaptive mechanism of culture, refers to the shared beliefs and values that enable humans to adjust to the

fluorocarbons
Organic compounds containing fluorine that are used in aerosol sprays, refrigerants, solvents, and lubricants.

technology, social organization, and ideology
The relationship between technology, social organization, and ideology and dietary practices will be discussed in Chapters 5, 6, 7, and 8.

metaphysical
Relating to a reality beyond what is
perceived by the senses.

supernatural and **metaphysical** environment. Everywhere, people
ask questions about the meaning of life, death, and the future. Our
ideologies attempt to provide answers to these questions. They help
define our relationship with the supernatural, and they justify our
particular way of life. Religion, folklore, art, music, literature, and
philosophy are all aspects of a society's ideology.

The relationships between food habits and a society's technology, social organization, and ideology are so strong that many nutritional anthropologists believe that the study of food systems
provides a prism through which they can learn about other aspects
of culture (Goodman, Dufour, and Pelto 2000). In the next four
chapters, we examine the relationship between food practices and
technology (Chapters 5 and 6), social organization (Chapter 7), and
ideology (Chapter 8) and the impact each aspect of culture has on
food habits. First, however, let's look more closely at the concept of
culture and its proper use in understanding human behavior.

Culture Is Learned

Like all people, you are born with the potential to learn any culture,
regardless of your genetic background. Anthropologist Clyde Kluckholn (1968) describes a young man he met in New York

> who did not speak a word of English and was obviously bewil-
> dered by American ways. By "blood" he was as American as you
> or I, for his parents had gone from Indiana to China as mission-
> aries. Orphaned in infancy, he was reared by a young Chinese
> family in a remote village. All who met him found him more Chi-
> nese than American. The facts of his blue eyes and light hair
> were less impressive than a Chinese style of gait, Chinese arm
> and hand movements, Chinese facial expressions and Chinese
> modes of thought. The biological heritage was American, but the
> cultural training had been Chinese. He returned to China. (p. 25)

The process by which we learn the beliefs, attitudes, behavioral
standards, and even body movements of a society is called *enculturation*. In all societies, the family is the major vehicle for training the
young to fit into society. But as many parents know, it is not always
easy to induce children to conform to social conventions. Parents
rely on formal instruction, subtle and not-so-subtle rewards and
punishments, and modeling correct behavior to educate children in
traditional ways of thinking and behaving (see Figure 4.2).

Industrialized societies have formal institutions—schools—to
enculturate their children. This formal schooling is combined with
informal teaching and modeling by relatives, friends, and the media
as a way to motivate people to adopt the society's values and goals.
Many positive characteristics such as pride in oneself, loving others,
and sharing may be encouraged in this way. Unfortunately, sexism,
classism, and other forms of prejudice may be transmitted to children in this way as well.

Because much of our enculturation takes place early in life and the culture we learn is shared by many people around us, certain behaviors may appear automatic or instinctual rather than culturally patterned. Many of the assumptions we learn to make about how the world works and what we were taught to value in life are unconsciously accepted as "just human nature" or the correct way of looking at things. Most of us are not aware that these conventional understandings are a product of our culture—one distinctive view among many. Because we take them for granted, many components of our culture, especially our value orientations and basic assumptions about how the world works, become *implicit*. Many Americans, for example, believe competition is instinctive; they believe it is natural for people who play and work together to compete. A look at some other societies, however, shows us that competition is *not* "just human nature." The Samoans of Malaysia, for instance, are rarely competitive when playing or working together. They value cooperation and look down on competitive behavior.

Culture as a Guide for Behavior

Culture has been likened to a mental map. As members of a social group, each of us uses this map to help us decide how to act and how to interpret what other people do. We are not saying culture determines human behavior or that people blindly follow their cultural maps. As noted in Chapter 1, a wide variety of biological, psychological, and environmental factors, and broad-reaching geopolitical forces, interact with culture to influence human dietary behavior. However, when people's mental maps or cultural understandings are similar, their behavior is more likely to be consistent and predictable, enabling the social group to function more smoothly.

The mental map that we call culture is made up of *norms, beliefs, values,* and other conceptions about what should take place. *Norms* refer to rules, principles, or standards for behavior; they are people's conception of what *should* occur in a given situation. People in each society make choices from a wide range of possible alternatives about how to behave. A comparison of several societies' norms for how male friends should greet each other after a few weeks' separation offers a good example. Such behavior probably takes hundreds of forms in the world at large, but in a given culture the alternatives are few. Irish American men will most commonly shake hands. Two Barabig males (eastern Africa) will clasp hands, then shift the grip to the thumb, and then back to the handclasp, all in rapid succession. Among a neighboring group, men spit into each other's right hand, a norm found offensive in most other parts of the world. In the Middle East, kissing may be more usual than a handshake, even in public situations, while in Japan the usual form of greeting is a bow. Each pattern of behavior reflects a shared rule, a cultural choice among conceivable alternatives. Each represents a cultural norm.

Throughout history, many people have questioned cultural norms and acted differently. In North America, some men have

Figure 4.2
Parents play an important role in teaching their children the society's food practices.

© Myrleen Ferguson Cate/PhotoEdit

rejected the traditional firm handshake as the only appropriate greeting for male friends and choose to hug each other instead.

In the dietary realm, there are many norms for how to prepare, cook, and serve food. In an urban intentional community in North Carolina, a household of 18 people follow many norms regarding food. Meat is prohibited, foods are cooked only in cast-iron cookware (aluminum cookware is prohibited), the person assigned to cook for the week serves food, and when the meal is over, each person washes his or her own plate and utensils. Their neighbors, just three doors down, expect meat at every meal and usually delegate the cooking to the mother and oldest daughter, and the two youngest sons do the dishes.

Even our most intimate bodily functions are governed by cultural norms. "Of all the regulations relating to table manners throughout the ages, that against breaking wind (or passing gas) has had the longest life. Surviving texts are not always specific about which aspects of the subject they are discussing . . . but it is clear that while a delicate burp has usually been acceptable in most societies, the audible release of digestive gases from the nether regions most certainly has not" (Tannahill 1973:231). Societies also share norms about people's appearance. Focus 4.2 explores variation in cultural norms about the body's weight and shape.

Cultures often allow several choices in deciding how a specific situation should be handled. In North America, we have a wide range of options for what is acceptable to serve to dinner guests: hamburgers, tacos, lasagna, barbecued chicken. However, it would be unusual for North Americans to serve dessert first or offer hot oatmeal as an appetizer.

Figure 4.3
Culture also dictates how you eat your food. In many societies, it is normative to eat with your hands.

FOCUS 4.2   Body Imagery: Beauty Is in the Eye of the Beholder

If you could make your body look any way you wanted, what would you change? Would you be heavier? thinner? more muscular? How much time and money do you spend on trying to alter your body's image?

Culture plays a significant role in how you answer these questions. The norms you have acquired about how people should look to be healthy and attractive come from many sources. Your parents encourage you to eat more or less and comment on your appearance. Peers may also discuss the body shapes they find most attractive or comment on your weight. The medical community and teachers tell you about the importance of maintaining a healthy weight. Fashion models, celebrities, and athletes provide visual reminders of what society says you should like, while the media persuades you to buy products ranging from diet aids to breakfast cereals that will help you look more attractive.

Definitions of the ideal body size and shape vary greatly from one culture to another. Members of many traditional societies value fatness as a sign of beauty, wealth, and health. A large, voluptuous body is typically valued in societies where food supplies are limited or unstable. In situations where only the rich have the means to eat enough to gain excess weight, a fat man is admired for his economic affluence and power, and a fat woman signifies that her husband is a good provider. Extra fat stores protect people from food shortages, bouts of illness, and parasitic infections. They also protect women's fertility in times of food scarcity. In some African countries with high rates of HIV/AIDS, thinness is associated with this deadly disease. Fatness is also considered beautiful and sexy in many social groups (see Figure 4.4).

In some societies that value a plump figure, members of the ruling class or other people undergo a special fattening process. Systematic overeating is especially common among women who are preparing to get married. Among the Annang of Nigeria, all girls, except the very poorest, were traditionally fattened in a special room—in either a separate room or a barricaded portion of her mother's room. While in seclusion the girl was supposed to refrain from all unnecessary physical activity and eat anything she wanted. She was completely naked so that others might observe her fatness. When she left the room, the village held a feast and danced in her honor. The fatter she had become, the higher the price her husband would be expected to pay her parents for the right to marry her. The length of stay and degree of obesity contributed to her family's status in the village. It was also believed to improve her ability to conceive, carry, deliver, and nurse her first child. Today, the practice of fattening girls for marriage is dying out or going underground among the Annang. However, it is still used to treat Annang women who are infertile (Brink 1995).

Although seclusion practices are more common among women, systematic fattening practices for males have also been documented. Igore de Garine (1995) gives a detailed description of a complex institution, the *guru,* used to fatten Massa men in Northern Cameroon. Massa men may participate in individual *gurus* for a 2-month seclusion during which time they may ingest as much as 10,000 calories a day. More commonly, however, men participate in a much longer, collective *guru.* During this period, the men consume about 3,500 calories per day while living a semiactive life in camp. They tend to

Figure 4.4

Many traditional Polynesian societies valued fatness as a symbol of beauty and prestige.

their cattle to prevent them from eating crops; they dance, sing, wrestle, and spend most of their time grooming and training to be beautiful. At the end of the *guru,* most participants have gained about 7 kilograms, giving their family lineage prestige and increasing their changes of securing a desirable wife because of their enhanced sexual attractiveness.

Many traditional Polynesian societies valued fatness in both men and women. A fat body was a symbol of beauty—and a source of pride for both the individual and those who helped bring it about. Obesity was also an important health advantage; it may even have had an important evolutionary advantage by protecting Polynesians during voyages across the cold Pacific Ocean. Even today, the love of food prevails in these societies and large caloric intakes are not uncommon; one report documented daily intakes as high as 7,500 calories for men and 5,000 calories for women (Pollack 1995). Recently, national attention has been drawn to the high rates of obesity, diabetes, and glucose intolerance among Hawaiians, Samoans, and Nurauans living in modernized settings. Among Hawaiians, for example, McGarvey

(continued)

Focus 4.2   (continued from previous page)

(1991) found that 75% of adult men and 80% of adult women were overweight and that 11.5% of males and 19% of females were severely obese. As Nancy Pollock (1995) explains, however, Westerners' criticism and efforts to fight obesity are viewed by some Polynesians as a culturally insensitive slur on their national image and as a failure to recognize that fatness is a manifestation of traditional cultural values and a survival mechanism in the face of many natural disasters and man-made disruptions (e.g., World War II) confronting these Pacific Islanders.

Cultural ideas about body shape and size are not static. Until relatively recently, obesity was also admired in many Western societies. A trip to the art museum reminds us that many Renaissance artists chose obese subjects to depict beauty; in Rubens' eyes, cellulite was beautiful! Books like *How To Be Plump* were published to help people gain weight, full-figured women like Lillian Russell were idolized, and wealthy men sported large bellies accentuated by fashionable vests and gold watch chains (Dyson 2000).

It wasn't until the late 1800s and early 1900s that North American body ideals began to change. Food supplies became more stable and abundant so that only the poorest families lacked access to enough food to become fat, and a large influx of immigrants—many of them heavy—moved into American cities. Fatness no longer offered a survival advantage, and slimness became a way for wealthier Americans to distinguish themselves from the poor and new arrivals. "When it became possible for people of modest means to become plump, being fat no longer was seen as a sign of prestige. . . . The status symbol flipped: it became chic to be thin and all too ordinary to be overweight" (Fraser 1997).

Since the flapper craze in the 1920s, slenderness has been firmly established as the norm. Although a more voluptuous female figure was popular in the 1950s, as evidenced by Marilyn Monroe's popularity, the ideal body weight has remained slim until present day. In fact, a study of body norms based on Playmate centerfolds in *Playboy* magazine, and Miss America Pageant contestants and winners, found a steady decrease in women's weight from 1959 to 1978. At the same time, there was an increase in the amount of space in six major women's magazines devoted to dieting. Since 1990, the ideal female body type has become stronger but remains lean (Fraser 1997).

While attractiveness, in general, and a slim body, in particular, are usually considered more important for women, men also respond to cultural norms about the ideal masculine physique. Men, however, appear to be more concerned with being muscular than thin. An analysis of men's magazines, for instance, found more advertisements and articles about changing the male body shape than losing weight (Anderson and DiDomenico 1992), and a study of *Playgirl* centerfolds found an increase in body density and muscularity from 1973 to 1997 (Leit, Pope, and Gray 2000). Boys' action toys—another reflection of ideal body types—have also become increasingly muscular over time (Pope et al. 1999).

Slimmer body ideals have had powerful implications for people whose bodies do not comply. Jeffery Sobal (1995a) describes three societal responses to those who deviate from today's body ideals. The first, and most powerful, response is stigmatization and discrimination. Stigmatization is based on the view that fat people are to blame because they lack willpower. Even today fatness is often equated with weakness, laziness, slothfulness, ugliness, immorality, and other negative stereotypical characteristics. Unfortunately, most obese people internalize these stereotypes and experience extreme dissatisfaction with their bodies and problems with self-esteem. Overweight people are also the subject of personal and institutional forms of discrimination: They encounter discrimination that makes it more difficult for them to get married, hired, promoted, or accepted into institutions of higher learning. On average, obese people receive lower wages, are subjected to more frequent prejudicial medical treatment, and are served less promptly by sales clerks (Sobal 1995a).

Another societal response has been the medicalization of obesity. Medicalization is a process by which physical and social conditions become defined and treated as medical disorders or diseases. As explained by Jeffery Sobal (1995a):

> *Medicalization of obesity occurred as medical people and their allies made increasingly frequent, powerful, and persuasive claims that they should exercise social control over fatness in contemporary society. Claims supporting medicalization occurred in many forms: naming, defining obesity as a disease, organizational activities, and applications of medical treatments. (p. 69)*

Terms like corpulence and fatness were replaced by the more scientific term, obesity. Obesity became designated as an official disease listed in the *International Classifications of Diseases (ICD-9-CM 1990)*, (U.S. Department of Health and Human Services 1990), putting it under medical responsibility rather than the moral control of the individual. Medical journals, professional organizations, and medical conferences were developed to legitimize claims that obesity is a disease, and a host of medical interventions—from prescription diets to surgery—were developed to treat it.

(continued)

> **Focus 4.2** (continued from previous page)
>
> Although medicalization of obesity has helped change the view of fatness as a moral problem, obesity is still stigmatized as a form of illness or disability. To counter negative stereotyping of overweight people and promote the acceptance and normalization of obesity, some people have reframed obesity as a political rights issue. Calling themselves "size activitists," they have encouraged the use of words like "full figured" and "ample" to refer to overweight bodies, applied litigation to protect fat people's rights, and established organizations such as the National Association to Advance Fat Acceptance and Association for the Health Enrichment of Large Persons to promote size acceptance and provide social support for the obese (Sobal 1995a). Despite these developments, it is unlikely that North Americans' preference for a slim body will yield to widespread acceptance of or preference for fuller figures in the near future.

In addition to norms, members of a society share information, or *beliefs*—conceptions of reality and propositions about how the universe works. Each society has a set of beliefs about food and conceptions about how it affects the human body. Some Southeast Asian peasants avoid fish because they believe it will give them worms. Traditional Australian Bushmen will not kill animals they believe to be ancestral spirits. **Hare Krishna devotees** shun meat because they believe it is dirty, while North Americans encourage their children to drink milk because it is believed to have special properties that promote health and strength. When describing our own beliefs, especially those that have gained widespread acceptance among experts, we may consider them "facts" or "knowledge." In this course, you are learning the beliefs held by your professor and the authors of this book about the relationship between food and society. Over time, many of the beliefs shared here will change, but the notion that our beliefs are "true" probably will not.

Hare Krishna devotees
Members of a religious sect—the International Society for Krishna consciousness—which is based on ancient Hindu philosophy.

Cultures are also made up of *values*—conceptions of what is desirable and undesirable, good and bad. As we acquire our society's values, we begin to react with approval and disapproval toward certain situations. In a social group that values honesty and generosity, people who display these traits are admired; those who do not are met with some form of disapproval.

Values such as North American concepts of democracy and freedom have an integrating, unifying effect that gives meaning to broad areas of life. Values are expressed in literature, art, religion, and other cultural institutions, including health practices. The value placed on youthfulness is seen in North Americans' obsession with vitality, health, and youthful figures. Commercial marketers use an understanding of our values, and the accompanying fears we have about how we look and smell, to sell us cosmetics, feminine hygiene sprays, deodorants, and a host of other products that promise to increase our vitality and youthful appearance. These practices might seem strange to many Chinese, who hold their elders in great esteem and value the wisdom and maturity of old age.

Some values are shared by many societies, whereas others differ significantly from one social group to another. A marketing research firm, Roper Starch Worldwide, recently conducted a study of human values throughout the world (Roper Starch Worldwide 2000). After developing a list of 60 personal values held by people around the globe, the firm's researchers asked a sample of 30,000 respondents aged 13 to 65 living in over 40 nations to rank them. When combining responses from all nations, "protecting the family" ranked first, with 47% of survey respondents ranking it as "extremely important." Honesty ranked second, followed by health and fitness, self-esteem, self-reliance, justice, freedom, friendship, knowledge, and learning (in that order). Many regions of the world, including North America, shared at least six of the world's top-ten values; however, important differences were also found between regions and countries in the study. In Canada and the United States, for instance, individuality, romance, excitement, adventure, and social tolerance received high rankings. But in parts of Asia that are not highly industrialized and where socioeconomic status is less stable, status, power, public image, wealth, and respecting one's ancestors were valued most highly. Values that exhibited the largest differences between nations included:

- Faith—ranked 1st in Indonesia, Egypt, and Saudi Arabia, but 45th in France
- Friendship—ranked 1st in Germany, but 26th in South Africa
- Freedom—ranked 2nd in Argentina, France, Italy, and Spain, but 40th in China
- Knowledge—ranked 2nd in India, but 34th in Egypt

Culture Is Expressed Through Behavior and Artifacts

artifacts
Anything produced or modified by humans.

We have said that culture is made up of norms, beliefs, values, and broader conceptions about life. Because these are concepts or ideas, they are not directly observable. We can observe, however, the ways in which culture is expressed in people's behavior and in the tools and other artifacts they produce. Archaeologists reconstruct the way ancient societies built homes from the potholes and other artifacts left behind. *Ethnographers*—anthropologists who compare and analyze cultures—infer norms and beliefs from how people act as well as from the explanations they give for their behavior. When people respond to a given situation in a repetitive, patterned way, they are likely to be following shared norms or other conventional understandings about how to act. By observing a group of people, ethnographers can begin to notice behavioral patterns. By talking with them, they are able to uncover the norms, values, and beliefs that guide individuals' behavior.

Take, for example, the case of an ethnographer observing a hypothetical tribe. She notices that men usually eat the food prepared by their wives but that they occasionally dine with their sisters. Her curiosity piqued, she begins to record exactly when the men eat with their sisters. After several months, she discovers a pattern: Most men dine with their wives for about 25 consecutive days and then eat with their sisters for the next 5 or 6 days. A few men, mostly older ones, deviate from this pattern and dine only with their wives. The regularity of this behavior leads the ethnographer to believe the tribespeople are following a norm of some type. When she discusses her observations with them, her suspicions are confirmed. The tribe believes women have magical powers during menstruation and that men may be harmed if they eat food prepared by menstruating women. When a man's wife is menstruating, he dines with a sister or other female relative whose food is not potentially harmful to him. Older men, whose wives are postmenopausal, eat at home throughout the month.

In this instance, the tribe's behavioral pattern corresponds closely with their normative pattern and accompanying beliefs and values. In other words, they act according to their norms and beliefs. But people do not always do what they think they should. Some Krishna devotees occasionally drink wine and some Mormons drink coffee or tea, so it is important to distinguish between what people say they should eat and what they actually consume. Even though you have acquired many beliefs and values to reinforce notions about what you should eat to stay healthy, you, too, are likely to deviate from these norms, especially when faced with stressful situations like final exams.

Another example comes from the village of Bisipara in the Orissa State of India (Bailey 1969, cited in Salzman 2001). Traditionally, the Pans caste in Bisipara was viewed as ritually polluting. Made up of landless agricultural laborers, musicians, beggars, and processors of dead animals, this "untouchable" caste was excluded from interacting with "clean-caste" members of priests, warriors, and herdsmen. According to traditional norms, they were not allowed to occupy leadership roles or attend the temples where clean-caste members worshiped. After government officials outlawed discrimination against the untouchable castes, many Pans took advantage of opportunities to get an education and obtained jobs as teachers, government workers, and other leadership positions. Some Pans, still unhappy with continued discrimination and stigmatization, tried to improve their social standing by avoiding practices seen as ritually polluting: They stopped eating meat, drinking alcohol, making music, and begging, and they adopted the dress and other habits of the cleaner castes. The ethnographer living with them during this time, F. G. Bailey, reported many incidents in which the Pans challenged norms that excluded them from religious and other village activities. Eventually, several Pans were elected to the Orissa State Legislative Assembly where they controlled access to important resources, for example, a new school for the village. As members of the clean castes called on the Pans for support in obtaining resources for the village, the Pans' inclusion in village life became important and was viewed as normal. Eventually, a new norm in which merit and achievement were used to select leaders replaced the old rule of excluding Pans members for ritual reasons. As individuals intentionally created new norms, considerable variability and tension within the system was generated.

Culture as a Functionally Integrated System

Norms, beliefs, and values tend to reinforce each other. Social groups share traits that are compatible with other traits that make up the overall cultural system. Ancestor worship would be unlikely to thrive in societies such as Canada or the United States where youthfulness is highly valued; the introduction of steak sauce into a Hindu society in which most people do not eat beef would make little sense. Cultures, then, are best viewed as *loosely integrated* systems of norms, beliefs, and values. When we refer to culture as an integrated system, we are saying that its parts have a special relationship to the whole.

Food practices are integrated into the overall culture and serve many functions: economic, political, recreational, social, aesthetic, religious, ceremonial, magical, legal, and medical. Consider, for example, the variety of ways food functions in your life. Like many North Americans, you may use food as gifts, host dinner parties to improve business relations, share meals with relatives to reinforce kinship ties, and cook special meals for aesthetic or creative gratification. Observance of food prohibitions or other special practices may allow you to express religious beliefs or prevent illness. Some foods serve as important symbols of prestige so their consumption enhances your social status, and occasionally you may eat special "comfort" foods to cope with boredom or anxiety.

Because food is integrated into a broader cultural system, attempts to change dietary practices must recognize the numerous functions food can play. Health care providers' immersion in the scientific aspects of nutrition may divert their attention from the important role food plays in social life. Health care professionals need to remember food is more than just something to eat; it plays many non-nutritional functions. When a health care provider advises a person to alter his diet, she may forget to consider the social significance of the changes she has recommended. The client, unwilling to sacrifice the many non-nutritional benefits associated with the practices being replaced, may reject the professional's scientifically sound advice. The goal, then, is to find ways to alter dietary patterns to make them more nutritious while preserving the other functions food plays in people's lives.

Awareness of food's many functions enabled one nutritionist to work successfully with a 53-year-old woman suffering from severe heart disease. This woman ate many meals high in calories, cholesterol, and saturated fat, including two potluck suppers held each week at her church. Recognizing that the church suppers were not just a source of nutrients, but also a way to reinforce social ties and reaffirm religious affiliation, the nutritionist was careful to make his advice compatible with this important social engagement. He advised the woman to limit her caloric and fat intake throughout the day so that she could afford to eat a little extra at the church suppers. He also provided her with recipes for tasty, low-calorie dishes she could take to the potlucks, like the one depicted in Figure 4.5. This advice allowed her to enjoy the social benefits of the potlucks without jeopardizing her health; it also encouraged her friends to do the same.

Despite the functional linkages between cultural traits, no system is in perfect harmony. All societies experience some degree of strain between inconsistent parts. This tension or disharmony is especially common in societies undergoing rapid change. Nutrition controversies, for example, are a regular feature in our current magazines and newspapers. When the U.S. Department of Agriculture and the U.S. Department of Health and Human Services issued the Dietary Guidelines for Americans and the USDA Food Guide Pyramid, the meat and dairy industries protested loudly. The debate over these recommendations and many other nutrition issues continues as we attempt to strike some compromise between inconsistent dietary beliefs, values, and customs.

Intracultural Variation

Although members of a social group share many cultural traits, important differences also exist between them: Not everyone holds the same values or acts the same way. No doubt your own dietary habits differ from those of your friends and relatives, particularly when daily routines are disrupted. In fact, cultures are not distinct, homogeneous, or set apart from others by clear-cut, fixed boundaries (Brumann 1999). Each culture represents a unique *combination* of traits, but very few societies have traits that are unique. Instead, they share most norms, values, and symbols with some other social groups. Cultural boundaries become blurred as members adopt ideas and meanings from other groups to whom they are exposed via personal contact, media, or other means. Immigrants hold on to certain traits from their culture of origin while adopting others from their new homeland. It is important, therefore, to acknowledge differences within as well as between social groups. These differences are called *intracultural variation.*

Intracultural variation may arise from a variety of sources. First, culture is not replicated perfectly as it is passed on to new members. We forget some ideas and reject or ignore others. You probably do not rely solely on family traditions or your parents' cooking practices when making a meal. Most likely you practice some traditional ideas but have rejected others. In addition to using family recipes, you might clip recipes from a magazine, try exotic fruits and vegetables available at the local supermarket, and swap advice about vitamin and mineral supplements with friends.

Exposure to other cultures and subcultures is another important source of intracultural variation. We may adopt food practices from countries we visit (via personal travel, travel magazines, or television shows) or from immigrants who bring new dietary practices with them when they move to Canada or the United States. In fact, North Americans have adopted a wide variety of cuisines from other countries. Not only are you likely to dine in a variety of ethnic restaurants (Chinese, Italian, Mexican, Indian, Mediterranean, Ethiopian, and others), but you may also cook many "ethnic" foods

Figure 4.5

Nutritionists who understand and respect the many social and cultural functions food plays in their clients' lives are better prepared to help them modify their dietary practices.

© Jeff Greenberg/PhotoEdit

at home. Consider the bagel. In the 1890s, bagels were consumed almost exclusively by Jews from Eastern Europe. In the 1900s, people from many other ethnic groups in the northeastern part of the United States discovered the bagel and it eventually became "an icon of urban, northeastern eating, a key ingredient of the 'New York deli'" (Gabaccia 1998). Today the bagel is mass-produced and marketed nationally where people living throughout the United States enjoy it.

Intracultural variation also arises from differences between families. When couples marry, they develop a new blend of dietary norms and beliefs, adopting some practices from each of their family traditions as well as developing new ones to pass on to their offspring. Perhaps you recall eating at a childhood friend's home and discovering different meal patterns or table manners from those practiced in your own family. In some instances, these differences represent new strategies for dealing with specific economic or other situations. For instance, families in which both parents work may rely more frequently on "meal replacements" or previously prepared foods than families in which a homemaker continues to prepare most meals. Divorce and other changes in family structure may also lead to the development of new food norms or practices.

Differences in access to food and the economic resources needed to purchase goods also create intracultural variation in dietary practices. Families with limited incomes may be unable to purchase some foods they believe will promote health and instead adopt strategies more compatible with their budgets. Despite economic constraints, many economically challenged families eat nutritiously. Although studies have documented increased rates of nutritional deficiencies among the poor, the U.S. Food Consumption Surveys in 1965 and 1990 found that blacks and whites in lower socioeconomic strata actually had a higher quality diet (based on a 16-point diet quality index) than upper-class whites in the 1960s and rather similar diets in the early 1990s (Last 1998). Of course, differences in socioeconomic status have a strong relationship with access to medical care and other preventive health behaviors that place the poor at increased risk for premature death and disability. In fact, longevity, mortality, and morbidity rates are strongly correlated with socioeconomic status.

Maternal education also affects dietary practices. Education increases a woman's productive capabilities and her contributions to the household income. Highly educated women usually have a better understanding of nutrition and how the body functions and, as a result, may place a higher priority on healthy lifestyles when feeding their families.

Occupation, age, household size and composition, and other situational factors also impact dietary beliefs and practices. When immigrants arrive in North America, they may find it difficult to purchase basic ingredients or special foodstuffs used to prepare traditional dishes. As they learn to substitute new foods and adopt new recipes from neighbors and coworkers, the differences between them and other members of their ethnic group increase. Infant feeding practices among Puerto Ricans in Dade County, Florida, vary significantly depending on the length of time the family has spent in the United States, the presence of the husband and his attitudes toward different feeding methods, especially breastfeeding, and the location of the mother's mother.

Intracultural variation also results from differential access to information. Many new food practices are acquired from television cooking shows, the Internet, news programs, and adult education courses. Women who work outside the home are exposed to ideas from their coworkers with whom they exchange recipes, home remedies, and shopping tips.

New ideas also come from groups who are trying to promote change. Most governments publish dietary recommendations and sponsor social marketing campaigns to encourage their citizens to adopt healthier habits. Government agencies also try to promote commodities like wheat in foreign countries to build stronger overseas markets and improve the national trade balances. At the same time, the food industry uses advertising to persuade you to purchase more of their products.

Individuals also create intracultural variation as they develop new information and solutions for coping with environmental challenges. We have already seen that people do not blindly follow cultural norms: They may intentionally break the rules, adopt differing values, or acquire new information. A good example of how individuals may exercise choices that do not conform to the normative rules comes from women's concerted effort during the second half of the twentieth century to fight gender-based discrimination. Some women consciously rejected traditional norms regarding their role as housewives and mothers and demanded different treatment from their husbands, employers, and others. Coexistence of these new norms and values has created considerable intracultural variation in women's roles and the lifestyles they adopt today. In many societies, new ideas are being adopted at such a rapid rate that we speak of a "generation gap": Many values and beliefs held by the younger generation differ from those held by their elders. Vegetarianism, for instance, is far more common among young people today than among older generations.

An interesting illustration of how these factors can work together to create intracultural variation comes from the Zumbagua, an indigenous region of highland Ecuador (Weismantel 2000). In recent years some young males have traveled to the capital, Quito, to work as construction workers. There they are socialized into a more modern, urban society than the women and older family members who stay behind. The men are also able to purchase bread to bring home as gifts or treats for family members and guests. As bread has become a symbol for status and modernity, their children—also exposed to the outside world through school—have begun demanding it for breakfast instead of the traditional porridge. These "cries for bread" have led to serious conflicts within families as well as variation between those willing and financially able to purchase it regularly for their children and those who cannot.

In sum, while *culture* is a powerful conceptual tool, it cannot explain all of the interesting diversity in human dietary practices today. Many factors—genetics, personality, the physical and social environment, and broad-reaching geopolitical forces—interact with culture to influence dietary behavior. Also, although culture provides us with rules and information about what we should eat, we can decide to break the rules and attempt to change the group's norms. As a result, cultures are neither static nor homogeneous so that, in many cases, the differences between members of a social group may be just as marked as those between members of different social groups. For this reason, it is dangerous to generalize about *all* members in a society or fail to realize that cultures are constantly changing. **Stereotyping** is another danger associated with misuse of the concept of culture and failure to recognize *diversity* within social groups. Finally, we hope that an understanding of the concept of culture will enable you to become more culturally relative and avoid the pitfalls of ethnocentrism, to which we turn next.

stereotyping
The ascription of a group's collective traits to all of its members.

Ethnocentrism and Cultural Relativity

Our cultural heritage is so much a part of our way of thinking that we sometimes find it difficult to understand or accept other people's behavior. We consider our own way of life and value system as natural, good, beautiful, or important, while puzzling over the strangeness of other societies' lifestyles. We have already discussed entomophagy, a practice that North Americans find repugnant but that many groups of people use as an important nutritional resource. Some societies find it most convenient to eat from a common pot using the fingers and hands as implements and feel comfortable sampling food from others' plates, whereas most North Americans view these practices as unsanitary, primitive, or bad manners.

The danger of ethnocentrism lies in its failure to view other people's practices as viable alternatives with potentially equal or even superior value. The uncritical acceptance of one's own value system and lifestyle as the most appropriate is called *ethnocentrism*. During colonial times the British viewed central Africans' reliance on cassava (a long sweet-potato-like root that contains various levels of cyanide) with skepticism. It was seen as having little nutritional value and negative ecological impact. Because its cultivation appeared to require relatively little labor, the British labeled it "the lazy man's crop" and discouraged its use. This ethnocentric view prevented the British colonialists from seeing the many advantages cassava offers as a dietary staple for this African population: It provides food security because it is highly adaptable to the difficult ecological conditions in Northwest Zambia; it produces high yields relative to caloric energy inputs; it can be stored for long periods of time; and it can be sold in the marketplace where it generates income needed to purchase other desirable products and enhance the social status of many groups within the society (Lenten 1999; von Oppen 1999). Another example closer to home is Western medical practitioners' reluctance until relatively recently to understand the benefits of acupuncture, acupressure, and other Chinese medical practices.

The reverse of ethnocentrism is a viewpoint known as *cultural relativism.* According to this viewpoint,

> *The standards of rightness and wrongness (values) and of usage and effectiveness (customs) are relative to the given culture of which they are a part. In its most extreme form, it holds that every custom is valid in terms of its own cultural setting. In practical terms, it means that anthropologists learn to suspend judgment, to strive to understand what goes on from the point of view of the people being studied, that is, to achieve empathy, for the sake of humanistic perception and scientific accuracy. (Hoebel and Frost 1976:23)*

Many anthropologists argue it is impossible to shed our own cultural lenses and look at the world objectively. Although it is certainly true that we will always be influenced by deep-seated values, cultural relativity reminds us to be more respectful and open-minded in analyzing and comparing cultural differences. It asks us to focus on the functions and meanings a custom has in another society and evaluate other cultures' practices using their own values and beliefs rather than our own. Using a culturally relative viewpoint, we can study the functions of cultural practices such as warfare, Hindu deification of the cow, or polygamy even though they run counter to our personal ethics and values. Cultural relativity does not imply that just because a cultural group behaves in a certain way, we should accept it as morally correct. We can recognize the economic advantages herders gain by having multiple wives and yet still find polygamy unacceptable. Nor does cultural relativity imply that we should reject our own lifestyle and adopt the cultural ways of other societies. We may challenge certain attitudes or practices such as imperialism in our own as well as other cultures, without sacrificing the benefits a culturally relativistic perspective affords.

By pointing out the conflicts between a narrow, ethnocentric point of view and a more culturally relativistic point of view, we wish to impress on students the need to approach each society and its people with an open mind, one that looks respectfully at new situations and examines critically the implicit assumptions with which they approach the world. Pharmaceutical companies, for instance, have learned they can identify powerful new medicinal agents by studying herbal remedies used by indigenous peoples in Africa and Latin America. (See Chapter 8's discussion of health beliefs and practices.)

Throughout this and other chapters we will emphasize the need to view food-and-health related practices as functional units of a wider cultural context, as well as to recognize the adaptive nature of certain food practices. At the same time, we will critically examine situations in which a range of interventions can result in improvement.

Implications for Health Care Professionals

Students of health and nutrition no doubt recognize the need to understand how people choose and prepare the foods that make up their diet, and how to use this understanding to help them improve people's nutritional status. Health professionals learn quickly that human dietary behavior is complex and that attempts to change it, more often than not, will be resisted, resented, or ignored.

We believe that an understanding of the concept of culture can be useful to nutritionists in five ways. First, by recognizing the pervasive influence culture has over all aspects of human behavior, we are less likely to make the common mistake of assuming that people are empty vessels into which we pour the truth. Being sensitive to the norms, beliefs, and assumptions people make about food and recognizing the need to introduce advice in ways that are compatible with their culture will result in more effective dietary changes. For example, nutrition educators who assume that their task is to tell people "the facts" about nutrition will soon find that little change takes place. A more effective strategy is to package information in such a way that it fits into clients' lives. At a meeting with a group of working mothers, a nutritionist began by asking them their concerns about feeding their children. When she learned that their major problems were due to lack of time and money, she abandoned her prepared lecture notes and led a discussion of ways to prepare nutritious foods inexpensively and quickly. The meeting proved effective in changing many of the mothers' practices because it was compatible with their needs and interests.

Second, recognition that much of culture is implicit can prevent health care providers from assuming that other people's way of life is just like theirs. With this perspective in mind, a nurse found that the way in which she ate her meals was very different from that of many of the clients she encountered. In her family, meals were served three times a day often with everyone gathered around the table. Had she assumed that her clients ate this way, her nutrition counseling would have been ineffective. By asking clients about their meal patterns, she found that some families rarely shared meals because of incongruent work schedules and food preferences. Family members snacked throughout the day and ate small meals when their work schedules permitted. If she had distributed copies of menus for family breakfasts, lunches, and dinners to these clients, the information would have been of little use. A more effective strategy proved to be a discussion of nutritious snacks and quickly prepared small meals that they could incorporate into their hectic schedules.

Third, appreciation of culture as an adaptive, functionally related system enables health care professionals to identify important linkages between food practices and other aspects of life. By recognizing the many functions food can play in people's lives, nutritionists are less likely to recommend changes that require unnecessary, and often unrealistic, sacrifices. Recognition of the relationship between food practices and other aspects of a culture also enables nutritionists to understand an important problem inherent in making dietary change: Modification of one cultural trait will often bring about concomitant changes in related traits. When these ties are not anticipated, resistance and/or negative effects may occur as the system achieves a new internal balance. Take, for example, the failure of the Eighteenth Amendment, which prohibited the sale of liquor in the United States. Reformers originally promoted the amendment as a means of abolishing the saloons used by political ward bosses to amass support for their corrupt political machines and uplifting the lower classes they felt were morally corrupted by drink.

Unfortunately, lawmakers, administrators, and reformers mistakenly viewed the sale and consumption of liquor as an isolated custom rather than as part of an integrated cultural system. When they abolished the legal sale of liquor, they were unprepared for the widespread repercussions created in family life, politics, law enforcement, and the economy. As hoped, blue-collar workers consumed less alcohol because they couldn't afford **bootleg** prices; however, the middle and upper classes began to support unlicensed saloons ("speakeasies") where homemade brews such as "bathtub gin" were sold. In fact, Prohibition caused a major economic shift. Because alcohol could still be produced for medical, industrial, and sacramental (religious) purposes, these industries expanded, with approximately one-third reaching bootleg channels at the expense of previous manufacturers. Wine making also prospered with the creation of a "harmless grape jelly" that could be converted to wine simply by adding water. Smuggling, another illicit source of alcohol, played a crucial role in the expansion of the Canadian liquor industry and contributed to the economic growth of other countries as well. Another unanticipated consequence of Prohibition was increased drinking among young members of the middle class and newly liberated women—largely because it was something disapproved of by society. Drinking bootleg booze and visiting a speakeasy became, for many, a way of flaunting convention and an expression of independence. A final and more tragic, unanticipated consequence of Prohibition was its impact on entrepreneurial crime in the United States. The emergence of bootlegging gangs, the consolidation of organized crime, the resulting gangland violence, and the general public's open contempt for the law finally led to the passage of the Twenty-First Amendment and reinstitution of legal liquor sales (Nelli 1976).

Fourth, recognition of intracultural variation helps us to respect diversity and avoid stereotyping. You can imagine feeling offended, frustrated, or perplexed if a health care professional assumed you used your grandparents' home remedies or shared other "old-fashioned" beliefs you rejected long ago. Because of the increased diversity found in the North American population, cultural competence is now considered an essential skill in many health care organizations.

Finally, cultural relativity encourages us to be receptive to the norms, beliefs, and customs of other groups. We are less likely to

bootleg
To produce or sell alcoholic beverages illegally.

offend them by responding with amazement or disgust at their beliefs and customs. By being nonjudgmental, nutritionists have a greater chance of developing the rapport needed to work effectively across social and cultural boundaries. While we may not be interested in consuming the diet of other groups, cultural relativity allows us to understand their food ways and recognize the value of their practices. We are less likely to mistake our beliefs for the "truth" and theirs as "wrong." As a result, we are less likely to try to change customs that are harmless or even have positive nutritional value. Even practices we find truly harmful can be changed more effectively if we first learn the reasons they came about and identify their functional relationship between those traits and other parts of the culture.

Summary

- Societies differ in what foods they define as edible and inedible, tasteful and distasteful.
- Foods considered delicacies in one society might not be considered edible in another.
- Differences in food choices can be traced to cultural factors as well as to biological and environmental factors.
- Culture refers to the knowledge, traditions, beliefs, and values that are developed, learned, and shared by members of a society.
- Culture influences how people respond to the physical environment (technology), the social environment (social organization), and the spiritual or metaphysical environment (ideology).
- Shared norms, beliefs, and values are part of a group's culture.
- Culture is manifest in behavior and artifacts.
- Culture is a functionally integrated system. Food, for example, fulfills many needs other than nourishment. It serves economic, social, aesthetic, and religious functions as well.
- While culture is a useful concept for describing a group in general terms, there is much intracultural variation. People from the same culture may have very different diets.
- The unexamined acceptance of one's own lifestyle and value system as the most appropriate is called ethnocentrism. A more effective approach is cultural relativism, a viewpoint that encourages approaching new situations and people with an open mind.

HIGHLIGHT Body Image and Health

A North American health educator working in an African village, concerned about the health consequences of obesity, launched an educational campaign to alert women to the benefits of weight loss. A poster was designed to teach women that they would feel more energetic and healthy if they lost weight. The poster showed a thin, energetic woman dusting a table next to a fat woman sitting listlessly on a couch. After displaying the poster throughout the community, an evaluation of its impact on local knowledge and attitudes was conducted. The results illustrate the importance of understanding cultural conceptions of body images: Instead of being convinced that physical activity is healthy, most villagers believed the poster showed a rich woman relaxing on a couch while a maid cleaned her house! In this village, as in many other parts of the world, fatness was valued as a sign of wealth, and thinness as evidence that women are poor and undernourished.

Cultural notions about the ideal body size and shape have important implications for the public's health. In this highlight, we examine the relationship between body image and health and look at current explanations for the growing number of North Americans who are obese or suffer from bulimia, anorexia nervosa, or other dangerous eating disorders.

The Obesity Epidemic

Despite the idealization of thinness, increasing numbers of North Americans have become overweight, and in the last three decades the problem has reached epidemic proportions. As seen in Figure H4.1, the proportion of U.S. adults who are about 30 pounds or more overweight increased dramatically from 1991 to 2000. States that appear in white did not provide Behavioral Risk Factor Surveillance data on obesity. Less than 10% of adults living in states that appear in the lightest shade of gray are obese. Between 10% and 15% of adults are obese in the next darkest shade of gray. And over 15% of adults are obese in states in black. States where over 20% of adults are obese are striped.

Today, approximately one-third of Canadians now exceed healthy weight standards (Health Canada 2002). The problem is even more widespread in the United States where 54% of adults and 25% of children are overweight (Flegal et al. 1998; Troiano and Flegal 1998). It is not uncommon for people from other countries visiting the United States to comment on the degree of obesity and large numbers of people who are overweight in this country. Approximately one in five people in the United States are now considered obese, and more children are obese now than at any other time in history.

Figure H4.1
Obesity trends among adults, 1991–2000.

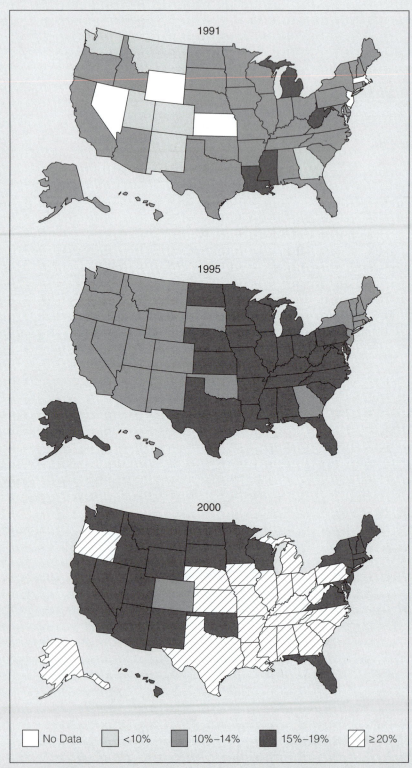

Source: Center for Chronic Disease Prevention and Health Promotion, Centers for Disease Control and Prevention, 2002. *http://www.cdc.gov/nccdphp/dnpa/obesity/trend/maps/index.htm*

The most accurate method for determining if you are over-weight or obese is to calculate your *body mass index* using the following formula:

$$BMI = [\text{Weight in pounds} \div \text{Height in inches} \div \text{Height in inches}] \times 703$$

You may also determine your BMI using the table or calculator on the Centers for Disease Control and Prevention (CDC) Web site: *http://www.cdc.gov/nccdphp/dnpa/bmi/bmi-adult.htm*. If your BMI is equal to or above 25, you are considered overweight. If it is equal to or above 30, you are considered obese (unless you are extremely muscular). Underweight is defined as having a BMI below 19.

Public health officials are concerned about the rapid rise in excess weight gain because of the increased risk it creates for coronary heart disease, hypertension, stroke, gallbladder disease, type 2 diabetes, **osteoarthritis,** some forms of cancer, and emotional problems. According to the CDC, more than 300,000 premature deaths are now linked to obesity annually in the United States. The impact of obesity on health and the rapidly increasing rates of obesity have motivated the Surgeon General of the United States to declare obesity as a national epidemic and the Centers for Disease Control and Prevention to launch a major initiative to prevent and control it.

osteoarthritis
A degenerative joint disease in which the connective tissue (cartilage) is lost.

Obesity is especially problematic for children. Sixty percent of overweight children already have at least one risk factor for heart disease, such as elevated blood lipid levels, insulin levels, or blood pressure (*www.cdc.gov/nccdphp/dnpa/obesity-epidemic.htm*). Excess weight gain has been linked to a sudden increase in type 2 diabetes and **hypercholesterolemia** among children. Also, the metabolic changes that accompany excess weight gain during childhood make it more difficult for them to lose weight later in life. About one-third of overweight preschoolers and one-half of overweight school-aged children will become overweight adults, making the magnitude of the consequences of the recent rise in childhood obesity difficult to estimate.

hypercholesterolemia
High cholesterol.

Many factors contribute to the obesity epidemic. The most obvious are easy access to an abundant food supply and a sedentary lifestyle. Environmental factors that encourage people to consume more calories include the increased availability and variety of inexpensive, calorie-rich food products and soft drinks (the largest single source of sugar in the North American diet), growth of the fast-food industry and corresponding provision of high-fat, calorie-dense food, and a growing tendency to socialize with food and drink (Centers for Disease Control and Prevention 2001). The "supersize" portions served at many restaurants also contribute to increased caloric intakes: An order of spaghetti and meatballs at the Olive Garden has enough meat and pasta to feed the average-sized man for an entire day, Applebee's New Steakhouse Salad contains 7 ounces of steak, and one pecanbon from Cinnabon has 1,100 calories (Hill and Peters 1998).

BOX H4.1

Excerpt from the *Lexington Herald Leader*

According to the USDA, per capita consumption of food in America increased 8% between 1999 and 2000. That translates to almost 140 extra pounds of food per person per year. Americans are not only eating more healthful foods like fruits and vegetables, but more of almost everything else. One reason for this increased food intake is the food industry's highly profitable practice of serving larger portion sizes. It may be a cultural compulsion to get the most for our money or it may be a biological imperative to store up in case of famine, but it seems that not many can resist the pull to buy supersize products, especially when they are so inexpensive. At 7-Eleven stores it costs only 37 cents more to purchase four times the soda. At Cinnabon a sweet roll nearly three times as large as the regular one costs only 48 cents more. And at McDonalds, 87 cents will buy you supersized French fries which contain almost three times as many fries as the regular size portion. The impact of supersizing portions is particularly pronounced because Americans eat out so frequently.

Source: Lexington Herald-Leader Wire Services, July 7, 2002, tag 0207070003, © 2002. *Lexington Herald-Leader.*

Unfortunately, school meals have also contributed to the rise of childhood obesity. Although school breakfasts and lunches must meet federal nutrition requirements, including guidelines restricting the amount of fat they contain, students who participate in the National School Breakfast and Lunch Program have been shown to have higher intakes of energy, fat ,and saturated fat than those who bring their lunches (U.S. Department of Agriculture 2001). A la carte foods and vending machines that sell soft drinks, cookies, and candies are not subject to these nutritional restrictions and give students mixed messages, as well as excess calories.

At the same time food consumption has increased, physical activity levels have declined. One in four adults in the United States are not active at all in their leisure time, and more than 60% do not get enough exercise to enjoy any of its many health benefits (National Center for Chronic Disease Prevention and Health Promotion 2000). Automation of workplaces, use of labor-saving machinery for household chores, and reliance on automobile travel instead of walking and cycling also contribute to sedentary lifestyles among adults (Centers for Disease Control and Prevention 2001).

Children have also become more sedentary. Nearly half of young people between the ages of 12 and 21 do not engage in regular vigorous physical activity. Their increased sedentary lifestyle reflects a growing reliance on television, computers, and electronic games for recreation. Television viewing, for instance, has been shown to contribute to obesity in two ways: (1) the more time children spend watching TV, the less time they spend burning calories on more active forms of recreation; and (2) children consume more energy during and after watching TV advertisements for food (Robinson 1999). It is also more difficult to exercise in unsafe neighborhoods and play areas, or communities that lack sidewalks and trails for safe walking. Children are also less likely to exercise today because many schools have eliminated or greatly reduced physical education in an attempt to save money and increase time for other activities. Interestingly, studies have shown that increasing physical activity during a school day actually increases students' ability to learn. Few opportunities exist for organized noncompetitive physical activities outside of school (Centers for Disease Control and Prevention 2001).

Disordered Body Image and Eating Behaviors

As more and more North Americans gain excess weight, it is not surprising that the proportion of people dissatisfied with their bodies' size and shape has also increased. Although very young children are rarely concerned about their weight, by age 6 or 7, many children start worrying about how they look and want to be thinner (Maloney et al. 1988). Body discontent typically increases with the onset of bodily changes during adolescence, with between 50% and 80% of teen girls in the United States dissatisfied with how their bodies look (Mossavar-Rahmani et al. 1996; Parker et al. 1995). Although girls worry about looking fat, boys tend to be more concerned that they don't appear big and strong enough (Sobal 1995b). This dissatisfaction usually continues throughout adulthood, and eventually declines during the later years when it is replaced with concerns about overall health.

Many people are dissatisfied with their appearance because they overestimate how much they weigh. In a study of female college students, only one-third of girls who thought they were fat actually exceeded healthy weight limits. Even among adults, only about half of the people who perceive themselves as overweight actually are (Centers for Disease Control and Prevention 1997). (See Figure H4.2.)

Unfortunately, body dissatisfaction often leads to unhealthy eating practices, such as chronic or "yo-yo" dieting, the use of diet pills, or vomiting. In a study of third through sixth graders in the United States, 37% had tried to lose weight, 20% had highly restrained their caloric intake, and 7% had distorted eating patterns suggestive of anorexia nervosa (Maloney et al. 1989). By grades 9 to 12, 58% of girls and 25% of boys had tried to lose weight, with 6% of the girls and 1.5% of the boys resorting to vomiting, diet pills, and laxatives (Story et al. 1998).

Important ethnic differences exist in perceptions of the ideal body size and responses to believing that one is overweight. However, recent studies suggest theses differences may be decreasing as the cult of thinness is adopted by many Hispanic, Native American, and to a lesser extent, African American youth.

In general, white Americans of European descent (Euro-Americans) are the most likely to overestimate their bodies' weights. They usually score substantially lower than African Americans and Native Americans on numerous measures of body satisfaction and self-perception. Dissatisfaction is particularly common among teenagers. As many as 90% of Euro-American teens are displeased with some aspect of how their bodies look (Parker et al. 1995), and

Figure H4.2
Many people, especially young women, see themselves as fat, even when they are not.

© Tony Freeman/PhotoEdit

over 60% of female high school and college students are trying to lose weight (Centers for Disease Control and Prevention 1997).

Research on women of Hispanic descent is relatively scarce and the findings are mixed, but it appears that, compared to Euro-Americans, members of many Hispanic groups are more accepting of a larger body ideal. In some studies, Hispanics are less likely than Euro-Americans to feel dissatisfied with their bodies, even if they are overweight. However, as Hispanics acculturate into U.S. society, they too begin to adopt thinner body ideals and disordered eating practices, such as chronic dieting, purging, and use of laxatives. Among young Hispanics, dieting is only slightly less common, and body satisfaction scores are the same as or only slightly higher than those of non-Hispanic whites (Harris, Walters, and Waschull 1991; Miller et al. 2000).

Native Americans have also been thought to prefer a larger body size than Euro-Americans. However, recent research conducted by Sally Davis and her associates in New Mexico (Davis and Lumbert 2000; Davis et al. 1999; Stevens et al. 1999) suggests Native American children may be adopting mainstream ideals for thinness. These researchers showed a wide variety of body sizes and shapes to a group of Southwestern Indians in the fourth and fifth grades and asked them to select the ones they considered ideal. Only 37% of the Native American students chose an ideal body shape that corresponded with their own size, and almost half the fourth graders and about one-third of the fifth graders said they wanted to be thinner. Many (38% of fourth graders and 61% of fifth graders) were attempting to lose weight (Davis and Lumbert 2000; Stevens et al. 1999).

As a group, African Americans enjoy the highest level of satisfaction with how they look. Compared to other ethnic groups, African Americans score highest on measures of body acceptance, appearance evaluation, and body esteem and satisfaction (Miller et al. 2000). They are also less likely to be preoccupied with weight loss, even if they are moderately overweight. In a comparison of elderly overweight women, African Americans were 1.6 times less likely to feel guilty after overeating, 2.5 times more likely to be satisfied with their weight, and 2.7 times more likely to consider themselves attractive than Euro-Americans (Stevens, Kumanyika, and Keil 1994). Focus group interviews with African American teens reveal that their ideals of beauty are more flexible than those of other groups, with less emphasis on body size and shape and more concern with projecting confidence, a positive attitude, and style—"making what you have got work for you" (Parker et al. 1995). Despite higher levels of satisfaction with their bodies, similar proportions of African American girls try to lose weight as Euro-Americans and Hispanic Americans, and increasing numbers of African American teens are using laxatives and diuretics to control their weight (MDConsult 2002).

Widespread body dissatisfaction and chronic dieting have drawn concern from health professionals, psychologists, educators, and sports administrators because of the impact they have on people's emotional and physical health. Body dissatisfaction is associated with obsessive-compulsiveness, anxiety disorders, and depression. Chronic or yo-yo dieting not only is ineffective in controlling excess weight, but also increases the chances of dying from cardiovascular disease, **osteoporosis,** and cancer (Cogan and Ernsberger 1999). Many weight-loss methods have short- as well as long-term risks, including muscle depletion, mood disorders, heart problems, and increased risk for more serious eating disorders—anorexia nervosa and bulimia.

osteoporosis
A degenerative bone disease in which the bones become brittle and porous.

Eating Disorders

Anorexia nervosa, self-induced starvation and a phobia of gaining weight, is found in approximately 1% of female adolescents in the United States, and estimates are even higher for teen girls from middle- and upper-class Euro-American families (Mehler 1996). Although still relatively rare, some researchers believe the problem is becoming more common among teen boys. Although self-induced starvation has been reported for centuries, it was rarely associated with a fear of being fat. Some religious aesthetics, for example, fasted to express their religiosity, and Chinese Taoists fasted to extend their longevity, separate themselves from the mundane world, and encourage mystical experiences. These forms of fasting appear to differ significantly from the type found in Western societies today (Rieger et al. 2001).

Table H4.1   Criteria for Diagnosis of Anorexia Nervosa

A person with anorexia nervosa demonstrates the following:

A. Refusal to maintain body weight at or above a minimal normal weight for age and height, e.g., weight loss leading to maintenance of body weight less than 85% of that expected; or failure to make expected weight gain during peirod of growth, leading to body weight less than 85% of that expected.

B. Intense fear of gaining weight or becoming fat, even though underweight.

C. Disturbance in the way in which one's body weight or shape is experienced; undue influence of body weight or shape on self-evaluation, or denial of the seriousness of the current low body weight.

D. In females past puberty, amenorrhea, i.e., the absence of at least three consecutive menstrual cycles. (A woman is considered to have amenorrhea if her periods occur only following hormone, e.g., estrogen, administration.)

Two types of anorexia nervosa include:

• Restricting type: during the episode of anorexia nervosa, the person does not regularly engage in binge eating or purging behavior (i.e., self-induced vomiting or the misuse of laxatives, diuretics, or enemas).

• Binge eating/purging type: during the episode of anorexia nervosa, the person regularly engages in binge eating or purging behavior (i.e., self-induced vomiting or the misuse of laxatives, diuretics, or enemas).

Source: Reprinted with permission from American Psychiatric Association, *Diagnostic and Statistical Manual of Mental Disorders,* 4th ed. (Washington, D.C.: American Psychiatric Asociation, 1994).

Anorexia is a devastating disease. The anorexic may lose between 15% and 60% of her normal body weight, losing significant amounts of muscle as well as fat. As her body loses weight, signs of chronic starvation appear: Menses stop, a covering of soft hair appears on the skin, the heart rate slows, and she feels cold even in warm surroundings. Excessive exercise combined with anorexia often leads to joint problems, especially in dancers and athletes. The skin may become dry and scalp hair thin. The stomach is often bloated. Memory may become impaired, and thinking may be confused. Anorexia can cause renal failure, serious gastrointestinal problems and abdominal pain, neurological complications, hormonal imbalances, osteoporosis, serious heart problems, blood disorders, brain damage, and a permanent reduction in stature. The psychological trauma and loss of self-esteem due to anorexics' obsession with their weight cannot be calculated (Fraser 1997). Between 4% and 20% of anorexics die, usually from heart failure or suicide (Herzog et al. 2000). The criteria for diagnosing someone with anorexia are summarized in Table H4.1.

Bulimia, purging by vomiting or the use of laxatives and diuretics, is far more common than anorexia (reported by approximately 3% of women and 10% of female college students). Bulimia usually starts in early adolescence when youth try to restrict their diets, fail, binge, and then purge. These binges often start with relatively small amounts of food (from 100 to 1,000 calories) but may go up to over 5,000 calories in just a few hours. Binges are followed by vomiting and use of laxatives, diet pills, and/or drugs to reduce water retention. Bulimics average about 14 binge-purging episodes a week, with some vomiting as many as four times a day. The criteria for diagnosing someone with bulimia are summarized in Table H4.2.

Victims usually develop severe tooth decay from destruction of the tooth enamel by the acid in their vomit. The strain of vomiting can break blood vessels in their eyes. Gums may become diseased, and pimples or rashes may break out on the face. Other common health problems include a constant sore throat; swollen glands near the cheeks; esophageal inflammation; liver, heart, and kidney damage; dehydration; and stomach rupture.

A third type of disordered eating includes people who binge-eat without purging, those who binge and vomit infrequently (less than twice a week for less than 3 months), repeatedly chew and then spit out food without swallowing it, or follow extremely restricted diets (MDConsult 2002).

In addition to teenage girls, dancers, wrestlers, skaters, gymnasts, and other competitive athletes, homosexual men, and men and women in the military have higher-than-average rates of eating disorders.

What Causes Eating Disorders?

If you or anyone you know has experienced "disordered eating patterns," you know how devastating this problem can be. What causes a person to starve herself or himself continually or binge and

Figure H4.3
Anorexics lose significant amounts of muscle as well as body fat.

© Tony Freeman/PhotoEdit

Table H4.2   Criteria for Diagnosis of Bulimia Nervosa

A person with bulimia nervosa demonstrates the following:

A. Recurrent episodes of binge eating. An episode of binge eating is characterized by both of the following:

   1. Eating, in a discrete period of time (e.g., within any two-hour period), an amount of food that is definitely larger than most people would eat during a similar period of time and under similar circumstances; and

   2. A sense of lack of control over eating during the episode (e.g., a feeling that one cannot stop eating or control what or how much is eating).

B. Recurrent inappropriate compensatory behavior in order to prevent weight gain, such as self-induced vomiting; misuse of laxatives, diuretics, enemas, or other medications; fasting; or excessive exercise.

C. Binge eating and inappropriate compensatory behaviors that both occur, on average, at least twice a week for three months.

D. Self-evaluation unduly influenced by body shape and weight.

E. The disturbance does not occur exlusively during episodes of anorexia nervosa.

Two types:

- Purging type: the person regularly engages in self-induced vomiting or the misuse of laxatives, diuretics, or enemas.
- Nonpurging type: the person uses other inappropriate compensatory behaviors, such as fasting or excessive exercise, but does not regularly engage in self-induced vomiting or the misuse of laxatives, diruretics, or enemas.

Source: Reprinted with permission from American Psychiatric Association, *Diagnostic and Statistical Manual of Mental Disorders,* 4th ed. (Washington, D.C.: American Psychiatric Association, 1994).

BOX H4.2

I feel like such a hypocrite. People look at me and see a small, healthy person. I see a person who gorges on food and is totally out of control. You wouldn't believe how much I eat. I shove food into my mouth so fast that I choke. Afterwards, my stomach feels like it will burst. . . . My boyfriend knows and is real supportive, but it hurts our relationship. I won't eat out with him. Sometimes I want him to take me home so I can binge. I'll make up an excuse to end our date . . . I hate to say this, but I'd rather binge than make out. I get real moody if anything interferes with my bingeing. (quoted in Pipher 1994:167–69).

then purge? A variety of theories have been offered to answer these questions. These theories are often categorized into biological, psychological, sociocultural, and feminist orientations (Banks 1992; Thompson et al. 1999b).

Biological explanations of eating disorders usually focus on abnormal hormonal balance and other biochemical changes observed in people with anorexia. Of special interest are changes in a tiny portion of the brain called the hypothalamus that regulates appetite and weight control. Brain imaging studies show that people with anorexia have high levels of certain proteins secreted during stress that may indirectly block appetite. Other substances that stimulate appetite and reproductive hormones are abnormally low in anorexics. Most experts believe excessive weight loss and restrictive dieting cause these changes. However, menstrual changes have occurred before weight loss began in some girls, suggesting that brain chemical abnormalities may contribute to the development of anorexia in some teens.

Many *psychological factors* have also been linked to eating disorders. Among the most common theories is the view that anorexia is a symptom of other psychiatric problems, especially anxiety disorders, and obsessive-compulsiveness. Emotional problems are very common among people with anorexia and bulimia; however, it is unclear if these cause eating disorders or result from them. Phobias and obsessive-compulsiveness usually develop before the eating disorders begin, whereas anxiety attacks tend to occur afterward. Approximately one-third of people with bulimia and two-thirds of those with anorexia suffer from obsessive-compulsiveness—persistent, recurring thoughts and repetitive routines associated with dieting and weight gain. They may exercise excessively, weigh every bite of food, cut food into small pieces, and/or chew each morsel a specified number of times. Other psychological characteristics associated with eating disorders are narcissism (sensitivity to criticism and the need for a great deal of admiration from others) and borderline personality disorder (difficulty controlling anger, fears, and impulses). Depression is also common among people with eating disorders, but it is more likely to be the result rather than the cause of the problem; treating depression rarely cures bulimia or anorexia, but weight gain often relieves depression (MDConsult 2002).

Psychologists have also examined parent-child relations in a search for clues about the causes of disordered eating. Bulimia and anorexia nervosa appear to be more common among children raised in a critical family environment, especially when parents encourage them to diet, tease them about their weight, or try to dominate them in other ways (Lunner et al. 2000). Teens who suffer from bulimia are more likely than teens without the disorder to report that their mothers invade their personal privacy and relate to them in a jealous or competitive manner. Bulimia is also more common among girls whose fathers are detached (perhaps making them feel worthless) or treat them in a sexual or seductive manner. And many studies show increased eating disorders among teens whose parents show excessive concern about their weight, eating, and appearance (Rorty et al. 2000). Children who are sexually abused are also much more likely than those who are not to develop disordered eating patterns (Neumark-Sztainer et al. 2000). In contrast, young people raised in cohesive families with good parent-child communication are less likely to develop these problems (Thompson et al. 1999).

Sociocultural theories propose that the value Western societies place on thinness and self-control leads to eating disorders. Sociologists and anthropologists point out that the cult of thinness and disorders such as anorexia nervosa and bulimia are found predominantly in Western societies. As described in Chapter 4, a slimmer body has been the norm in the United States and Canada for almost a century, and this idealization of thinness may serve as a counterbalance against the tendency to gain weight in societies with abundant, stable food supplies where people can gain weight easily and obesity is common.

According to sociocultural theorists, the media plays a powerful role in transmitting the cult of thinness. Television, radio, magazine, and other print messages use anorexic models whose pictures have been airbrushed and computer-altered to create unachievable standards of beauty. Numerous studies have shown that exposure to images of ultrathin models can lead to body dissatisfaction and unhealthy eating behaviors. In one randomized experiment, college women viewed 10-minute videotapes of commercials that either contained thin and attractive images of people or contained images that were neutral and unrelated to people's appearance. After viewing the tapes, the women who saw the videotape stressing the importance of thinness reported greater amounts of depression, anger, weight dissatisfaction, and overall dissatisfaction with their appearance than those exposed to the neutral message (Heinberg and Thompson 1995).

BOX H4.3

When I was 12, I stopped eating. . . . I didn't eat lunch at all any time. I wouldn't eat much breakfast or dinner. . . . (My parents) sort of thought I just didn't feel well with breakfast and dinner. . . . They didn't know about me not eating lunch till one day I left about two days' lunches in my bag and Mum found them and I got blasted. . . . I do that regularly now. . . . Sometimes I think I'm fat, which I'm not, I think. I just get these feelings where I think I'm revolting in the public eye. . . . You see on TV lots of people who are tall and glamorous and skinny, and that's probably what you're meant to look like, so you feel odd and you don't fit in and just feel revolting. (12-year-old Australian girl quoted in Wertheim, Koerner, and Paxton 2001.)

Some social scientists blame a capitalistic society for the increase in anorexia and bulimia (Nichter and Nichter 1991). They note the obvious benefits of creating a demand for weight-control products for the $33 billion diet industry, $20 billion cosmetic industry, $300 billion cosmetic surgery industry, and rapidly growing fitness industry (Colditz 1992).

> *These industries have developed a "sure-fire" formula for success: Standardize a thin ideal of beauty that the majority of women can never attain, but make it look so appealing that they actually seek it out. The pursuit of thinness and the subsequent failure of most women in this pursuit construct an indefinite marketing of consumers. (Germov and Williams 1999:121).*

Feminist scholars have also entered the debate, proposing that eating disorders are a natural response to pathological societal pressures to be thin rather than a manifestation of psychiatric illness (Bordo 1993; Thompson et al. 1999). Although recognizing that many men are concerned about their weight, they argue that body imagery has special meaning for women. Women grow up knowing that attractiveness is a key to attracting men. They expect others to evaluate themselves in terms of their appearance and begin to watch their own bodies as outside observers. Women also learn that dieting and appetite control are necessary to achieve society's standards of beauty (McKinley 1999).

Feminists note that today's women are conflicted because they are socialized to seek equal opportunity in education and employment but, at the same time, must maintain the sexual identity of a pleasing wife and mother. The result is confusion, body dissatisfaction, low self-esteem, and ambivalence about eating. By restricting their weight, women gain a sense of control that is otherwise thwarted in a male-dominated society (Beardsworth and Keil 1997; Thompson et al. 1999). Anorexia nervosa is considered a rebellion against misogynistic societal norms that demean women by objectifying their bodies, and the anorexic's decision to lose weight is seen as a symbolic protest against male domination, a rigid sexual division of labor, and the traditional female role (Banks 1992).

Other feminists argue that eating disorders represent anxieties about success and femininity. Starvation stops menstruation and delays their bodies' maturation, allowing them to return to the less threatening prepubescent body and avoid becoming a sexually mature, autonomous person in a world they see as dangerous for women.

Of course, each of these theories has its shortcomings. No single theory, for instance, explains why only a small proportion of women develop eating disorders, whereas others gain excess weight or stay within healthy limits. Even in randomized experiments, not all women respond negatively to media messages promoting thinness. Most likely, a combination of factors, such as young people's susceptibility to social pressure or their feelings of worthlessness or shame, mediate the effects of media exposure to cultural ideals of thinness (Banks 1992; Murray, Waller, and Legg 2000; Thompson et al. 1999).

Other evidence that a combination of biological, psychological, and sociocultural factors work together to create eating disorders comes from studies of twins who are raised together and twins who are raised separately. These studies suggest that a combination of genetic factors and environmental factors (parental influence, peers, etc.) contributes to the development of eating disorders (Bulik et al. 2000). As a result, some researchers are now testing comprehensive models that examine the combined effects of people's global psychological makeup (e.g., self-esteem, emotional problems, susceptibility to external feedback), societal pressure (media exposure; teasing by relatives, peers, and friends), body image, and levels of dissatisfaction on people's actual weight and disordered eating practices (Thompson et al. 1999).

Applications for Nutritionists

Health care providers are frequently requested to assist people in achieving their ideal weight. Success requires an understanding of both the medical and cultural definitions of the term "ideal weight," as well as the ability to reconcile the differences between them.

Medically, ideal weight is usually determined in terms of height, as given in the body mass index, Metropolitan Life Insurance tables, or other indexes. Because the relative amounts of fat and muscle tissue and bone size vary greatly from person to person, ideal weights are given as ranges. Many experts argue that these ranges should be made even more flexible and point out that health hazards are associated only with *extreme* overweight and underweight (that is, 20% over or under the ideal range). Recommendations to diet, therefore, should be directed only to those who are extremely over- or underweight.

Flexibility in medical definitions of ideal or acceptable weights gives the health care professional an opportunity to adapt his or her recommendations to the client's cultural notions of body size. When working with people who value fatness, the professional can recommend weights that fall at the upper end of the range, and focus on the other end of the range when working with people who value thinness. Both groups can be made aware that a range, rather than a specific weight, will allow them to maintain good health.

It is important to recognize that even well-intentioned actions of health promoters may do as much harm as good if their efforts reinforce unrealistic standards of beauty and thinness. Health care professionals working in North America and other settings where thinness is highly valued need to be careful to avoid contributing to the barrage of media messages portraying an ultrathin figure as the only way to be attractive and healthy. Particularly when working with groups prone to anorexia nervosa and bulimia, obesity prevention and control efforts should promote protective behaviors, such as increased fruit and vegetable consumption and physical activity, rather than dieting. Promotion of healthy eating and physical activity can focus on a wide variety of benefits they offer people, rather than solely on weight loss. When working with Native American youth, for example, Davis and her associates (Davis and Lambert 2000; Davis et al. 1999) recommend that public health professionals build obesity prevention programs on traditional cultural values by promoting exercise as a way to become strong and healthy eating as a way to reinforce ethnic pride.

It is also important to avoid blaming the victim by focusing on environmental factors rather than just individual knowledge and attitudes. Some programs are now making it easier for people to maintain their weight by lowering the fat content of school cafeteria food, offering physical education programs during or after school, and building walking trails in safe neighborhoods.

An understanding of the social and cultural factors that impact body imagery and dieting should also help nutritionists avoid the common mistake of blaming the victim when counseling or designing health promotion interventions. Because few people succeed at long-term weight loss, many nutritionists become frustrated and impatient with their clients. To avoid blaming the victim, most nutritionists have learned to reframe weight-loss goals so that even small weight losses are considered successful and cause for celebration. (For example, the loss of just 10 pounds has been shown to help normalize blood sugar levels in some diabetics.) Debate continues on the feasibility of significant, long-term weight loss for severely overweight patients, and some medical professionals have changed their approach radically, recommending a "nondiet" approach to eating moderately. Others focus their efforts on prevention of obesity rather than weight loss (Parham 1999).

An understanding of the multiple factors that contribute to eating disorders should also make it clear that anorexia nervosa, bulimia, and other severe forms of eating disorders are *not* nutritional problems and may require assistance from other professionals. Most experts believe success requires a multidisciplinary team including psychologists, nutritionists, physicians specializing in the relevant medical complications, and other medical personnel skilled in treating eating disorders.

Food Technologies: How People Get Their Food in Nonindustrial Societies

The next time you go into a supermarket, take a few moments to consider the technological achievements behind today's food system. As you hunt and gather fresh fruit and vegetables from the produce section, notice the amazing abundance before you. Oranges, grapes, tomatoes, corn, and broccoli—most foods, in fact—are now available year round. Just 50 years ago, the same items were missing from the grocer's shelves during much of the year. Today, modern transportation not only enables you to eat your favorite produce in winter, but also brings you many new and exotic foods: California artichokes, Australian kiwifruit, grapes from Chile, star fruit from Asia, avocados and mangoes from Florida, and the tropical papaya and plantain. Access to foods from such a diversity of regions is unprecedented in human history. It may make it necessary to ask the produce manager for advice in preparing, or even identifying, the store's newest selections.

Consider also how this produce reaches your supermarket. The transportation system that makes this abundance possible is not without its costs. In fact, over 500 million gallons of gasoline are used annually to move fresh vegetables and produce to market; 1,300 million gallons are needed to move manufactured food products from processors to warehouses and supermarkets.

As you leave the produce section and push your cart down the aisles, past the meat and dairy departments, stop again—this time to appreciate the tremendous variety of food products from which you may choose. Your supermarket has over 49,000 different items stacked on its shelves, with 16,390 new flavors, colors, or varieties of products introduced in the year 2000 alone (Food Marketing Institute 2001). More than 50% of these items were not there just 10 years ago. Ask yourself how these new products are made. Can you, for example, name the major ingredients in Cool Whip, give the formula for 7UP, or explain how Grape Nuts and Cheetos are processed? If you are among the vast majority of North Americans who were not raised on a farm, you may also recall the first time you realized that hamburger meat, drumsticks, and pork chops came from those lovable animals pictured in your storybooks.

When your shopping is finished and you line up at the checkout counter, stop to compare your experience in getting food with that of people in other places and times. There are several striking differences.

CHAPTER 5

First, unlike the majority of people who have inhabited the earth since human beings emerged, you are probably not directly involved in collecting, hunting, or raising your own food. Because each farmer in industrialized societies can produce enough food to feed 35 to 50 people, you are free to pursue other occupational interests, investing only a small proportion of your time and money for food.

A second difference is that your access to a food supply is set apart from most contemporary and past people because of its remarkable consistency, quality, quantity, safety, and convenience. Whether you are in New York City or Carmel, California, you are likely to find a similar selection of breakfast cereals, canned vegetables and juices, and meat and dairy products. When you purchase a carton of strawberry yogurt, you expect to get a product flavored in a consistent way and free from the risk of food poisoning.

Finally, because you are not directly involved in food production and you have access to such a predictable food supply, it may often appear as if modern industrialized societies have achieved technological mastery over the environment. When you are eating tropical fruits during winter, it is sometimes hard to recognize how environmental conditions influence your diet and health. Even the economic effects of droughts, untimely frosts, and other climatic disasters are cushioned by national and international pricing and trade policies. Nevertheless, warnings of vanishing topsoil, acid rain, shrinking farmland, and the danger of pesticides should remind you that our society, too, is struggling to achieve a harmonious balance between its technology and the environment.

The wide array of fruits and vegetables and other foods available in North American supermarkets is the result of the development of a global industrialized food and nutrition system, which has its roots in the "age of exploration" and the industrial revolution (see Chapters 3 and 6). This chapter describes how groups of people adapt to the challenges posed by the physical environment in which they live. Every society has developed tools, techniques, and strategies for obtaining food, clothing, shelter, transportation, and protection from natural predators, and for processing food for consumption. We refer to this part of culture as **technology.**

At the core of each culture's technology is food-getting, or **subsistence activities.** A society that lacks effective methods for obtaining food or other energy resources cannot survive. Subsistence activities are cultural because they are shared by the society and take place in an organized, systematic manner based on decisions the group has made about the use and distribution of food and energy. People do not simply go out and hunt, plant, raise animals, build irrigation channels, or establish factories as if they were independent workers. Subsistence activities, and the distribution of the resultant food and wealth, require that the group make decisions about the directions hunters will take in pursuit of animals, how land will be divided, which fields will be cultivated, and where factories will be built and who will work in them.

Because the provision of an adequate food supply is imperative to survival, a society's subsistence activities are intricately tied to

technology
Tools, techniques, and strategies used by a group for obtaining, producing, and processing food for consumption, and for providing clothing, shelter, transportation, and protection from natural predators.

subsistence activities
Those parts of technology that are used for obtaining, producing, and processing food.

other aspects of its culture. In fact, when societies are grouped according to the type of food-getting techniques they employ, those falling into the same subsistence categories often share many similar social organizational and ideological features as well. Furthermore, they share some common challenges to health and nutrition. The complex of social and cultural conditions that are associated with different sets of tools and techniques for food getting is called a culture's **adaptive strategy** (Cohen 1974). The food system described in the first few paragraphs of this chapter is associated with the adaptive strategy of *industrialism.* It is characteristic of highly industrial and postindustrial countries such as the United States, Canada, and Europe, but is also becoming increasingly characteristic of the urban areas of less industrialized and developing countries. However, industrialism is still not the only way in which groups of people adapt to their environments in order to meet their nutrient needs.

As noted in Chapter 1, Yehudi Cohen (1974), and a number of writers who followed him, identify five major types of adaptive strategies with a number of variations:

- Foraging (hunting, gathering, and fishing)
- Extensive agriculture (horticulture)
- Pastoralism
- Intensive agriculture (plow agriculture)
- Industrialism (industrialized agriculture)

The characteristic social organization, ideological features, and their consequences for diet and health of each of these adaptive strategies are described in the sections that follow. As with any attempt to group societies into categories, we must overlook many important differences between the societies grouped together in order to focus on the outstanding characteristics shared within each category. Keeping this in mind, we nevertheless find it useful to compare societies by how they obtain their food: foraging, horticulture, pastoralism, intensive agriculture, and industrialized agriculture. In this chapter we focus on the first four of these strategies, strategies in which people use nonindustrialized technologies. These strategies are comparatively *localized,* that is, they are systems in which a large number of people are involved in food production and in which foods are consumed relatively close to where they are produced. In the following chapter (Chapter 6), we will discuss the global industrialized food system that characterizes industrialized countries and urban centers in less developed countries.

Foraging

Foraging, or **hunting and gathering,** is a subsistence technology that relies almost totally on human energy and simple tools to collect wild plants and kill free-roaming animals. As we discussed in Chapters 2 and 3, foraging was followed by all pre-human and human groups until about 12,000 years ago. Today there are few groups in which more than 75% of the food comes from the gathering of wild plants, the hunting of wild animals, and fishing.

adaptive strategy
The complex of social and cultural conditions that are associated with different sets of tools and techniques for food getting.

foraging (hunting and gathering)
An adaptive strategy that relies almost totally on human energy and simple tools to collect wild plants and kill free-roaming animals.

Today's foragers can be found in remote regions of the forests of Madagascar, Southeast Asia, including Malaysia and the Philippines, and islands off the Indian coast (Kottak 2000). However, a larger number of foraging groups were observed and described in the twentieth century. These are the contemporary foraging groups to which we will refer in this section.

In contemporary times, foraging groups are found in marginal lands not suitable for agriculture, such as the desert fringes of Australia, southern Africa, and North America; in the forests of Asia, South America, and Central Africa; and in the frozen wastelands of Siberia and the Arctic. Figure 5.1 shows the location of groups that used foraging as an adaptive strategy in the twentieth century. Among these groups are the Ju/'hoansi (sometimes called Bushmen, or !Kung) of the Kalahari Desert of Botswana and South Africa; the Mbuti people (sometimes called Pygmies) of the equatorial rainforest of the Republic of the Congo; the Inuit and Indian peoples living in or just below the Arctic Circle; the Nootka and Kwakiutl peoples of the Northwest Coast of North America; and the indigenous (aboriginal) peoples of Australia. Foraging groups also exist or existed recently in South America, the Philippines, Southeast Asia, and South Asia. However, as we discuss the diets and lives of contemporary foragers we should keep in mind that, in large part, contemporary foraging groups have been pushed onto these marginal lands by competing social groups that have more complex tech-

Figure 5.1
Shaded portions of the map indicate regions where foragers lived at the time of first contact with Europeans.

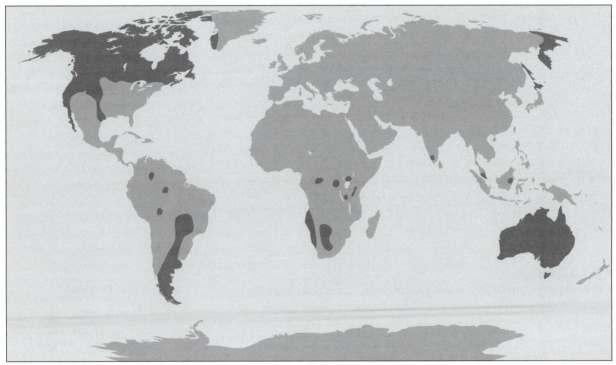

Source: Authors

nologies. Before the Neolithic revolution, human groups in a wide variety of ecological settings relied on foraging for their livelihood. The lives of foragers living in more abundant landscapes would have been a good bit different from the lives of contemporary foragers inhabiting marginal lands.

The subsistence tools used by foragers tend to be simple: digging sticks for digging up roots, spears and bows and arrows for killing animals, pounding sticks for preparing tough foods, spits for roasting, and baskets and slings for carrying tools and children. The number of tools that foragers can accumulate is limited by the need to pack them up frequently and move. Most foragers either roast, bake, or steam foods or eat them raw. It appears the use of pots for boiling foods for foragers is very recent and is the result of borrowing the technology from neighboring groups of cultivators or pastoralists. This means that foragers may well lose much of the fat and juices in meats during the cooking process. It is only people with pots for boiling that can capture fat and other meat juices in broth (Hayden 1981).

One factor that accounts for some of the differences between groups of foragers is an aspect of the physical environment—latitude. That is, foraging groups living at different latitudes (closer to or farther from the equator) face different physical environments and different health and nutrition challenges. Most importantly, ecosystems at higher latitudes (Arctic and sub-Arctic) have fewer plants available for food. Foragers at high latitudes, such as the Arctic, must rely almost exclusively on animals for foods, whereas foragers nearer the equator rely more heavily on plant foods.

The dietary challenge to contemporary foraging groups is to find enough food to feed themselves, without overexploiting the resources available, in environments in which they live off of the locally available plants and animals. This aspect of contemporary foragers may be different from that of Paleolithic foragers, as contemporary foragers generally find themselves in restricted environments, hemmed in by cultivators and others, with little chance of migration. Paleolithic foragers had the opportunity to move on to new areas if they overexploited a particular environment. The need to use available resources without overexploitation results in a common demographic characteristic of foragers: They live in small groups that move frequently in their search for food.

Demographic Issues of Foragers

While there is a good deal of variation in the size of foraging groups, population density rarely exceeds one person per square mile. The exact size of a group and its density are limited by the resources available and the nomadic lifestyle. The average population density for contemporary hunters and gatherers is about one person for every 20 square miles. In harsh environments, camps may be limited to two or three families that move frequently within a large area. For example, in the rainy season, the Ju/'hoansi (Ju) of the

Dobe region of the Kalahari Desert in Botswana live in groups of 3 to 20 huts near seasonal water holes. They occupy a single site for anywhere between 3 weeks and 3 months and then move on. According to anthropologist Richard Lee (1993), "The Ju typically occupy a campsite for a period of weeks and eat their way out of it" (p. 48). The length of time they spend in a particular site is dependent on the size of the groups. Smaller groups have less of an impact on their immediate environment and can remain in one place for a longer length of time than larger groups.

The Inuit (Eskimo) peoples of the Arctic Circle face a different kind of ecosystem. In the high latitudes in which they live, there is little in the way of plant resources. The Copper Eskimo obtained about 55% of their food from hunting and about 45% from fishing. When following an exclusively foraging adaptive strategy, the Copper Eskimo moved their camps about once per month. The Mistassani Cree of Canada, at a similar latitude, with 75% of the diet coming from hunting and 25% from fishing, have a population density of about one person per one hundred square miles and spend only short periods in the same place (Hayden 1981). The primarily hunting/fishing foragers of the North need to be more mobile and to work harder than the gatherer/hunters of lower latitudes, and they face more frequent problems with food shortage and famine.

In foraging societies in lush surroundings, especially near rivers or other sources of fish, 50 to 100 or more families may reside in a village for much of the year. The Kwakiutl and the Nootka peoples of the Northwest Coast of North America live in an area of abundant resources. They occupy an area of the coast of British Columbia and southern Alaska not suitable for agriculture. However, the Japan Current brings warm water from the South Pacific to this area. The warm current moderates the climate and brings abundant supplies of fish, especially salmon, to the coast and up the rivers. Until recently, the runs of salmon in the rivers were so heavy that the rivers seemed to be alive with fish. It was so easy to spear, net, and trap salmon that during a salmon run a person could gather enough fish in a few days to provide enough dried fish for a year. A number of other marine resources including shellfish, hair seals, sea lions, sea otters, porpoises, and whales were available, as were a number of land animals, such as deer and mountain goats, and waterfowl. And although there were a limited number of plant foods available, several varieties of berries were abundant seasonally. With these especially plentiful resources, Nootka villages might have had populations of 1,500 people and population density was about 200 persons per 100 square miles (Gross 1992; Hayden 1981).

The problem of managing resources faced by the peoples of the Northwest Coast was different from what it was for the peoples of the Kalahari Desert. Rather than being spread out over the year, most of the foods were abundant only seasonally. Therefore, foods needed to be processed for long-term storage. The Kwakiutl dried and smoked fish and stored them in baskets. They dried berries and meat. They boiled candlefish and seals, skimmed off the edible oil, and stored it in jars made of dried kelp. All of this required a much more elaborate technology and tool set than that needed by the Ju/'hoansi. Whereas the Ju/'hoansi solved the problem of seasonal food availability by moving across a wide territory and sharing food, the Nootka and Kwakiutl solved the problem by developing a comparatively elaborate set of tools that allowed for preservation and storage of large quantities of food, but it kept them from moving about. The result was sedentary villages that established territories along salmon streams, which the villagers defended. Few contemporary foraging societies lived in regions of such abundance. However, the idea that most foraging peoples live under extreme hardship has also been revised. Until recently, foragers were believed to live on the brink of starvation, working long hours each day to eke out a meager existence. Studies have shown that this is true for only a few groups living in extremely harsh regions such as the Arctic.

The majority of foraging tribes enjoy good nutritional status and leisurely work schedules. The Dobe Ju/'hoansi (Lee 1968, 1972a, 1993) offer a good example. Typically, a Ju/'hoansi woman collects enough food in 2 days to feed her entire family for a week. The rest of her time is spent cooking, fetching wood, embroidering, dancing, storytelling, resting, and visiting with friends. This steady mixture between work and leisure varies little throughout the year. The Ju man's schedule is more uneven. Hunting may occupy an entire week followed by several leisurely weeks. As with the Ju woman, 2 or 3 days of hunting produce enough food for the entire week. Even the most avid hunter in the group studied by Lee worked only 32 hours per week. While this kind of work week is common for foragers at low latitudes, as we noted earlier, foragers at high latitudes need to spend much more time hunting and fishing to feed themselves.

Social, Political, and Ideological Features of Foragers

In foraging groups, labor is divided according to age and gender. Young children assist their mothers in gathering food, and young boys also hunt small animals. Women collect most of the plant materials and hunt animal species that are relatively abundant, distributed uniformly over the landscape, and can be found close to camp. Men usually hunt bigger game, especially species that are unevenly distributed over the landscape and require significant amounts of time away from camp to track. However, in some cultures women, especially young women without children, may do a significant amount of hunting. Among the Agta of the Philippines, a horticultural group in which hunting is an important activity, both women and men are considered capable and effective hunters. Anthropologist Agnes Estioko-Griffin (1986, 1992) describes a day of hunting with Agta women:

> On Christmas morning I accompanied Abey, a grandmother about 55 years old, her 12-year-old niece, and her three dogs to the edge of the forest. Abey carried her hunting knife, which she preferred to bow and arrows, and her betel pouch [a small bag for carrying betel nut]. Another team of two sisters, Littawan and

Taytayan, and their five dogs chose to go up river. The other women and children remained in camp.

We spent six hours walking on slippery trails and chasing after the dogs through thickets of thorns and vines. This was interrupted by only a few minutes of rest to chew betel and catch our breath. Finally, Abey decided that it was another unlucky day and we returned to camp.

We arrived in camp tired and hungry. The men, who had been away for two nights, also returned empty-handed. Later in the evening, however, Littawan and Taytayan walked into camp, each carrying a pack containing about 20 pounds of butchered wild pig. They appeared tired after having been gone for nearly 10 hours. The meat was immediately divided among all of the families in the camp, including us. Although it was evident that everyone was pleased with the sisters' success, no one immediately expressed praise for their accomplishment. (Estioko-Griffin 1992:273)

Because plants are found in greater abundance than game and are more evenly dispersed across the landscape, in most ecosystems women contribute a much larger proportion of the diet than men, often a two to three times greater volume of food. Meat provides only 20–40% of the calories, but is more highly valued than plant foods.

All members of foraging societies are food-producers. Children learn early to collect plants and small animals, although they are not expected to contribute regularly to the food supply until they are adults. Virtually all men hunt game, and all women hunt and gather plant foods. There are no other specialized occupations.

Most hunters and gatherers place a high value on generosity and sharing. Because no single hunter has consistent success and meat spoils quickly without refrigeration, sharing ensures that most people in the camp get an adequate supply of protein. Typically, foragers have cultural rules regarding how the meat from game animals will be distributed among family members, hunting parties, and visitors (Hayden 1981; Lee 1993; Marshall 1976). Estioko-Griffin, in the passage above, describes how game provided by two sisters was shared among all of the families in camp. The value placed on generosity and the regular exchange of food also strengthens ties within the group, enhancing cooperation and social cohesiveness. The words of Nisa, a Ju/'hoansi woman, have been recorded by anthropologist Marjorie Shostak. About the value of sharing food, Nisa says:

"When we were living in the bush, some people gave and others stinged [were stingy]. But there were always enough people around who shared, people who liked one another, who were happy living together, and who didn't fight. And even if one person did stinge, the other person would just get up and yell about it, whether it was meat or anything else. 'What's doing this to you, not giving us meat?' When I was growing up, receiving food made my heart happy. There really wasn't anything, other than

stingy people, that made me unhappy. I didn't like people who wouldn't give a little of what they had." (Shostak 1981:88)

The high value placed on sharing raises the issue of how groups deal with freeloaders. Even though almost every healthy adult contributes to the food-getting process, there is the occasional freeloader—a healthy adult who does not gather or hunt his or her fair share. In more abundant ecosystems, freeloaders are supported with little concern by others. In harsher, Arctic and sub-Arctic climates, nonproductive adults are unlikely to survive.

The basic unit of social organization in foraging societies is the **band,** a small group of people related by real or fictional kinship ties. Members are often expected to marry outside the band and usually live near the husband's or wife's parents. Most foraging groups have no recognized social groups outside of the family—no clubs, no classes or other types of social stratification within the group. In most foraging societies (the Northwest Coast of North America is an exception), there is no formal leadership. Some individuals may have more influence on decisions because of their personal characteristics, but groups are essentially **egalitarian.** Some groups, such as the Ju/'hoansi, have part-time ritual specialists—healers and shamans—but not other types of occupational groups.

Many foraging groups have mechanisms to keep people from gaining more prestige than others. The situation described by Estioko-Griffin regarding the successful sisters is not uncommon among foragers. Successful hunters are not praised, and may even be teased about their poor performance as hunters. Among the Ju/'hoansi, Lee (1993) describes a custom he calls "insulting the meat" in which successful hunters are subjected to criticism about the meat they have captured and hunters themselves downplay the importance of their success. Customs such as these help keep even more skilled individuals from gaining prestige and power over their companions.

In most foraging societies, hunting is the focus of religious belief and ritual. Supernatural favor is sought to attract game and increase the hunter's chance of success. For example, during the whale-hunting season, the Eskimo hunter carries out rituals, wears special amulets to bring him good luck, and adheres to special **taboos** such as the avoidance of raw meat.

Diet and Health of Foragers

As mentioned earlier, the diet of foragers varies with the ecosystem in which they live and is highly correlated with the degrees of latitude from the equator. At lower latitudes it is comprised largely of plant materials, typically 60–80% by weight. The Dobe Ju/'hoansi get about 70% of their food by weight from gathered plant resources and about 30% from hunted animals. The aboriginal peoples of Australia typically ob-tained about 60% of their food from gathered plants and 40% from animals. However, the peoples living above about 50° north or south of the equator average less than 20% of the diet coming from plants, with the Copper Eskimo, at 69°N and well within

band
A type of egalitarian social/ political organization composed of a small group of people related by real or fictional kinship ties.

egalitarian
A type of societal group that lacks formalized differentiation in access to, and power over, basic resources among its members. There is no formal leadership and all members are more or less equal.

taboo
A prohibition against doing something for fear of harm from a supernatural force.

the Arctic Circle, traditionally obtaining none of their food from plant sources. When all foragers are considered, the majority of people following a foraging strategy obtain more of their diet and calories from plant sources than from meat. For Arctic groups, such as the Copper Eskimo, exclusively meat and fish diets provide more than enough high-quality protein, but they are low in fat, and, in order for essential vitamins and minerals to be obtained, the entire animal (entrails, organs, etc.) must be consumed (Hayden 1981). For groups such as the Copper Eskimo, some vegetable foods are provided through the consumption of the stomach contents of grazing animals such as caribou.

To get a better idea of what a forager's diet is like, let's take a closer look at the diets of Dobe Ju/'hoansi living in the traditional manner. As noted above, the Ju/'hoansi get about 70% of their diet from plant foods, which are collected by women using digging sticks and carrying bags. Fifty percent of the plant foods consumed come from a single source—the mongongo nut (*Ricinodendron rautanenii*). Mongongo nuts are abundant but scattered in the Kalahari (see Figure 5.2). Mongongo trees grow in groves, and camps are often located near a grove. The mongongo tree is highly drought resistant and reliably produces a crop even in drought years. The nuts have a very hard shell and can be picked up off the ground up to 12 months after they have fallen from the tree. Even in drought years, in which tens of thousands of pounds a year are gathered and consumed, thousands of pounds more are left to rot on the ground. The mongongo nut has a high protein and fat content. Mongongo nuts are consumed daily throughout the year. The average Ju/'hoansi consumes about 300 nuts/day. This yields about 7.5 ounces of nut meat, which provides 1,260 calories and 56 grams of protein. This is as many calories as 2.5 pounds of rice and as much protein as 14 ounces of lean beef. The supply of mongongo nuts is so reliable that Richard Lee (2000) reports that when a Ju/'hoansi man was asked why he did not plant crops, he replied, "Why should we plant, when there are so many mongongo nuts in the world?" (p. 38). Mongongo nuts are gathered by women and brought back to camp. There they are roasted and cracked with a stone, and the roasted meats are consumed.

In addition to the mongongo, 13 other plants are considered major foods and provide another 25% of the plant foods consumed. These foods are widely available on a seasonal basis. They include four species of fruits, berries, and melons, including the fruit of the *baobab* (a wide-trunked tropical tree), a species of wild bean, and a species of roots. These foods are eaten on a daily basis when they are in season. Minor foods are eaten several times per week when in season and provide abut 15% of the plant foods consumed. Among these the most important are 7 species of roots and bulbs that the women dig from the ground. Most fruits and melons and many roots are consumed raw. Some roots, beans, and stalks may be roasted.

About 30% of the Ju diet comes from animals, hunted primarily by men who use bows and arrows (with and without poison), spears, knives, hooks, rope snares, digging sticks, fire-making equipment, and a carrying yoke (a wooden bar laid across the shoulders in order

Figure 5.2
Ju/'hoansi women gather mongongo nuts while the men hunt.

to carry a load in two equal portions). While some hunting is done by stalking animals and shooting them with poisoned arrows (e.g., kudu, gemsbok, and wildebeest), more animals are killed by using dogs and by pursuing burrowing animals underground. Dogs are used to hunt warthogs, steenboks, duikers, and hares. The dogs bring the animals to bay, and the hunter kills the prey with a spear. Antbears, warthogs, and porcupines are pursued into their burrows and killed. Springhares are killed in their burrows with a special hook, and the burrow is then excavated with a digging stick. In some cases, hares, guinea fowl, and small antelopes are snared.

In a 28-day period, Richard Lee (1993) observed hunters bringing 454 pounds of meat, enough to provide an average of 9.1 ounces of meat per person per day. On average, a Ju will consume between 175 and 200 pounds of meat in a year. Plant foods are not usually shared, but most meat kills (except for small animals) are shared among all of the people in a camp. Men skin and butcher the animals, distribute the meat, and roast it over open fires.

On an average day, a Ju/'hoansi will eat about 300 roasted mongongo nuts, a bit more than one-half pound of roasted meat, and about a pound of raw and roasted roots, beans, leaves, edible gums, berries, fruits, and melons. This diet of meat and plants provides about 2,300 calories and 96 grams of protein, and it apparently provides adequate intakes of essential vitamins and minerals. What the diet doesn't provide is refined carbohydrates and, indeed, it is a bit low in fat.

Food Preferences of Foragers

Foraging groups clearly have preferences for some foods and methods for preparing them. The Ju/'hoansi consume only about half of the plants they recognize as edible in their environment. A large number of potential plant foods are ignored because they are not preferred, and these are consumed only when other foods are not available. Hunters in relatively abundant ecosystems seem to avoid the meat of carnivores, which tends to have a stronger flavor than herbivores. Hunters in sparser environments, however, eat virtually any animal and consume all edible parts of an animal. The Ju/'hoansi never eat vultures, dogs, or hyenas, and only rarely eat lions, whereas the Mistassani Cree of subarctic Canada consume almost every animal available.

A large number of foraging groups have been described as having a strong preference for fat (Hayden 1981). The diets of foragers generally provide little fat. Most plant foods (mongongo nuts are an exception) have low fat content. Game animals also have low fat content. The carcass of a 97-pound kangaroo, an important source of meat for aboriginal Australians, has only about 4 ounces of fat (Gould 1966). The land animals of North America are also quite lean. A condition known as "rabbit starvation" has been described for hunters in North America who starve despite consuming a large number of rabbits, presumably because there is almost no fat in the diet (Hayden 1981). For African hunters, the eland is highly prized because of the fat content. Australian foragers deem the intestinal fat of animals as a delicacy and may abandon especially lean carcasses. Although we have a stereotypical image of the Inuit peoples as heavy consumers of the fat of sea mammals (blubber), in fact, most of the fish and animals they consume are very lean, and blubber is a necessary source of calories and essential fats. Some researchers suggest that a preference for fat may be a human physiological adaptation to past human diets low in fat. It is certainly clear that many contemporary foragers have a stated preference for fat and work to include it in their diets.

Nutrition and Health of Foragers

As we saw above, the Ju/'hoansi diet appears to be quite good and the Ju people live in what appears to be a relatively stable and productive ecosystem. They are not atypical. Nutritional surveys of other contemporary gatherer/hunters show them to be generally free from severe protein-calorie malnutrition, scurvy, rickets, and vitamin B deficiency diseases. Bitot's spots, indicative of vitamin A deficiency, were noted in a small percentage of children. Although evidence of severe deficiency is rare among most gatherer/hunters, loss of subcutaneous fat during lean times of the year has been noted, suggesting that the amount of energy available at those times is not optimal. Some Inuit, and other peoples living at high latitude, who live traditionally do appear to have experienced periodic severe food shortages, especially during the harsh winter months. Some ethnographers have reported that old or unhealthy members of the Inuit community have left the camp and wandered off to the ice floes to die in order to take some of the pressure off the community created by the limited food supply.

The children of some foragers who have left hunting and gathering as a way of life and settled with agriculturists have grown taller and heavier than the children of those living traditionally. This suggests that energy intake, overall, may be somewhat restricted or that energy expenditure in a nomadic lifestyle is very high. However, as Frederick Dunn (1968) and other researchers have noted, foragers rarely experience the periodic times of famine and starvation that can accompany crop failure among agriculturists. The wide variety of the wild diet means that few specific nutrient deficiencies are found.

The low population density and nomadic lifestyle have other beneficial health impacts as well. As noted by a number of analysts (Black 1990; Black et al. 1974; Cohen 1989; Dunn 1968), the small size of the foraging group protects them from many of the infectious diseases that plague people living in crowded conditions. For example, illnesses such as measles, mumps, polio, influenza, and even the common cold require a large, dense population to be maintained endemically. In small isolated populations, an infectious disease that confers at least a short period of immunity cannot be maintained. When such a disease is introduced, all susceptible people become infected, they are then immune, at least for a time, and the disease dies out. The disease will not present a problem unless it is introduced again from the outside. It takes a large population, perhaps more than a million people, in order for some infectious disease to have enough susceptible people to infect. In modern times, such diseases have been introduced into small isolated foraging communities by outsiders. The result is that all community members get the disease quickly, but once there are no more susceptible people, the disease dies out until it is reintroduced by an outsider. Our Paleolithic ancestors probably did not have to contend with most of these diseases, including the common cold. It is only in the last 5,000 years or so that human populations have been large enough to suffer from measles, influenza, mumps, chicken pox, and colds.

Because foragers are nomadic or seminomadic, they usually move away from their wastes. This means that they have less contact with sources of parasite infection. The number of parasites in the intestines tends to be lower among foragers than among the sedentary societies we will describe later. As they have few animals, often only dogs, foragers are not often in close contact with the domesticated animals that share diseases with humans, such as pigs. However, other parasitic and infectious diseases, such as those transmitted by insects—especially those that have wild animals as hosts such as sleeping sickness, typhus, and tetanus—may be important afflictions. Without modern medical care, there is a high mortality rate among children and fewer children reach adulthood than in sedentary societies. Children who survive early childhood and adults are healthier, however, than in many peasant agricultural communities.

The chronic diseases we see in modern societies are rare in foraging societies. This is partly because fewer foragers reach the ages at which these illnesses occur, although a significant number of them do reach 60 years of age and beyond. However, the foraging lifestyle is also protective of many of the chronic diseases faced by people following an industrialized adaptive strategy. Few foragers suffer from obesity, an important contributor to chronic disease. Most foragers have a very active lifestyle. Lee (1972b) estimates that the average Ju/'hoansi woman walks 2,400 kilometers (about 1,500 miles) per year, often carrying a small child, food, and equipment. Diets tend to be varied and provide a wide variety of nutrients. Minimally processed diets provide large quantities of dietary fiber.

In sum, the foraging strategy has proven to be a successful adaptation to a wide range of environments and was characteristic of humans for thousands of years. Although now restricted mainly to marginal areas of the world, hunters and gatherers generally enjoy an adequate diet, good nutritional status, and good health.

At the present time, few, if any completely independent foraging groups remain. Most have some contact with settled, technologically more complex groups. For some, steel tools have replaced stone, bone, and wooden tools. Most Ju/'hoansi, for example, spend some time working for their agriculturist neighbors, using the cash they have earned to purchase foods. For centuries the Mbuti Pygmies of the Ituri forest of the Republic of the Congo, in Africa, have traded meat for agricultural products with their neighbors. They have essentially become commercial hunters who hunt animals for trade for grains and other plant foods. In competition for land and resources with more complex societies, the technologically simpler foragers are at a disadvantage.

Horticulture

horticulture (extensive agriculture)
A nonmechanized system of food production that relies solely on human labor to cultivate plants in small garden plots. It is characterized by reliance on human energy and a limited inventory of simple tools.

Horticulture, or **extensive agriculture,** is a nonmechanized system of food production that relies solely on human labor to cultivate plants in small garden plots. It is characterized by reliance on human energy and a limited inventory of simple tools. Digging sticks and wooden hoes wielded by people are used rather than plows drawn by animals. In more recent times, steel machetes and steel-tipped digging sticks and hoes obtained in trade have become common items in the horticulturalist's tool kit. Neither irrigation nor terracing of fields is used.

The defining characteristic of horticulture as an adaptive strategy is that it is *shifting* cultivation. That is, forest land is cut down and often burned to remove vegetation and leave an ash that enriches the soil. A crop is planted in the resulting "garden" for 1, 2, or 3 years. After 1 to 3 years, the soil nutrients are exhausted and/or the weed problem becomes too difficult to deal with using simple tools. The garden is then abandoned to return to forest and regenerate, and new fields are cleared. A long fallow period allows the soil to regain its fertility and be available in the future for cultivation. Depending on the environment, fallow periods may last from a few years to a few hundred years. A common name applied to horticulture is "slash and burn" agriculture, which involves cutting down the forest or brush and burning it.

As we saw above, the principal problem of foragers is managing hunting and gathering activities so that the environment is not degraded and continues to provide sufficient food. The problem of horticulturalists is similar. They must manage their food production activities in a way that allows them to continue to find sufficient productive land for their crops close to home. Horticulturalists, then, tend to be semisedentary or seminomadic. They build semipermanent villages and cultivate the land around the village. After several years, most of the land has been used and is in a 10- to 30-year period of regeneration into forest or brush. When the travel distance needed to get to fresh fields becomes too great, the village moves to a new site. For the Northern Kayapo of Brazil, the cycle of village movement appears to be about 10 to 12 years. Part of the seminomadic nature of the lives of horticulturalists is the need to find new areas to hunt and gather.

In traditional, independent-living horticultural groups, food production is intended for home consumption rather than commercial sale, each farmer has control over his or her own production, and there is little interdependence between groups.

Horticulture is still practiced by a large number of people throughout the world. In Southeast Asia, for instance, one-third of the total land area used for agriculture is cultivated with this type of subsistence technology. It is a common adaptive strategy in Africa (see Figure 5.3) and the rainforests of South America and Central America (Gross 1992). Horticulture appears to have been a common cultivation technique in Europe during the Neolithic time, but has been replaced in Europe with more intensive methods of cultivation such as plowing and fertilizing (Gross 1992). We can speculate that during the Neolithic revolution, the first attempts at agriculture around the world would have relied on shifting cultivation, as the large draft animals important for plow agriculture were not yet available. In contemporary times, horticulture as an adaptive strategy is found primarily in tropical areas with high rainfall and forest vegetation.

The shifting of garden plots in and out of cultivation means that at any one time a great amount of potential land is not in cultivation. The environment surrounding a village is more "natural" than that around intensive agricultural villages; wild plants and animals are still available for use as food. The "wild" and "tamed" parts of the environment are very close to one another.

Today, few horticulturists are independent. Many people using the simple techniques associated with horticulture are peasant farmers, living within nation-states and producing crops for sale as well as home consumption. In Central America, farmers cultivating hillsides with a machete or digging stick look out over mechanized agriculture being practiced in the valley below.

Horticulture demands relatively little investment of labor to produce an adequate diet as compared with intensive cultivation

Figure 5.3
This basket of food has just been harvested from a field in Sudan. It contains sorghum, millet, watermelon, okra, and hibiscus flowers (used as a tea). This Sudanese farmer also grows sesame as a cash crop.

© B. R. DeWalt, 1982

strategies. Among sago palm farmers in New Guinea, one day of work is sufficient to produce food that will meet 85% of the family's caloric needs for 4 or 5 days (Jerome, Pelto, and Kandel 1980). In a similar study of another group, the New Guinea Tsembaga, Rappaport (1968) found that only 380 hours per year per food producer were needed to grow enough taro, yams, and other crops to meet the caloric needs of the community. Studying Mekranoti, a village of the Northern Kayapo in the Amazon rainforest of Brazil, Gross et al. (1979) found that only an average of about 8.5 hours/week of work in the gardens were needed to provide the staple foods of a household. Horticultural tasks also have a seasonal rhythm requiring large inputs of labor at planting and harvesting times, interspersed with more leisurely periods at other times of the year.

Social, Political, and Ideological Features of Horticulturalists

As in foraging societies, the tasks associated with getting food are assigned according to gender. In most horticultural systems, men clear the brush, fence off the land, care for large domesticated animals, hunt and fish, and defend the villages from outsiders. Men may also help with the harvesting. However, the bulk of the tasks associated with gardening are carried out by women. Women are usually responsible for cultivating, planting, weeding, most of the harvesting, raising small animals, including pigs in some of the Pacific horticultural groups, some gathering of wild plants and small animals (including insects), and preparing food. DeWalt observed several instances in which Kayapo husbands and wives traveled to a garden together. While the wife cultivated and weeded the field, the husband sat in the shade with the baby. Because of the need for much land to be left fallow, horticulture cannot support dense or completely sedentary populations. Although the population density of horticulturists is higher than for most foragers, it is much lower than for intensive and industrialized agriculturalists and does not produce enough surplus to support a nobility or specialized, nonagriculturalist categories of workers. However, villages of several thousand people were reported for the Amazon, before introduced disease decimated the populations (Dufour 2000a).

It does appear that several civilizations were supported for a time by horticulturalists, principally the Classic Mayan civilization and several small pre-Inca states in the Andes. Horticulturalists, unlike foragers, may claim exclusive rights over land. Garden plots may be owned by individual families or kinship units, and people are expected to recognize the rights of the owner to the food produced in her or his plot. The increased value of land as compared with foraging, and the large amount of land needed as compared with intensive cultivation, places emphasis on territorial protection and warfare. Horticultural tribes tend to be in frequent conflict with their neighbors. The horticulturalists of New Guinea and the Amazon rainforest, which are relatively densely settled, are often described as warlike, although the technology of war (weapons) historically available to these peoples, such as bows and arrows and spears, means that mortality from war is much lower than it is in societies that use mechanized warfare.

Most horticulturists produce enough food to store for future needs and trade with other groups for items not produced locally. Because they are able to produce some surpluses, horticulturists often place some value on the accumulation of private wealth, creating differences between families. It should be kept in mind that in this context, wealth may mean more animals, such as pigs, or more ceremonial goods. Households with the most possessions frequently command the most respect and have the greatest influence in the community. However, this gap is not large enough to create major inequalities in control over resources or lead to the formation of social classes.

Another feature of horticultural societies is the increased complexity of social organization. Because most settlements are at least semisedentary, relationships between households are close and long-lasting. In addition to the family group that shares a common household, larger kin groupings are recognized. In most societies, people are organized into groups that trace their descent from a common ancestor. A leader of the descent group is expected to settle disputes and make decisions on behalf of the group's members.

Political organization varies depending largely on the size, density, and permanence of villages. Contemporary horticulturalists living independently tend to live in **tribal societies,** that is, sedentary or semisedentary societies in which there is little concentrated power or authority. In tribal societies there are often leaders, usually (but not always) males, who are referred to as *headmen* or *big men* by anthropologists. The leader or headman's power may be relatively weak, based primarily on respect and persuasion, or he may have some formal authority and power, but typically can influence the actions of his neighbors only through his powers of persuasion. Often, the headman has acquired more wealth and wives than other men in the community and is selected for personal attributes such as generosity, eloquence, bravery, or perceived supernatural powers. In some horticultural societies, the headman has the obligation and authority to regulate conflict and enforce rules. More commonly, however, he lacks the right to issue orders and must persuade people to take his advice.

Headmanship in horticultural societies is not a full-time occupation. Headmen are neither relieved of their farming and hunting/fishing tasks nor supported through taxes and tributes, although they do enjoy some special privileges not found in hunting and gathering societies, such as the use of special insignia or clothing.

Finally, whereas foragers rarely have multiple spouses, horticulturalists may have several. Most commonly the pattern of marriage is one in which **polygyny,** that is, one husband being married to two or more wives, is allowed and even encouraged. Men who can acquire several wives may be at an advantage in circumstances in which women do much of the work of cultivation. In the film

tribal society
A type of social/ political organization in which a number of households live together in permanent or semipermanent villages and in which there is little concentrated power or authority.

polygyny
A form of marriage in which one husband is married to two or more wives.

Onkga's Big Moka, Onkga, is a "big man" in a New Guinea horticultural group who is seeking to improve his influence by putting on a big feast in which a number of goods, including a large number of pigs, will be given as gifts. In order to manage the extra pigs, Onkga takes an older widow as a fourth wife. This wife's chief attribute is that she is good at raising pigs and the sweet potatoes that they eat. (Onkga needed another woman in the household to help in managing the pigs for the feast.) In cases in which women do a good deal, even most, of the work of cultivation and animal husbandry, polygyny is attractive for men who can manage larger households. However, it is also more difficult to manage and support several wives and their children. While polygyny is often allowed and even desirable, in most settings only a few men will be able to attract and support more than one wife. The majority of marriages are monogamous.

Much of the religious or ritual activities of horticulturists are aimed at improving food supplies. Fertility rites and other cultural artifacts are associated with horticultural societies dating back to the agricultural revolution 10,000 years ago.

Diet and Health of Horticulturalists

Historically, some grain cultivators have followed a strategy of shifting cultivation. For instance, in parts of Mexico and Central America, where population densities were low in the early twentieth century, corn and sorghum were cultivated using slash-and-burn technology. However, most contemporary horticulturalists live in tropical, high-rainfall environments in which grains are not successful crops. Due to high rainfall and poor soils, areas such as the Amazon rainforest and the rainforests of equatorial Africa and the Pacific, cannot maintain intensive cultivation, nor are these areas suitable for growing the **cereal grains** and legumes that historically supported the growth of cities and civilizations in more temperate climates. Contemporary horticulturists tend to rely on root and some tree crops for their basic staples. In the Amazon and Africa, these are manioc and sweet potato. In New Guinea and parts of the Pacific, the main crop tends to be yam. In Polynesia, taro has been the principal staple crop. In other parts of the Pacific, sago palm, which produces a starchy pith, is cultivated. Bananas are found as staples in parts of Africa and South America.

The root and tree crops characteristic of horticultural systems are much lower in protein and fat than the grains grown commonly in intensive agriculture—maize (corn), wheat, barley, rice, sorghum, millet, and legumes. The starchy staples must be supplemented with other foods to provide sufficient protein and fat. For this reason, most contemporary horticulturists are also hunters, fishers, and gatherers, relying on wild food to provide some of their nutrients. Wild fruits and berries, honey, and insects, along with some wild vegetables, are also collected and provide needed vitamins and minerals. Men frequently spend a good deal of time hunting and fishing. In New Guinea and parts of Southeast Asia and the Pacific, pigs are raised to supplement the diet of root crops.

cereal grain

A crop used for its seed that is a member of the grass family. The most common cultivated cereal grains are maize, wheat, barley, rice, sorghum, and millet.

In recent years, researchers increasingly have come to see the gardening systems of horticulturalists as more complex than originally thought. The food systems of the indigenous peoples of the Amazon rainforest are a case in point (Dufour 2000a). The Tukanoan people of the northwestern Amazon grow bitter manioc (cassava) as their staple crop. Tukanoan households clear one or two new gardens per year. They choose garden sites on the basis of what they know about drainage and soil conditions. They tend to clear roughly equal amounts of primary forest, land that has not been previously cultivated and secondary forest, land that has been previously cultivated and left fallow for a number of years. After the forest is cut and burned, bitter manioc is planted densely over most of the plot. Other crops are also planted, including taro, sweet potato, arrowroot, pineapple, chili peppers, *mafafa* (a root crop), *lulo* (a member of the potato family), bananas, and plantains. Coca, a shrub, and several tree crops, *guama, uvilla,* and peach palm, may also be planted.

Each plot is **polycultural,** that is, a number of different species of crops are planted. The plots are also **polyvarietal,** or planted with different varieties of the same species of crops. However, there is much more variation in the types of manioc, the staple crop, planted than there is in other species. Dufour (2000a) reports that in a survey of Tukanoan garden plots there were between 2 and 16 different species of crops in each plot, but between 17 and 48 different varieties of manioc. Tukanoan women plant different varieties of manioc for different uses, including a sweet variety that is attractive to a large rodent, which is killed and used as food. At any one time, a Tukanoan household will have access to gardens at various stages of development. Some will be newly planted with immature crops, others will be producing manioc, and one or more older plots, in the process of reverting to forest, will have fruit trees, fish poisons, and medicinal plants. The transition between cultivated and fallow plots is not a sharp one. Cultivated plots fade into fallow, often serving as orchards and grasslands on the way to reverting to forest.

Processing Manioc We noted above that the Tukanoa and many other South American and some African horticulturalists rely on the highly productive, bitter manioc plant for the bulk of their calories. It is called bitter manioc because it contains cyanogenic glucosides that break down into glucose (a sugar) and toxic prussic acid (hydrogen cyanide) when the roots are disturbed by harvesting, cutting, and processing. All manioc varieties contain some cyanogenic glucosides, but the amount varies from minute amounts in "sweet" manioc to the relatively high levels in bitter manioc. Sweet manioc can be peeled, cooked, and eaten without further processing. Bitter manioc must undergo additional processing to remove the prussic acid before it is consumed. Dufour (2000b) has described the process used by the Tukanoan people to detoxify bitter manioc (see Box 5.1).

The hydrogen cyanide is washed out of the manioc in the dewatering process and becomes a gas when the liquid is boiled. As the fiber and the starch are left for up to 48 hours, they ferment slightly.

polycultural
A planting strategy in which single plots are planted in a number of different species of crop.

polyvarietal
A planting strategy in which a number of different varieties of the same species of crop are planted together.

Detoxification of Bitter Manioc

First the roots are rasped to remove the outermost layer of peel, and then they are grated. Grating reduces the roots to a fine watery mash, which is then separated into three fractions: liquid, starch, and fiber. The separation is accomplished by washing (with water and extracted juices) and squeezing the mash in a basketry strainer to remove the starch and liquids from the more fibrous portion, and then allowing the starch to settle out of the wash water. Once the starch has settled, the supernatant [the substance floating on the surface] is decanted off and boiled immediately to make the beverage *manicuera*. The other two products, starch and fiber, are stored at least overnight but preferably for 48 hours, and are then recombined and baked as *casabe* [manioc bread, baked in a pit oven]. To prepare *casabe,* the fiber is de-watered in a *tipiti* [basketry sleeve press], lightly roasted, and then mixed with the starch. The starch is also used to thicken beverages (*mingao*) and fish porridges (*pune*).

Source: D. Dufour, "A Closer Look at the Nutritional Implications of Bitter Cassava Use," in A. H. Goodman, D. Dufour, and G. H. Pelto (Eds.) *Anthropology: Biocultural Perspectives on Food and Nutrition* (Mountain View, CA: Mayfield, 2000), p. 165.

The resulting manioc bread (*casabe*) has a characteristic sweet/sour flavor. The Tukanoa also make two other dishes from manioc. *Fariña* is made from a particular variety of manioc with firm yellow-fleshed roots. They are trimmed and soaked in a stream for 2 to 4 days until they are soft, and then peeled, grated, allowed to ferment for at least 3 days, de-watered in a *tipiti* (manioc press), and sifted to produce an even-textured flour. This is roasted into *fariña,* which is sprinkled on any number of dishes. If kept dry, *fariña* can be stored for an extended period of time. Finally, the Tukanoa make "fresh" manioc bread by de-watering and baking the grated manioc mash. It is denser than regular *casabe* and does not have a fermented taste. The Northern Kayapo process manioc in very similar ways. In both cases the method of detoxifying manioc results in a set of flavors and textures that have become highly preferred and are identifiably Amazonian. Indeed, *fariña* is incorporated into the diets of most Brazilians and give Brazilian cuisine a set of characteristic textures and flavors.

Noncultivated Foods Horticulturalists also rely on noncultivated foods. Plots left fallow remain productive areas. Some game species would not occur in large numbers in the forest if it were not for the availability of fallow gardens reverting to forest, which provide food for them (Posey 1982). Deer, tapir, and collared peccary may reach higher densities in secondary growth than in primary forest. Vickers (1988) carried out a long-term study of the game available around a village of Siona-Secoya people in the Amazon of Ecuador. He found that some species of game did seem to disappear over time as a result of hunting. These included woolly monkey, a large forest bird called the curassow, and a large ground-dwelling bird—the trumpeter. However, a number of species appeared to remain abundant. These included peccary, tapir, deer, other monkeys, birds, rodents, and reptiles used for food.

Although game animals are eaten, most of the foods of animal origin consumed by Amazonian peoples come from fishing. Both the Tukanoa and the Kayapo fish with bow and arrow, hook and line, and fish poisons.

Finally, insects provide protein and fat in the diet. The use of insects is widespread and includes ants, termites, larvae of both moths and butterflies, and palm grubs (Dufour 2000a).

Kayapo harvest certain kinds of palms, principally for the fatty white grubs they contain. A number of plant foods are also gathered from the wild. These include palms, which provide hearts of palm, medicinal herbs, and, for the Kayapo, large quantities of Brazil nuts, which provide high-quality protein and fat.

Horticultural societies that consume a wide variety of wild and domesticated plants and animals enjoy a well-balanced diet. The Tsembaga of New Guinea, for example, combine 36 species of domesticated and wild plants with domesticated pigs and small wild animals. Nutritional studies of the Kayapo found them to have excellent nutritional status. Although Kayapo adults are somewhat shorter than European adults, Kayapo children grow similarly to North American children, surpassing North American growth standards by 5 years of age.

However, for some horticulturalists in areas with high population density and less opportunity for hunting and fishing, the picture is not as good. The people of the highly populated highlands of New Guinea appear to suffer from low intakes of both calories and protein. A common problem for manioc cultivators in West Africa is the disease **kwashiorkor,** which is the result of insufficient protein intake, but adequate calorie intake. In kwashiorkor a child loses muscle and fat, but may appear plump as a result of edema. As a result, children with kwashiorkor often exhibit a characteristic round face called "moon face." In one of the languages of Ghana, the word *kwashiorkor* means "the illness the toddler gets when the new baby is born." It is a common affliction of toddlers weaned from breast milk and fed porridge made from manioc.

The diets of many contemporary horticulturists (such as the prehistoric horticulturists at Hardin Village described in Chapter 3) rely primarily on one or two crops that are high in carbohydrate and low in protein. Crops such as yams, sweet potatoes, plantain, and manioc are the dietary mainstay of many horticulturists. Because their diet contains minimal amounts of protein and sometimes other nutrients as well, and "because they have more of their eggs in one basket, they live a more precarious nutritional existence" (Pelto and Pelto 1976:4) than hunters and gatherers. In particular, they are susceptible to widespread protein malnutrition and are more vulnerable to starvation when crops fail. Kwashiorkor, beriberi, and other deficiency diseases are commonly found in societies that rely on a single crop as a major dietary staple.

Horticulturists' health is also affected by their pattern of living in permanent or semipermanent villages. Villages provide new breeding places for many forms of disease. Domesticated animals such as cattle, pigs, and fowl can transmit anthrax, Q fever (a tick-borne disease), brucellosis (a disease of cattle), tuberculosis, and intestinal infections caused by the salmonella bacterium or the ascarid worm.

kwashiorkor
A nutritional disease in which a child loses muscle and fat but may appear plump as a result of edema. It is thought to be the result of a diet sufficient in calories but low in protein.

Horticulturists in some regions also face increased threats from typhus and malaria when clearing new ground for cultivation.

Livingstone (1958) has illustrated the relationship between the spread of agriculture, malaria, and sickle-cell anemia. As the West African agriculturists expanded into the forest and destroyed the trees in the preparation of ground for cultivation, they encroached on the environment of the pongids (apes). The pongids, which were the primary host of malaria carried by the mosquito, were exterminated or forced farther into the forest. The mosquitoes quickly transferred to the hominids for their meals. Livingstone points out that agricultural activity, which provides new breeding areas for mosquitoes as well as a large population for the mosquitoes to feed on, led to malaria becoming an endemic disease (Armelagos and Dewey 1970). At the same time, the relatively small size and low population density of horticulturalists living in relatively isolated areas means that, like foragers, they are unlikely to maintain a number of infectious diseases.

Pastoralism

pastoralism

An adaptive strategy based on a set of technologies through which food energy is extracted from herds of domesticated animals.

Pastoralism is a set of technologies that extracts food and other forms of energy from large herds of domesticated animals. Like horticulture, it is an adaptive strategy that is extensive, that is, it uses large amounts of land. In some intensive agricultural systems, animal husbandry is important and the people are seminomadic. However, here we are discussing a system in which there is no cultivation and the economy and diets are centered on animal husbandry alone. Predominantly found in the Near East and Africa, pastoralism is a cultural adaptation to semiarid, open country where the land will support ruminant animals but not agriculture (Goldschmidt 1968). By specializing in animal husbandry, pastoralists are able to exploit harsh environments that cannot be irrigated and are too dry to support other modes of subsistence.

Pastoralism developed as a distinct way of life in Africa, Asia, and Europe. Five different zones of pastoralism, relying on different types and mixes of animals, have been noted (Gaisford 1978). In the desert and semidesert regions of the Sahara, East Africa, Arabia, Iran, and Baluchistan, the camel is best adapted to herding. Cattle are also well-suited to the grasslands of Central Africa, while sheep and goats are kept on the desert's fringes. In the more temperate mountains and valleys of Southwest Asia and the Mediterranean, sheep are herded. Mongolian nomads have traditionally herded horses in Central Asia and have also raised Bactrian (two-humped) camels, sheep, goats, and cattle. In the subarctic tundra of northern Europe and Asia, only the reindeer can be herded. Pastoralism appears to be a successful adaptive strategy for arid and cold environments.

Because they must move from place to place to keep their grazing animals well-fed, some degree of nomadism is involved in this lifestyle. Fully nomadic pastoralists do not occupy permanent dwellings, and they do not practice agriculture nor depend on hunting and gathering for food. Seminomadic pastoralists differ in that part of the group, usually women and children, set up seasonal settlements near water and cultivate crops (Jerome et al. 1980). The lives of pastoralists are somewhat different depending on the species of animals they herd. For example, because horses are well adapted to the winters on the steppes of Central Asia and can find food even through snow, the contemporary horse herders of Central Asia are less likely to be nomadic than the camel herders of North Africa where both forage and water are sparser (S. Olsen personal communication, May 2001).

Some herders of cattle, sheep, and goats are *transhumants* rather than nomads. In **transhumance,** pastoralists move their herds seasonally from lower altitudes to higher altitudes in the same region, much like Heidi's grandfather in the well-known children's book. The Nuer, of East Africa, for example, move their herds of cattle to higher ground in the rainy season, when much of the low land is flooded, and move them back down to lower areas during the dry season when water is scarcer in the high grounds (see Figure 5.4).

Often, pastoralists depend on neighboring agriculturists for grain and other produce. Regional arrangements are established for trade and collective defense. Because pastoralists are often mobile and can steal the agriculturists' grain surpluses, they can force neighbors to acknowledge them as overlords. The Mongols and Arabs are two examples of pastoralists who gained control over huge civilizations in this way (Harris and Johnson 2002).

Even today, the reluctance of pastoralists to observe laws and national boundaries has resulted in efforts to settle them. Pastoralists are being driven from their lands and relocated on sedentary ranches by encroaching agricultural societies that have acquired the equipment and skills to cultivate arid terrains (Jerome et al. 1980). In some areas, improved water and fodder supplies have allowed pastoral nomads to settle while maintaining pastoralism as a way of life.

transhumance
A form of nomadism in which herds of animals are moved seasonally to different locations, often different altitudes within a region.

Figure 5.4
A Maasai woman herding cattle.

© Adrian Arbib/Anthro-Photo File

Social, Political, and Ideological Features of Pastoralism

Labor among the pastoralists is divided according to gender. Adult women in seminomadic groups usually stay in camps to look after the children and collect and cultivate crops, while men and older children tend the herds. When the pastoral way of life is fully nomadic, women and children accompany the men. The major social correlates associated with pastoralism have been outlined by Walter Goldschmidt (1968):

- Because of their mobility, pastoralists own few tools or material objects.
- Pastoralists are usually considered militaristic, frequently raiding neighbors' food supplies and fighting to protect their herds.
- Pastoral societies are frequently organized into male-dominated groups. When a woman marries, she usually leaves her family to join her husband's group.

- In pastoral societies, inheritance of wealth, like descent, is usually passed down though male lines.
- There is little or no concept of land ownership in pastoral cultures.
- Because water is scarce, pastoralists travel throughout the year depending on the availability of water. Often, they spend several months camped near watering holes, then move on.

Many of these characteristics appear to be obvious from requirements of pastoral life. To feed their animals, pastoralists must be on the move. They come into frequent conflict with others over the use of grazing lands and agricultural products. They have a military advantage over nonnomadic groups in that they frequently have animals that can be ridden. The social conditions of patriliny (inheritance through the male line) and the higher social power of men compared to that of women may reflect a desire to keep herds together among brothers and male relatives. The religious traditions of Judaism, Christianity, and Islam with strongly patriarchal deities and an institutionalized value on maleness have clear pastoralist roots. As we will discuss in more detail in Chapter 8, some specific dietary restrictions imposed by religious beliefs such as Hindu avoidance of beef and Jewish avoidance of pork may be partially explained by the realities of a pastoralist way of life.

Some of the gender inequality found in many pastoral societies may be related to the distribution of labor. In most groups, men do the more rigorous and migratory tasks of herding and warfare, and women milk animals, make animal products such as cheese, prepare food and clothing, and take care of children. Unlike the foraging and horticultural groups we have discussed, women tend to be less involved in the food-producing process. Boserup (1970) and many others have argued that when women have less to do with primary food production than men, they also have less social power than men. The Tuareg are one of the few nomadic pastoralist groups that are matrilineal. In this society, the inheritance of group membership and goods is passed through the female line, primarily from the uncle (mother's brother) to the nephew (sister's son).

Diet and Health of Pastoralists

The pastoralists' diet is usually based on renewable animal products such as milk, milk products, and blood rather than meat. In most, but not all, systems meat is eaten on a limited basis because the animals' milk and blood are more valuable than their meat. In these cases, only old or sick animals or superfluous young males are slaughtered.

Drinking the blood of live animals is extremely practical in nomadic groups. Blood is naturally stored and will not spoil as long as the animal remains alive. In their sweep across the steppes, the Mongols relied on the blood of their horses as a major food source, drinking it fresh. Horses may be bled about once every 10 days with no ill effect. Blood may be obtained by thrusting a sharpened hollow needle or other instrument into a blood vessel in the animal's neck and catching the blood that is released. Marco Polo described the Mongols on campaign noting that each soldier traveled without provisions and fire, with a string of 18 horses, living only on the blood of his horses (Tannahill 1988). Most groups that consume the blood of their animals, however, cook it into a coagulated form before eating it (Tannahill 1988).

In some pastoralist societies, milk products are the primary foods. The Khurgiz of Afghanistan use milk from yaks in a variety of ways. Yogurt, cheese, and ghee (clarified butter) enable these people to preserve milk as well as enjoy it in a variety of forms. Finally, some societies rely on a combination of blood and milk products. The Maasai of Africa, for example, mix the blood and milk of their cattle and drink the mixture.

Milk cannot be kept fresh without refrigeration for more than a few hours in warm climates. Groups that rely heavily on milk have developed a number of ways to preserve milk. The simplest is to allow milk to sour in the presence of beneficial bacteria. In this case, bacteria that convert lactose, the principal sugar in milk, into lactic acid are introduced to the milk. The resulting acid medium inhibits the growth of other spoilage bacterial and the soured product can be kept for as long as several days. Beneficial bacterial are introduced to milk through the addition of active bacteria cultures. If you have ever used a spoonful of commercial yogurt to start a batch of home-made yogurt, you have done exactly the same thing. However, even putting fresh milk into a container that has previously contained soured milk gets enough bacteria into the mixture to do the job of souring. Cow's milk and camel's milk are routinely soured to keep them edible. The type of soured milk that results depends on the specific type of bacteria used to start the souring process. The yogurt we usually eat is made from a mixture of *Lactobacillus bulgaricus* and *Streptococcus thermophilus*. Buttermilk includes the bacterium *Leuconostoc citrovorum,* which converts citric acid to diacetyl, a molecule that gives a buttery flavor to the buttermilk (McGee 1985). The original soured milk products were likely the results of ubiquitous wild bacteria getting into fresh milk left to sit in warm temperatures.

Some groups allow milk to go through an ethanol fermentation process rather than lactic acid fermentation. The result is a mildly alcoholic beverage that can be kept for days, even weeks. *Koumiss* is mildly alcoholic fermented mare's milk made by Kazakhs and other horse pastoralists of the Eurasian steppes.

Finally, milk can be processed into cheese, a product that under the right conditions can be kept for years. A wide variety of milks, including the milk of cows, goats, sheep, mares, yaks, and water buffalo are made into cheeses, especially when pastoralists are sedentary or transhumant rather than nomadic. Cheese takes up only about a tenth of the volume of the original milk, and because it is drier and slightly acidic, it is much more resistant to spoilage.

The making of cheese requires more equipment than the simple souring of milk. Cheese is useful in times of scarcity, but it must be carried. Cheeses are commonly found among agricultural-pastoralists (who farm as well as raise animals) and transhumant pastoralists (who move animals to seasonal pastures) often leaving some community members behind.

Although pastoralists have been characterized as relying exclusively on animal products, pastoral groups also eat some vegetable foods, primarily cereal grains. Frequently they trade or sell animals and animal products, such as milk and cheese, for grains. When they are not able to do this, they may raid the grain fields of neighboring agriculturists (Harris and Johnson 2002).

The Turkana and the Borana are two groups of pastoral people who live in East Africa. Although they live in the same region of the world, the Turkana and the Borana occupy somewhat different ecological zones with differing climate and vegetation (Galvin, Coppock, and Leslie 2000). The Turkana live in the Gregory Rift Valley of northern Kenya at elevations of between 1,800 feet and 3,330 feet.

The most notable characteristic of the diets of contemporary East African pastoralists is the overall low energy content of their diets. Galvin (1992) found that the energy intake of the Turkana was only 1,340 calories/day. (Compare this with Lee's estimate of 2,300 calories/day for the Ju/'hoansi.) For Turkana women, the average energy intake was only 1,009 calories/day. Comparable data on the Maasai show a daily energy intake for women and children of 1,080 calories (Galvin 1985, 1992; Nestel 1986). This is a diet high in good-quality protein from animal sources, but low in calories. Not surprisingly, the body mass index (BMI), a measure of weight compared with the height of an individual, of the Turkana, Borana, and other groups is very low. A BMI below 16.9 in adults likely shows chronic or severe energy deficiencies (James, Ferro-Luzzi, and Waterlow 1988). Approximately 23% of Turkana adults had BMIs below 16.9. Fifteen percent of Borana adults had BMIs suggesting severe or chronic energy deficiency. At the same time, 35% of adult Turkana and 53% of adult Borana showed BMIs above 18.5 suggesting adequate energy reserves in these people. The height of Turkana men averages about 5'8", and Turkana women are on average 5'4". Borana tend to be a bit shorter with men averaging 5'7" and women 5'2" to 5'3" (Galvin et al. 2000).

When Europeans first encountered East African pastoralists during European colonial expansion into East Africa, the Africans, at 5'8", seemed to tower over the Europeans. However, the African pastoralist populations have maintained about the same average height for the past 100 years, suggesting a stable, barely adequate diet, while Europeans have increased in height dramatically over that time, suggesting increasingly good nutritional status for Europeans (Bogin 2000). It seems that pastoralists in East Africa suffer chronic low energy intake, but seem to rarely suffer severe periodic shortages of food (except under conditions of war and extreme drought).

In recent years, attention has been directed toward other aspects of the health of pastoralist groups, most notably the Maasai. The Maasai's heavy reliance on blood and whole milk makes their diet especially high in saturated fat and cholesterol, both of which have been implicated in the development of atherosclerosis and coronary heart disease. The Maasai, however, seem to have little or no heart disease. In an attempt to understand this phenomenon, Mann, Sperry, and Gray (1971) examined the hearts and aortas from 50 deceased Maasai. Although none of the hearts showed any signs of myocardial infarction, the amount of atherosclerosis present was equal to that of Americans of comparable ages. The saving factor appeared to be that the blood vessels of the Maasai were far larger than those of the Westerners. In fact, the blood vessels of the Maasai were large enough that the atherosclerosis that was present did not affect their functioning. Mann et al. attributed the increased size of their blood vessels to the Maasai's extensive exercise. However, there is another possible explanation for the low cardiovascular disease rates. Despite high intakes of cholesterol and saturated fat, the Maasai, as we have seen, have low energy intake and are quite lean. As contemporary thinking about the impact of diet on cardiovascular health focuses more on the impact of high energy intake and low energy expenditure, rather than on fat and cholesterol, it would seem that the lean physique and active lifestyle of contemporary African pastoralists account for the lack of heart disease among the Maasai.

Intensive Agriculture

Intensive agriculture, or **plow agriculture,** is a form of plant cultivation that requires intensive labor and land use. In contrast to horticultural systems, intensive agriculture relies on animal and mechanical labor to supplement human labor. Irrigation, terracing, and manure or other fertilizers and elaborate systems of crop rotations make it possible to cultivate the land continuously. Hoes, plows pulled by draft animals, and large inputs of human labor are needed to maintain irrigation systems and care for animals, setting this subsistence mode apart from horticulture. The increased labor investments of agriculturists are rewarded by the long-term yields gained from the land. Unlike horticulturists, who must allow their fields to lie fallow, intensive agriculturists can produce one or two crops a year for many consecutive years. This, in turn, allows populations to form dense, permanent settlements around their fields. Intensive cultivation allowed the development of large, stable cities about 5,000 years ago.

In Third World countries, there continue to be large numbers of farmers who use land intensively but rely on technology that is based on nonmechanized animal and human labor (see Figure 5.5). These semisubsistence producers produce food and other agricultural products for home consumption, as well as for sale. Some societies use technologies that appear to be virtually the same as those used thousands of years ago. In other societies, contemporary technologies such as the use of chemical fertilizers and commercial animal feed are incorporated into a system that continues to rely on human and animal labor.

Contemporary semisubsistence, nonmechanized farmers can be found on virtually any continent. In some areas of Latin America and Africa, these forms of cultivation predominate. Intensive agriculturalists have a somewhat different set of issues regarding their

intensive agriculture (plow agriculture)
An adaptive strategy based on the intensive use of land. Crops are planted year after year in the fields. Plows may be used to till the soil, and soil fertility is maintained through use of manure and other fertilizers and crop rotations.

Figure 5.5
Man using animal traction to plow fields in the Andes of Ecuador.

ability to get enough of the right kind of nutrients in their diets. We discussed some of these in Chapter 3, in the discussion of the Neolithic revolution. When sedentary communities and households cultivate the same land over long periods of time, they quickly eliminate all of the game animals and may find themselves in competition for food with their domesticated animals. For this reason, sedentary, nonmechanized agriculturalists tend to be able to include only small amounts of foods of animal origin in their diets.

Social, Political, and Ideological Features of Intensive Agriculture

The maintenance of irrigation and terracing systems requires organized labor. Farmers must work together to build irrigation channels and regulate the water supply. Life in large, permanent villages requires cooperative social relations among community members. Perhaps more importantly, intensive agriculture can generate a large surplus that can be taxed, commandeered, or purchased to support groups of people who specialize in other non-food-producing tasks such as craftspeople, religious leaders, health specialists, and nobles. Although there were some complex societies with many different specialized occupations that appear to have been supported by horticultural production, such as the Classic Mayan civilization and several early states of the Andes, these were fleeting and probably collapsed precisely because extensive agriculture did not generate enough of a surplus to support a sedentary complex society. Only with the technologies that allow intensive cultivation can complex societies such as chiefdoms and states be maintained. Although not all of the agricultural societies that have existed were associated with the state level of political organization, no contemporary states and almost all of the ancient states were based on intensive agriculture.

All ancient and modern cities have been based on agriculture as the means of food production. Intensive cultivation allows not only for the extraction of a surplus to support other occupational classes and an elite, but also for the concentration of populations to very high levels. As noted earlier in the chapter, this also allows for the maintenance of a number of infectious diseases in populations that could not have existed in less dense populations.

In contrast to egalitarian foraging and horticultural groups, most agricultural societies have developed highly stratified social systems with rulers and nobles, religious and merchant elites, and a large group of food producers from whom a surplus is extracted through taxation. The nutritional implications of stratification are profound and will be discussed below.

Land is much more valuable in intensive agricultural societies than in foraging, horticultural, or pastoral societies. In these other systems, land is used but not "owned" in the sense in which contemporary North Americans think of owning land. The notion of ownership of land and land as private property is meaningful only in intensive systems. Other goods also take on new meanings. The nomadic forager and pastoralist, and the seminomadic horticulturalist, have little interest in accumulating goods, even luxury goods they have to drag along with them when they move. Pastoralists with beasts of burden can accumulate more, foragers relying on their own labor, less; but in the end, surplus goods are at best a big nuisance. It is not so with sedentary agriculturalists. Acquiring and owning goods may be a key goal of the elite. Sharing among equals becomes philanthropy from the elite to the poor. And, more importantly, the need to defend private property from others results in the need for formal judiciaries to resolve disputes and police forces to defend individuals' rights to property. Although horticultural and pastoral groups frequently have individuals who can be called on to help resolve disputes over homicides and property issues (especially over animals in pastoral societies), the development of formal judicial systems and police forces is associated with intensive cultivation.

Finally, stratified societies often rely on a religious system or system of political ideology that supports an elite or noble class. A political or religious ideology of superiority for the elite is a hallmark of the complex societies that developed with intensive agriculture. These were also sometimes codified into the legal system as sumptuary laws—that is, laws allowing only the elite or nobility to consume some goods, including foods. Both the Inca and the Aztecs, for example, reserved the consumption of some foods to the nobility. More to the point, the ability of rulers to extract a surplus, sometimes forcibly, from a large food-producing class results in a difference in diet and nutritional status even with laws restricting the consumption of some foods. In sum, agricultural societies often have the following social organizational features:

- Specialized occupations, with some individuals performing skilled tasks such as engineering, toolmaking, and tool repair.

- Significant differences between the rich and the poor; society is stratified into social classes.
- Formalized political and legal systems to protect land rights, protect irrigation systems, and resolve conflicts.

Today these states are organized into nations tied together by international laws and trade agreements. Yet most nation-states are organized into a ruling elite, most of whom live in urban centers, and a much larger group of rural agriculturists. Thus, with the exception of a few isolated economically independent societies, today's agriculturists live in regular contact with the market towns and cities. They participate in the social, political, and religious life of the nations in which they live and are dependent on the nations' markets as outlets for their surplus crops. Agriculturists who make up part of a larger economic and political unit are referred to as *peasants*.

Contemporary Peasant Societies

peasants
Rural cultivators who raise crops and livestock primarily for household consumption, not primarily as a business enterprise, but also raise some crops and animals to sell to an urban elite.

Peasants are rural cultivators who raise crops and livestock primarily for household consumption, not primarily as a business enterprise. They also produce some cash (nonfood) crops and enough surplus food to support an urban elite. Peasants use their surpluses to pay tax or rent and buy necessities and luxury items produced outside their communities. Thus, unlike traditional or independent horticulturists, who distribute their surpluses to members within their own group, peasants are linked through trade and other economic exchanges to an elite or wealthy group outside the community. Typically, the elite use peasants' surpluses to enhance their own standard of living. This link between peasants and the elite is a key factor distinguishing peasants from more autonomous agriculturists. It also is essential in understanding the peasants' lifestyle and the social inequities that exist between them and the larger society (Jerome et al. 1980).

The crops grown by peasant agriculturists to sell for cash often end up on the international market. Mayan farmers, each with less than 2.5 acres of land, living in the highlands of Guatemala, produce snow peas and broccoli that appear in the frozen-food sections of North American and European supermarkets. Small farmers in Ecuador produce bananas that are shipped to Europe and North America. Much of the coffee that we drink is still produced by small, non- or lightly mechanized farms in Central America, South America, and Africa. However, in many areas, commercial, mechanized agriculturists produce crops for export while peasant agriculturists, using simple technology and household labor, produce the basic grains that feed the people of the area.

Peasant agriculturists are characteristically small farmers. That is, they farm less than 3 acres, rarely more than 12. Most rely on hand labor and simple tools: the hoe, machete, cutlass (a short sword with a curved blade), and ax. Some use larger machines such as plows pulled by animals or humans to aid in planting, cultivating, and harvesting. They use little fertilizer and few chemicals.

Women provide over half the labor on small farms in planting, cultivating, and postharvest food processing. In some regions of Africa, women do most or all of the farm labor and almost everywhere they do all the cooking. In addition, women usually transport produce to rural markets where they work as vendors. Older children often assist with agricultural duties or collection of wild food. Practically the entire family must work to obtain food needed to sustain its members.

This form of agriculture is referred to as *subsistence* agriculture because the farmer's main interest is in raising food for his or her family. It also is known as *traditional* agriculture because of the dependence on traditional farming practices handed down for many generations. Specific practices vary from one society to another and even between regions. Usually, however, the methods and systems used are well-adapted to the environment. In some areas of the tropics, for example, farmers use land to grow crops for several seasons and then allow it to lie fallow for a period of time; they may plant pasture or fruit trees if lengthy fallow periods are required to replenish the soil for crop cultivation.

Other methods that get optimal use of the land include mixed cropping and intercropping. With mixed cropping, several plant species are grown together in the same field. Individual plants can be positioned in the plot to yield maximum growth for each species. Intercropping is similar: Two or more crops are planted in rows side by side. Several crops can use sunlight, soil nutrients, and water more effectively than one. Plants with leaves at different heights enable a small, rapidly growing crop to temporarily use empty spaces between larger, slower growing plants. Also, crops with varying root lengths can more effectively intercept water as it percolates down through the soil. In this way, nutrients can be absorbed that would otherwise dissolve and wash below the roots. Erosion is reduced. If legumes or other nitrogen-fixing crops are planted, nitrogen is made available for other plants as well. Finally, diseases and pests are minimized with mixed cropping by making it harder for a fungus or pest affecting one species to reach other plants of the same type in the field. Because the various crops sharing one field often are not planted or harvested at the same time, most work must be done manually. For the farmer with a small amount of land, this is not a major drawback, however, because the plot is small; labor is readily available; and machines for planting, cultivation, and harvesting are beyond his or her financial reach anyway.

Farmers in the Andean region of northern Ecuador follow a system of mixed cropping that incorporates many of the techniques mentioned above. In the province of Carchi, near Ecuador's border with Colombia, small farmers in the high Andes have developed a system of cultivation that mixes a number of crops with animal husbandry to produce foods for home consumption and for the market. They still use plows pulled by oxen, horses, or mules, and human labor to till the soil and plant and harvest the crops.

Not all small farmers live harmoniously with the land or produce enough to feed their families well. Population pressures and increased production of exportable crops, combined with destructive agricultural practices, have thrown traditional land-human relationships out of balance in many parts of the world. Increased population pressure, for instance, has forced some farmers in tropical rainforests to shorten the period during which land is allowed to lie fallow. Subsoil is eroded when it is overcultivated, and land productivity declines. Flooding may result from the soil's inability to hold water, further damaging the subsoil. Overcropping in fragile ecosystems such as the Sahelian area of Africa is contributing to the process of desertification. In fact, the desert is expanding in parts of the Sudan at the rate of 6 to 7 kilometers per year due to overuse of the land.

Soil erosion and declining soil fertility have placed many peasant groups in nutritional jeopardy. The Bapedi, a Bantu-speaking group in the northeastern Transvaal, South Africa, traditionally ate a diet that was based on thick porridges made from maize, sorghum, or millet. The stiff boiled porridge was rolled into balls and dipped into a gravy or relish made from meat or vegetables. Loss of soil fertility, soil erosion, and overhunting of game mean that the ingredients for the nutritionally important supplementary relishes and gravies are less available. As a result, use of these dishes has declined and so has the nutritional adequacy of the Bapedi diet (Waldmann 1980).

The Transition to Market Economies

Today, most peasant societies are experiencing a shift from subsistence food crops to cash crops and a cash economy. Focusing on the production of subsistence crops and raising most of the food they eat, peasants grow tobacco, jute, cotton, sugar, and other crops that are sold for cash instead of being consumed by their families. Mexican farmers replace some maize production with sorghum for the Mexican feed market or tomatoes and avocados for European and American supermarkets. Ghanian farmers grow rice for the world market rather than sorghum for home consumption. Large sugarcane or cotton schemes encourage and sometimes force small farmers to produce cotton or sugar for the world market. In these cases, some of the cash they earn may be used to buy foods they don't produce themselves, but food produced at home tends to be eaten, whereas cash produced through the sale of cash crops can just as easily be used to buy television sets, cigarettes, and alcoholic beverages or foods of lower nutritional value, such as soft drinks. It can also be used to purchase more secure housing, health care, and education.

In some cases, it appears that cash crops are more likely to be managed by men than by women. With this change, control of income in a household may shift from women to men. When income is controlled by women, it is more likely to be spent on food and health care than when it is controlled by men. Finally, there has been some concern that the production of cash crops takes more human energy than subsistence crops, requiring households to consume more healthful foods to sustain the hard work.

In a case in Mexico studied by Hernandez et al. (1974), 13 years of agricultural development resulted in a 600% increase in production among small farmers. At the same time, the population doubled. The crops changed somewhat. While corn production increased at about the same rate as the population, bean production did not keep pace with population growth. The greatest increases in crop production came in coffee, cocoa, sugar, and bananas: all export crops. Over this time the diets of townspeople (representing merchants and middlemen) improved, while the diets of the rural cultivators who experienced the tremendous increase in production remained about the same. About 22.5% of children under 5 years of age showed moderate to severe malnutrition. It seems that the unequal status of peasant producers in relation to the wealthier merchants means that peasants are denied access to the fruits of their agricultural labor.

There are, in fact, numerous examples of peasant farmers who abandon subsistence crops for cash crops, either on their own or as a result of pressure from outside. In many instances, the economic benefits of this shift are not seen by the farmer, who must use his or her cash to buy a diet that is less nutritious than the one previously produced.

Because of concern over the potential impact of shifts from subsistence to cash cropping, a number of studies have been done to monitor the nutritional status of people making this shift (Dewalt 1993a; Kennedy, Bouis, and Braun 1992). The picture painted by these case studies is not as grim as we might think. In some circumstances the nutritional status suffers, particularly in children, when small, nonindustrialized (peasant) farmers shift from subsistence to cash cropping. This is especially likely when the crop is high risk and generates little or no income. Nutritional status is also compromised when all, or most, subsistence production is abandoned and when women lose control over food and income. However, when cash crop production is successful, subsistence production is at least partially protected, and women continue to control food and income, the impacts are more often neutral or involve a slight improvement of nutritional status. The majority of the cases studied did seem to show a nutritionally neutral or slightly beneficial effect. When the growing of cash crops results in increased income, then there is also more of an investment made in education and health care (DeWalt 1993a).

Where Do Cuisines Come From?

In this chapter we have talked about the kinds of strategies that humans use to procure food. The kinds of foods that exist in a particular environment, and the ways in which people extract or produce those foods, lay the groundwork for cuisines. As we have noted, the simple technologies used by foragers result in diets that are based on the specific edible plants and animals available in the environment in which they live and which are consumed raw or baked or roasted over open flame. Food producers were initially dependent on the presence of domesticatable plants and animals in their environments. Later, as domesticated plants and animals circled the globe, they were dependent on the plants and animals that would produce well under local environmental conditions.

For foragers, foods are minimally processed. Their cuisines are largely dependent on the physical environment and the amount of equipment that a highly mobile group of people can haul along with them. The cuisine of sedentary foragers, such as the native peoples of the Northwest Coast of North America, includes flavors and textures that result from the processes used to preserve fish and other foods for seasonal storage.

Pastoralists confront the problem of how to keep foods that spoil quickly, such as milk, fresh for at least a short period of time. Soured and fermented milk products and clarified butters lend characteristic flavors to food. Indeed, these are flavors that people from outside the culture may find unappealing. *Koumiss* is clearly an acquired taste.

Agriculturalists, both extensive and intensive, process foods to improve nutritional content or make them more palatable. As noted earlier in the chapter, bitter manioc must be processed to remove toxins, giving the resulting breads a characteristic flavor. The proteins in sorghum are more available after mixtures of water and flour are allowed to sour. Maize is more nutritious after *nixtamalization,* and the processing of soybeans into tofu removes the trypsin inhibitors (see Figure 5.6).

Many elements of cuisines come directly from the foods that are available with a particular set of technologies in a particular environment. Other elements of cuisines come from the use of processing techniques that make foods more nutritious and impart characteristic flavors and textures. Finally, some of the elements of cuisines come from the improvement in nutritional efficiency that results from combining foods, such as cereals and legumes, in dishes and in diets. In subsequent chapters we will discuss the effects of aspects of socioeconomic organization, social structure, and ideology on cuisines. But one of the most fundamental rules of cuisines is that they must provide enough nutrients to keep human beings alive and allow them to reproduce.

Figure 5.6

This Honduran woman is making tortillas. The characteristic flavor and texture of tortillas comes from the nixtamalization *process in which maize is soaked in an alkaline solution to soften it and allow for the removal of the pericarp (seed coat).*

© Kathleen DeWalt 1983

Summary

- Humans use a range of technologies to get and process their foods. Some are more complex and others simpler in terms of the tools and techniques used. Analysts suggest that the types of technologies used to get food are part of larger "adaptive strategies," that is, complexes of social and cultural conditions that are associated with different sets of tools and techniques for food getting. Cohen and other researchers identify five broad types of adaptive strategies: foraging, horticulture, pastoralism, intensive agriculture, and industrialized agriculture.

- Each strategy is associated with a set of challenges to be solved. The dietary challenge to contemporary foraging groups is to find enough food to feed themselves in environments in which they live off the locally available plants and animals without overexploiting the resources available. Horticulturalists must manage their food production activities in a way that allows them to continue to find sufficient productive land for their crops close to home, and they must find ways to supplement starchy staples with higher protein foods. Pastoralists must find sources of starchy crops with higher calorie content within a context of diet fairly high in animal products and must find ways to preserve highly perishable animal products. Peasant agriculturalists must find ways to produce enough subsistence crops to feed themselves or generate sufficient income from small-scale agriculture to buy food.

- There is a rough gradient of number of hours needed to work to obtain food from forager to agriculturalist.

- The social and demographic consequences of foraging and gardening are such that population densities are small. As a result, peoples following these strategies suffer from fewer infectious diseases. Indeed, population densities great enough to sustain some diseases were possible only after 5,000 years of food production allowed the formation of cities.

- The diets of foragers differ depending on the environment, and foragers at high latitudes include more meat in their diet than those at lower latitudes. However, foragers tend to have a highly varied diet that provides sufficient nutrients. Depending on the environment, horticulturists may have low protein intake and fairly monotonous diets. Pastoralists may have diets rather low in calories, but high in animal products and protein. Contemporary peasant agriculturists can be at high risk for poor nutrition.

- Intensive agriculture, however, has the potential to generate surpluses of food that can be used to support nonfarming classes of people such as religious specialists, elites, and artisans.

- For contemporary peasant agriculturalists, the transition from subsistence to commercial agriculture may be a particularly vulnerable process in terms of food and nutrition. A decade of research shows that under conditions in which some subsistence production can be maintained, women's income is protected, farmers get good prices for their commercial crops, and workloads are not significantly increased, the increased income from cash cropping can be beneficial to households and to the nutrition of vulnerable household members, such as children. When these conditions do not exist, nutrition may suffer in the transition from subsistence to commercial agriculture.

- Cuisines—that is, characteristic foods, preparations, and tastes—result, in part, from the foods available and the particular flavors and other aspects that prepared food has as a consequence of the technology of food getting, preservation, and preparation.

HIGHLIGHT

Farming Strategies in the Andean Region of Ecuador

In this Highlight, we'll take a closer look at the adaptive strategies of farmers of the Andean region of the province of Carchi, Ecuador, a contemporary intensive agricultural society.

The most important food crop from the point of view of these farmers is the potato. Potatoes are an indigenous Andean crop, domesticated about 4,500 years ago, a hundred miles south of Carchi in what is now Peru. Virtually every farmer grows potatoes, and no farm household buys potatoes. Households aim to be self-sufficient in potato production. When they have too little land to produce enough potatoes to support the household, household members will work on the land of larger landholders for payment both in potatoes and in cash. There are also elaborate potato-borrowing and payback systems in which relatives and neighbors lend potatoes to each other. These are later paid back when the original potato receiver has sufficient potatoes.

Small farmers in the Andes grow literally hundreds of different varieties of potatoes. Only a handful would be familiar to North Americans. Figure H5.1 shows eight different varieties harvested from a plot of land only about an acre in size. Different varieties have distinctive colors, textures, and flavors. Some have white flesh with red skins, some have yellow flesh with white skins, some are mottled red and white, some have black skin with blue flesh. Some are russeted and some are smooth. Some are good for mashing, some for boiling, and some for frying. Some make better soup; some make better potato pancakes, both common dishes in the Ecuadorean Andes. And some are easier to sell in urban markets.

Farmers plant a variety of potatoes in their fields in order to have appropriate potatoes for the different dishes they prepare and serve at home and other varieties to take to market. In this system, potatoes are a cash crop as well as a subsistence crop. In 1988 the average farm household in the potato-growing area of Carchi consumed approximately 22 pounds of potatoes per person per week, or about 3 pounds of potatoes per person per day (DeWalt, Uquillas, and Crissman 1988). For a household with six people, this amounted to about 120 pounds of potatoes per week. Kitchen tables hold a bowl of boiled potatoes that are left out for snacking all day long, and toddlers are often seen walking about with a half-eaten boiled potato grasped firmly in their hands.

Potatoes are a good food for humans. They are about 3% protein, and the protein they contain is of relatively high quality; that is, they contain most of the amino acids humans need in about the right proportions. Potatoes are also good sources of vitamin C, potassium, and fiber.

Figure H5.1

These eight different varieties of potatoes were all harvested from a one-acre field in the province of Carchi, Ecuador. They have different tastes and cooking characteristics. Each variety has a different purpose. Some are good for making potato pancakes, some for soup, some for boiling. Some are grown exclusively for sale in the market.

© Kathleen DeWalt 1983

The importance of potatoes in the cuisine was evident in a meal consumed by Kathleen DeWalt in a farm household in Carchi in 1987. The first course was *locro de papa* (potato soup) with a slice of homemade cheese melting on top. The second course was a dish of boiled rice with fried potatoes and a fried egg on top. The third course was a bowl of plain boiled potatoes in their skins.

However, as culturally and nutritionally important as potatoes are to Andean farmers, the small farmers of Carchi also grow a series of other crops in their fields (DeWalt et al. 1988). In order to maintain the fertility of their soils, they rotate several crops. They also keep one or two cows and often a few sheep. The other crops grown include barley, wheat, fava beans (English broad beans), peas (which are used as dry peas), lupine beans, quinoa, and three other species of tubers found only in the Andes: the *oca, melloco,* and *mashua.* These are tubers that were domesticated in the Andes and produce crops at high altitudes in cold weather. They are rapidly disappearing from farmers' fields, displaced by more productive crops such as improved varieties of potatoes. This is also true for the **pseudo cereals** *quinoa,* which is a member of the chenopodium or goosefoot family, and *amaranth,* which Americans know as pigweed. Quinoa and amaranth are very high in protein but do not produce as well as the cereal grains. Finally, "wild" turnips grow as weed in fields. The leaves and stems of the wild turnip provide greens and vegetables for

pseudo cereal
A crop that resembles the cereal grains and is used for its seeds, much like the cereal grains, but which is not a member of the grass family. Quinoa and amaranth are common pseudo cereals.

Carchi households. When weeding fields, farmers allow wild turnips to remain in the fields. They are then "harvested" as needed.

In 1999 we estimated that about 38% of the calories consumed in the households of peasant farmers in the Andean region of Carchi came from grains (principally barley). About 16% came from tubers. These were mostly potatoes, but about 1% of all calories came from traditional Andean tubers (*oca, melloco, mashua*). About 7% came from legumes (principally fava beans and dried peas). Finally, about 5.5% of the calories came from dairy products: milk and cheese. The average Carchi household consumed enough food to meet the energy and protein needs of the household (DeWalt et al. 1999).

Potatoes are a crop that takes a lot of nutrients out of the soil. If potatoes were grown year after year in the same place, the soil would soon be exhausted. To avoid this situation, Carchi farmers practice an elaborate system of crop rotation. They grow potatoes for 2 or 3 years in a field. They follow potatoes with a grain such as wheat, barley, or maize. The cereal grain crop uses up the leftover fertilizers in the field. After the grain crop, farmers usually grow a legume such as fava beans, peas, lupine beans, or lentils. Because legumes are nitrogen-fixing plants, the legume crop begins the renewal of the soil. Finally, farmers allow the field to go back to grass for a few years and pastures his or her animals on it. The manure produced by the animals also helps renew the fertility of the field.

In the current system of cultivation in Carchi, farmers also use some chemical fertilizers, especially on the potatoes they grow for sale in the market. They also use pesticides, principally insecticides and fungicides. In fact, some varieties of potatoes will be fumigated every 2 weeks to control fungi and insects. Farmers say that they themselves do not like to eat potatoes that have been grown with chemical fertilizers and pesticides. If they have enough land, they will plow previously unused, very fertile land to grow potatoes for their own consumption, and use chemicals only on the potatoes they intend to sell.

Food Technologies: How People Get Their Food in Industrialized Societies

When you think of a modern American farm, what comes to mind? Checkered fields of corn, oats, hay, and clover; chickens scrambling for food in the backyard; cattle grazing in the fields; and horses in their stalls? While these kinds of farms do still exist, they do not produce the bulk of our food and other agricultural products. By and large our food comes from large farms that are likely to produce only one or perhaps two crops. Cattle are fed grain in crowded pens and draft horses have been replaced with tractors, some equipped with front wheels that reach 7 feet in height and have air-conditioned cabs. Chicken farms consist of long buildings that house thousands of birds in small cages. Conveyor belts deliver their food and water and remove their eggs and waste.

In industrialized societies, agriculture is a large-scale business enterprise requiring large amounts of capital and energy. Industrialized agriculture can be thought of as the adaptive strategy of the industrial and postindustrial world, including North America, Europe, and Asia, as well as some of the agriculture in many industrializing countries. Industrialized agriculture is significantly different from the four adaptive strategies described in the previous chapter. It is more capital intensive, replacing human labor with capital. As we shall see later in this chapter, it is, in fact, "an energy sink." It is also highly productive in terms of output per unit of labor.

One of the most profound differences between industrialized agriculture and the other systems is that food is produced far away, both geographically and socially, from where it is consumed. The industrialized food system is a global, "delocalized" system. Only about 2.4% of the population of the United States is directly involved in food production. Compare this to the fact that all adult Ju/'hoansi, Kayapo, and Turkana are food producers, and more than 50% of the populations of preindustrial states, such as the Aztecs or Persians, were food producers. On top of that, the United States is a *net food exporter*. It is one of the few countries that export more food than they import. In the United States, each person involved in food production produces enough food, on average, to feed at least 40 people engaged in other pursuits.

Industrialized agriculture is the commercial production of food using mechanization, chemical pesticides, and fertilizers, coupled with scientifically generated knowledge to increase productivity. It has a local, national, and global marketing system to distribute food. Barlett (1989) defines industrial agriculture as a system of food production that "uses the products of industry in its own production process" (p. 254). This capital-intensive system substitutes fossil fuel, machinery, and purchased inputs such as chemical fertilizer for human or animal labor. Food systems in industrialized societies are technologically complex, with a number of components making up the chain from field to table. These components show the extent of this complexity:

- *Farming industry:* Farms and farmers, family farmers, corporate farmers, ranchers, fishermen.
- *Agro-technology industries:* People and industries that produce farm machinery, agrochemicals (fertilizers, pesticides), and biological inputs (hybrid seeds, genetically modified organisms).
- *Purveyors of factors of production:* Capital, credit, financial services, information, training, suppliers of seeds, chemicals, and equipment.
- *Intermediate industries:* Wholesale sellers, import/export concerns, storage, transportation, and marketing boards.
- *Food industries:* Processing, manufacturing, packaging, wholesale distribution, catering, and retailing.
- *Scientific research:* The international agricultural research centers of the Consultative Group for International Agricultural Research (CGIAR), researchers in national agricultural research systems (NARS) such as the U.S. Department of Agriculture, researchers in universities, and research and development sections of private seed-producing and food-processing corporations.
- *Regulation:* State health and nutrition policies, state and national food quality and safety monitoring, food security programs (food stamps, commodities).
- *Food consumption:* Household labor in food purchasing and preparation, food habits and culture. (Adapted from Atkins and Bowler 2001:11.)

This complex of activities that occur to develop new technology, produce food, and process, transport, and deliver it to consumers—along with the economic, agricultural, and energy policies that regulate the production, distribution, and consumption of food—is referred to as the "food system." Industrialized agriculture characterizes developed countries and exists alongside less technologically complex agriculture, and even foraging, in many developing countries. For an in-depth examination of industrialized agriculture, we look at the food system in the United States, a key component of the global food production system and one of the most industrialized systems.

Features of Industrialized Agriculture

monoculture
The practice of planting only one species or variety in a field.

Industrialized agriculture in the United States and elsewhere relies predominantly on **monoculture,** the practice of growing one crop within a given land area. In some places, such as the American corn belt, entire regions are planted with seeds that produce plants genetically identical to each other. Each plant responds to the environment in much the same way. Farmers can then plant seeds of a corn hybrid that was specifically developed to give high yields within their geographic location. This is advantageous because hours of daylight, average daily temperature, and total days suitable for growth and maturation of the crop can differ greatly between geographic regions.

The fields of industrialized agriculture look very different from the complex fields of horticulturalists and many nonindustrialized agriculturalists. They are uniform and neat looking. The neatness and uniformity of corn fields in Iowa, and tomato fields in California, make it easier to use chemicals to control weeds, insects, and diseases and to use mechanical harvesting equipment and chemicals. Because the crop can be bred to produce fruit that will form at the same place on the plant and that is firm enough to be handled mechanically, a machine operated by one or two people can be used to harvest the crop (see Figure 6.1). Planting, cultivating, and fertilizing also can be done mechanically, often by a single person.

Today, the farmer can even plant the seeds of crops that have had genes from other organisms, such as bacteria, inserted into their DNA to make them resistant to pests, diseases, and herbicides. These are **genetically modified (GM) crops.** *Bacillus thuringiensis* (Bt) is a bacterium long used in both conventional and organic agriculture to attack insect pests of food crops. The inclusion of the gene that produces the Bt protein Cry9C helps protect the corn from attack by the European corn borer and, to a lesser extent, the corn earworm, the southwestern corn borer, and the lesser cornstalk borer. However, there are some concerns that Cry9C can cause allergic reactions in some people, and the use of Bt corn as a human food has not been approved by the U.S. Food and Drug Administration. The Environmental Protection Agency (EPA) approved Bt corn in August 1995, and its use grew from about 1% of planted corn acreage in 1996 to 19% in 1998. It peaked at about 26% in 1999 before falling to 19% in 2000 as a result of a scare caused by the identification of Cry9C in taco shells sold for human consumption (USDA/ERS 2001a).

Extensive agriculturalists move to new land when the land they are cultivating loses its nutrients. Nonmechanized agriculturalists use crop rotation, animal manure, and **intercropping** (growing more than one crop in a field) to help maintain soil fertility. However, in industrial systems, soil fertility is enhanced principally through the use of chemical fertilizers. Chemical fertilizers contain phosphorous, potassium, nitrogen, or some combination of these. Other minor nutrients may also be added.

Soil is especially vulnerable to nitrogen depletion. Unless crops that contain nitrogen-fixing bacteria (e.g., clover, alfalfa, soybeans) are planted to replenish the soil, its natural fertility drops. With the development of chemical processes that synthesize atmospheric nitrogen into inorganic compounds, it has been possible to greatly boost soil fertility and crop yields. Between 1960 and 1995, the

Figure 6.1
Family farmer and hired hands on his farm near Mound Bayou, Mississippi. The family-owned farm is still the most common farm type in the United States.

Courtesy U.S. Department of Agriculture

genetically modified (GM) crops
Although the term "genetically modified" can refer to changes in the genetic makeup of plants as a result of either conventional breeding or genetic engineering, it has come to refer most commonly to crops that have had the genes of bacteria, animals, or other plants mechanically inserted into their DNA in order to include characteristics that make the modified crop more desirable. *Transgenic* refers specifically to the movement of genes from one species into another.

intercropping
The practice of growing different species of plants in the same plot. Some of the most common combinations include one or more grain crops with one or more legume crops. Legumes, as nitrogen-fixing crops, help provide more nitrogen to grain crops that are heavy users of nitrogen.

amount of nitrogen fertilizers used in the United States grew from 2.7 million tons of nitrogen per year to 11.7 million nutrient tons per year (Anderson and Magleby 1997). The greatest part of this almost fourfold increase took place in the 1960s and 1970s as farmers noted the increase in crop yields when more nitrogen was applied. The use of fertilizer per acre of land has remained steady since the 1970s. The United States is a net importer of nitrogen fertilizer.

Although animal manure and municipal wastes can be used as fertilizer, the use of these materials is less than it could be. Both manure and municipal wastes pose problems for U.S. agriculture. The use of manure is economically feasible only when the manure is produced close to where it could be used as fertilizer. Relatively few farmers also have animal production operations that could serve as a source of manure for their farms. Problems with the use of municipal wastes include the lack of basic research on how to best manage, compost, and mix municipal wastes for use in agriculture (USDA/ERS 2001b).

Finally, one of the most dramatic aspects of the use of chemical fertilizers in U.S. agriculture is that there has been a consolidation of fertilizer producers and sellers. There are fewer providers now than even 20 years ago. For example, between 1976 and 2000, the number of companies producing anhydrous ammonia (a nitrogen fertilizer) declined from 58 companies to 27 companies, and the number of factories that make anhydrous ammonia declined from 113 to 39. Perhaps more importantly, the biggest four companies sell almost 50% of this nitrogen fertilizer. The industry that provides fertilizers to the U.S. farmer has become an oligopoly. That is, a small number of companies provide the product with a smaller and smaller amount of competition among them (Kim et al. 2001). As we will see below, this same consolidation is taking place in other aspects of the system, including farms themselves.

Other practices associated with industrialized agriculture are **multiple cropping** (planting more than one crop on the same land in the same year); the use of chemical pesticides and herbicides to control insects, weeds, and other pests; and irrigation to provide controlled amounts of water to a crop throughout the growing season.

In 1995 U.S. agriculture used 72.9 million pounds of chemical insecticides, 358.7 million pounds of chemical herbicides (weed killers), and 181 million pounds of chemical fungicides (Fernandez-Cornejo and Jans 1999). The reasons for adopting genetically modified crops such as Bt corn, soybeans, and cotton is to reduce the use of chemical pesticides. The adoption of Bt and herbicide-tolerant (Ht) crops does seem to have had the result of decreasing the amount of chemical pesticides on crops being grown in conventional ways (USDA/ERS 2002b). Later we will discuss the impact of the growing interest in organic approaches to producing food.

One of the most outstanding features of the U.S. farm is its production efficiency. It is recognized throughout the world for its efficiency and ability to increase crop yields, while keeping inputs relatively constant. In fact, productivity, the yield obtained from all

multiple cropping
The practice of planting more than one crop on the same land in the same year. After the first crop is harvested, the field is quickly replanted with a second, and sometimes a third, crop. This is possible in areas in which the growing season is very long.

inputs (fertilizer, seed, etc.), has risen over 150% since 1948. In just one decade (1970 to 1979), productivity increased 16.7%. Currently, average productivity in crop production is increasing about 1.89% per year, and in animal production it is increasing about 1.7% per year. However, actual gains in farm productivity depend on how efficiency is defined. As N. Omri Rawlins (1980) explains:

> *There are many different ways to measure farm efficiency, and the degree of efficiency depends on the type of measurement used. One of the most common indicators used is the number of persons supplied by the average farm worker. Presently, the average farm worker supplies over 55 persons with food and fiber, compared with only 20 in 1955 and 15 in 1945. The ratio of total farm output to farm inputs may also be used as an indicator of efficiency. Overall output per unit of input has increased over 20 percent since 1960. Also, output per labor hour for farm workers has increased much faster than that of nonfarm workers since 1969. The result of this amazing productivity is the reliable abundance of the United States food supply. (p. 43)*

Barlett (1989) points out that the increasing use of technology, including mechanization, use of chemical inputs, and improved seeds and animals (see Focus 6.1), places industrialized agriculture on a **"technology treadmill."** A technological change that increases production or lowers costs gives those who adopt the process early a competitive advantage over others, and a period of higher profits. Eventually, even farmers who resist adopting new technology are forced by competition to do so. The introduction of bulk tanks to hold and transport milk was an innovation that had a significant impact on the way milk was produced and processed in Wisconsin. Before the use of bulk tanks, dairy farmers in Wisconsin collected their milk in 10-gallon cans that were placed in a milk cooler. Private truckers picked up the cans and transported them to a processing plant every day. In the 1950s, a new technology became available—refrigerated bulk tanks that were connected directly to the milking system and transported milk directly from cow to tank with a minimum of contamination and with much lower need for labor. Bulk tanks were served by special milk tank trucks. The technology needed a large capital investment, but it reduced labor and costs for both the dairy farmer and the hauler. Also, as bulk tank technology became available, sanitation laws were changed to require the use of this technology because of the sanitation advantages it offers.

Larger farms adopted this technology more quickly than smaller farms. However, "late adopters" found that they could not find truckers that would haul their 10-gallon cans. The greatest benefits of the technology were gained in herd operations that were large enough to fully utilize the larger bulk tanks (economy of scale). Further, the more efficient technology lowered the price of milk. Farmers who resisted adoption found they could not sell their milk, and even if they could, they could not get a price for their milk that

technology treadmill
The situation in which early adopters of a new technology have an advantage over other producers. The competition forces others to adopt the new technology. Often, large producers can adopt a technology first, forcing smaller producers to get bigger in order to compete.

FOCUS 6.1 Food Processing

The United States' food supply is the product of technological achievements in food processing, manufacturing, and transportation. Every culture has its own way of processing and preparing foods. However, in the United States, food technology is also big business: "The fourth largest industrial user of energy, the employer of about one-seventh of the working population, is a complex network with its parts spanning the globe. (It's no wonder that few of us know where our food comes from, how it gets to us or what processes it passes through on the way." [Katz and Goodwin 1980:149].)

The number of food items available in U.S. supermarkets continues to increase, with thousands of new products added every year. In 1980 the average number of products on a supermarket's shelves was about 14,000. In 2000 that number was a bit over 49,000. In 2000 alone, 9,249 new food products were introduced. Of those new products, 1,379 were sauces, dressings, and seasonings; 1,271 were beverages; and 1,146 were confectionery products (Food Marketing Institute 2000). Clearly, very few of these represent new food crops or domesticated animals, although some fruits and vegetables from tropical areas of the world continue to enter North American markets. Most of what are called "new" foods are simply different ways of processing and packaging the same ingredients that have characterized the North American diet for the past century.

Thanks to technology, consumers have more food available than ever before. As we noted above, they pay comparatively less for it than they did 25 years ago. The price includes all the built-in "maid-service" features in our modern food supply: frozen foods, baked products, and other processed foods. However, the convenience of highly processed food is purchased at the expense of valuable nutrients and home-cooked flavors. Natural texture, taste, and aroma are replaced with artificial color, flavor, and even a smell to simulate natural food. Preservatives are added to extend shelf life (Meyerhoff and Tobias 1980).

North Americans have a love-hate relationship with food technology. Throughout Canada and the United States, people appreciate the convenience and abundance of their food supply, but condemn its impact on their social and physical environments.

The food industry's corner on the North American market grew explosively in the second half of the twentieth century. With women entering the labor force in higher numbers, less time is available for menu planning and preparation of home-cooked meals. At the same time, women's earnings allow them to buy more processed foods. These same factors have led to the rise of another component in industrialized nations' food systems: the "away-from-home food market." This market consists of restaurants, cafeterias, fast-food chains, taverns, clubs, hotels and motels, trains, boats, planes, vending machines, and all other facilities that sell food. According to figures published by the USDA, the total expenditure for meals and snacks eaten away from home was $394 billion in 2000. This represents 47 cents of each dollar spent on food in the United States. Approximately one-third of those sales take place in fast-food franchises (USDA/ERS 1998). Most towns in the United States have the familiar golden arches of McDonald's, Kentucky Fried Chicken, and Long John Silver's Seafood Shops. Conveniently located near shopping centers, theaters, and busy thoroughfares, these restaurants offer a standardized, predictable menu at moderate prices. The menu also reflects the very features of the developed nations' diet that are under attack: foods high in fat, cholesterol, salt, sugar, and preservatives. This concerns many nutritionists because of the increasing reliance on these fast foods: 25% of all of the meals that children consume are away from home, and close to 50% of these are at fast-food outlets. Studies show that American children consume more fat, more saturated fat, more sodium—and less calcium, iron, and fiber—than is recommended (Lin, Guthrie, and Blaylock 1996).

allowed a profit. The introduction of bulk cooling tank technology pressured Wisconsin farmers to increase the size of their herds or leave dairy farming. "Farmers were forced to get bigger or get out" (Barlett 1989:257).

Another outcome of the productivity of industrialized agriculture is a reduction in the real price of food. That is, compared to wages and the cost of other items, food has become less expensive. In 1929 (the first year the USDA reports data), 23.5% of the average family's total expenditures was for food. By 1970 that percentage had dropped to 13.5%, and in 2000 it was only 10.6%

Farm Size

Another consequence of the technology treadmill may be the consolidation of farms (see Figure 6.2). As we saw above, increasing technification creates economies of scale. Larger operations can produce each unit of production at a lower unit cost. Competition among farm operations focuses on the ability to produce food at a higher level of productivity. While the family farm continues to be the basic production unit of industrialized farming, accounting for about 98% of all farms, the number of farms and the number of families engaged in farming have been declining since 1935 (see Figure 6.3). The number of farms in the United States reached its peak in 1935 when there were 6.8 million farms. In 1974 this number had dropped by two-thirds to 2.3 million farms. Between 1950 and 1970, over 100,000 small farms were lost each year, nearly 2,000 per week. Since the 1970s the number of farms has continued to decline, but much more slowly. In 1997 there were about 2 million farms. As the number of farms decreased, the average size of farms increased from 155 acres in 1935 to 487 acres in 1997 (Hoppe et al. 2001).

Averages can be deceiving, however. The farms that remain are very diverse and most are quite small. Sixty-eight percent of the land used in farming is in farms with incomes under $10,000 per year. At the same time, large (annual sales between $250,000 and $500,000) and very large (annual sales over $500,000) family farms,

Figure 6.2

Contemporary American farms in York County, Pennsylvania.

Courtesy U.S. Department of Agriculture

Figure 6.3

Changes in the size and number of farms in the United States, 1850–1997. The number of farms has decreased since 1935 while the size of farms has increased.

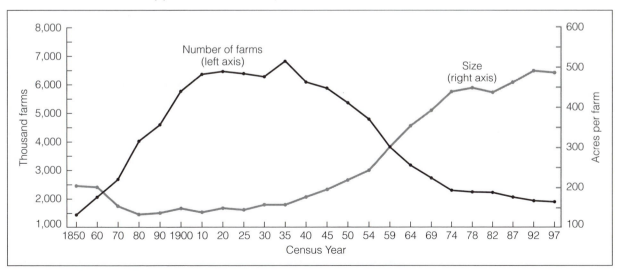

Source: U.S. Bureau of the Census, 1900–1992 Censuses of Agriculture; U.S. Department of Agriculture, National Agricultural Statistics Service, 1997 Census of Agriculture.

which make up about 8% of all farms, produce 53% of all the agricultural products produced. There has been a consolidation of farms and a continuing concentration of production in large and very large family farms. Family farms are also getting more "corporate." Single proprietorship (noncorporate family farm) is still the most common form of ownership with about 90% of all farms in this category, a proportion of farms are owned by family partnerships and corporations, and only about 2% are owned by nonfamily corporations. However, nonfamily farms account for over 13% of all agricultural production.

Farms vary in their size and profit-making potential. Many farmers, especially those with small landholdings, produce less income and must rely on off-farm jobs for themselves or family members to supplement the farming income. Larger farmers do much better and account for most of the nation's agricultural production. In 1998, for example, about 8% of the farms had sales of $250,000 or more; these accounted for 68% of total farm sales. On the lower end of the scale, about 50% percent of farms sold less than $10,000 worth of farm products, accounting for only 1% of farm sales (Hoppe et al. 2001).

The optimal acreage and the amount of capital and labor needed for a profitable operation differ both by crop and by geographic area. In general, however, small farmers do not have sufficient land or capital to withstand inflation (e.g., increased costs of fertilizer, seeds, and pesticides) and wide fluctuations in crop prices. In large part, this accounts for the continuing decline in farm numbers and the increase in the size of farms. As noted in the example of bulk milk storage, in industrialized agriculture there is often an **economy of scale.** Larger farms are more able to invest the capital needed to acquire and maintain equipment, purchase diverse inputs, and weather hard times. Corporate farms tend to be found in some parts of the country and not others, and to focus on particular food industries such as feedlots (where cattle are fattened, slaughtered, and processed) and citrus fruit operations.

Contract Farming, or Vertical Integration

Finally, the process known as **contract farming,** or **vertical integration,** is changing the farm family's relationship with the food system and agribusiness dramatically. Contracting, or vertical integration, refers to a system of food production in which a processing or marketing company contracts with farmers to produce crops and animals to the company's specifications. Formerly, farmers made the major decisions about what to produce and how to produce it. When the crops were harvested and the animals ready for sale, the farmer sold them to the highest bidder. In contracting, the farms are operated by the farm families, but inputs (seed, animals, chemicals) may be provided by the company and the farmer paid a fee for producing the animals or crops, or the farmer has a contract to sell all of the production to the company. Contracting is a system that ties independent producers into close relationships

economy of scale
A situation in which the mass production of an item lowers the average cost of production of each unit. For example, for the production of some crops, after the initial investment in machinery, the cost of cultivating more land is small compared to the initial investment.

contract farming (vertical integration)
A marketing arrangement in which the producer agrees to produce crops or animals to the specifications of a buyer, who agrees to buy the products. In some cases, the buyer also provides the seeds or animals, and other inputs, and the farmer provides land and labor.

with agribusiness. Food processors or other agribusinesses contract with farmers in order to produce known quantities, with specific and standardized characteristics. For example, General Mills uses a specific variety of wheat to produce the cereal Wheaties. This wheat produces flakes that curl up in milk rather than getting soggy. General Mills contracts with farmers to produce this specific variety of wheat and agrees to buy the production (USDA 2001a). In all, about 11.5% of all farms have one or more contracts, and about 35% of the total dollar amount of farm production is produced under contract arrangements (Harwood et al. 1999; USDA 2000).

While a number of different products are produced through contract arrangements, the most vertically integrated food production system is poultry production (see Figure 6.4). Ninety-five percent of all poultry production, both meat and eggs, is carried out under contract arrangements in which farmers produce poultry for specific processing and wholesaling companies. The poultry industry has been vertically integrated since new techniques for raising and feeding animals were developed in the 1950s. About 50% of fruit and dairy production and 42% of hog production are carried out under contracts (Harwood et al. 1999; USDA 2000).

Contract farming has the potential to bring about fundamental changes in both the production and marketing of food, but also in the rural communities in which farmers reside. It is a way in which family-owned-and-operated farms industrialize, not only through the mechanization of production, but also through the corporatization of agriculture. Contract farming can stabilize pricing of inputs

Figure 6.4

Baskets of day-old chicks are delivered to a broiler house near Magee, Mississippi. The chicks will be tended until they are large enough to be processed into whole broilers and broiler parts. Ninety-five percent of the poultry products produced in the United States are produced under contract farming arrangements.

Courtesy U.S. Department of Agriculture

and production. It can relieve the farmer of some of the risk involved in producing a crop or an animal. However, while remaining independent businesses, farmers give up a good deal of control over the production process through contracting. In addition they give up the ability to look for the best price at which to sell their products at harvest. There is concern that as contracting arrangements increase, farmers will be less able to purchase inputs in the local communities and will bypass local middlepeople in the sale of products. This is likely to further weaken the local economies of rural farming communities (USDA 2000). It is clear, however, that the practice of contracting is increasing and likely to continue to increase as U.S. agriculture becomes more and more corporate and, at the same time, more delocalized.

Industrialized agricultural technology appears at first glance to represent a major advance in people's mastery over nature. Never before have humans produced such high yields per farmer or controlled such a wealth of knowledge about plant and animal domestication and the preservation of foods. Modern technology allows us to grow food in areas once considered uncultivable and to synthesize completely new food products.

Despite these achievements, many scientists are concerned. As with all technologies, new practices affect the environment and create new environmental conditions that require even newer technological responses. Industrialized agriculture creates stress on soils, water supplies, flora, and fauna. Unlike less mechanized strategies, the effects of industrialized agriculture are not confined to the immediate environment. Demands for gasoline to run tractors and for oil to produce fertilizers have contributed to oil shortages and price increases felt around the world. North Americans complain about high prices at the gas pump while Latin American peasants find that they can no longer afford the fertilizer made from natural gas that helps make their farms profitable. Some scientists fear that industrialized agricultural techniques, combined with other technological advances of Western societies, may jeopardize the environment to such an extent that humans may be working themselves out of existence. Others claim that the environment is not in grave danger and advocate the spread of industrialized agricultural techniques throughout the world. In either case, energy shortages, smog alerts in major cities, and contaminated rivers and lakes have made members of industrialized societies aware of the ongoing struggle to adapt to the physical environment.

Energy Used in Food Production

The U.S. agricultural system is the most productive in the world, that is, if you consider only the amount of production per unit of human labor. However, there is another way to look at the efficiency of a system of production. If you count calories, you no doubt are aware of the energy you get from eating food. But how often do you consider the amount of energy that went into producing it? You may be surprised to learn that in the United States it takes 9 calories of energy to produce just 1 calorie of food. You may also be surprised to learn that this is a much less efficient energy ratio than food produced in nonindustrialized societies or even in the United States 20 years ago (see Figure 6.5). In 1910 more calories were produced from agriculture than were invested in it (Harris and Johnson 2002). By 1970 it took 8 calories of energy to produce 1 calorie of food energy.

Why does it take so much energy to produce our food? First, industrialized agriculture is highly mechanized and farm machines use large amounts of gasoline. Second, chemical fertilizers are made from natural gas, and the production of pesticides also requires fossil fuels. Third, food processing is a major consumer of natural gas as well as petroleum products. And fourth, the transportation of food from farms and processors to warehouse retailers consumes millions of gallons of gasoline and diesel fuel. The average food item travels 1,300 miles from farm to table. Almost every state in the United States gets over 85% of its food supply from outside the state.

Figure 6.5

Energy subsidies for various crops. The energy history of the United States food system is shown for comparison.

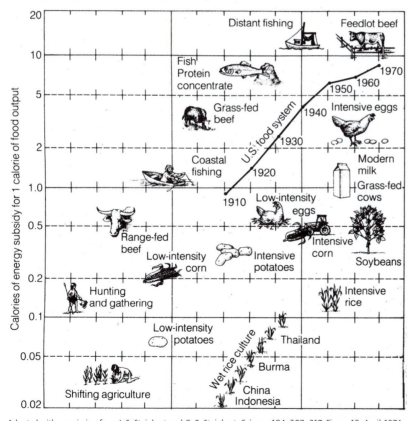

Adapted with permission from J. S. Steinhart and C. E. Steinhart, *Science:* 184, 307–317, Figure 19, April 1974. Copyright © 1974 by the American Association for the Advancement of Science.

Finally, retailers, especially fast-food restaurants, are significant consumers of energy. One study estimated that the equivalent of 12.7 million tons of coal was used in 1971 by McDonald's hamburger franchises alone (Perelman 1976). Pimentel et al. (1980) estimated that when you take into consideration the 15 tons of machinery, 22 gallons of gasoline, 203 pounds of fertilizer, and 2 pounds of chemical pesticides invested in 1 acre of production in 1975, a total of 2,890,000 calories of nonfood energy was expended. And this does not include the energy cost of transportation and packaging. When the fuel bills of all components in the food system are totaled, the agricultural industry is responsible for 10–20% of the nation's energy consumption (Rawlins 1980). Compare this with the energy ratio of 11 calories of food energy produced from 1 calorie of energy expended by the Ju/'hoansi, or the 18:1 ratio of energy produced to energy expended by the horticultural Tsembaga of New Guinea, or the 54:1 ratio of nonmechanized agriculturalists of China, the Luts'un (Harris and Johnson 2002).

Industrial agriculture is a huge net energy sink when all of the sources of energy are included. This reliance on nonrenewable energy resources creates serious problems for farmers who must pay higher prices. It also has national and international significance as we attempt to decrease our dependence on foreign oil and avoid future energy shortages.

Soil Erosion

A problem that is not unique to industrialized agriculture, but nevertheless important for today's society, is soil erosion (see Figure 6.6). Most soils have the capacity to regenerate between 2 and 5 tons of topsoil per acre annually.

> *If erosion does not exceed the rate of regeneration, the soil resource is maintained. Most soils, therefore, incur some erosion without significant adverse impacts. A soil tolerance loss of five tons per acre has come to be accepted as the threshold at which erosion becomes excessive. When this standard is applied to the available information on erosion soil loss, the problem seems to be relatively concentrated, that is, the most severe erosion occurs on a small proportion of the acreage. (Knutson, Penn, and Boehn 1983:331)*

In 1997, 108 million acres of land were considered to be eroding excessively resulting in an overall loss of 1.3 billion tons of topsoil (USDA 2001a).

Most erosion reaches excessive proportions on croplands, particularly those used to grow cotton and sorghum. Rangelands and pastures experience considerably less erosion. The recent expansion of U.S. agriculture into foreign markets, resulting in higher prices, has encouraged many farmers to plant crops year after year and from "fence to fence, frequently at the expense of the shelter belts and windbreaks" (Knutson et al. 1983:331). The 1970s and 1980s, therefore, brought increased rates of soil erosion and new concerns for soil conservationists. These concerns continue today.

Figure 6.6
Vast areas of soil under mechanical plowing are exposed to erosion, and those that are irrigated may, over time, become salty and unusable. This photograph shows the result of wind erosion.

AP/Wide World Photos

Water Pollution and Shortages

For many years, water was taken for granted and treated as an unlimited resource. But as irrigation uses expanded from surface to ground supplies, water supplies dropped and fears of water shortages have spread throughout much of the country (Knutson et al. 1983). Soil erosion also has an impact on water quality. Many surface water sources have become polluted with chemical fertilizers, pesticides, and topsoil. In fact, with the control of other sources of water contamination, agricultural sources of water pollution have become the most important threat to surface water and a significant threat to ground water (Crutchfield, Feather, and Hellerstein 1995).

Social Costs

Industrialized agriculture has its social costs, too. The decline in the number of farms and people employed on farms and the expansion of corporations into many segments of the food system have raised questions about the potential survival of the family farm. When crop prices drop, cash flow becomes an acute problem and many farmers are forced to seek supplemental jobs or face bankruptcy.

Small-farm closures and mechanization that displace human labor force people to migrate to urban areas in search of employment. When the nonfarming sector cannot absorb them, problems arise. "The transition or movement of people out of agriculture is often extremely painful, imposing a severe burden on those who are displaced" (Knutson et al. 1983:169).

Alternative Strategies

The concerns raised by the environmental impacts of farming practices and the cost in nonrenewable forms of energy has resulted in the development of a number of alternative strategies. Included among these are organic farming techniques, systems to increase sales of locally grown foods, food co-ops and buying clubs, and personal dietary changes.

Organic or Sustainable Farming

One of the most popular alternatives available to help solve some of these problems is organic or sustainable farming. Organic farming has been gaining in popularity among both farmers and consumers. From 1992 to 1997, the amount of cropland dedicated to certified organic production more than doubled. The organic production of eggs and milk increased even faster. However, the total amount of organic cropland is still low compared to all U.S. cropland. In all, 1.3 million acres in 49 states were certified organic in 1997. This represents only 0.2% of agricultural land. Two-thirds of this land was in crops and about a third in organic pastures and rangeland. Over 30% of the production of herb and mixed vegetable farms is now organic. About one-third of the production of spelt and buckwheat is organically produced. However, much less than 1% of the production of wheat, corn, rice, and barley, and only about 1% of oat production, is grown organically.

BOX 6.1

What Does a "Certified Organic" Label Mean?

Certified organic production means that agricultural products have been grown and processed according to the specific standards of various state and private certification organizations. Certifying agents review applications from farmers and processors for certification eligibility, and qualified inspectors conduct annual on-site inspections of their operations. Inspectors talk with operators and observe their production and processing practices to determine if they are in compliance with organic standards that, for example, virtually prohibit synthetic pesticide use in crop production and require outdoor access for animals in livestock production (USDA 2001b).

The U.S. Department of Agriculture and a number of state departments of agriculture are heavily promoting organic production. Several states have begun subsidizing conversion to organic farming systems. For example, in Iowa, organic crop production is eligible for cost-share support from the USDA's Environmental Quality Incentive Program. Since 1999 the Minnesota Department of Agriculture reimburses Minnesota producers for up to two-thirds of the cost for organic inspection and certification. The USDA also promotes the export of organic crops (USDA 2001b).

Organic farmers have greatly reduced or eliminated their reliance on chemical fertilizers with the use of organic techniques. This is done by planting hybrid crops that are less dependent on fertilizer, rotating the growth of legumes that return nutrients to the soil, and using animal manure and crop residues to enrich soils. Also, some small farmers are replacing monoculture with the more traditional practice of diversified farming. A farmer may grow corn, oats, hay, clover, a family garden of vegetables and fruit trees, and raise cows, chickens, and pigs. Such a system creates a healthy ecosystem and allows for recycling of wastes.

Relocalization of Food Systems

There is also an increasing promotion of relocalizing food systems through farmer's markets and the sale of locally produced foods, particularly fresh fruits and vegetables, in supermarkets. Nineteen thousand farmers nationwide report that they sell all of their produce through farmer's markets. Many towns have formed or revitalized farmer's markets. Fairgrounds, public squares, parking lots, and closed-off streets provide space where farmers can sell their locally grown food directly to consumers.

In 1976 the New York City Council on the Environment began a small program to help preserve the few farms in the area and provide consumers with locally grown fresh fruits and vegetables. Since then, the program has grown to over 25 weekly farmer's markets, with 16 in Manhattan alone. As an outgrowth of this project, the council recently initiated a farm survival program to expand its efforts to help local farms. This project works to improve regional marketing channels, dealing with issues such as land preservation, zoning legislation, and small farmers' tax problems.

Many "local food systems" are right in America's own backyard. In 1979, 33 million American households (42%) grew some of their own food. The total value of this produce reached $13 billion—an average yield of $386 for each garden.

Another strategy gaining popularity is **community-supported agriculture (CSA)** (University of Massachusetts 2000). CSA originated in Japan about 30 years ago when a group of women, concerned about rising food imports in Japan, began to develop direct purchasing agreements with local food producers. In Japanese this arrangement is called *teikei,* which means "putting the farmer's face on food." Community-supported agriculture is a partnership of mutual commitment between a farm and a community of supporters that provides a direct link between the production and consumption of food. Supporters cover a farm's yearly operating budget by purchasing a share of the season's harvest. CSA members make a commitment to support the farm throughout the season, and they assume the costs, risks, and bounty of growing food along with the farmer or grower. Members help pay for seeds, fertilizer, water, equipment maintenance, labor, and so on. In return, the farm provides, to the best of its ability, a supply of seasonal fresh produce throughout the growing season. Becoming a member creates a responsible relationship between people and the food they eat, and the land on which it is grown and those who grow it (University of Massachusetts 2000).

As of January 1999, there were over 1,000 CSA farms across the United States and Canada. The prices consumers pay for food are usually less than retail, but more than wholesale, so both the farmer

community-supported agriculture (CSA)

An arrangement in which consumers contract directly with farmers to purchase their products over a growing season. Consumers generally buy a share in the output of a farm and then receive weekly deliveries of produce.

and the families benefit. CSA farms are a strategy that serves to relocalize at least part of a food system and may help to build a stronger sense of community and support the local farm economy.

Foods Co-ops and Buying Clubs

Food co-ops and buying clubs are nonprofit structures in which consumers buy directly from wholesale dealers or farmers, allowing them to save the money normally paid to middlepeople. Buying clubs usually consist of several households that purchase bulk wholesale goods collectively. The clubs operate out of homes, churches, or community centers, with members contributing labor and transportation. Co-ops have more members and a greater variety of food than buying clubs, and permanent quarters. While some co-ops resemble large supermarkets, such as those in Berkeley (California) and Hyde Park (Chicago), more frequently they are small enterprises that emphasize wholesome foods sold in bulk, rather than prepackaged. An estimated 250 and 350 storefront co-ops now exist in the United States.

Diet and Health of Industrialized Agriculturists

The industrialized, globalized food system of North America provides the most diverse, the most abundant, and probably the safest food supply in the world. The abundance and variety of food supplies in industrialized societies enable most families to consume sufficient protein, calories, and other essential nutrients to maintain good health. Widespread shortages are no longer a problem, although the poor often experience **hunger** and **food insecurity** (see Chapter 7 for a discussion of hunger and poverty in the United States). A 1995 survey suggests that about 11.9 million households, or about 12% of U.S. households, experience food insecurity, that is, they do not have "assured access at all times to enough food for an active healthy life" (FAO 2001b). Of these, 7.78 million households were classified as *food insecure **without** hunger.* They reported concerns about the adequacy of their food supply, substituted cheaper and fewer foods, and reduced the quality and variety of diets, but did not have to significantly reduce food intake. An additional 3.34 million households were classified as *food insecure with **moderate** hunger,* that is, they reported some reduction in food intake due to inadequate household income. And, finally, 817,000 households, or about 1% of U.S. households, were identified as *food insecure with **severe** hunger.* In this case, reductions in food intake were observed for both children and adults and one or more adults were reported to have experienced a substantial reduction (Carlson, Andrews, and Bickel 1999).

However, while an unacceptably high proportion of American families experiences hunger or is food insecure, clinical undernutrition, as measured by deficiency diseases or growth failure in children, is extremely rare. Among the classic nutrient deficiencies, only

hunger
The physiological phenomenon of not having enough food as a result of not being able to afford enough food; the actual experience of not having enough food.

food insecurity
Condition in which a household has limited or uncertain availability of food, or limited or uncertain ability to acquire acceptable foods in socially acceptable ways (i.e., without resorting to emergency food supplies, scavenging, stealing, or other unusual coping strategies). Many food-insecure households are worried or unsure whether they will be able to get enough to eat, and most of them reduce the quality, variety, or desirability of their diets. They may resort to emergency food sources or other extraordinary coping behaviors to meet their basic food needs.

iron deficiency continues to be a problem. Mild iron-deficiency anemia is relatively common among children and women, especially in economically disadvantaged families. Other forms of malnutrition are rarely found.

Industrialized agriculture in the United States exists in a context of great social and economic diversity. Obesity and hunger exist side by side, sometimes in the same households. In a recent study of a sample of poor women who were classified as food insecure or experiencing hunger, the more food insecure women were more likely to be obese (Townsend et al. 2001). In fact, obesity and the health problems associated with it are clearly the most important nutritional problems in North America and, increasingly, in the affluent segments of populations worldwide.

Since the beginning of the twentieth century, the food consumption of Americans has changed in important ways. Consumption of meat has increased and continues to increase, with current consumption at about 201 pounds per person per year, about 25 pounds per person more than in 1970[1] (see Focus 6.2). However, the types of meats consumed are changing with consumption of poultry replacing consumption of beef. Americans consumed an average of 21 pounds less red meat (mostly beef), 31 pounds more poultry, and 3 pounds more fish and shellfish in 1997 than in 1970. As a result of substituting poultry for red meat, the meat eaten by Americans is leaner than it was in 1970 (Frazao 1999). Another category of food that is increasing is carbohydrates. Americans now consume about 491 grams of carbohydrate per day. This is up 27% from 1970. Americans' consumption of cereal grains (wheat flour, rice, corn, etc.) has been increasing since 1970, and rather dramatically since the mid-1980s, but it is still below their consumption of grain products in 1909.

However, the most dramatic increase in carbohydrate-containing foods is the increase in caloric sweeteners, sugars coming from table sugar and corn sweeteners. The average American consumes more than two-fifths of a pound (53 teaspoons) of caloric sweeteners per day or 154 pounds per year. That 154 pounds of sweeteners includes 240 candy bars and almost 300 twelve-ounce cans of soda. Some teenage boys consume as much as 400 pounds of sugar per year due to high-sugar snacks and beverages. (The USDA's Food Guide Pyramid recommends that people consuming 2,200 calories per day consume no more than 12 teaspoons of sweeteners/day.) Sugar is known to promote dental caries and may contribute to obesity and related conditions such as cardiovascular disease.

On the whole, Americans consume 580 pounds of milk per year as fluid milk, milk products, cheeses, and butter, which is only about 15 pounds per year more than in 1970. However, the types of products we consume have changed. Whereas the consumption of

---

[1.] The data on food consumption are based on food-disappearance data. That is, they represent the food that was available to Americans in the marketplace. It is assumed that the amount of wastage has remained steady over time.

FOCUS 6.2   Americans and Beef

Today, Americans eat more meat per capita and pay less for it in terms of proportion of total income than any other people in the world. Although the United States has one-fifteenth of the world's population, it eats one-third of its meat. Beef has always been America's favorite meat.

In the mid-1800s, steak was the most popular meal in the United States, with Americans eating twice as much of it as their English counterparts. Even today, as overall beef consumption falls, beef is still the most heavily consumed meat. Each American consumed about 63 pounds of beef in 1999. The next most commonly consumed animal products were pork and chicken, each at about 48 pounds per person (Putnam and Allshouse 1999).

The history of beef consumption in the United States is, in some sense, the history of America. When Europeans arrived in the Western Hemisphere, they found a land that, compared to Europe, was sparsely populated. With the deaths of the Native Americans as a result of introduced diseases, vast stretches of land became available for European settlement. As farming communities became established in early colonial America, the demand for beef was easily met. Each settlement was able to raise its own cattle. Even though cattle required three times as much land as grain, this posed little problem because at that time land seemed unlimited. But as the population of the East Coast grew, the pinch for space began to be felt and the large tracts of land needed to raise cattle were employed for other purposes.

If Americans were to continue to eat meat in large quantities, more grazing land would be needed. The western United States provided the vast lands required. When cattlemen arrived in Texas to set up their business, wild cattle were already there. Texas cattle were a tough breed whose ancestors had been imported by the Spaniards in the sixteenth century and abandoned in Texas. Within a generation of the cattlemen's arrival, these feral cattle had become domesticated as the notorious Texas Longhorns.

In the mid-1800s, the western United States began supplying beef to the eastern part of the country. One drawback was the long distance separating producers from consumers. Cows had to walk many miles to reach the marketplace. A Longhorn drive from west to east might consist of 2,500 animals with 1,500 miles to travel, advancing at a rate of 10 to 20 miles a day and requiring 2 to 5 months of steady plodding. Some drives ended in the Midwest, where cattle were sold and slaughtered. Others took a short break in a Chicago feeding pen to fatten up the animals, then went back on the road to New York. When the first herds of Texas Longhorn arrived in New York in 1854, citizens flocked to see them arrive (Root and de Rochemont 1976). Cattle drives became more difficult in the late 1800s as land was enclosed. The open range was closed off as farms were fenced and cities grew up. Railroads offered a solution, revolutionizing the fashion in which beef was transported from the western plains to the eastern market. In 1867 the first shipment of cattle left by rail from Abilene, Texas. By the end of the nineteenth century, cattle drives came to a virtual halt and railroad transport took over.

The advent of the refrigerated rail car in 1882 also contributed to the availability of safer, fresher meat. Cattle could now be carried short distances to packing centers to be cut into carcasses and sent by refrigerated rail car to butcher shops in the East.

Beginning in the 1940s, a new system of feeding cattle was developed in the United States. Instead of

Figure 6.7
Beef cattle on a feedlot.

© Courtesy U.S. Department of Agriculture.

(continued)

Focus 6.2   (continued from previous page)

going from pasture to slaughter, with a possible short stop in a feedlot, most U.S. cattle now stay in feedlots for extended periods (see Figure 6.7). Here they are each fed over 2,500 pounds of grain and soy products (about 22 pounds a day), hormones, and antibiotics (Lappe 1983). Prior to this time, relatively few cattle were fed grain, but by the early 1970s approximately three-fourths were raised on grain feed (Pimental 1997; USDA 1961). Currently, 90% of the soybeans, corn, oats, and barley grown in this country is fed to livestock.

Controversy surrounds the current method of raising beef in the United States because cattle convert grain into protein inefficiently. Sixteen pounds of grain and soy are needed to produce a single pound of meat. In contrast, cattle are well equipped to convert many fibrous plants (grass, wood pulp) that are inedible for humans into protein that we can eat. Therefore, many people object to the current practice of feeding grain to cattle—which would be better used to feed the world's hungry people—and recommend instead that cattle be pastured on grasslands. Such ecological considerations have attracted considerable attention recently in light of Third World food shortages and have motivated some concerned citizens to adopt vegetarian diets.

Between 1910 and 1976, beef consumption increased 72% in the United States. In 1976 the average citizen ate 180 pounds of meat, the vast majority of which (100 pounds) was beef. Americans' passion for beef peaked in 1976 and then began dropping. According to USDA food consumption surveys, beef consumption dropped 6.1% from 1976 to 1981. Consumption has continued to drop. In 1997 Americans consumed about 190 pounds of meat, poultry, and fish per person. Only 63 pounds were provided by beef (Putnam and Allshouse 1999). As both health concerns over the impact of high levels of saturated fat became more prevalent, and the relative price of poultry, especially of chicken, became lower, Americans have replaced some of the beef they had consumed with chicken.

beverage milk has continued to decline, the consumption of cheeses has increased. The consumption of beverage milk has been declining since 1945. In 1945 Americans drank four times as much milk as soft drinks. By 1975 the consumption of milk and soft drinks was about equal. Since 1975 the consumption of soft drinks has outstripped the consumption of milk, and in 1997 Americans drank 2 1/2 times as much soda as milk. (Clearly that is where some of the sugar intake comes from.) Consumption of cheeses by Americans increased from 11 pounds per person in 1970 to 28 pounds per person in 1997. One of the more amazing changes, however, is in the consumption of ice cream. Ice cream production and consumption in the United States grew from 860 million gallons per year to 960 million gallons/year (12%) just in the 5 years from 1995 to 2000.

After remaining virtually unchanged since 1909, the total number of calories consumed by Americans began to rise in the 1980s, rising about 15% from 1984 to 1997. Based on food-disappearance data, Americans now have an average of 3,800 calories/day available to them. Estimates of individual intakes based on recalls of foods consumed by individuals suggest an average consumption of just over 2,000 calories/day, which also represents an increase since the mid-1980s.

On the more positive side, the per capita consumption of fruits and vegetables has risen by 21% since 1970. The consumption of saturated fats and cholesterol has declined 4% and 13% respectively, although the overall consumption of fat has increased slightly. Because the number of calories consumed has increased, the data show an overall decline in the percentage of calories coming from fat from 41% in 1977 to 33% in 1996 (Gerrior and Bente 2002; USDA/ERS 2002a). Table 6.1 summarizes changes in food consumption since 1970.

North Americans' dietary patterns present problems. Intakes of calories, fat, carbohydrates in general, cholesterol, salt, sugar, additives, and preservatives have been linked in the United States to major health problems including cardiovascular disease, hypertension, some cancers, dental caries, and chronic liver disease (National Research Council 1989). But in recent years, the problem of positive energy balance—that is, obesity—has come to the forefront as the most common and most dangerous nutritional problem in North America. Obesity and the diseases closely associated with it (diabetes, heart disease, and high blood pressure) affect more than 40% of the U.S. population. A sedentary lifestyle, combined with a food supply that offers literally thousands of calorically dense products, contributes to North Americans' failure to balance food intakes with energy expenditures. And the problem is getting worse. Obesity doubled in children and tripled in teens between 1980 and 2000. While this may in part be a result of less physical activity and lower energy expenditure, it also appears that Americans are eating more food. In the 1950s, Americans reached the lowest level of food consumption in the twentieth century. Energy intake has been rising ever since, but began to accelerate after about 1985. At the end of

Table 6.1 Average Annual *Per Capita* Food Consumption—Selected Years, 1972–1997

	Unit	**1972–1976**	**1977–1981**	**1987–1991**	**1997**
All meat, poultry, and fish	pounds	175.3	179.2	185.0	190.3
Red meats	pounds	128.6	127.2	115.4	111.0
Poultry	pounds	34.3	39.2	54.3	64.8
Fish and shellfish	pounds	12.5	12.8	15.3	14.5
Eggs	unit	284.0	270.1	241.1	238.7
Dairy products					
Beverage milk	gallons	30.0	28.1	25.9	24.0
Cheeses	pounds	14.1	17.1	24.3	28.0
Ice cream	pounds	17.8	17.5	16.8	16.2
Total fruits and vegetables	pounds	579.5	597.0	651.9	710.8
Flour and cereal products	pounds	134.4	142.9	177.4	200.0
Caloric sweeteners	pounds	122.9	124.4	134.5	154.1
Carbonated soft drinks	gallons	28.1	34.4	45.2	53.0
Regular	gallons	25.1	29.6	34.7	41.4
Diet	gallons	3.0	4.9	10.5	11.6
Coffee	gallons	32.8	26.8	26.5	23.5

Source: Data taken from J. J. Putnam and J. E. Allshouse, *Food Consumption, Prices, and Expenditures, 1970–97* (Washington, D.C.: U.S. Government Printing Office, 1999).

the twentieth century, the consumption of meat, cheese, caloric sweeteners, and carbohydrates was at an all-time high (Putnam and Gerrior 1999).

Sodium intakes also exceed what most experts consider advisable levels. The average U.S. citizen consumes 4,000 mg to 8,000 mg of sodium (2 to 4 teaspoons of salt) each day. The recommended intake for sodium for adults is 1,100 mg to 3,300 mg, but some people have lived well on as little as 220 mg of sodium per day. Excessive intake of sodium is linked to hypertension and its potentially fatal consequences, heart and kidney disease and stroke.

Of equal concern are the additives and other chemical compounds found in many processed foods. In addition to colorings, flavors, and preservatives added by food technologists, many foods may contain pesticides, antibiotics, and residues of other chemicals used in the production phase. This has raised many new questions about the safety of our food supply and how it should be maintained. If a substance causes cancer in rats, should we assume it is dangerous for us as well? What are the health risks associated with antibiotics in meat? Should the need to control weight be considered in a decision to ban artificial sweeteners? How should we evaluate the relative value of production cost savings versus health hazards in legislating guidelines for food production?

Ultimately, the environmental, social, and health consequences of industrialized agriculture must be viewed as an issue of conflicting goals and values. "The key to food selection for a large segment of the U.S. population is convenience and fun (for a growing number it is also maintenance of good health), while the food industry's major goal is production, diversification, and profit" (Jerome, Pelto, and Kandel 1980:37).

Agrarian values such as ownership of land and family farming must be reconciled with consumers' desires for a safe, nutritious food supply at reasonable prices. And conservationists' efforts to protect the soil and water must be considered in light of economic realities that make it more profitable to use environmentally destructive tools and techniques.

Resolution of these conflicts requires the involvement of farmers and consumers as well as federal and state governments. Because the issues are complex and the need for resolution urgent, emotions tend to interfere with objective analysis and policy formation. As Knutson et al. (1983) conclude:

> *The basic volatility of agriculture and the importance of food to the survival of humankind complicate the problems of farmers, consumers and policy makers in arriving at mutually acceptable agricultural and food policy. This initially requires an understanding of the policy options, their consequences, and their interrelationships. Secondarily, it requires a willingness on the part of those affected by agricultural and food policy to recognize each other's interests and seek compromise solutions. This is not an easy task, but it is as important to agriculture as food is to life itself. (p. 379)*

Summary

- Industrialized agriculture is a highly complex food production system that uses machinery and fossil fuel to replace human and animal labor, as well as chemical pesticides and fertilizers and complex irrigation systems.
- It is an essentially delocalized system, with production and consumption taking place far from each other.
- Consolidation in both farms and the manufacturers of inputs means we are relying on an ever-decreasing number of farmers and companies.
- The development of family corporations, an increase in the number of nonfamily corporations in farming, and the increase in contract farming also result in food production becoming more corporate.
- While this technology is highly productive, many people are concerned about some of the environmental, economic, social, and health costs associated with industrialized agriculture.
- There are a number of movements aimed at relocalizing at least some parts of the food production system.
- The abundance and variety of industrialized societies' food supplies enable most people to consume a diet with adequate calories, protein, and other essential nutrients. Such groups, however, often have health problems associated with obesity, the overconsumption of calories, saturated fat, cholesterol, sodium, and sugar.

HIGHLIGHT

Genetically Modified Foods: Friends or Foes?

"BEWARE OF FRANKENFOODS!" "GENES FROM GENETICALLY ENGINEERED FOODS COULD BE DETECTED IN BRAIN" (http://www.dorway.com/franken.html) Headlines like these are common on Web sites condemning the development and use of genetically modified (GM) or transgenic plants and animals. The article that followed the second headline went on to describe research in which scientists were able to find gene fragments from GM plants that had been included in the diet of mice as part of an experiment to study the impact of consuming GM plants. What the story didn't say was that one is likely to find gene fragments in the brains of mice from *any* food they consume. The development of transgenic food plants and animals is perhaps one of the most controversial developments of industrialized agriculture. There are at least three sets of issues that are debated regarding the development and use of genetically modified organisms. The first is the discussion concerning the potential health impacts that GM foods have for humans. The second set of issues has to do with the possible environmental impacts of GM organisms. Finally, there are a number of questions regarding the impact of laws protecting intellectual property rights on the potential of GM plants and animals to aid resource-poor farmers in developing countries in which, arguably, the greatest need for improved production and reduced use of inputs exists.

What Are Genetically Modified Foods?

In some sense, all domesticated plants and animals are genetically modified. In the process of domestication, the domesticators selected plants and seeds for characteristics very different from the original wild ancestors. As discussed in Chapter 3, domesticated maize is so different from its wild ancestor that one would not recognize them as close relatives. Farmers have continued to change the genetic characteristics of domesticated plants through the process of selection, creating new varieties by selecting for a set of characteristics that meet specific needs. For example, the Tukanoans (Chapter 5) select different varieties of manioc (cassava) for many different reasons. Some varieties mature earlier, some later; some have more dry matter content, some less; and so on. The process of selecting plants and seeds for desirable characteristics is a process that farmers carry out all the time. These are all kinds of genetic modification through selection of desirable characteristics.

When geneticist Gregor Mendel published the papers that outlined the basic laws of heredity, the scientific basis for controlled breeding was laid. Scientifically based controlled (conventional) breeding was little different in practice from the selection process that farmers and pastoralists had been using for millennia. An understanding of genetics and the use of explicit scientific experimentation made the process more rapid and effective. Even today, much of the work of genetic change in plant and animal populations is carried out using conventional plant breeding—selecting existing variations or mutations that are deemed desirable for a variety of reasons. The resulting seeds or animals "breed true," that is, successive generations look much like the parent plants and animals. Once a new variety produced through selection is released to the public, farmers can use seeds from one harvest to plant for the next.

Much of this conventional selective breeding was, and still is, carried out by public institutions such as governments, universities, and publicly supported research organizations. The main reason for this is that, although there are some laws that protect the rights of institutions and individuals to sell new varieties, these are weak in comparison to the patent laws that can be used to protect rights to transgenic crops, as we will discuss below. The result is that private companies cannot expect to make much profit from the work they have to do to produce new varieties using conventional selective breeding. Also, once the seed is purchased by a farmer, she or he will not have to purchase seed again, but can use the seed from the previous harvest. Private seed companies are not much interested in investing in the development of these kinds of varieties because there is little potential for profit. They have left that work to the public sector, which uses public (tax) monies to support it, or to private philanthropic organizations. In Chapter 10 we will discuss at greater length the role of an international network of agricultural research institutes that make up the Consultative Group on International Agricultural Research (CGIAR). These institutions do much of the conventional breeding work for the crops and animals used by resource-poor farmers in developing countries. They are funded by a mix of philanthropic organizations, multilateral donors such as the United Nations, and contributions from individual countries. For-profit companies are much less interested in the crops and animals that are used by poor people in less developed countries.

In the first half of the twentieth century, a new tool was added to plant breeding: the development of the techniques for hybridization. In hybridization, two different varieties of the same plant are crossed to produce a new variety in the hope that the new plant will have the desirable characteristics of both parents. Although most hybrid plants are the result of crossing different varieties of the same species, some hybrids have been made by crossing closely related species. Triticale is a grain that is the result of a cross between wheat and barley. It combines some of the best qualities of both species. Triticale has a high yield and good grain quality like wheat, and the ability to withstand very cold weather and resist disease like rye. However, few species will cross well enough to produce viable seeds.

Many hybrids turn out to not be useful, but a number show dramatically improved characteristics. They have much higher yields or better growth characteristics. Some plants, like maize, are easier to hybridize than others. Maize hybrids were the basis of the "green revolution" of the 1960s and 1970s (see Chapter 9). Indeed, the ease of hybridizing maize as compared with wheat is part of the reason that maize yields have more than doubled globally, while wheat yields have not increased as much.

Seeds of hybrid plants do *not* "breed true," that is, the offspring from a hybrid plant do not look much like the plant that produced the seed. Only a small percentage of the resulting plants look like the hybrid parent. Most of the offspring look more like the original parents of the hybrid plant, usually with undesirable characteristics. For this reason, farmers using hybrid seed must buy new seed each year. Hybrid seed is something that for-profit companies were interested in. Because farmers would have to buy the seeds every year, the potential for profit was higher than in selective breeding for varieties that bred true. The introduction of hybrid seeds to resource-poor farmers has meant that they could not plant seeds from their own harvest. One of the results was that in developing countries hybrid plants were adopted first by affluent farmers, who could afford to buy seed. However, the much higher yields of some hybrid varieties made the new seeds attractive to poor farmers as well, and many of them found that increased yields covered the costs of buying new seed each year.

In 1953 the discovery of the double helical form of the DNA molecule opened another powerful avenue of scientific research. The understanding of the molecular nature of genes laid the basis for research leading to the ability to move single genes from one cell to another and then from the cells of one organism to the cells of another organism. By 1973 actual recombination of DNA from the genes of one organism into the DNA of another organism was carried out. Recombinant DNA again sped up the process of inducing variation, but also allowed scientists to select very specific characteristics to include in new genetic configurations. Moreover, it allowed for the recombination of genetic material from organisms that were very different from one another, such as genes from one type of plant (say, the daffodil), into the genes of another very different plant (say, rice; see discussion of golden rice below), or genes from bacteria into genes of plants and animals. Technically, the resulting organisms are "transgenic" organisms; that is, they contain genes that have been transferred from one organism to another. However, they are most often referred to as genetically modified (GM) organisms (or GMOs) even though this term can also be used to refer to plants and animals that are the result of domestication, conventional selective breeding, and hybridization. To the activist community, these are "frankenfoods." Recombinant DNA technology dramatically speeds up the process of creating new plants. It is also very expensive and requires scientific technology that is much more sophisticated and expensive than that required by either selective breeding or hybridization—technology that is often not available in developing countries.

GM plants and animals have several characteristics that make them attractive to private companies. At least initially, they have to be purchased by farmers, although many GM plants that are not also hybrids breed true. In a number of countries, including most notably the United States, they can be protected by patents. That is, a university, a company, or an individual can patent the variety produced through recombinant DNA methods and restrict its use or sale by others. This, perhaps more than any of the other characteristics of GMOs, makes their use by poor farmers in developing countries problematical. For some companies, such as the chemical giant Monsanto, the appeal is that a GM plant can be created that is resistant to agrochemicals produced by the company, as in the case of Ht crops discussed below.

What Kinds of GM Foods Are Available?

We have mentioned several different types of GM plants that are currently available. There are several others currently available and a number of potential GM plants. Recombinant DNA approaches have been used in several broad categories:

- To improve resistance of plants to pests and diseases
- To make plants resistant to herbicides
- To improve the nutrient content of plants

Resistance to Pests and Diseases

Some of the work on transgenic plants is aimed at increasing the resistance of plants to insects and, potentially, diseases such as those caused by plant viruses. To date, the most advanced work has been in incorporating genes that act as insecticides. Bacteria produce a number of proteins that act against other organisms. *Bacillus thuringiensis* (Bt) is a naturally occurring soil bacterium that produces crystal proteins (Cry proteins) that selectively kill specific groups of insects. Different strains of Bt attack different insects. Some attack larvae of moths and butterflies, others attack the larvae of mosquitoes and flies. In general, Bt does not harm useful or friendly insects such as honeybees. Theoretically, the use of plant varieties with insecticidal properties built into the plant should improve yields and decrease the use of chemical insecticides.

Bt was first isolated in 1901 in Japan from diseased silkworm larvae. There are several strains of Bt, each with differing crystal proteins. Research in the 1960s identified Bt Cry9C protein among other strains of Bt protein that have insecticidal activity against the European corn borer, a common pest of maize (Conway 1997; Gianessi and Carpenter 1999).

In the development of Bt maize, generally three different sections of genetic material are inserted into maize cell nuclei using a "gene gun." The three sections are the gene that codes for a specific Cry protein, a promoter gene that controls where and how much of the protein the plant will produce, and a genetic marker that allows for the identification of successful transformations. This last is often an antibiotic that demonstrates that the other genes are present when tested. Successful genetic recombinations are called "events." Events vary in what genetic information is transferred and where it is inserted into the maize cell's DNA.

The U.S. Environmental Protection Agency (EPA) must register events before they can be used in commercial seed production. To date, the EPA has registered five unique Bt events for maize: 176 (Novartis Seeds and Mycogen Seeds), BT11 (Northrup King/Novartis Seeds), Mon 810 (Monsanto), DBT418 (Dekalb Genetics), and CBH351 (AgrEvo).

The Mon 810 and BT11 events are used in production of YieldGard maize. Event 176 is sold as KnockOut by Novartis and Nature Gard by Mycogen. The DBT418 event is sold as BT-Xtra. The CBH351 event is sold as Starlink by AgrEvo. These events vary in how much Cry protein is produced in the plant and where it is produced, thus affecting European corn borer control. The Starlink maize hybrid contains the Cry9C protein while all the other Bt hybrids contain the Cry1A(b) or Cry1A(c) protein. It was the presence of the Cry9C protein in taco shells that alerted consumers and the EPA that Starlink maize was being used in food for humans before it was approved for human use. There is concern that Cry9C (as well as other proteins) may cause allergic reactions in some people (Gianessi and Carpenter 1999).

Studies of the insecticidal attributes of Bt maize show that it does reduce damage by some insects, especially the European corn borer (Gianessi and Carpenter 1999). However, the degree to which the use of Bt corn has resulted in reduced pesticide use is under debate. In their review of the impact of GM plants on the environment, Wolfenbarger and Phifer (2000) estimate about a 1% reduction in the use of pesticides. However, some proponents suggest the percentage is higher (Gianessi and Carpenter 1999). USDA data do show a decline in the use of pesticides.

The most common Bt varieties are available for maize, soybeans, and cotton. The assumption is that these are all meant for nonfood purposes, for either animal feed or fiber. But, as noted above, Starlink maize was detected in taco shells.

Virus resistance has been incorporated into several plant varieties, including potatoes, squash, and papaya. However, virus-resistant plants account for less than 1% of the land planted with GM crops worldwide (James 2000).

Herbicide Resistance

Although resistance to pests and diseases has been an important focus of GM crops, the most successful transgenic crops are those that have been engineered for tolerance to specific pesticides, most commonly glyphosate, which goes under the trade name Roundup. Roundup is a product of Monsanto, a large drug and chemical company. Roundup is an herbicide that kills just about any plant. It is used to control weeds worldwide and is frequently used in conservation-tillage (con-till) agriculture, a technique used in industrialized agriculture in which a powerful herbicide is used before planting to eliminate weeds. The field is planted and not tilled or weeded again. Con-till agriculture helps prevent soil erosion. It also saves on the use of fuel and on labor costs. Roundup is the most common herbicide used in con-till agriculture (Barboza 2001).

In the last two decades, Monsanto has invested hundreds of millions of dollars in developing plants that are herbicide tolerant (Ht), that is, tolerant of glyphosate. In 1996 Monsanto released the first Roundup Ready seeds. The plants grown from these seeds do not die when exposed to Roundup. The advantage should be obvious. Roundup Ready plants can be "weeded" chemically with Roundup without damage to the crop. To date there are Roundup Ready varieties of maize, soybeans, wheat, and cotton. The combination of Roundup and Roundup Ready seeds have made Roundup the most successful agrochemical in the world (Barboza 2001).

The combination of Roundup and Ht Roundup Ready seeds has been very profitable for Monsanto and has allowed the company to shift from a "chemical" company to a "life sciences" company. As the Roundup patent has run out, Monsanto has been able to lower prices to be more competitive with companies manufacturing the generic glyphosate by shifting some of the costs to Roundup Ready seeds, for which Monsanto holds a number of patents (Barboza 2001).

Improving Nutrient Content

One of the most exciting prospects for GM foods is the potential for the improvement of the nutrient content of key foods. The improvement of the nutrient content or bioavailability of nutrients in staple foods could mean improved nutrition for poorer consumers who cannot afford a varied diet (although one could also argue that all people should have access to a varied enough diet to provide all required nutrients).

A number of different strategies have been pursued to improve the nutrient content of foods using genetic engineering. In one case, the iron content of rice has been improved by moving the entire gene sequence for production of ferritin (an iron storage protein found in animals, plants, and bacteria) from soybeans to rice. The transgenic rice seeds stored up to three times as much iron as normal seeds (Bouis 2000). In another strategy, scientists moved a gene sequence from the common bean to rice to double its iron content (Bouis 2000). Lysine content of canola and soybean seeds has been increased by transferring genes from several different bacteria. One of the more interesting stories is the story of attempts to increase the beta-carotene (precursor to vitamin A) content of rice. As we will discuss in more detail in Chapter 10, vitamin A deficiency, along with iron and iodine deficiencies, is one of the most widespread micronutrient problems in the world.

Golden Rice Rice is the world's most important grain. It is especially important in areas of the world in which deficiencies of vitamin A are common. All green plants contain carotenoids—vitamin A precursors with a yellow color—in their leaves. Some grains, such as yellow maize and yellow sorghum, contain carotenoids in their seeds (which we eat as grains)—but unmilled rice contains very little, and milled rice, our common white rice, contains virtually no vitamin A precursors in the grain. Scientists had long been looking for a rice variety with higher carotenoid content in the grain, with no success. Therefore, conventional breeding for higher carotenoid content was not possible.

Molecular analysis of rice, compared with yellow maize, showed that rice lacked several enzymes in a particular synthetic pathway that would lead to higher beta-carotene content. Beginning in 1985, the Rockefeller Foundation committed about one-half of all of its agricultural funding to studies of rice biotechnology. About two dozen high-priority traits were targeted by the program, selected because they (1) would benefit poor farmers and consumers and (2) were not readily achievable through conventional breeding. Beta-carotene production in rice endosperm was one of these targeted traits (Toenniessen 2000).

But it was not until the early 1990s that two European scientists, Dr. Ingo Potrykus of the Swiss Federal Institute of Technology in Zurich and Dr. Peter Beyer of the University of Freiburg in Germany, with Rockefeller Foundation funding, developed a strategy to insert genes for the missing enzymes into the DNA of rice. They took two genes from the daffodil plant, and one from a bacterium, and shot them into rice cell nuclei. The resulting rice has a yellow (golden) grain and a high beta-carotene content—enough to make a difference to people with vitamin A deficient diets. However, several of the techniques used to create golden rice were patented in some countries, and some of the materials used were obtained under legal agreements that restrict further dissemination. Potrykus and Beyer also wanted to assess the biosafety of golden rice. The genes that have been added to the rice endosperm affect a small portion of a complex pathway. It is important to analyze the effect of these changes on other chemicals in rice to determine if there are health or environmental hazards—or, perhaps, additional benefits. It is possible that daffodils are responsible for what has been called "daffodil pickers' rash," so the allergenicity of the rice needs to be assessed to be sure that neither the daffodil genes nor the bacterial gene have introduced a new allergen to rice.

In order to subsidize the very expensive research needed to test the grain for human safety and for nutritional effects, and to deal with issues of intellectual property rights, Potrykus and Beyer entered into an agreement with the Zeneca company to further develop and market golden rice. They have retained the right to share golden rice with public sector agencies that will make it available to low-income farmers. This is known as the Humanitarian Project. Zeneca has retained all commercial rights in all countries and will donate support to Potrykus and Beyer in the Humanitarian Project (Toenniessen 2000). The International Rice Research Institute in the Philippines, a member of the CGIAR (see Chapter 10), is now testing golden rice.

The delivery of golden rice from the inventors' laboratories in Europe was possible as a result of the donation of intellectual property licenses from Syngenta Seeds AG, Syngenta Ltd., Bayer AG, Monsanto Company Inc., Orynova BV, and Zeneca Mogen BV. Each company has licensed free-of-charge technology used in the research that led to the invention of golden rice. Subject to further research, initially in the developing countries of Asia, as well as local regulatory clearances, golden rice can then be made available free-of-charge for humanitarian uses in any developing nation (Rockefeller Foundation 2001).

Beware of Frankenfoods!!!

The potential for GM foods to reduce pesticide use and to improve the availability of nutrients to consumers, especially poor consumers, is high. But what are the costs? Are these "frankenfoods" unnatural organisms that pose important risks to human health and the environment? There are both activists and scientists who believe that the answer is yes. In addition to the potential health and environmental impacts, there is a question about the impact of protection of intellectual property rights on the availability of GM technology for poor farmers.

What Are the Questions About Human Health?

Both scientists and activists voice concern over the potential impacts of GM foods on human health (see Figure H6.1). Chief among the possible effects is the potential of unknowingly introducing allergens into foods. Food allergies affect about 1–2% of adults and 5–8% of infants. They are usually triggered by proteins and peptides (parts of proteins). There is concern that moving genes

Figure H6.1
Protesters uprooting genetically altered plants.

coding for proteins from one organism to another would cause a previously harmless food to provoke allergies in some people.

A second area of concern is that introducing new genes into foods would create food intolerances, that is, toxic responses as a result of small-molecular-weight compounds that are not tolerated by humans and cause illness. Food intolerances are more common than allergies. As noted in Chapter 5, chemicals that have these effects are often found in plants like manioc (cassava), which need to be processed in order to render them harmless to humans. Genetic modification may introduce chemicals into foods that cause toxic responses (Conway 1997; de Kathen 2000).

A third issue is related to the practice of using antibiotic genes to serve as markers for genetically modified organisms. There is concern that the massive release of antibiotic-resistant plants into both the food and feed systems may increase the likelihood of widespread development of antibiotic-resistant forms of bacteria in the humans and animals that consume them. In fact, some of the antibiotics used in the production of GM crops are commonly used in human and veterinary medicine.

All of these concerns are very real. Although there are no specific cases to show that any of these potential hazards have taken place, as both activists and scientists point out, there is not even a good methodology available for realistically assessing all of the risks for specific GMOs. As a result, there has not been the degree of testing needed to either uncover problems or assure the consuming public of the safety of GM foods. The lack of information available means that both sides of the debate make claims that cannot be well tested.

What Are the Questions About the Environment?

In addition to the concerns about human health, there are important questions concerning the impact of GMOs on the environment. It is not clear whether the genes that code for resistance to pests and herbicides can escape from the GM plants to wild populations. A number of cultivated plants have weedy cousins with which they regularly intercross. For example, sorghum intercrosses with Johnson grass to produce weeds that are especially aggressive. Maize is open-pollinating, which means the pollen is carried in the air to other plants. In Mexico, maize frequently intercrosses with related weedy grasses. Although rice is primarily self-pollinating—that is, most of the rice pollen stays on the same plant that produced it— some does spread to other cultivated rices and noncultivated wild rice relatives. The fear is that pollen from resistant plants will pollinate related weeds, creating "super weeds" that are resistant to herbicides and to the insects that normally keep the populations of these weeds under control.

Another concern is that virus-resistant genes will transfer to existing viruses, creating new strains of viruses or an increase in the variety of plants that they attack (Conway 1997). As noted previously, there is concern that the inclusion of antibiotic resistance as marker genes may result in increased resistance of bacteria to drugs commonly used for both humans and animals.

Scientists are also concerned that the widespread introduction of GMOs will result in a decrease in biodiversity. The transgenic crop itself might "escape" from cultivation and replace wild species, reducing overall biodiversity. As noted earlier, pollen might "escape" and be incorporated in wild species that are closely related to the cultivated species (de Kathen 2000). Both of these processes already happen with varieties developed through conventional breeding. However, conventional breeding takes a long time and some of the impacts can be anticipated. The rapidity with which genetically engineered change takes place leaves little time for problems to emerge slowly.

Finally, there has been concern that the escape of Bt pollen and plants can have a detrimental impact on beneficial insects. Several years ago, scientists demonstrated that monarch butterfly caterpillars that had fed on milkweed dusted with pollen from Bt maize died. This raised a concern that the monarch butterfly population, and perhaps other butterflies and moths, might suffer in areas in which Bt maize is grown. Several research projects carried out subsequently showed that the mortality of monarch caterpillars in Bt maize fields was not significantly different from that in other fields. However, this was, in part, the result of the already very high mortality due to predation. Most caterpillars (90%) get eaten anyway. The research to date still leaves open the possibility that use of Bt crops may have a detrimental impact on some nontarget insects.

In general, both the human health and environmental impacts of transgenic crops are poorly understood (Wolfenbarger and Phifer 2000). In many cases, the issues concerning the potential impacts are similar to those found in conventional plant breeding. Genes and plants "escape" from cultivated plants to "pollute" wild relatives or become pests. Cultivated plants have become invasive and aggressive weeds, effectively reducing biodiversity since the time in which domesticated maize replaced wild maize in the areas of domestication. However, the pace and magnitude of GM change is different. In some ways, the precision used in identifying and transferring specific characteristics and genes may make the impacts more predictable, but, at the same time, the novelty of the recombinations makes predictions more difficult. There is need for much better research and information in a rapidly moving field. That research is very expensive. A number of countries, without the resources to carry out the testing, have found it easier to ban the use of GM crops.

What Is the Role of Laws Protecting Intellectual Property Rights?

For some critics, the impact of the issues surrounding intellectual property rights is the most important problem. Intellectual property rights have to do with the rights that inventors (or writers, or filmmakers, or musicians) retain over the materials they create. Because a good deal of the work in developing the technologies for gene cloning and transfers was carried out by private companies, or researchers working for universities, many of the key components of the development of transgenic seeds are protected by patents. While there are mechanisms to protect varieties of plants and animals developed with more conventional techniques, these protections are not as strong as patent protection. In the golden rice example earlier, without special agreements, the technology could not be made available for adaptation to meet the needs of poor people.

Gordon Conway (1997), president of the Rockefeller Foundation, sees this as the most critical issue for the potential use of GMOs to meet the needs of poor farmers and the undernourished. As he notes, the cost of developing GMOs is so high that the public sector cannot afford the cost, and the private sector needs to be assured that it will be able to profit from the advances enough to pay for the investment in research and make a profit. Because of the huge costs involved, companies carefully guard patent rights. Agracetus was the first company to develop a technique to put genes into soybeans using a chemical "gun." This company has been granted the patent rights in Europe to all genetically engineered soybeans, regardless of what technology was used to develop them (Conway 1997). The old process by which new materials were taken by public sector and nonprofit institutions to be adapted to the needs of poor farmers is not viable when the original materials are protected by a number of patents.

There are a number of new types of partnerships being developed. One of them is the arrangement with private and public sector institutions to make golden rice available to poor farmers at low cost. Agracetus is involved in a relationship with the University of Wisconsin and the International Center for Tropical Agriculture (CIAT) in Colombia to develop bean seeds that are resistant to bean golden mosaic virus. Agracetus keeps the process of gene transfer secret but makes the resulting material available to CIAT, which then passes it on to public-sector seed suppliers in less developed countries (Conway 1997).

Can GMOs Help Poor People?

There is good potential for GMOs to help poor people. In Chapter 9 we will discuss the global impact of conventional breeding on the cost of foods to poor consumers. To the extent that improved production and lower use of inputs as a result of genetic modification lower the cost of foods to poor consumers, GM foods can have a positive impact on food security. Also, enhancing the nutrient content of staple foods so that they provide more nutrients to poor consumers can be beneficial. The biggest questions have to do with whether or not the technologies are available to poor farmers both domestically and in less developed countries. In Chapters 9 and 10, there is further discussion of the problems with technological transfer for resource-poor farmers.

CHAPTER 7

Food and Social Organization

Take a moment to consider the many people you come in contact with each day. During the course of a single day you might interact with roommates, friends, family, coworkers, teachers, store clerks, waiters, and new acquaintances made on the Internet. Your ability to get adequate food, find shelter, and meet other basic human needs requires interacting with these and many other people.

Over 30 years ago, Gerhard Lenski (1970), a famous sociologist, observed that the amount of social interaction in modern societies had reached levels that

> *would stagger the imagination of the members of simpler societies. Never before have people had so much contact with so many other people. To a large extent, this is the natural result of increasing urbanization: communities are larger, people live closer together, and increased contact is inevitable. Industrialization also has an effect: the home is no longer the workplace except for a small minority of men; children spend half their days in crowded schools; and women are increasingly drawn outside the home by a variety of responsibilities and opportunities. (p. 251)*

Today, globalization—fueled by advanced transportation, communication, information technology, and trade liberalization—brings you in touch with an even greater number and diversity of people.

Social organization refers to the way a social group organizes its members into families, social strata, communities, and other groupings. Social organization also includes a complex set of norms that regulate relationships, provide templates for the way work is organized within households, and influence how the day is organized.

Food practices are intricately tied to social organization. As people interact, they develop guidelines to help them divide food-getting tasks, distribute food, eat together, and determine what people occupying different positions in society should and should not consume. In turn, food practices can reinforce a society's social organization, as when people cooperate to produce food or share a meal. In other cases the social customs may be divisive, as when people are denied access to food or other resources by more powerful segments of the society. Through the production, distribution, and consumption of food, people act out some of their most important social relationships. As we discussed in Chapter 1, the way a group of people organize themselves to produce, distribute, and consume food has an important influence on other aspects of their social organization. Because of this **reciprocal** interaction, some nutritional anthropologists believe food practices are a type of language that guides our social relations and encodes patterns of inclusion and exclusion from social groups. Food practices are symbolic of a particular social order and communicate a great deal about who we are and how we distinguish ourselves from others. In this chapter, we examine the many ways in which food practices are tied to social organization.

reciprocation
An exchange between two or more people.

Food as a Means of Solidifying Social Ties

People often use food in building and maintaining human relationships. Food brings people together, promotes common interests, and stimulates the formation of bonds with other people and societies. Feeding is one of the first and most powerful ways in which parents establish a relationship with their children. Almost everywhere, a food offering is a sign of love, affection, or friendship, whereas withholding it may be seen as an expression of anger or hostility or a form of punishment. Likewise, to accept food from someone signifies the acceptance of his or her offer and the reciprocity of the feelings expressed, whereas refusal may be a sign of enmity and a rejection of their kindness. The Old and New Testaments abound with references to the use of food to seal **covenants,** eating together to solidify relationships, and refusing food to signal anger or ruptured social ties. Then, as now, people in most societies felt most secure when eating with friends and loved ones. We rarely dine with our enemies, and when we do so, it usually signifies an attempt to put our antagonisms aside, at least for the moment.

covenant
A binding agreement or contract made by two or more people.

Kinship and Familial Alliances

Around the world, the family unit is the basic building block of society: Almost everywhere families are a major source of physical and social nurturance, meeting individuals' basic needs for food, safety, shelter, love, and affection. Family roles in many societies are organized around the way they produce and distribute food.

The concept of "family" has been defined in various ways. The U.S. Bureau of the Census defines it broadly as a group of two or more people residing together that are related by birth, marriage, or adoption. Many social scientists prefer to talk about family systems as "small, dynamic, semiclosed groups made up of members who have mutual obligations to provide a broad range of emotional and material support. Families have structure, functions, assigned roles, modes of interacting, resources, group culture, and a history that forms part of the developmental or life cycle of the group" (Coreil, Bryant, and Henderson 2000:123).

Families come in many forms. The most common "types" or labels used to describe families are nuclear, attenuated, or extended. The *nuclear* family refers to a family made up of two parents and their children. *Attenuated* families are smaller, typically one parent and children, while *extended* families include members outside of the nuclear unit, such as uncles, aunts, and, grandparents. In some cases, families may include fictive kin—people who are not related by blood or marriage but have been "adopted" into the family. Focus 7.1 shows how the composition of families who reside together in the United States differs from the nuclear family pattern that is still considered the norm.

Regardless of the way you define a family, families are usually an important economic unit responsible for producing and distributing the food needed to sustain its members. In North America, family members usually pool their income to buy food for the entire family, not just for the individual members who earn it. Food is stored in the house and made accessible to all. Grocery shopping, meal preparation, and food preservation are typically done with the entire family in mind.

Because the extended family plays such a limited role in industrialized societies, we are often surprised by the importance it plays in carrying out domestic activities elsewhere. Let's look at some examples of how extended families are structured in societies in other parts of the world and how these structures are tied to food habits.

The Ashanti of West Africa were traditionally organized into *patrilineal* units, or extended families formed around men (the grandfather, father, sons, and grandsons) and the men's sisters. Ashanti men shared meals with their mothers, sisters, nephews, and nieces rather than their wives and offspring. Each wife did the cooking for members of her extended family unit, so before each meal there was a steady stream of traffic as children took food prepared by their mothers to their fathers' sisters' homes.

Families in the Trobriand Islands were traditionally organized into extended family groups organized around the women. Referred to as *matrilineal* groups, these families consisted of a woman, her offspring, sisters, brothers, mother, and grandmother (see Figure 7.1). Often the women lived near each other and shared common property and economic obligations. The husbands lived with their wives but belonged to their mother's and sister's matrilineal group. When a man married, he did not become a part of his wife's group, but always remained somewhat of an outsider. Traditionally, women's brothers rather than their husbands carried out male responsibilities in the family. A Trobriand household composed of the husband, wife, and offspring grew yams primarily for use by the husband's sister's family. In return, the wife's brother supplied a large proportion of the household's yams. Because yams were distributed to members of the matrilineal group rather than the nuclear unit, a household might be working to produce food for families who were not even living in the same village. At harvest time, the biggest and best yams were transferred and displayed in front of the recipient's home, where the producer received recognition for his gardening achievements and generosity.

Food is also intricately tied to another important function of the family—socialization and control of its members. Because the

FOCUS 7.1 U.S Household Composition

Although most individuals (95%) have experienced marriage by the time they reach 65 years of age, household composition in the United States differs dramatically from the normative ideal form—the nuclear family (Fields and Casper 2001). Data provided by the U.S. Bureau of the Census continue to illuminate this gap between our perceptions of the ideal family form and our actual household composition patterns. Its studies have revealed several interesting demographic trends such as the decreasing proportion of families with children under 18 years of age and the increasing proportions of nonfamily households, grandparents raising their grandchildren, and nontraditional family forms such as single-father and same-sex households.

Census 2000 data revealed that, from 1990 to 2000, nonfamily households increased proportionately at a faster rate than family households (Simmons and O'Neill 2001). As seen in Table 7.1, married couples occupy only about half of the nation's 105 million households, and less than half of these couples have children under the age of 18 (Simmons and O'Neill 2001). The proportion of married-couple households continues to decline, with a decrease from 55% in 1990 to 52% in 2000. The greatest decline is seen in the proportion of married-couple households with their own children, which dropped from 40% of all households in 1970 to only 24% of households in 2000 (Fields and Casper 2001). In terms of racial trends, only 48% of African American households in 2000 were married-couple households, compared to 68% of Hispanic, 80% of Asian and Pacific Islander, and 83% of non-Hispanic white households (Fields and Casper 2001). The number of unmarried couples has increased steadily over the last 40 years, with an increase from 3.5% in 1990 to 5.2% in 2000. Approximately 11% of unmarried-partner households in 2000 were composed of same-sex couples.

Table 7.1 Household Types

Total Households	105, 480,101	100%
Family Households		
Married couples	54,493,232	51.7%
[With children <18 years	24,787,824	23.5%]
Female-headed household, no spouse	12,900,103	12.2%
[With children <18 years	7,594,567	7.2%]
Male-headed household, no spouse	4,394,012	4.2%
[With children <18 years	2,215,082	2.1%]
Family household total	71,787,347	68.1%
Nonfamily Households		
Householder living alone	27,230,075	25.8%
Living with nonrelatives	6,462,679	6.1%
Nonfamily household total	33,692,754	31.9%
Multigenerational households	3,929,122	3.7%
Unmarried-partner households	5,484,965	5.2%
Average number per family	3.14	
Average number per household	2.59	

Source: U.S. Bureau of the Census, *Profiles of General Demographic Characteristics: 2000 Census of Population and Housing.*
Note: In some cases, numbers and percentages do not coincide because of rounding. Also, although the U.S. Bureau of the Census classifies families into two major types, nonfamily and family, data about the nature of relationships between members of a household were collected to estimate the number of intergenerational households and unmarried-partner households. These estimates, in addition to the number of people per family and household averages, provide additional information about U.S. household composition.

(continued)

Focus 7.1   (continued from previous page)

In contrast to the declining proportion of nuclear families, the proportions of female-headed, male-headed, and single-person households have increased. The number of single-mother families more than tripled (3 to 10 million) from 1970 to 2000, and the number of single-father families quintupled (393,000 to 2 million) during this same period (Fields and Casper 2001). Since 1970, the number of divorced adults has more than quadrupled and the number of never-married adults has more than doubled (U.S. Bureau of the Census 1998). Almost half of the mothers who lived alone with their children in 1998 had never been married (National Marriage Project 2001).

In addition, the proportion of single-person households (i.e., individuals living alone) has increased from less than 8% in 1940 to 26% in 2000. In 2000, multigenerational households accounted for 3.7% of all households, with most composed of the householder with his or her child and grandchildren. The number of grandparents raising or responsible for their own grandchildren under 18 years of age has increased dramatically since the 1970s (Casper and Bryson 1998). Supplementary U.S. Bureau of the Census (2001b) data suggest that over 5.6 million grandparents live with their grandchildren under 18 years of age and that almost half of them are responsible for their grandchildren.

Figure 7.1
This matrilineal Hopi family centers around a core of women. A husband moves to his wife's household, in which he has important economic responsibilities. Even after marriage his most important relationship is to his sisters' sons.

© Terry Eiler/Stock Boston

family unit is usually responsible for producing, socializing, and caring for the next generation, it has a powerful formative influence on individuals' food preferences and habits. Compared to most mammals, humans are dependent on adults to feed them for many years, creating an extended period of socialization and strong emotional ties. As a result, parents are a major source of information on what is and is not edible, which foods are "good" and which are "disgusting," and norms regulating food production, preparation, sharing, and table manners (McIntosh 1996; Rozin 1988).

Parents influence their children's dietary practices through a variety of mechanisms: modeling, making certain foods readily accessible in the home (e.g., placing them in a visible location easy to reach), selecting places to eat outside the home, transmitting beliefs, norms, and values that guide food selection, teaching their children self-management skills, and rewarding desired behaviors and punishing others. Food is also used to punish and reward children for other behaviors, though the wisdom of this has been questioned.

In turn, children may influence their parents' and siblings' behavior through requests to try new foods or restaurants, cooperation with or resistance to parents' persuasive efforts, and introduction of new ideas from outside the family unit (Baranowski 1997). Many families, for example, rely on their children to select breakfast cereals and snack foods and to decide where they will eat when they dine out. Marketers are well aware of this phenomenon and target youth with extensive advertising and promotions.

The ways families divide labor and define members' roles have an important impact on food practices. Much of family life revolves around domestic work, and culture influences how that work should be divided among family members. From foraging societies to modern industrialized societies, the division of domestic labor is typically unequal, with women carrying out a larger share of the duties and contributing more to the family's caloric intake than men.

Figure 7.2
*Ju/'hoansi boys help their father butcher a small animal, which he will
share with others in his band.*

© Lee/Anthro-Photo File

Recent studies of domestic work in Great Britain (Charles and
Kerr 1988; Warde and Hetherington 1994; Warde and Martens
2000), the United States (DeVault 1991; Harnack 1998), Sweden
(Ekstrom 1991), and South Wales (Murcott 1983) show that food
remains predominantly women's responsibility in these urban settings. Although some men help more than others, relatively few
men in these industrialized nations assume the central tasks of grocery shopping and cooking. In a study of 599 couples in Great
Britain, Warde and Martens (2000) found that men's contributions
to food work were significantly greater if they reported taking a personal interest in food and cooking. Men with university degrees also
contributed more than those with less formal education. In households where both partners worked full time or children were present, men were more likely to help, but only with ancillary tasks,
such as setting the table, clearing up, or washing dishes. A similar
picture is painted by a recent U.S. survey: Only 23% of men compared to 93% of women were involved with meal planning; 36% of
men compared to 88% of women shopped for food; and 27% of
men compared to 90% of women cooked. Even in families with children and a mother who was employed full time, only 31% of men
participated in meal planning, 39% assisted with food shopping,
and 39% prepared meals (Harnack 1998).

Children also contribute to production and food preparation. In
foraging societies, children help collect food and may hunt insects
or small game, but usually make a relatively small contribution to
the group's caloric intake (see Figure 7.2). In North America, as children become old enough to carry out kitchen duties, they, too, may

contribute to food production by carrying out tasks such as husking corn, setting the table, washing the dishes, or taking out the garbage. Over half of American teenage girls (ages 12 to 19) shop for part or all of the family's groceries each week (Kraak and Pelletier 1998).

Food practices play a key role in maintaining ties between its members and linkages with the outside community (Beardsworth and Keil 1997). For most families, mealtime is an important event, perhaps the only time during the day when family members have a chance to sit down and talk. In some societies, the family meal is a metaphor for the family itself—its importance evident in the strong sentiments attached to favorite family dishes and customs. Perhaps after being away for several months, you have returned home to be served one of your favorite meals and became aware of the feelings associated with family dinners.

The importance of mealtime in creating and maintaining familial ties is nicely illustrated by a study of 200 families with young children in northern England (Charles and Kerr 1988). Drawing on in-depth interviews with the mothers in these families, this study described the norms they shared for what should be served and how meals should be prepared, as well as when, where, and how they should be eaten. To these mothers, a "proper meal"—the main meal of the day—must be made up of a meat dish (or occasionally fish) and two vegetables. To be "proper," the meal should be cooked by the wife for her partner and children, and enjoyed together, usually at the family dinner table. The inability to provide a proper meal is considered a real deprivation, not only because it indicates an inability to feed their families adequately, but also because it reflects on the status of the family. The researchers concluded that the main meal of the day is served to nurture and nourish the family and demonstrate that it is behaving properly. It is, in short, a way to literally produce family.

Family meals may be especially important for youth and adolescents. Teens who eat with their families five times per week or more tend to be at decreased risk for poor adjustment, for example, depression, drug use, and low school motivation (Bowden and Zeisz 1997). When adolescents leave home, some miss home cooking and family meals, which they see as a form of love and nurturance, but others welcome the autonomy to make their own food choices.

Of course, mealtimes can also be a time of stress and conflict. Food may become a battleground between parents and children—a source of criticism and conflict—or a time for parents to fight with each other or their children. In extreme cases, mealtime may contribute to eating disorders, disruption of spousal ties (Murcott 1983), even domestic violence, as when men beat their wives for failing to prepare meals (Counihan 1999).

Mealtimes can also tie families to the outside world, providing linkages between the nuclear unit and extended family members or the wider community (Beardsworth and Keil 1997). In Charles and Kerr's study (1988), extended family members were frequently invited for Sunday meals and holiday feasts, couples often dined together, and women occasionally shared non-main meals with friends. In each case, food sharing was an important way to strengthen relationships. The type of food and drink served at these meals signaled a great deal about the status of the family and importance of the relationship. A three-course meal with wine was much more likely to be served to dinner guests in upper- and middle-class homes than in working-class homes.

Because of the importance of the family in strengthening ties and socializing its members, many scholars and social pundits in the United States and Great Britain have lamented the demise of the family mealtime. One complaint centers on the declining numbers of families who regularly eat meals. In 1996 only 49% of Americans reported eating three meals a day. Fifteen percent ate just two meals, and another 26% ate two meals and one to two snacks. The others ate one meal and snacks or other combinations of meals and nonmeals (Mogelonsky 1998).

The number of meals cooked and eaten at home has also decreased dramatically. Once reserved for special occasions and travel, the typical U.S. citizen over the age of 7 now eats an average of 4.2 commercially prepared meals per week (National Restaurant Association 2000). In a 1999 survey of shoppers, the Food Marketing Institute (2000) found that only 66% of American shoppers eat most of their meals at home. As a result, the restaurant-industry share of the food dollar spent grew from 25% in 1955 to over 45% in 2000. Restaurant sales were expected to reach $399 billion in 2001 (National Restaurant Association 2000).

Fast-food sales have also increased. Between 1988 and 1998, fast-food sales grew by 56% to reach $102 billion (Clauson 1999), and Americans now spend more money on fast foods than they do on cars, higher education, or personal computers (Cummings 1999). To meet North Americans' time-starved, mobile lifestyle, most fast-food chains now emphasize convenience by offering a narrow range of "handheld" products so that busy consumers can combine mealtime with travel, work, and other activities (Jekanowski 1999). Other timesaving measures include the sale of food in discount stores, shopping malls, and gas stations.

Even when Americans eat at home, they often rely on meals prepared elsewhere. According to the National Restaurant Association (1996), Americans pick up a take-out dinner to eat at home 61% more often than they did in 1986. Some retailers, such as Boston Market, offer home-style meals complete with entrées and side dishes. Drive-through purchases account for more than half of fast-food sales, and many companies now offer take-out delivery. Supermarkets also offer a wide selection of home-replacement meals—"meals like mom used to cook at home that are quick and convenient" (Price 2000:26).

In addition to ready-to-eat items, many shoppers buy partially prepared foods. In a national survey of shoppers, approximately two-thirds of shoppers say they buy ready-to-cook vegetables, 50% buy ready-to-serve, bagged salad, 44% buy frozen side dishes, and 40% buy ready-to-serve main dishes at least once a month (Food Marketing Institute [FMI] 2000). Consumer research has shown that

Figure 7.3

Time-starved Americans are buying more than twice the number of home-replacement meals than they did in the mid-1980s.

© Novastock/Index Stock Imagery

many "Americans want to do a bit, but just a bit, of actual cooking" (Eig 2001:A1). To meet this demand, food companies are developing products that take only one pan and less than 15 minutes to prepare, but still look, smell, and taste like a complete, home-cooked meal (see Figure 7.3). These home-replacement meals, or "convenient-involvement products," include Hamburger Helper, Green Giant Complete Skillet Meals, and Kraft dinner mixes that can be prepared in 5 minutes and baked or fried for another 15 to 30 minutes to fill the house with the aroma of a home-cooked meal and give the chef credit for cooking.

Nutritionists are concerned about the impact foods-away-from-home and meal replacements are having on dietary quality. Food purchased outside the home contains more of the nutrients overconsumed and less of the nutrients underconsumed by North Americans than home-cooked food. Not surprisingly, people who eat out frequently consume more calories, fat, and sodium, and fewer fruits and vegetables, than those who eat out less often (Lin, Frazao, and Guthrie 1999).

Building Relationships with Neighbors and Friends

When you ask someone you have just met to come over for something to drink or go out for pizza, your offer has important social implications. Asking a neighbor over for a cup of coffee, for example, can be seen as an invitation to establish a relationship of recurring exchanges of coffee.

Among the Sherpa in Nepal, food offerings to friends, neighbors, and even casual visitors are considered mandatory. The Sherpa consider it rude to allow anyone to leave their house without an invitation to drink and eat. Although guests are expected to refuse initially so that the host can insist, they must eventually consume the proper amount of food and drink—or risk censure. "All transactions begin with a hospitable offering—of cigarettes, tea, milk, food, and other gifts, but, above all, of "beer" and "whiskey"—that must be accepted and must be reciprocated. An offering is effective if or because it creates an ambience of amicable feelings" (March 1998:46).

Another example comes from an in-depth study of two Italian-American households in a North American city (Theophano and Curtis 1991). In these households, foods were exchanged in multiple forms: hospitality, preparation of favorite Italian foods given as gifts or payment for services rendered, sharing meal preparation and division of leftovers, and exchange of raw as well as cooked foodstuffs. The authors of the study conclude with these remarks about their findings and the implications for health providers:

In addition to representing social links and differences, [food exchange] constitutes a significant means of provisioning households. Shopping, distributing leftovers, and preparing large quantities of food are part of the system of exchange and are often important means for provisioning other households. Participa-

tion in such a cooperative network has implications for social and nutritional policy concerns. Most social and medical programs are based on a model that identifies the individual or the family as the locus of decision making about food and health. Here, as for other groups in variable economic situations, integration in social networks has been shown to be significant in household decision making and provisioning. (Theophano and Curtis 1991:170).

The type of food and drink shared with friends often conveys meaning about the nature of the relationship desired or already achieved. Potluck dinners and cookouts suggest a closer degree of friendship than an invitation to morning coffee. Sit-down dinners preceded by drinks are served to close friends and honored guests, whereas cocktails alone may be offered to acquaintances (Farb and Armelagos 1980).

For North Americans working in non-Westernized societies, the use of food to strengthen friendship ties can sometimes be problematic. Imagine you are designing nutrition programs for a rural community in the Peruvian Andes and would like to build a friendship with the local teacher who has impressed you with her understanding of the community and her pleasant personality. When she invites you to her house for dinner, you eagerly accept. Upon arrival at her home, your enthusiasm is heightened by the warm hospitality and interesting conversation. Then suddenly you are faced with the menu: an appetizer of **ceviche,** pickled pigs' feet, and a whole, baked guinea pig as well as several unfamiliar vegetables. What would you do?

When faced with such a situation, it is important to recognize the strong association between food and social intimacy. If you choose not to eat the foods, it would be helpful to know the local conventions by which invitations and food are properly declined. If you choose to eat the foods, you might strengthen your ties with your hostess. You might also find that you like the foods.

ceviche
Marinated, raw fish and other seafood.

Food as a Means of Strengthening Economic and Political Alliances

Throughout history, food practices have been closely tied to economic and political alliances between groups and to political and economic power. In many societies around the world, it is customary to seal an economic contract or exchange by sharing a meal and/or drink. Access to food connotes power, whereas hunger—the inability to satisfy one's most basic needs—is the most absolute form of powerlessness. In the words of David Arnold (1988), "Food was, and continues to be, power in the most basic, tangible and inescapable form" (p. 3). Because of its importance, food has been used as currency in some societies and is a symbol of power almost everywhere. Sidney Mintz's (1985) discussion in *Sweetness and Power* is an excellent example of how control of sugar production and consumption contributed to colonial domination and the accumulation of wealth.

Trade

Food trade has been used in all societies as an important means of economic distribution. Trade networks have always existed among foragers, horticulturalists, and agriculturalists. Today virtually all nations are involved in trade for some desired, if not needed, foods.

Trade gives people access to resources unavailable in their immediate environment, including basic necessities in times of famine. The importance of trade is evident in the U.S. **embargo** against Cuba and the symbolic as well as nutritional value that accompanied the Bush administration's decision to sell a large shipment of food to the island immediately after Hurricane Isadore.

embargo
A government order prohibiting trade with another country or governmental entity.

Sometimes, groups with very different lifestyles and technological adaptations become interdependent in ways that strengthen ties and enhance the nutritional adequacy of the diets of both groups. For example, in some regions of Mexico and Guatemala, a number of villages form a loosely knit socioeconomic community in which they trade goods and interact socially. In this system, villages specialize in making a particular product (e.g., bread, ceramics, flowers, or other commodities). Because the villagers have specialized in a specific commodity, they are forced to trade, and thus interact socially, with people in other villages (Hendry 1999).

Today, all nations are economically interdependent because of their reliance on international trade. Globalization has liberalized trade agreements and created many new forms of economic and political interdependence between nations. With few exceptions, most nations import much of their basic food and rely on exported foodstuffs and manufactured goods to pay for imports and internal development. In Chapters 9 and 10, we look at how the interdependence between developed and developing nations affects the nutritional as well as economic status of both trading partners.

Food as a Gift

Another form of distributing wealth and building economic alliances is gift giving. Despite the giver's contention that it is a free presentation of goods that involves no obligation, gift giving is part of a network of distribution that is often intentional. Gifts serve to solidify social ties and build economic alliances. Friendship, kinship, and other relationships are reinforced and validated by the exchange of gifts.

Food is an appropriate gift for many occasions and, as such, serves to distribute wealth and strengthen social ties. Asking the boss over for dinner has economic as well as social implications. Soft drinks, beer, and snacks are offered to friends who help with residential moves or other labor-intensive projects. Grateful patients sometimes supplement their cash payments to physicians with food from their gardens or kitchens.

One example of the important role gift giving can play comes from a study of food security among a rural elderly population in North Carolina. Among this population, food gifts are one of the few socially acceptable ways of getting food when it is scarce. "Food gifts are given child to parent, neighbor to neighbor, anonymously, and in many cases explained by a pattern of generalized reciprocity: 'I used to help people when I was young, and now people help me'"(Quandt et al. 1999:8). One of the most important aspects of gift giving is the ongoing system of exchange and reciprocity it creates (Mauss 1967). People are expected to give, to receive, and to repay. "Refusal to give or receive is a vast insult that severs relationships. Refusal to repay signifies inability to do so and loss of sudden face. Giving to others is the basis of power, for recipients are beholden to donors" (Counihan 1998:102).

Political Alliances

Closely associated with the economic function of food is its ability to enhance political alliances. Elaborate state dinners for visiting dignitaries are a common occurrence in North America as well as in many other societies. Sharing a meal is symbolic of an effort to establish ties and may dispel tension when attempting to reach political decisions.

A classic example of how food exchanges are used to create political alliances comes from the Yanomamo, slash-and-burn horticulturalists living in the Amazon (see Figure 7.4). The Yanomamo live in a loose network of villages surrounded by gardens where they grow **plaintains,** bananas, peach palm fruits, manioc, yams, sweet potatoes, and other vegetables. From the surrounding forest, they collect insects, hunt game animals and small birds, and gather fruits and honey.

plaintains
A large fruit like a banana used as a staple food in many tropical regions.

Political alliances are essential for survival among the Yanomamo because of the ever-present risk of attack by stronger groups attempting to steal their women. Raids frequently force weaker

Figure 7.4
Food exchange plays an important role in strengthening political alliances among Yanomamo villages in the Amazon basin.

© Napoleon Chagnon/Anthro-Photo File

groups to leave their villages and gardens. Because the forest cannot feed or hide large groups, the defeated villagers must take refuge in an ally's village for a year or longer until they can establish new gardens to support an independent existence. Of course, the defeated Yanomamo Indians cannot simply arrive at any village and declare they need help, without risking attack. They must have already built a strong alliance with the village. Ceremonial feasts, held only when entertaining the members of another village, are an important vehicle for building these political ties.

During Yanomamo feasts, hundreds of pounds of plaintains and meat are consumed—enough to sate the guests and send them home with a generous supply of leftovers. The feast gives each man an opportunity to adorn himself in gawdy decorations, dance, spin around, and prance as he displays himself "like a glorious peacock that commands the admiration of his fellows" (Chagnon 1997:102). It also gives the hosts great prestige from guests who flatter them for their generosity. Most importantly, however, these feasts are political events that "reaffirm and cultivate intervillage solidarity in the intimate, sociable context of food presentations, thereby putting the ally under obligation to reciprocate the feast in his own village at a later date, bringing about another feast and more solidarity" (Chagnon 1997:97).

The U.S. and Canadian governments also use food aid to build political ties. Unfortunately, this type of "food diplomacy" often fails because recipient countries resent being asked to support U.S. views or activities in order to qualify for food shipments. The reverse situation, prohibition of food shipments, is also used for political purposes with equally erratic success. Consider the embargoes against Cuba and Iraq; neither country has adopted a democratic government nor overthrown its leaders.

Food and Social Status

Social status denotes the place an individual holds with reference to other members of society: child or parent, husband or wife, boss or employee. Status is one's position within the group's social organization. This definition of status as a position within society is somewhat different from the more common use of the term as a synonym for prestige. Social status does not necessarily imply superiority or rank; rather, it is a set of rights and duties associated with the person occupying a certain position within a cultural group. Social status is based on gender and age within almost all societies. In the remainder of this chapter, we examine the relationship between dietary practices and one's gender and socioeconomic status. (See Highlight 7 for a discussion of the relationship between food and people's position in the life cycle.)

Food and Gender

In most cultures, food practices are linked to gender in several ways. First, food tasks are typically divided between men and women within families. As already noted, labor is often divided in distinct and complementary ways that reinforce male and female identity and relationships (Counihan 1999). Meal preparation has also been viewed as a symbol of women's domesticity and femininity—and her inequality (Beardsworth and Keil 1997).

In many societies, women's identities and domestic roles are changing as they move into the workforce outside the home. Carole Counihan's (1999) ethnography of women in Florence shows that some women who work outside the home rarely cook traditional meals like their mothers did. They know how to cook, admit it is their responsibility to cook, and do not expect their husbands to do it, but still have difficulty fitting it into their busy lives. As a result, they accept gifts from their more traditional mothers, purchase processed foods that have become increasingly available in local stores, and rely on fast foods, snack bars, self-service cafeterias, and restaurants.

> *Women clearly felt ambivalence and conflict about their declining role in food provisioning. They wanted to control their family's foods, but did not have time because they also wanted to work. Cooking was still essential to their femininity, but working women could not achieve the standards of their mothers. They ended up dissatisfied and at war with themselves, insecure about this part of their femininity. They suffered identity conflict and lacked the security of their mothers, who were more able to fulfill the roles they had internalized as necessary for themselves. . . . Yet on the other hand, the definitive entry of modern Florentine women into the workplace introduced them to an economic power their mothers never had. (Counihan 1999:57)*

Culture also influences how food is distributed between men and women. In many societies men eat first, best, and most; children eat next; and women consume the leftovers (Adams 1990). Taboos prohibiting consumption of rich protein sources are directed primarily at women of childbearing age. As women gain rights, such food restrictions become less common. These prohibitions may play a role in diverting scarce food sources to men in the event they become available and are commonly explained as a way to ensure adequate intake for men whose workload and caloric needs are greater, even though women may be pregnant or lactating (Ross 1987). However, they may also contribute to lower rates of malnutrition found in men in relation to women and children. (See Chapters 9 and 10 for a more detailed discussion of the malnutrition that affects women and children in developing countries.)

In North America, gender discrimination is often more subtle: The father's preferences may be given highest priority in deciding which foods to serve and he may be given special privileges, such as getting the best cut of meat or having the first chance at second helpings. Studies of domestic life in industrialized societies (Charles and Kerr 1988; DeVault 1991; Warde and Martens 2000) show that many women make an effort to serve meals they know their partner will enjoy, rarely serve him foods that he has rejected, and invest considerable effort in preparing and presenting meals to please him. As McIntosh and Zey (1998) point out, responsibility for food preparation is not equivalent to control of what is served; that often resides with men.

Some feminists and other scholars say that women's need to serve, satisfy, and defer to their partners reflects and reinforces their subordination (Beardsworth and Keil 1997; DeVault 1991). They note that in egalitarian societies, these disparities are less pronounced or absent. For instance, among the Wamirans of Papua New Guinea (Kahn 1998) and the Culina of western Amazon (Pollock 1998), gender roles are distinct but equal, helping to establish and reinforce men's and women's interdependence through complementary roles in food production and preparation.

A third way food is tied to gender identity concerns the types of foods considered appropriate for men and women. In Western societies, alcohol and red meat are widely linked to masculinity and salads are associated with femininity. Among the Wamirans, taro is a masculine food, considered the "men's children," while among the Culina, women are associated with vegetables and men with meat (Kahn 1998).

Finally, dieting and diet disorders are more common among women than men, reflecting different norms and values placed on body imagery and other aspects of gender identity. Although explanations of the high incidence of anorexia and other eating disorders among Western women abound (see Highlight 4), many scholars believe they reflect cultural values on thinness.

Food and Socioeconomic Position

The study of socioeconomic differences is one of the most enduring concerns of the social sciences. Public health professions, for example, have long recognized that **socioeconomic status (SES)** is one of the strongest and most consistent predictors of illness or premature death. Everywhere, poor people suffer disproportionately from most diseases and die younger than their wealthier counterparts. SES has been associated with infant mortality, cardiovascular disease, many types of cancer, arthritis, diabetes, and numerous infectious diseases (Coreil et al. 2000).

Socioeconomic position is closely tied to dietary patterns and nutritional status, with the diets of families in the higher social strata being more closely in line with recommended dietary guidelines and the poorer families suffering from a larger number of dietary inadequacies. These differences reflect differences in the amount of money available to spend on food, purchasing power, and food norms and beliefs.

Around the world, food spending varies according to household income. As income rises, a smaller proportion of food expenditures is spent on food, caloric intakes increase, diets become more diverse, and nutritional content generally improves. In the United States, for example, household income has a strong influence on weekly gro-

socioeconomic status (SES)
A person's position in a society's social and economic hierarchy or ranking in terms of prestige. Occupation, education, and income are the most common measures or components of a person's SES.

cery spending. In 1997 the average weekly household spending ranged from $66 for families earning under $15,000 per year to $107 for those earning more than $75,000 per year (FMI 1997). A comparison of nations reveals similar differences: Citizens living in high-income countries (per capita gross domestic product exceeds $10,000) spend an average of 16% of their income on food compared to 35% in middle-income nations (per capita gross domestic product between $700 and $10,000), and 55% in low-income countries (per capita gross domestic product less than $700) (Meade and Rosen 1996).

Many poorer families also have to pay more for food. Their purchasing power is diminished because they have limited access to supermarkets and large grocery stores that typically charge 10% less than convenience outlets or smaller "mom and pop" grocery stores. A 1999 survey by the U.S. Department of Agriculture found that many low-income families living in rural communities have difficulty traveling to large retail outlets because they do not own a car or cannot afford to travel long distances. Because they have to rely on small grocery stores that offer fewer selections and higher prices, they pay 2.5% more for food than the wealthier families in the rural communities and 3.1% more than households in suburban neighborhoods (Kaufmann 1999).

Finally, some studies have also examined class differences in food norms and beliefs that may contribute to socioeconomic inequalities in food consumption. A study of 849 families in three European cities showed that upper-class mothers took health into account more often when purchasing foods for their families and restricted their children's intake of sweets and other non-nutritious foods more often than their counterparts with lower incomes. In contrast, poorer families considered their family's taste preferences more often and tended to agree that their children could snack as long as they ate their main meals (Hupkens et al. 1998). One explanation for the difference comes from another study of mothers in the United States that showed families on a tight budget were less likely to impose healthy foods on their children or partners because they could not afford to purchase foodstuffs that might be wasted if the family refused them (DeVault 1991). (See Focus 7.2 for information on how the U.S. government responds to the dietary needs of the poor.)

Food as a Symbol of Prestige

Some foods are used as "class markers" or symbols of prestige. Which foods do you associate with high status and which ones with low status? The use or display of expensive foods is used in many societies to communicate a person's social importance. Among Latin American peasants, prestige is acquired from sponsoring a celebration of the village saint's birthday that includes the presentation of a ritual meal. Some peasants are known to use most of their wealth—even go into debt—in order to benefit from the prestige associated with such sponsorship.

FOCUS 7.2   Poverty and the U.S. Government

Because socioeconomic status reflects different concentrations of wealth, SES or social class has a powerful impact on diet and nutritional well-being. The chance for an adequate diet diminishes as the amount spent for food decreases. Below a certain point, it is no longer possible to purchase the foods needed for an adequate diet. The use of poverty indexes or a poverty line to identify families at nutritional risk is based on the government's recognition of the relationship between income levels and nutritional well-being.

Notions of a "poverty line" date back to the turn of the twentieth century. In the late 1800s, nutrition surveys revealed higher rates of malnutrition in working-class families than among those who were more affluent. One in-depth study conducted in Great Britain concluded that regardless of how wisely some poor families managed their budgets, they simply did not have enough income to purchase an adequate diet. The researchers calculated an income level, known as the poverty line, below which malnutrition eventually would result no matter how wisely or economically the family shopped. Supporting evidence came from studies that showed that the infant mortality rate for families below this line was 247 out of 1,000 births, compared to only 173 per 1,000 in families above the line. Further confirmation of the relationship between poverty and nutritional status came from two follow-up studies in the same country.

Since 1963, the U.S. government has also used a poverty line to evaluate the extent of poverty and identify people in need of federal assistance. In calculating the poverty line, the government has relied heavily on estimates for the cost of a minimal, but nutritionally adequate, diet. Beginning in the mid-1930s, the U.S. Department of Agriculture (USDA) developed "food plans" as a way of determining the cost of providing a family with adequate nutrition. These plans have been updated periodically and are now set at four cost levels: thrifty, low-cost, moderate-cost, and liberal. The lowest cost plan, the Thrifty Food Plan, represents a nutritionally adequate diet at minimal cost. To develop the plan, the USDA examined low-income families' food consumption patterns, modified the diet pattern to make it nutritionally balanced, and then calculated the cost of the items in the plan. Minor revisions were made in 1999 to bring the plan into compliance with current dietary guidelines—for example, by adding more fruits and vegetables and decreasing the amount of fats and oils—but the actual cost of the plan (In 2002 a working family of four received $289.00 per month.) did not change significantly (Center for Nutrition Policy and Promotion 1999).

The Thrifty Food Plan is used to calculate poverty thresholds—the amount of money that marks the point below which a family is considered poor. These thresholds are used to compare trends in the number and proportion of people who are poor in the United States and determine eligibility for government assistance programs. To allow for non-food expenses when calculating poverty thresholds, the cost of the Thrifty Food Plan is multiplied by three. The multiplier of three was chosen in the early 1960s because the food budget for a family of three or four persons—all families, not just those who are poor—represented about one-third of its after-tax expenses. Each year, poverty thresholds are adjusted to account for the family's size and composition as well as price inflation. Although slight adjustments are made for Alaska and Hawaii, the thresholds are the same for Americans living in the 48 contiguous states.

To determine eligibility for programs such as Food Stamps and the **Special Supplemental Nutrition Program for Women, Infants, and Children (WIC),** a family's before-tax income is compared to the threshold for a family with

(continued)

Special Supplemental Nutrition Program for Women, Infants, and Children (WIC)
A program that was established in 1972 to provide nutritious food, nutrition education, health checkups and referrals, and infant feeding advice and support to economically disadvantaged families. The program serves women who are pregnant, breastfeeding, or in the postpartum period and infants and children up to the age of 5.

Focus 7.2   (continued from previous page)

the same number of adults and children. If the family's income falls below the designated amount, they are eligible for assistance. In June 2002, for instance, a single person was considered "poor" if her or his before-tax annual income was $8,860 or lower; a family of four was classified as "poor" if their before-tax annual income was $18,100 or lower. The same person would be eligible for the Food Stamps Program in 2002 if his or her before-tax annual income fell at or below 130% of the poverty line: $11,518 for a single person, $15,522 for two people, $19,526 for three persons, and $23,530 for a family of four. Slightly higher thresholds are set for people living in Alaska and Hawaii where food costs are much higher (U.S. Department of Health and Human Services 2002). (For more information on the poverty guidelines and the thresholds for any given year, visit the Department of Health and Human Services Web site:*http:// aspe.os.dhhs.gov/poverty/02poverty.htm.*)

Critics of the current poverty measure claim it has always had weaknesses and that these shortcomings have grown more serious as the economy has changed during the last 40 years. They note that the plan has always been monotonous, failed to allow for eating outside the home, and required its followers to have a working refrigerator, freezer, and stove and be able to carefully select, store, and manage foods. Support for their concern is evident in a pilot test of the revised Thrifty Food Plan: Two of the eight "test" families required assistance in preparing some of the foods included in the plan (Center for Nutrition Policy and Promotion 1999).

The Panel on Poverty and Family Assistance convened by the National Academy of Sciences has also criticized the poverty measure, noting its failure to keep pace with the economy (Citro and Michael 1995). In an extensive study of the measure, they concluded that the current system needs to be replaced because it fails to account for geographic differences in the cost of living, increased costs of nonfood items (e.g., child care, transportation, and housing), Social Security and other payroll taxes paid by the poor, and increased standards of living. They also recommended the revised poverty measure include the value of government assistance benefits (e.g., Food Stamps and public housing), as well as use after-tax income levels.

Development of a reliable and valid measure of poverty is important for several reasons. First, the measure of poverty is an important social indicator used to compare economic conditions between states and regions and to monitor trends over time. An accurate poverty measure may enable policy makers to assess the impact of public policies and economic changes. Second, it is important to correctly identify families who are eligible for government programs so that they can obtain the assistance needed to maintain good health. Finally, it is important to protect the notion of a poverty line as a way to dispel the myth that poor people are able to meet their nutritional needs if they would just use good sense in how they spend their money.

Specific foodstuffs, dishes, or cuisines usually become symbols of prestige because they are made of expensive or rare ingredients, take a great deal of time to prepare, and/or are consumed only on special occasions or by members of privileged groups (see Figure 7.5). Caviar, truffles, and prime cuts of red meat are often associated with the wealthy, whereas legumes, peas, beans, and canned meats carry less prestige.

As supplies and prices change, so do the foods seen as class markers. Today, lobsters are associated with affluence in the United States, but in the seventeenth century, the situation was reversed. At that time, lobsters were in such abundance that they were considered "fit only for the poor, who could afford nothing better." When a group of new colonists arrived in Plymouth in 1633, Governor William Bradford was deeply humiliated because his colony was so short of food that the only dish they could present their friends was lobster "without bread or anything else but a cup of fair water" (Root and de Rochemont 1976:51). Similarly, chicken was once reserved for guests on Sundays and special occasions. In the United States, Calvin Coolidge's promise of a "chicken in every pot" represented affluence

Figure 7.5
Special dishes that are made from costly ingredients, or take a great deal of time and skill to prepare, may carry a great deal of prestige.

© Robert Brenner/PhotoEdit

to his listeners. Today, with highly efficient production methods, the price of chicken has dropped and so has its status. Now, chicken is often featured in cookbooks and magazines as a money-saving meal.

As a prestige marker, food can be used to communicate one's social status or facilitate movement up the social ladder. In medieval Europe, food supplies were not stable and the powerful could distinguish themselves from peasants by consuming gargantuan feasts. Obesity was a visible sign of wealth. Beginning in the seventeenth and eighteenth centuries, food supplies became more secure and consumption of large quantities of food was insufficient to distinguish the rich from the poor. (The nobility had reached the limits of how much more they could eat than the peasantry.) While the rich continued to eat lavishly, their social status was expressed in their consumption of a variety of refined, exotic, and delicately prepared dishes. Good taste, good manners, etiquette, and self-control, also became important ways to establish their superior status.

The switch from food quantity to quality as a prestige marker was accompanied by parallel changes in the appetite and norms pertaining to body image. By the nineteenth century, self-control and moderate eating began to be considered important social virtues, contrasted with the gluttony associated with lower classes (Bell and Valentine 1997; Mennell, Murcott, and van Otterloo 1992). As we have already seen, a slimmer body was idealized as evidence of the restraint and refinement associated with success. Even today, slimness is more common among the wealthier classes, except perhaps in some societies where food supplies are still limited (Corrigan 1997).

Because food is such a conspicuous prestige marker, many people adopt the dietary practices of the social group to which they aspire. Today, easy credit, in the form of credit cards, enables many people to consume luxury items associated with the rich. Eating out in restaurants is one way that people may temporarily rise above their social status. "A person does not have to belong to an exclusive club to dine on high cuisine, only have the means (cash or credit) to do so. Many exclusive restaurants restrict access, thus increasing their desirability by limiting the number of patrons they can serve. Others do so by simply elevating their prices. The diner gains both the cuisines and the elegant circumstances thought to characterize the everyday lives of those with higher social status" (McIntosh 1996:48).

One of the best examples of conspicuous dining is the presence of a special chef's table in the kitchen of the Citronella, an upscale Washington, D.C. restaurant. Patrons pay dearly for the privilege of eating a fixed menu selected by the chef while seated behind a glass window where all the other patrons can watch them dine.

The adoption of prestigious goods is also one of the first dietary changes to occur among wealthier citizens as a nation industrializes. In areas of Africa where sorghum and millet are traditionally grown, expensive wheat is now imported to meet the urban elite's demand for white bread. Red wine has replaced sour milk as the traditional offering at funerals in Senegal and in Chad, and sacrifices are now made to guardian spirits with red wine and lumps of sugar. As people switch to prestige foods—white flour, packaged bread, white rice, sugar, canned foods, baby formula, and soft drinks—their nutritional status suffers. Such substitutions can have tragic effects. The replacement of brown rice with polished white rice, for instance, created an outbreak of the vitamin B-1 deficiency disease, beriberi, causing the debilitation and death of thousands of people in nineteenth-century Asia.

Summary

- Social organization refers to the cultural traditions that people develop about how to interact with one another and organize themselves in order to achieve common goals. Food and eating habits are tied to social organization in many ways.
- People often use food in building and maintaining human relationships. Throughout history, food has been used to establish social ties with friends and neighbors and to build economic and reinforce political alliances between groups.
- Food plays an important role in creating and maintaining familial ties. Food ways are intricately tied to socialization and control of family members, role delineation and division of labor within the family, and linkages with people outside the family system.
- Socioeconomic status has a powerful impact on diet and nutritional well-being. Poorer families must devote a larger proportion of their income to food expenditures, often pay higher prices for food, and may make different food choices for their families.
- Food is often used as a class marker or symbol of prestige.

HIGHLIGHT Food and the Life Cycle

As people grow older, they pass through a series of recognized statuses within the life cycle. Most societies recognize these status changes with ceremonies or *rites of passage,* such as christenings, puberty initiation rites, marriages, and funerals. These ceremonies proclaim the status changes and dramatize the new roles the individual is expected to occupy. They also help other society members adjust to new ways of relating to him or her. In this highlight, we look at the special dietary practices and rituals associated with major phases in the life cycle: birth, infancy and childhood, adolescence, reproduction or parenthood, old age, and death. Before reading on, think about the associations you make between food and each of these phases in the life cycle. Are there foods or special dietary practices you associate with children, teenagers, elders?

Birth, Infancy, and Childhood

The first phase in the life cycle is birth. Many societies mark the arrival of a new member with naming ceremonies or other rituals. In places where infant mortality rates are high, naming ceremonies may be delayed until the most dangerous period has passed. Among the Gurage of southwestern Ethiopia, the godmother ritually administers the first food to the child, but a feast to celebrate the infant's arrival is delayed for 5 days (Farb and Armelagos 1980). Catholics in the United States hold christenings to welcome the arrival of new members into the church and formally designate the child's godparents.

Most families view infancy and childhood as a vulnerable period requiring special nutritional concern. In North America, children's diets are selected to provide sufficient quantities of protein, calories, and other essential nutrients needed to support growth. Foods are also chosen based on their digestibility by the still-developing digestive tracts of infants and children. Because infants and children are considered too young to tolerate some items, they are not given the entire repertoire of adult foods. Parents are advised to withhold solid foods until the baby is 4 to 6 months old and to withhold potentially allergenic cow's milk, egg white, and orange juice until the baby's first birthday. Foods that could get caught in the windpipe, such as grapes and chunks of hot dogs, are often denied to children under 3 years of age. Coffee and alcohol are usually avoided until adulthood or at least until late adolescence. Of course, not all parents follow the prevailing nutritional advice. Some parents view the consumption of solid foods and cow's milk as a sign that the baby is healthy and maturing faster than other children.

Childhood is also viewed as an important time for establishing future food habits that will sustain good health. Children are encouraged to "clean your plate" and "eat your vegetables." Because children prefer foods that are familiar, parents often struggle to increase the varieties of foods they consume. Researchers have found support for the old adage—"Take three bites of everything"; they conclude that children will accept new foods but only after repeated exposure, perhaps as many as 10 or more times. Experts also recommend that parents use modeling to instill desirable food habits in their children and warn them against the use of rewards and punishment to encourage children to eat a certain type of food. Rewarding children for eating a certain food has been shown to actually decrease their acceptance of that food (Escobar 1999).

The Teen Years

About the time of puberty, the rights and responsibilities of children begin to change. In much of the world, adolescence is a brief period with a clearly defined and ceremonially observed end. Puberty rites are held to mark the youth's transition from childhood to adulthood and announce the changes in rights and responsibilities that the group must recognize.

The more important a woman's contribution to the family food supply, the more likely it is the society will mark her passage into adulthood with a special ceremony (Crapo 1993). Among the Shoshone Indians, females' readiness for adulthood was traditionally announced by menarche. At the time of her first menstrual cycle, the Shoshone girl was isolated in a special hut where she was prohibited from eating meat. At the end of the isolation, her mother brought her new clothes—women's clothes—to mark her new status as an adult.

Puberty rites are especially dramatic in societies where passage to adulthood is expected to be difficult. Among men, adult status is often marked by **circumcision,** scarification of the body, tattooing, or other bodily changes. Secret knowledge and a change in diet almost always accompany this rite of passage (Farb and Armelagos 1980). In some African and Middle Eastern societies in which males dominate women, the female genitals are mutilated as a way to control their sexuality and/or increase fertility.

circumcision
Removal of the foreskin of the penis or the clitoris.

In contemporary North America, adolescence is a long and, for some, difficult period, during which rights and responsibilities are ambiguously defined. Some bars recognize the transition to adulthood at age 21 by offering a free alcoholic drink to customers celebrating this occasion; however, in general, changes in age status and the accompanying role expectations are not socially marked.

Due to the rapid growth and sexual maturation that occurs during the teen years, nutrient needs may be greater than at any other time of life. A rapidly growing, active 15-year-old boy may need 4,000 calories or more just to maintain weight. Along with increased

Recommended Dietary Allowances (RDAs)

The levels of intake of essential nutrients considered to be adequate to meet the nutritional needs of practically all healthy people. RDAs are derived by a group of nutritional scientists who advise the Food and Nutrition Board, a committee of the National Academy of Sciences' National Research Council.

calorie needs, **Recommended Dietary Allowances (RDAs)** for protein, the B vitamins, vitamins A, E, C, calcium, phosphorous, iodine, magnesium, and zinc also increase for teenagers. Teenagers of both genders need more iron; boys have an additional requirement for building up large quantities of red blood cells, and girls need more iron because of the onset of menstruation.

As teenagers become more independent, they assume greater control over what they eat. Parents and teachers no longer make dietary decisions for teens whose mobility allows them to eat when they want, where they want, and what they want. In some households, people in this age group are responsible for shopping and preparing foods for themselves and their families.

A study of Canadian teenagers (Chapman and Mclean 1993) found that they divide food into two categories: good food and junk food. Good food is associated with home and the family, while junk food is associated with peers and having fun. Although teens are aware of the nutritional deficits of junk food and fear its ability to make them gain excess weight or cause skin problems, they eat it because it represents autonomy from parental control and good times with friends (see Figure H7.1).

In an attempt to express their individuality, teenagers sometimes deliberately try new things that are different from their parents' ways of life. In some cases, dietary habits such as vegetarianism offer adolescents a way to identify or distinguish themselves. Typically, however, behaviors that are different from those of the larger society are in strict conformance with the teen's own subculture. Teenagers are generally sensitive to peer pressure and often conform to norms shared by other members of their subgroup.

In North America, the fast-paced, independent lifestyle of teenagers affects their dietary habits significantly. Skipped breakfasts, increased snacking on empty calories, irregular meals, and frequent dining at fast-food restaurants are characteristic. Between 30% and 40% of teens surveyed skip breakfast most days, and 25% of those who do eat breakfast, eat it away from home, usually at a fast-food restaurant. Teens who skip breakfast are two to five times more likely to have an inadequate diet and score significantly lower on tests than those who eat breakfast regularly (American Dietetics Association 1998a). In the United States, 43% of elementary, 74% of middle/junior high, and 98% of senior high schools have a vending machine or a school store, canteen, or snack bar where students can purchase food or beverages. The majority of food sold is considered to be "junk food" (**School Health Policies and Programs Study** CDC 2000). These sales provide schools with a way to make money for various programs and activities. This practice is being challenged by nutrition advocates through state legislation that would set nutritional guidelines on the foods sold to students at school.

Given the special nutritional and social circumstances of adolescence, it is not surprising that the diets of many North American teens fall below the RDA for some nutrients. Deficiencies in calcium, iron, vitamins A and C, and the B vitamins are not uncommon. Due to the increased consumption of empty-calorie snack foods and

Figure H7.1

Enjoying food with friends is an important social activity during the teen years.

© Mary Kate Denny/PhotoEdit

School Health Policies and Programs Study (SHPPS)

A national survey periodically conducted to assess school health policies and programs at the state, district, school, and classroom levels.

processed meals, the diet of some teens is also high in fat, cholesterol, sodium, and sugar, while it lacks fiber. The diets of some
teenagers also contain large amounts of fast foods. As a result, they
ingest too much fat, calories, and sodium, which can lead to health
problems (Kraak and Pelletier 1998). Of course, not all teenagers eat
poorly. Some people become particularly interested in nutrition during this time in their lives and eat quite well.

Reproduction and Parenthood

Reproduction is another phase in the life cycle that is associated
with special dietary habits. Most societies recognize that eating
habits of pregnant women affect the development of their babies,
and many prescribe special foods or avoidance of certain foods for
the prenatal period. In the United States, women are advised by
health care providers to increase their calories, protein, calcium,
phosphorous, magnesium, folacin, and other vitamins and minerals
to meet the increased nutrient needs of pregnancy and ensure a
birth weight greater than 5 1/2 pounds. Milk is presented by many
nutrition educators as a "super food" necessary to obtain the Recommended Daily Allowances for calcium, while alcohol and coffee
are to be avoided.

Food beliefs, poverty, lack of prenatal care, and lack of exposure
to nutrition advice, as well as a host of other circumstances, can
interfere with the observance of these dietary recommendations. In
general, however, the pregnant woman is considered to be eating for
two, weight gain is desirable, and her cravings for specific foods are
accepted.

In contrast to North Americans who increase their food intake
to produce large babies, women in some societies restrict weight
gain and fetal size in hopes of minimizing the dangers of childbirth.
This is especially important in populations that experience stunted
growth and development due to nutritional inadequacy. The short
stature and small pelvis of many poor women makes the birth of
large babies difficult.

Many societies observe food taboos during pregnancy as a way
to protect the child or mother. By reducing the uncertainty associated with the pregnancy's outcome, food taboos may also reduce a
mother's anxiety. Often these taboos focus on foods that resemble
the feared outcome: The Shoshone avoided eating trout because it
flops and could cause the child to get entangled in the umbilical
cord. Javanese women are advised to avoid the kepel fruit with its
horizontal seeds for fear their fetus will present in a transverse position (Hull 1986). The Mbum Kpan women of Chad in equatorial
Africa avoid meat from antelopes with twisted horns, which might
cause them to give birth to deformed offspring. And some North
Americans still avoid looking too long at strawberries because they
believe it can cause a similar-shaped birthmark on the fetus. Pica, the
practice of eating nonfood substances, is discussed in Focus H7.1.

Food restrictions and proscriptions are even more common during the postpartum period. One of the most widespread practices is

FOCUS H7.1　Pica

You may have heard that some pregnant women crave pickles, ice cream, and strawberries; but did you know that others crave and eat clay, laundry starch, and ice as well? The consumption of nonfood items is called *pica*, the Latin word for magpie—a bird known for carrying off and eating unusual things. Although clay, laundry starch, and ice are among the most frequently craved items, the desire for chalk, burned matches, and the inner tubes of tires has also been reported.

Geophagia, or clay eating, is practiced by a wide variety of peoples around the globe—West Africa, New Guinea, the Philippines, Guatemala, and the Amazon Basin. In the United States, geophagia is thought to be most prevalent among young children and women in the rural Southeast. Geophagia among poor African Americans living in the South has been studied for many years. It has also been reported among non-Hispanic whites, Mexican Americans, and Native Americans as well. It also has been observed in other primates.

For most people who eat clay, not just any clay will do. In Ghana, for example, certain sites are held in high regard for the quality of their edible clay. The clay is formed into egg shapes and marketed throughout East Africa. In the United States, clay from Georgia is considered to be especially good by some. Clay was once sent to farmer's markets, where it was sold in shoeboxes (Farb and Armelagos 1980). Some people eat clay plain; others add salt and vinegar and bake it to impart a smoked flavor. Unable to obtain good clay, some women have used laundry starch or cornstarch as substitutes. Despite its resemblance to clay (the texture is similar), laundry starch has none of the minerals found in the clay. Other nonfood substances reported to be consumed include baking soda, cigarette butts, cloth, coal, coffee grounds, crayons, dirt, hair, mothballs, paint chips, paste, and plaster (Brown 2002).

Table H7.1　Some of the Many Nonfood Substances People Reported Eating

Animal droppings	Coal	Laundry starch	Paste
Baking soda	Coffee grounds	Leaves	Pebbles
Burnt matches	Cornstarch	Mothballs	Plaster
Cigarette butts	Crayons	Nylon stockings	Sand
Clay	Dirt	Paint chips	String
Cloth	Hair	Paper	Wool

© PhotoDisc

Why do people practice pica? A review of the literature provides a wide variety of conflicting information on the subject. The six major hypotheses are outlined below.

1. *Pica is the body's response to the need to obtain certain nutrients.* Samples of clay eaten by people in West Africa have been analyzed for mineral content. The calcium, magnesium, potassium, iron, copper, and zinc levels have been shown to compare favorably with the mineral supplements given to pregnant women in modern societies. Some researchers speculate that the craving for clay is due to mineral deficiencies or increased mineral needs.

2. *Pica is a response to hunger.* Pica is prevalent among people with limited food available to them. Clay is free and filling. A one-pound box of laundry starch provides 1,800 calories, making it an inexpensive source of energy, though it supplies no other vital nutrients.

(continued)

Focus H7.1   (continued from previous page)

3. *Pica is a cultural phenomenon passed from generation to generation.* The stimulus for eating clay may be associated with women's early roles as gardeners and potters, which placed them in a position where consumption of clay was normal. Identity with the female community passed the tradition along.

4. *Pica is a response to physiological changes.* For example, dry substances such as clay and laundry starch absorb the excessive amounts of saliva produced during pregnancy. Some people who practice pica say that eating clay or starch reduces nausea during pregnancy just as soda crackers do.

5. *Pica is a way of seeking attention.* Pregnant women who ask friends and family to help them secure hard-to-get items are expressing their need for support, love, and understanding. Along these lines, clay is taken as a gift to new mothers while they are recovering from childbirth.

6. *Pica is a means of protecting the body from toxic foods and other substances.* Perhaps the most compelling explanation for pica is a theory proposed by Timothy Johns (1991). Johns examined clay eating by men, women, and children in a variety of societies. He believes clay eating originated as a way to detoxify foods containing poisonous substances. Because clay binds to the toxins, rendering them less harmful, clay eating provides people with an important adaptive advantage—greater flexibility in the foods they consume. In the Andean mountains, natives eat wild species of potatoes that contain toxic **glycoalkaloids**, which cause stomach pains, vomiting, and even death. These natives are able to consume large quantities of these potatoes by dipping them into slurry clay gruel that binds to the glycoalkaloids. Similar practices have been observed elsewhere. The Pomo Indians of northern California mixed clay with the ground meal of acorns, and the Sardinians mixed clay with boiled acorns to make bread. In these cases, the clay doesn't bind to the toxic tannins but, rather, alters the structure of its molecules rendering them incapable of binding to the gut wall. Johns believes that people may initially practice pica as a way to eliminate the bitter taste; however, it has an important survival advantage because the clay's ability to actually detoxify foods makes them available for consumption. Over time, Johns argues, the practice of eating clay becomes a cultural norm and may be practiced by later generations after the diet no longer contains toxic substances. Even when other members of the society no longer practice geophagia, pregnant women may continue the custom because it helps them counter the nausea and vomiting that result from hormonal metabolites released into the small intestine during pregnancy. Indeed, many women say they eat clay or laundry starch (Argo in the Blue Box), to settle their stomachs. (Worthington-Roberts and Williams 1989)

Nutritional Consequences of Pica

Clay eating and other forms of pica usually have relatively little nutritional impact because limited amounts are eaten. Ingestion of some types of clay may decrease iron levels by binding to dietary iron, making it unavailable for absorption. Although laundry starch does not bind to iron, if eaten in large quantities, it can replace more nutritious foods. Some people have reported eating up to 2 pounds of laundry starch per day (a total of 3,600 calories), leaving little room in the diet for nutrient-rich foods. Fortunately, most people who practice pica do so on a limited basis and serious

(continued)

glycoalkaloids
Naturally occurring chemicals found in potatoes, especially in the peelings and near the outside edges. High amount of glycoalkaloids can occur in green, unripe, damaged, or poorly stored potatoes. They create a bitter, musty taste that can give people a burning sensation in their mouth or cause diarrhea and gastrointestinal problems when consumed in large quantities

Focus H7.1   (continued from previous page)

complications are uncommon, with one exception. Over a million children in the United States have elevated serum lead levels. Each year, many children die from lead poisoning from eating lead paint chips and inhaling lead-contaminated dust and soil near buildings (Brown 2002).

Because of potential complications, it is helpful for health care providers to know if unusual items make up a large part of a woman's diet. But many women who practice pica are embarrassed about it and not likely to tell health workers. Interviewers have found that if they mention that they have talked to others who eat nonfood items, the respondents are more likely to share their practices. A question a nutritionist might ask is, "I've talked to some pregnant women who say they crave things like clay, and laundry starch, and ice. Do you ever crave or eat them?"

La Leche League (LLL)
An international organization founded in the United States in 1956 by seven women who wanted to make breastfeeding easier and more enjoyable for mothers and children. The organization sponsors mother-to-mother support groups that offer information and encouragement to those who want to breastfeed. The Information Services Department of LLL also provides information to health professionals.

to seclude the new mother and child together for a specified period of time, typically 20 to 40 days, during which time she rests and consumes "strengthening" or "protecting foods." Called *la cuarentena* in traditional Hispanic cultures, this custom had a dual function. First, it exempted the woman from work and sexual intercourse during the seclusion period so that she could rest and focus on her baby. The assumption of cooking and other household tasks by family members and friends allowed the new mother to eat well, establish lactation, and adjust to new sleeping schedules. Second, it reinforced ties with relatives and neighbors who extended their services to the family—services the new mother was expected to reciprocate on a similar occasion. Similar practices still exist today in the United States. For example, in some communities, members of the **La Leche League (LLL),** a network of women dedicated to promoting and supporting lactation, take turns bringing nutritious meals to a new mother and her family, usually for a period of 1 or 2 weeks.

Although many food restrictions are potentially harmful because they limit protein and other nutrient sources, few studies have documented serious harm resulting from these practices. If the prohibited foods are rarely consumed anyway or alternatives are abundant, the nutritional impact of food taboos will be limited. Balancing practices exist in many places to help offset the potentially negative nutritional implications. Some Mexicans, for example, restrict pork, beef, red beans, milk, cheese, and eggs, as well as all types of fruits and vegetables. However, families go to great lengths during *la cuarentena* to provide new mothers with chicken, black bean broth, soups, boiled milk, toasted tortillas, and other nourishing foods. Researchers also have found that many food taboos are not always obeyed, particularly when food is scarce. In a survey of food practices of 2,000 mothers in East Java, only 10–15% of women reported observing food restrictions when pregnant or lactating (Hull 1986).

Old Age

The degree of respect and power afforded the elderly varies a great deal across cultures. Older people are most likely to be highly esteemed in societies where they maintain their roles as family leaders into old age. This pattern is especially common in societies requiring married couples to live near one spouse's parents and those organized into patrilineal or matrilineal extended family units. The elderly also occupy higher-ranking statuses if they remain economically productive (Crapo 1993). Among small foraging groups, bandleaders are chosen from the oldest men in the group, and older women are respected for their wisdom and extensive life experience. Elders also tend to be held in high esteem in societies where they control inheritable wealth or are believed to exert control over the living after death (as ancestral spirits). In China, the best foods are often reserved for the elders and food offerings are made regularly to ancestors who have died.

In contrast, older people are less likely to maintain high-ranking statuses in industrialized societies organized around nuclear family units. For many years, older people in the United States were stereotyped as sickly, poor, and lonely, receiving little respect and assistance from younger people. Today, the continued economic productivity and increased wealth of American seniors has shattered this stereotype and raised their social ranking. Americans are not only living longer today, but they are more active. The percentage of 60-plus seniors who report having a sedentary lifestyle declined from 34% in 1985 to 28% in 1998, and a growing number are starting new careers, taking university courses, learning new skills, and volunteering for social causes. In one study, 44% of elders reported that they worked for pay at some point after retirement, usually because they wanted to stay active (Raymond 2000).

Older North Americans' dietary habits are influenced by a complex set of biological, social, cultural, and economic factors that make it difficult for some to meet their nutritional needs. Aging brings many physiological changes that impact dietary behavior and nutritional status. As metabolism slows, caloric needs decrease. By age 70, men require an average daily intake of 1,875 calories compared to 2,501 calories during their 20s. For women, daily caloric needs drop from a mean of 1,634 calories to 1,386 calories. Many older people also produce less saliva, digestive acids, and enzymes than younger people, making some foods more difficult to swallow and digest. Dairy products may cause discomfort if eaten in large amounts because of the reduced activity of lactase, the intestinal enzyme that breaks down lactose. The gastrointestinal tract's motility (its involuntary folding and unfolding to move food along) is often decreased, allowing food to remain in one place much longer, producing harder stools and constipation. Some older people increase their use of salt and sugar to adjust to distorted taste sensations or diminished taste acuity. At the same time, physical disabilities and chronic diseases (e.g., hypertension, diabetes, cardiovascular disease, kidney disease, and osteoporosis) require many older people to make dietary modifications, including the reduction of salt and sugar (Schlenker 1998).

The later years are often marked by adjustment to changes in the social environment brought about by changes in family composition (e.g., children leaving home, death of a spouse), retirement, or new living accommodations that impact dietary habits. Elders learn to shop and cook for a smaller household, and they may also need to adjust to a fixed income or learn to eat alone.

Older people's diets also reflect traditional food habits and preferences. As with any age group, older people are most comfortable eating familiar foods. When a person moves into a nursing home or encounters other dramatic lifestyle changes, eating familiar foods can make the transition easier. Of course, while some elders may focus on an increasingly narrow array of valued foods, others find they now have the time to experiment and try new foods.

In addition to lifelong dietary patterns, many older people modify their diet habits to improve their health. As with other age groups, many food beliefs reflect an awareness of physiological changes and special nutritional needs. In other cases, they result from notions about what foods are appropriate for older people. Some foods are eliminated because they are perceived as "too strong" or "too hard" for the elderly, others are eaten in greater quantities because they are considered healthy for older people. Prunes and foods high in roughage may be eaten to prevent constipation. In a survey of 194 older Canadians, perceived health benefit was a stronger predictor of food choice than convenience, cost, or prestige value of foods (Schlenker 1998).

Food availability comprises another set of factors that influence elderly people's ability to meet their nutritional needs. For those who live below the poverty line, food availability is a major problem. When chronic illness requires special foods, many senior citizens living on fixed incomes find it hard to purchase some products, such as the more expensive low-sodium foods. Because most are shopping only for one or two people, the elderly usually purchase the smallest food package, for which they must pay the highest unit cost.

Loss of mobility and lack of transportation also create problems for some elderly people. With the flight of large chain stores from the central city to the suburbs, older people living in high-rise buildings or apartment complexes located in downtown areas are increasingly disadvantaged. Only small markets, with a limited variety of generally expensive items, are located in their immediate vicinity.

Food choices may be severely restricted for elderly people who live in nursing homes or other long-term care institutions. Although this group is small (less than 5% of people over 65), they are almost exclusively dependent on what these institutions serve (U.S. Government Accounting Office 2001). Studies show a tremendous range in the dietary adequacy and variety of foods served in nursing homes. Even when menus are designed that offer residents 100% of the RDA for all essential nutrients, many residents do not consume enough to meet their nutrient needs.

A third major influence on elderly nutritional status is the social environment in which the older person lives. Because eating is as much a social activity as a biological activity, some people lose interest in eating after a spouse dies or friends and other relatives move away. Grocery shopping, storage, cooking, and cleanup may appear too demanding when cooking for just one. These tasks appear particularly meaningless to the older adult who is depressed, grieving for loved ones, or lonely. Men who have relied on their wives for meal preparation, as well as mealtime companionship, usually have more difficulty in dealing with the loss of a spouse than women.

Some older persons overcome these problems by sharing meals with friends and neighbors. A strong social network—a number of close friends, attachments to relatives, ties with neighbors or members of your religious group—has been linked to improved nutritional status. In one study by McIntosh and his associates, social support was shown to help the elderly cope with stressful life events and physical disabilities. Companionship and mealtime companionship were associated with regular meal patterns, improved appetite, and better nutritional status (McIntosh 1996).

With the elderly population in most industrialized nations growing rapidly, policy makers, politicians, and others are under increased pressure to protect the interests of this increasingly powerful group. A variety of assistance programs such as Social Security, Supplemental Security Income, and the Senior Nutrition Program for the Elderly have been developed to benefit elderly Americans, and funds have been allocated for research institutes and special grants to study the problems of aging in our society (see Figure H7.2). Recognition of the important links between nutritional well-being, overall health, and the social environment is critical in developing new ways to extend life and improve the quality of those additional years.

Figure H7.2
The Senior Nutrition Program provides midday meals in a congregate setting so that elderly people can enjoy companionship along with nutritious food.

© D & I MacDonald/PhotoEdit

Death

With rare exceptions, societies mark the last transition of the life cycle ceremonially. Funeral rites provide a means of disposing of the body and helping survivors adjust to the person's absence. Kinship ties are reshaped, accumulated wealth is distributed, and duties are reassigned. Food often plays an important part in funeral rites and other events that surround a person's death. "Recognition by the community of this upheaval has its effect on the one activity common to everyone: the preparation and distribution of food. The disruption of community life is often symbolized by basic changes made in the customs of eating—fasting, temporarily extinguishing the hearths, placing new taboos upon foods, and special offerings of food to the gods" (Farb and Armelagos 1980:93).

Some societies believe that the dead require nourishment after they leave this world. Surviving kinsmen make regular food offerings to their ancestors' spirits in exchange for the spirits' help with harvests, offspring, or other earthly matters. In traditional Chinese society, food played a major role during the funeral and for years afterward. Food was presented immediately upon death and then again when the body was placed in the coffin to protect the spirit on its journey. After the body was placed in the coffin, food offerings were presented twice daily until the body was buried. A funeral banquet was held immediately after the burial, and additional offerings were made to the deceased. Later, an ancestral tablet was installed in a special altar room or hall, where additional food offerings and prayers were made on a regular basis (Lindsay 1998).

As in other phases of the life cycle, some dietary practices associated with death have important nutritional implications. The Tsembaga of New Guinea mark the passage of a warrior with a ritual pig slaughter. Several pigs are sacrificed and only close relatives consume the pork. Although the Tsembaga perform these rites primarily for ideological reasons, the consumption of a high-protein food such as pork during times of stress makes good nutritional sense as well in this essentially vegetarian population.

Moving closer to home, in North America friends and neighbors often take food to families who are mourning the death of a relative. When a newcomer to a rural eastern Kentucky community gave birth to a stillborn baby, the community responded by sending an entire van full of food to her home. Her refrigerator, freezer, countertops, and tabletops were filled with this bountiful display of support. This custom provides relief from cooking and shopping when family members are preoccupied with making funeral arrangements and grieving.

Worldview, Religion, and Health Beliefs: The Ideological Basis of Food Practices

At a vegetarian restaurant in Washington, D.C., one of the authors took an informal survey asking diners if they were vegetarians and, if so, why they did not eat meat. Responses included a wide variety of interesting reasons. A man from India cited the Hindu religious doctrine that prohibits meat consumption as one way to achieve the ideal of nonviolence. An American woman with a family history of heart disease chose the low-fat, low-cholesterol vegetarian diet for disease prevention. Respect for animals combined with disapproval of the inhumane practice used in livestock production was another reply. And several people referred to today's world food crisis and stated that they avoided meat as a way to protest the use of the large quantities of grain to feed livestock while people in other parts of the world go hungry.

Each of these reasons reflects some aspect of the consumer's ideology: religious beliefs, health beliefs, ethical standards, ethnic identity, or political positions. When we think of ideology, we usually think first of beliefs, knowledge, religion, folklore, art, dance, literature, and language—symbolic expressions of meaning and values. But ideology also embraces the entire realm of socially shared concepts and understandings about the universe and people's place in it. It includes goals for human existence and beliefs about the supernatural and its relationship to people. Ideology also consists of justifications for the social rules governing food production, division of labor, distribution of resources, marriage, child rearing, maintenance of law and order, and political relations with other groups. It is the way we think about the activities of everyday life, how we categorize them, and how we fit them into wider understandings of the world around us. We are also including here the symbolic meanings of objects and actions. Finally, ideology consists of "thoughts about thoughts," as in formal philosophical or scientific (**ethnoscientific**) systems (Harris and Johnson 2002). Several of the systems of shared understanding and knowledge that are included in ideology are particularly relevant to the study of food. They are *the symbolic meanings of foods, worldview, religion, ethnic identity,* and *health beliefs.* Their relationship to food and diets is explored in this chapter.

ethnoscience
The study of the systems of classification and ways of knowing about natural phenomena held by members of a distinct cultural group.

Food in a Forest of Symbols

Anthropologist Victor Turner (1967) has effectively argued that humans live in a "forest of symbols." For humans the physical world

symbol
An object that can be perceived by the sense and refers to or represents nontangible concepts such as ideas, emotions, identity, or religious beliefs; carries meaning.

mucilaginous
Slimy or mucus-like.

is full of **symbols**—objects that not only exist in and of themselves, but also are imbued with meaning by humans. A rose is not just a rose, but also a symbol of romantic love, or the finer things in life (as in "bread and roses"), or death. Similarly, food is not just a vehicle for nutrients; it also has meanings beyond those associated with the need to sustain life. In one culture the consumption of dog meat is a symbol of uncivilized behavior, in another a celebration of ethnic identity. For people in the United States, a turkey dinner denotes an important holiday.

The definition of what is a "food," what constitutes a "meal," and how meals can and should be sequenced are all part of the basic assumptions about foods that are culturally influenced and, in turn, influence diet. Dog is food in one culture and decidedly not food in another. In much of Asia, the **mucilaginous** excretion of a cow's udder is not considered appropriate food for humans. As anthropologist Marvin Harris (1985) points out, "The Chinese and other Eastern and Southeast Asian peoples do not merely have an aversion to the use of milk, they loathe it intensely, reacting to the prospect of gulping down a nice, cold glass of the stuff much as Westerners might react to the prospect of a nice, cold glass of cow saliva" (p. 130). (Author K. M. DeWalt has a similar response to milk: Yuck!) As we discussed in Highlight 2, aversion to milk may be, in part, the result of low lactase levels in adults in food systems that have not included dairying as a key component. We have also pointed out that North Americans tend to eschew insects as food. In Mexico and Thailand, among other places, they are delicacies. No cultural group defines as "food" every edible item in their environment. Even the Ju/'hoansi do not consider as food all of the potentially edible items in their environment. And some foods seen as appropriate for adults may not be considered appropriate for children.

We also have rules about what collection of edible items is considered a "meal." A hamburger, fries, and a drink may constitute a meal to North Americans. However, Japanese eating in McDonald's restaurants do not consider this a meal. Without rice, the eating event that includes the 'burger is a "snack," not a proper meal (Ohnuki-Tierney 1998). In Korea, however, this is not a problem. Koreans consider an eating event at McDonald's to be a meal and will even take guests to McDonald's restaurants as special treats (Bak 1998). Think about what constitutes a meal for you. What elements must an eating event have to be defined as a meal? What constitutes a snack? What things go together? Does corned beef *need* cabbage?

Finally, the number and content of meals in a daily, weekly, or yearly cycle is also something that tends to follow cultural rules. When we carry out a statistical analysis of American food consumption, we find that the consumption of bacon correlates highly with the consumption of eggs. Bacon and eggs go together and are generally eaten early in the day. Some North Americans may like cold pizza for breakfast, but they find the Japanese custom of soup for the first meal of the day to be rather strange. The pattern of three meals per day with the largest family meal in the evening is a product of the industrial revolution and the extended school day. Farm families used to consume the largest meal at midday to help the workers sustain themselves through a long working day. "Supper" was a light meal at the end of the working day. Many cultures define two meals per day as ideal; others may have four or more.

Many plastic and paper plates come with three distinct compartments, a more-than-symbolic representation of the "ideal" American meal of meat, starch, and vegetable. It is so central to our assumptions about what constitutes a meal that we construct plates to accommodate it. Meal cycles are also part of our basic assumption about food. For many North Americans, Sunday dinner is the central meal of the week. Special foods are served at this meal. In the Italian-American community, studies by Goode et al. (1984a, 1984b) found that Sunday dinner included more of the ideal elements of meal than any other meal during the week. It was generally based around a "gravy" dish that combined a tomato-based sauce with pasta. Meals during the rest of the week were much less elaborate and might be American-style "platter" meals with meat, starch, and vegetable.

Special occasion, holiday, or feast-type meals also have expected content (see Figure 8.1). North Americans consume turkey for many holidays. Holiday meals in the Philadelphia community studied by Goode et al. have an even more elaborate pattern than Sunday meals and include an antipasto course, a soup course, a whole-roast course, a multidish dessert course, and an ultimate course of nuts and fruits. DeWalt's Pennsylvania Dutch mother-in-law would serve pork on New Year's day—to start the year out right. Pigs "root" ahead and are forward-looking. Chickens and turkeys were inappropriate because they "scratch backward" to get their food. A person who ate poultry on New Year's day would start the year looking backward rather than ahead. Our Kentucky friends always included black-eyed peas and greens in their New Year's day meal.

The Meaning of Food

Cultural rules and expectations about food and meals can be important influences on what we eat and when. As Claude Levi-Strauss (1969) has pointed out, food is not just "good to eat," it is also "good to think." The symbolic value of food can exert a powerful influence on diet. In a study of food in the Andean region of Ecuador, Mary Weismantel (1988, 2000) found that indigenous people continued eating toasted barley flour (*machica*) for breakfast because it reaffirmed their difference from the Spanish-speaking *mestizos*, who ate bread with their morning coffee; that is, *machica* serves as a symbol of ethnic identity. At the same time, their children clamor for bread because they see it as a more "modern" breakfast food.

Like many symbols, food is **polysemic**—it carries many meanings simultaneously. To elite people, the consumption of chitterlings, collard greens, and corn bread may denote the low status of the consumers. For others, however, "soul food" celebrates their identity as African Americans with Southern roots.

polysemic symbol
A symbol that carries several meanings simultaneously.

Figure 8.1

Kwanzaa, the African American holiday celebrated from December 26 through January 1 each year, culminates with a feast featuring dishes from throughout Africa, the Caribbean, the U.S. South, and other regions where Africans now live.

© Merritt Veincent/PhotoEdit

The meanings of foods also shift with changing circumstances. When sugar first became available to Europeans, it was a luxury food, symbolizing affluence. Its use was a sign of conspicuous consumption. When sugar became inexpensive and a staple of the diet of the working classes in England, it quickly lost its value as a status symbol (Mintz 1985). And you might recall from Chapter 1 how the meaning of pork to the Scots changed considerably over time.

In some societies, specific foods have such important meanings that their use or consumption is restricted to only a few people, often the elite. Such rules are called *sumptuary laws*. For example, the Inca Empire of Peru allowed only the nobility to eat certain foods. As noted in Chapter 7, some foods have a symbolic affiliation with gender. We often give students a set of cards with the names of common foods on them and ask them to sort the cards into categories. The most common set of categories students use looks very much like the "four food groups" commonly taught in North American schools in the 1970s and 80s. However, when we ask students to put the foods into categories along a continuum of "masculine" to "feminine," they can do so with some consistency. Although some individuals will categorize some foods differently, two foods are always classified consistently. Steak is uniformly classified as a "masculine" food, and strawberries are uniformly classified as "feminine." We are not sure why steak and strawberries carry such consistent associations with gender in North America, but this is a result that has held up in classroom testing over more than a decade in several different parts of the United States.

Kahn (1986, 1998), writing about the meaning of food for the Wamira people of New Guinea, describes how men use their knowledge, spiritual power (*anona*: essence), and magic to produce taro (*Colocasia esculenta*). Taro is a root crop commonly found on Pacific islands. A man's taro plants are considered his "children." Men may try to harm others' taro crops using "taro sorcery." The quality of the taro that is harvested, then, represents men's virility, power, and essence. "When men harvest their crops, the appearance of their taro informs them of all they need to know: their horticultural potency, the accuracy of their magical knowledge and the intentions of their friends and enemies" (Kahn 1986:98). For the Wimira, taro is not only food; it is also a key representation of the social and political power of individuals.

Food as an Ethnic Marker

One of the most important ways food is used for its symbolic value is as a marker of ethnic identity. As mentioned earlier in this chapter, several generations away from Italy, Italian Americans in Philadelphia continue to have "Italian" dinners on Sunday, centered around pasta and sauce dishes when family members gather together (Goode et al. 1984a, 1984b). In Chapter 5 we discussed the subsistence system of the Kayapo of Brazil. The Kayapo use more sweet potatoes and less manioc than their neighbors. One of the names the Kayapo use for themselves translates into "Sweet Potato Eaters," which distinguishes them from their neighbors. The term Eskimo is actually a pejorative name given to the Inuit people by their non-Inuit neighbors. It means "Raw Blubber Eaters."

Food is sometimes used to identify individuals following a particular belief system and to differentiate between followers and non-followers. Many religious systems prescribe or proscribe particular foods for their followers. Later we will discuss the impact of religious beliefs on diet using the examples of the Jewish prohibition on eating pork and the Hindu prohibition on eating beef.

Worldview

A major component of every society's ideology is its **worldview,** which refers to the way a group defines, categorizes, and explains physical objects and living things. It is a complex set of beliefs, values, and norms describing "how the world works." Because of varying worldviews, people from two cultures looking at the same thing may describe it very differently. For instance, a simple description of some of the most basic observable components of this world—trees, land, and water—by a member of one society might prove to be completely unintelligible to a member of another. The Navajo Indian facing the East sees the world in the following way:

> *The sage-covered earth is changing, one of the most benevolent of the gods, who grows old and young again with the cycle of each year's seasons. The rising sun is himself a god who with Changing*

worldview

A system of shared basic understandings of how the world works and people's place in it.

Woman produced a warrior that rid the earth of most of its evil forces and who is still using his power to help people. The first brightness is another god Dawn-Boy. . . . The cone-shaped mountains have lava on their sides, which is the caked blood of a wicked tyrant killed by the Sun's Warrior offspring. (Leighton and Leighton, cited in Honigman 1959:591)

A white man looking at the same landscape sees the yellow day coming up over miles of sage, copse of pinyon, three or four yellow pines in the soft light, distant blue swells of mountains, with here and there a volcanic cone. (Honigman 1959:591)

Although everyone's view of the universe is greatly influenced by their society's particular worldview, this aspect of ideology tends to be so implicit (taken for granted) that people are often unable to describe it. If interviewed by an anthropologist from another culture, for example, how would you describe the United States' worldview? Take a minute and try to list some of the major assumptions North Americans make about how the universe operates.

Most likely you found this exercise difficult. Because your basic approach to the world is so implicit, you are unlikely to mention basic conceptions of the world and the fundamental values that you believe should guide humankind. Because our basic assumptions about food and the world are so implicit, we often do not recognize that they differ among people. We often don't even discuss them in cultural terms—that is, as one distinct way of looking at life that differs from the views held by others. Instead, you might assume that the values and beliefs that make up your worldview are the "natural" or "correct" explanation of how things occur. **Ethnocentrism** is a concept that applies here. Ethnocentrism refers to the belief that your own patterns of behavior and thought are good, beautiful, or important. It also includes the idea that other, different, patterns of behavior and thoughts are less good, less beautiful, less important, less civilized, and generally inferior. Ethnocentrism often also includes the idea that your patterns of action and thought are "natural," that is, they are basic to all humans and are the way all people do or should think and act.

Because our assumptions about food and life are implicit, they seem "natural" to us. We may not be well prepared for how other societies' assumptions differ from ours. When living and working in other countries or with other ethnic groups, people are often perplexed by their hosts' behavior; when attempting to introduce changes, they are frustrated by the resistance and other confusing behavior they encounter. Like cannibalism to Westerners, some of these behaviors may seem to be "wrong" and "unnatural." As we discussed in Chapter 4, the alternative to ethnocentrism is **cultural relativism.** The letter reprinted in Box 8.1 illustrates how differently groups of people may look at acceptable sources of food. As Mr. Piao points out, there is no reason to believe that the consumption of the meat of any animal is any more natural, good, or civilized than that of any other animal.

ethnocentrism
The tendency to interpret or judge other cultures in terms of one's own culture. It often includes the assumption that the behaviors and understandings that one has from one's own culture are "best," most "proper," most "natural."

cultural relativism
The position that others' culturally influenced behaviors and understandings should be viewed from the point of view of the people inside the culture. It often includes the assumption that there are a number of alternative ways to understand the world and behave in it.

(B·O·X) 8.1

Eating Dog in China
Letter to the Editor of the New York Times, September 13, 1994

To the Editor:

I read David M. Raddock's September 2 letter on China with amusement. Yes, we are still eating dogs in China. What's wrong with that? I used to feel the same as May-Lee Chai (whose Aug. 26 letter Mr. Raddock replied to) whenever I read or heard any Americans talk about this Dog Thing. I do not feel angry anymore. This is not a matter of who is more civilized or less civilized. It is simply a matter of different tastes.

Why is eating dog meat less dignified (at least that is how it is portrayed in the media here) than eating beef? Does one kind of animal enjoy more rights than another? . . . I am not ashamed of eating dog. I have a different color of skin, I speak a different language and I come from a different cultural background so I sometimes eat different food (meat in this case). This is the beginning of my seventh year in the United States. America has been a perfect country for me except for one thing: I (miss) having dog meat back home.

I like Mr. Raddock's letter. It is factual and non-judgmental. This kind of perspective is what we need when looking at different cultures. In today's world, linked by common interest, it is perhaps wise and beneficial not to impose our own values on others.

James Piao
New York

(Quoted in Johnson and Harris 2000:12)

We discuss worldview here as an important concept when we address ways to change food habits and diets. Generations of Western-trained nutritionists have looked at the diets and local food knowledge of non-Western people and tried to change them just because they were different from Western food habits. The attempt to introduce the consumption of cow's milk in either liquid or powdered form in many parts of the world is one example of this. In Mexico the Spanish attempted to displace the local maize-based bread—tortillas—with wheat, arguing that corn was clearly an inferior food to wheat, even after scientific studies in the twentieth century showed this to not be true (Pilcher 1998). Goode et al. found that nutritionists tried to change the food habits of the Italian Americans they worked with because they were thought to be unhealthy. An analysis of the nutrient content of Italian-style "gravy" meals and American-style "platter" meals found that, on the contrary, the gravy meals were lower in fat, cholesterol, and calories than the platter meals. All approaches to change need to include a careful analysis of the actual current nutritional conditions and an assessment of the likely impact of changes in diet on nutrition. They also need to include a deep-seated respect for local knowledge and understandings as possible alternatives to "conventional wisdom" with respect to food and diet.

Religion

And you shall observe the feast of unleavened bread, for on this very day I brought your hosts out of Egypt, therefore you shall observe this day, throughout your generations, as an ordinance for ever. In the first month, on the fourteenth day of the month at evening, you shall eat unleavened bread, and so until the twenty-first day of the month at evening. For seven days no leaven shall be found in your houses; for if anyone eats what is leavened, that person shall be cut from the congregation of Israel. (Exodus 12:17–19)

religion
A component of culture that includes a system of beliefs expressed through rituals and symbols and which is concerned with the supernatural.

Religion is a system of beliefs that is expressed through rituals and symbols and that is concerned with the supernatural. Religious behavior includes chants, prayers, myths, talks, ethical standards, and concepts of supernatural beings and forces. Systems of religious beliefs provide participants with an intellectual and emotional commitment to the ordered belief system on which social life is based. There are several things to note in this discussion. First, religious systems are about *belief*. Some elements cannot be "proved." They are taken on the basis of faith. This makes religious beliefs especially important to consider in discussion of food and diet. They can be important influences on diet.

Note that we have not mentioned the need for belief in a supreme being, nor have we mentioned moral codes. Not all belief systems that deal with the supernatural include belief in the existence of one or more discrete beings. Not all religious systems have clear moral codes. The Ju/'hoansi religious system suggests the existence of sources of spiritual power, and specialists who can access that power, but does not have much to say about a god or gods or good and evil. It is important to keep in mind that some religious systems do not have elements that are integral to Western religious belief. Indeed, some of the basic assumptions of humans' place in the cosmos are not shared among many religious traditions.

People everywhere relate to supernatural forces or entities, perform religious rituals, and incorporate the supernatural into their daily lives. The specific ways in which humans define and relate to the supernatural vary widely, but despite this diversity all religions serve many of the same functions. First, religion provides people with an organized picture of the universe and various supernatural forces or entities and their relationship to humans. Religion offers explanations for illness, death, and natural disasters, and ritual practices for controlling supernatural forces. Weather magic, fertility rituals, curing ceremonies, and rituals for attracting game offer the comforting illusion of control where scientific and technological mastery are lacking (Schwartz and Ewald 1968). Thus, by establishing an orderly relationship between humans and the supernatural, people use religion to make sense of the inexplicable and to reduce anxieties associated with the unknown.

Religious rituals that involve proscribed foods or food avoidances are often used in this way. For example, in most societies pregnant and lactating women avoid certain foods believed to be harmful. These food proscriptions give the woman a sense of security that she is protecting her unborn child from harm. Some ritual practices are nutritionally detrimental. Religious dietary restrictions placed on West Bengali children if the rice-feeding ceremony is delayed aggravate an already limited diet. Many of the foods considered ritually dangerous are high in protein and other nutrients needed by the growing child. However, these negative nutritional effects may, in part, be balanced by emotional benefits (e.g., the sense of spiritual protection) gained by family adherence to this practice.

Fasting is a common component of ritual. Some Christians fast during Lent, Judaism has several fast days, and Moslems fast from dawn to dusk as part of the observation of the holy month of Ramadan. The native peoples of the American plains used fasting as part of the ritual of the vision quest and the sun dance. Several contemporary organizations that assist people in making vision quests also emphasize fasting as part of the experience. For example:

"At [name withheld], we assist individuals who seek to find their true nature in nature. In our vision quest programs the participant spends 4 days and nights alone, fasting on the earth. The first 4 days are spent in preparation with the staff, including instruction in the three-phase dynamic of initiation; the four shields model of Self and psyche; emergency and survival procedures; archetypes and allegories of the heroic journey; the dynamics of fasting; and instruction in self-initiated ceremony. Preparation includes individual counseling and instruction in passage rite forms and ceremony relevant to the specific status of the individual." (http://www.questforvision.com/)

Religious ritual practices are also used to relieve anxiety about adequate food supplies. Communal hunting and agricultural rituals are practiced to ensure fertility of animals or soil, bring rain, or make game more willing to be captured. The Christian farmer, for example, may pray for good weather and a bountiful harvest. Although all people recognize the need for hard, efficient work to obtain food, the level of technological know-how often leaves sufficient room for uncertainty. The hunter knows that the arrow kills and plants need rain, but he is also aware that the arrow may go astray and that the rain may come too late. In the face of such uncertainty, rituals are seen as a way to enlist cooperation of supernatural forces and ensure a successful outcome. These rituals serve an important function. They reduce the food producer's anxiety about his ability to secure adequate food. And reducing this anxiety may actually improve efficiency: The hand is steadied, movements become more skilled, behavior becomes more flexible, and members cooperate more willingly in team endeavors. "Thus the improvement of confidence in the likelihood of success may very well be an important ingredient in achieving that very success. Ritual, by reducing anxiety and increasing confidence in hunting, agriculture, and other important activities, may very well be of material aid in direct proportion to the magnitude of the real risks involved" (Wallace 1966:175).

Religious ritual also enhances a feeling of unity and social solidarity. Religious ceremonies bring people together to participate in a common activity in an atmosphere heavily charged with emotion. Through this communal activity, people renew and reinforce their identification with the social unit and gain a heightened sense of social cohesion. Totemism, a feature of Australian aborigines, and some other societies' religions, exemplifies this function well. A **totem** is a certain plant or animal species considered sacred by a particular tribe or social group within a tribe. Members of the totem group identify in a special way with the totem, often through the belief that they are descended from an original totemic ancestor. Great respect is shown for the totem. It is neither killed nor eaten, except in ceremonies where it has a ritual role in feasting. Social scientists view totems as symbols by which groups express their unity and identity. Among Australian aborigines, each sex, clan, and local group has its own totem. Totemic feasts bring people together to worship a symbol (the totem) of the group and, thereby, enhance their solidarity.

Religious ceremony and feasting is a familiar theme in Christian tradition as well. The Last Supper and its derivative, Holy Communion, are sacred meals in which participants share the symbolic flesh and blood of the Divinity. By partaking of Communion, religious followers reaffirm their faith and reinforce their sense of oneness with each other. Church picnics, potluck suppers, and sharing prayer before meals are other examples of activity practiced in the Christian religion that serve to solidify the group.

A study of the relationship between religious participants and diet patterns of the aged found that elderly people who participate frequently in religious activities have higher essential nutrient intake and eat meals more regularly than those who do not (McIntosh and Shiffleet 1984). The nature of the resources exchanged and the ways in which these affect eating patterns, dieting, cooking, and shopping were not studied by these researchers. However, data collected in rural Kentucky (Bryant et al. 1981) suggest that the elderly benefit significantly from church dinners and picnics, assistance with shopping, invitations to share meals, and gifts from fellow church members.

totem

An animal or a plant that represents the founding ancestor of a group or clan.

Figure 8.2

Typical Seder meal. Seder is a ceremonial dinner held on the first night of Passover by Jews. Special foods are symbolic of the Israelites bondage by Egypt and of the Exodus.

© Michael Newman/PhotoEdit

Religion also reinforces other cultural values and standards of behavior. Common goals and rules of conduct are embedded in religious codes such as the Ten Commandments. Rituals and symbols, many of which involve food practices, are used to evoke a sense of commitment to the moral order. Almost everywhere, religious codes restrict some foods, while several (e.g., Hinduism, Judaism, and Islam) include elaborate dietary laws (see Focus 8.1). The Seventh-Day Adventist religion restricts meat, tobacco, and alcohol and prescribes simple diets containing a balance of whole grains and other fresh foods.

FOCUS 8.1 Jewish Dietary Laws

Perhaps the most elaborate set of dietary laws associated with religious beliefs are Jewish laws governing food and its preparation. Based on the Talmud of the Old Testament, as interpreted by scholars, these laws are still practiced by many Orthodox Jews as a central part of their religion, but not all Jewish families practice these traditional dietary customs.

Jewish dietary laws can be divided into three elements. First, all animal and vegetable components of food must be derived from approved species. Approved species are defined by the laws of *Kashrut,* which deal primarily with animal foods. Approved animal species include four-footed animals that both ruminate (chew cud) and have cloven hoofs, birds that do not eat carrion, and fish with scales. All other animals, as well as their milk and eggs, are ruled unfit. These include the swine (which has a cloven hoof but does not chew cud), the rabbit and camel (which chew cud but do not have cloven hoofs), as well as carnivores, rodents, shellfish, birds of prey, eaters of carrion, and reptiles.

(continued)

Focus 8.1   (continued from previous page)

Most plant species are approved. Exceptions include untithed and crossbred (hybrid) grains and the fruit of trees that have born fruit for less than 3 years.

A second dictate requires that approved species must be prepared correctly. Warm-blooded animals, for example, must be ritually slaughtered. The correct method of slaughtering animals is based on a principle of "pity for all living things" requiring that they must be killed with a minimum of pain. Hunting, which may inflict death in a cruel way, is forbidden. The *shoket*'s blade, used in ritual slaughter, must be so keen that there is no brutal tearing of the flesh or skin. A single, swift, almost painless stroke is used to sever both the trachea and jugular veins.

Also, the blood of all slaughtered animals must be removed by ritual soaking. This involves an alternating sequence of soaking, rinsing, and draining that frees the meat of all blood, for consuming blood is strictly forbidden. "If any man . . . eat blood I will set my face against his soul and will cut him off from his people" (Leviticus 17:11).

Finally, milk and all dairy products may not be mixed with meat products. All approved, or "kosher," foods are divided into three categories: *pareve, milkhk,* and *fleyshik. Pareve* foods include all kosher foods of plant, chemical, and animal origin, honey, eggs, kosher insects, and kosher fish. These cold-blooded animals are not subject to the ban on blood, and rules for slaughter require only that the animal be spared unnecessary pain. *Pareve* foods cannot pollute other foods and are ritually neutral. If a *milkhk* or *fleyshik* ingredient is added to a pareve food, it ceases to be *pareve. Milkhk* (dairy) foods include those products from the milk of kosher animals, milk, butter, cream, cheese, whey, as well as their derivatives. Any food that contains a trace of these items is defined as *milkhk. Fleyshik* (meat) foods are those produced from the meat of kosher mammals and birds, which must be slaughtered by a ritual expert and drained of blood. In the home, the housewife then ritually salts and bathes all exposed surfaces of the meat to remove remaining traces of blood. Only then does it cease to be *treyf* (impure) and is fit to be cooked and eaten (Regelson 1976:124).

The taboo placed on mixing kosher *milkhk* food with kosher *fleyshik* food is based on the biblical command "Thou shalt not seethe the kid in the milk of its mother." This has been translated by Jewish scholars to mean the total separation of the two food categories. If they are brought together at any time, they both become impure. Every effort is made, therefore, to keep them separate through the use of separate storage vessels and separate eating utensils, dishes, and cutlery. "These [foods and vessels] must be stored separately, and when washed separate bowls (or preferably sinks) and separate dishcloths (preferably of different colors to avoid confusion), must be used. If the meat and milk foods are cooked at the same time on a cooking range or even an open fire in a closed oven, care should be taken that dishes do not splash each other and that the pans are covered" (Rabinowicz 1971:40). The kosher household, then, has two sets of dishes and cookware, one for meat and one for dairy food, used on regular occasions. They also have two additional sets of dishes reserved for Passover.

Meal planning is made less difficult by the inclusion of *pareve,* or neutral, foods that can be consumed with meat or milk. In many traditional communities, these foods (fish, flour, eggs, vegetables, fruits, salt, sugar, condiments, and beverages) made up the bulk of the diet. Meat, in fact, is an uncommon item reserved largely for the Sabbath, holidays, or when someone is sick.

It is not just Judaism that prohibits or restricts the consumption of some foods. Islam prohibits consuming pork and alcohol. The emphasis on the sacredness of animal life and the value of vegetarianism is a feature of Buddhism. Fish is avoided in a number of African groups. In general, prohibitions on consumption of certain foods for either religious or other reasons (e.g., ethnic identity) are almost always directed against flesh foods (Simoons 1994).

Christian groups such as the Seventh-Day Adventists follow a rule of vegetarianism. The Church of Jesus Christ of Latter-day Saints (Mormons) prohibits using foods seen to contain stimulants, such as coffee, and alcohol. These practices have proven adaptive in modern North America. The Mormons' view of the human body as a temple, not to be defiled by poor diets and unnecessary drugs, has produced decreased incidences of cancer, heart disease, diabetes, and alcoholism among their followers.

Finally, many religious practices contribute to a society's adaptation to the physical environment. The adaptive functions may not be obvious to members of the society. People follow these practices for *religious,* not *ecological,* reasons. They are, nevertheless, important to the professional who introduces changes that might inadvertently disrupt the ecological balance established by traditional religious customs. Food taboos offer a good example. Before the arrival of Christian missionaries, Bantu-speaking peoples throughout southern Africa drank a fermented sorghum beer. In the late nineteenth century, the Bantu chief was converted to Christianity and, as a result, forbade his people to prepare or drink beer. Unwittingly, the chief placed them at nutritional risk. The sorghum beer had been one of their few sources of B-complex and C vitamins, effective in preventing beriberi, pellagra, and scurvy. It also served as a sanitary source of drinking water because ingredients were boiled several times to make the beer. Thus, acceptance of a religious innovation—the prohibition of beer—had several unintended, negative consequences for the group's adaptation to their environment.

Food taboos among peoples living in tropical forests of South America have also been shown to have important adaptive value. The religious restrictions placed on hunting and eating certain game animals protect otherwise endangered species and discourage hunters from expending energy on the animals that are most difficult (least "cost-effective") to hunt.

The classic example of supposedly irrational dietary behavior, which actually has adaptive value, comes from India, where Hindu doctrine forbids the slaughter of cows and the consumption of beef (see Figure 8.3). This taboo has long been cited as responsible for the creation of large numbers of aged, useless cattle that roam aimlessly across the countryside and throughout the city streets, defecating, blocking traffic, and stealing food. Many Western economic planners have recommended animal husbandry techniques that would improve the size and quality of India's cattle herds. Some analysts have also claimed that the Indians are ignoring a valuable protein and energy food source and suggest that they replace this religious practice with a more nutritionally sound policy allowing beef consumption. Because the Indian population has inadequate protein and calorie intakes and suffers from frequent famines, it appears as if these recommendations are well founded and that the Hindu doctrine is foolishly endangering the society's chances for survival.

Quite the contrary, says Marvin Harris in his study of cattle within the context of the Indian ecosystem. Harris (1974, 1975, 2000) found that the way Indians treat their cattle "increases rather than decreases the capacity of the present Indian system of food production to support human life" (1975:568). In fact, the Hindu doctrine making cattle sacred has several important adaptive functions. Most Indian farmers, for example, rely on cattle to pull plows and carts. By serving as cultivating machines, cattle play an essential role in traditional Indian agriculture.

Even though the cattle are scrawny because they live off the countryside, they can pull plows and carts well enough and do not

Figure 8.3
Hindus consider cattle sacred and believe that they should not be killed. But the meaning and importance of cattle and their products in India is very different than it is in the United States.

Henry H. Bagish/Anthro-Photo File

require as much food as larger breeds. This is important to the low-technology, resource-poor farmer who has limited land and food and cannot afford to divert arable land for grazing space or take grain from his family to feed his animals.

Cows, even the most scraggly and barren beasts, provide manure. Some of the dung is collected and used as cooking and heating fuel (a highly valuable product in a largely deforested country) and as fertilizer for crops. The rest remains in the countryside and is swept by rains into fields where it, too, fertilizes the poor soil.

Indian cattle serve another adaptive function. By scavenging throughout the streets and countryside, they convert many items of little or no direct human value (such as grass) into products of immediate human utility: milk, dung, and draft labor. Moreover, they provide a considerable amount of beef to those castes who eat cattle that have died from natural causes. Finally, the taboo placed on cows protects this valuable source of agricultural power, fuel, and fertilizer from being consumed during famines. If Indians were to eat their cows, they would have no way to produce crops when the next planting season began. Contrary to being useless, then, the sacred cow plays an important economic role in Indian society. In the words of Mohandis K. Gandhi (1954): "Why the cow was selected for apotheosis [elevation to divine status] is obvious to me. The cow was in India the best companion. She was the giver of plenty. Not only did she give milk but she also made agriculture possible" (p. 3).

In sum, it is important to be sensitive to the wide-reaching functions that some religious-based food practices can have for a society. Attempts to change a client's diet or modify an entire community's diet patterns should include a careful assessment of the context in which religious food practices are found.

Ethnicity and Ethnic Identity

Foods are also used as markers of ethnic identity. In the United States we commonly talk about "ethnic foods" and can name many of them: Mexican tacos, Japanese sushi, pizza (Italian? Perhaps.). In terms of food, America is definitely a melting pot. Food is so tied up with ethnic identity that, like the term "Eskimo" for the Inuit, and the name "Sweet Potato Eaters" applied to the Kayapo, foods are often used as shorthand to identify groups, usually pejoratively: the French become "frogs"; Germans, "krauts"; Italians, "spaghetti benders"; Mexicans, "beaners." The strength of these terms suggests that foods and differences in food and diet are powerful makers of ethnic identity.

In his historical analysis of food in the construction of Mexican identity after the conquest of the Americas by Europeans, Jeffrey Pilcher (1998) chronicles the ways in which both the conquering Spanish and the indigenous Mexicans used foods and diet to establish separate, then blended identities, and as symbolic tools in establishing social classes. First the Spanish and then, after Mexico's independence from Spain, the Mexican elites argued that the inferiority of maize as a food, compared to wheat, accounted for the poverty and "backwardness" of the indigenous peoples of Mexico. There was an active program to eliminate maize production and consumption among indigenous peoples that lasted into the twentieth century. For indigenous peoples, maize symbolized their identity as the first Mexicans, but more importantly it was a food crop that came from their own fields and that they controlled, unlike wheat, which was commercially grown and had to be purchased. The development of cookbooks celebrating a "Mexican" cuisine that was a blend of indigenous and European foods and cooking techniques was one of the key steps toward establishing the Mexican identity as a *mestizo*—a "mixed," or blended, identity that includes both Native American and European elements. In the end, the *tamal* (a dish that blends native and European foods) as a symbol of the Mexican national identity prevailed.

Ethnicity is one of the many ways that people categorize themselves and others. An **ethnic group** is a group of people who share a common identity as a result of common ancestry, region or nation of origin, language, and customs. Shared identities are developed and maintained through face-to-face interaction and the use of local, national, and, increasingly, global media such as native language newspapers, radio, and television broadcasts. While some ethnic groups are localized, for example, Italian Americans, others are transnational, such as world Jewry. Some ethnic groups exist only in an international sense. Armenians no longer have a discrete homeland, but have an international community of people who identify themselves as Armenian. Ethnic identity is maintained, in part, by common dress, language, religious beliefs, and of course, food.

It is important to keep in mind that we define ethnic groups only in comparison with the dominant culture of a nation or region. Being "American" in the United States is not an ethnicity. It is only when we wish to contrast minority groups with the dominant culture that we think in terms of ethnic groups. For example, we think of Italian Americans, Mexican Americans, Armenians, Native Americans, and so on, distinguishing these groups of people as different from "American." We need to be careful in talking about ethnic groups, because sometimes "ethnic" differences are really only class differences. Ethnic groups generally fit into the social hierarchy of a nation or region, and ethnicity may be used as a marker for higher or lower status and class. Many of the characteristics of the diets of poor Puerto Ricans or African Americans are the result of being poor in America, not necessarily the result of being African American or Puerto Rican.

Ethnicity is important to food and diet in two different ways. In some ways, ethnic backgrounds give people an alternative set of ways of thinking about food and diet. There is "cultural content" that influences food preferences and health and nutrition beliefs. But apart from the cultural influence, food is also used as a symbol of identity. It is used as a "boundary marker" between ethnic groups. "*We* eat beef, *they* eat pork." Indeed, some theories about the development of food prohibitions, such as the proscription against pork for Jews and beef for Hindus, suggest that these began as ways to

ethnic group

A group that shares a social identity based on common region or nation of origin, shared language, shared religious beliefs, or racial characteristics. Some ethnic groups are localized in a single country or region, such as Italian Americans.

distinguish a particular cultural group (Hebrews and Indo-Europeans) from their neighbors and were later codified into religious belief. Some analysts (Kalcik 1984) even talk about ethnic foods as elements of the "performance" of ethnic identity, implying a conscious manipulation of food and other aspects of ethnicity much as an actor in a play "performs" a role.

Both of these views of ethnicity are helpful in understanding how ethnicity may influence foods and diet. People in different cultural groups do have very different notions about what constitutes "proper" foods and meals. These ideas are important parts of cultural or local knowledge, and influence what people prefer to eat and what they actually eat. However, it is also clear that in the face of new and alternative information, people do change both their ideas about food and their diets. At the same time, food is used as an element of the celebration (or performance) of ethnicity even many generations after cultural change has taken place. Some preferences for ethnic foods persist. Ethnic foods used as heritage foods provide a feeling of belonging and comfort, even when consumed only occasionally. Actual day-to-day diets, however, become more and more diverse. For many immigrating groups, the diversity of food use within the group is continually increasing. While all the members of the Italian-American community in Philadelphia studied by Goode and her colleagues identified with Italian foods, the daily diets of individual members were very different, one from another. Knowing a person's ethnic identity tells us something about what the possible sources of diet and food preferences are, but it does not tell us much about what that person's actual daily diet is likely to be.

Health Beliefs and Local Knowledge

Three people with increased body temperature and nasal congestion may interpret the causes and treat the symptoms quite differently. A person who believes in a biomedical health model might describe it as a viral infection, caused by contact with an infectious person, and treat it by staying in bed, drinking plenty of fluids, and taking an analgesic. A proponent of complementary medicine might chalk it up to stress and use herbal remedies and visualization as the road to recovery. A person who explains it in spiritual terms might search for a violation of supernatural norms and perform a ritual dance to relieve the symptoms.

Health belief systems include notions about how the body works and explanations about what makes it healthy and ill. We can understand health belief systems by looking at the way in which symptoms and illnesses are perceived and the preventive and curative practices used, as well as the characteristics of the health care provider. The term **ethnomedical system** refers to the medical knowledge and practices associated with distinct cultural groups.

It is important to note that, in part as a result of ethnocentrism, those who subscribe to biomedical, scientific explanations and treatments (modern medicine) tend to think of their system of beliefs about health and illness as "knowledge" while considering alterna-

ethnomedical system
A system of medical knowledge, beliefs, and practices that is associated with a particular cultural or ethnic group. Biomedicine is the ethnomedical system of Western cultures.

tive systems to be merely "beliefs." A quick review of the ways in which "scientifically" based nutrition advice has changed over the past century would lead us to conclude that much "scientific" advice is influenced by Western worldviews and fundamental values, rather than completely objective "knowledge." And, conversely, much of the ethnomedical knowledge of other cultural groups is based on generations of observation and even experimentation.

Perceptions of Symptoms and Disease

Cultures vary in what symptoms they see as indicative of poor health. A set of symptoms considered serious in one society may be ignored in another. Primary and secondary yaws (an infectious tropical skin disease caused by a spirochete, a type of bacterium) was so common in parts of Africa, until the last generation people did not regard it as a disease. Similarly, in the 1930s, when goiter was widespread throughout parts of the north central United States where the soil is iodine-deficient, the development of goiter during puberty was considered by some to be a normal part of maturation.

The wide range of responses people may have to a set of symptoms is illustrated in the case of spirochetosis. Known as *pinta* (meaning "spotted"), this skin disease is caused by the bacterium *Treponema careteum*. It begins as scaly plaques in childhood. These become blue, later turning pink, brown, blue, and black. Eventually the affected areas become completely discolored, leaving the person with a multicolored skin. Symptoms seem limited to the skin, although they also provide the person with resistance to syphilis and yaws. Different groups of people have reacted to these symptoms in different ways that vary from scorn to admiration. In some places, people with the signs of *pinta* have been shunned, whereas in others they have enjoyed high status. In the early sixteenth century, Montezuma, the Aztec emperor, selected *pintados* ("the spotted ones") to transport his carriage because of their attractive skin coloration. Among some South American Indian tribes, the disease is so common that people who have it are considered healthy, while those without it are regarded as ill (Dubos 1980; Wood 1979).

Another example spotlights malaria. Now regarded as serious, this disease was not considered pathological during the nineteenth century when it affected most of the settlers moving into the upper Mississippi Valley. "As the frontier pushed westward, the cutting of forests, travel by riverboat, settlement in river bottoms, poor sanitation, and stagnant water increased the incidence of the *Anopheles* mosquito and people's exposure to it. In the beginning the 'chills' were regarded as a necessary element of the inevitable 'acclimatization' and after having 'shaken' for years people got so used to it that they hardly paid attention to a little 'ague'" (Foster and Anderson 1979:40–41).

Preventive and Curative Practices

Health belief systems everywhere offer people a variety of preventive and treatment measures to maintain health. The practices used in a given culture correspond closely with how illness is perceived. When illness is believed to result from magic, people avoid offending their neighbors, who might direct evil forces against them. When people believe an illness such as kwashiorkor is caused by the breach of postpartum sexual taboos, they observe strict sexual abstinence. And when disease is believed to be caused by bacteria, sanitation is practiced carefully. Both of these practices help prevent kwashiorkor.

Dietary practices are used by virtually all peoples to treat illness—and by most to prevent it. In North America, nutrition has become exceedingly popular as a preventive measure, linked in the minds of some people to almost all diseases. As noted in Chapter 6, recent surveys of food consumption show that a significant proportion of North Americans have decreased their intake of red meat, eggs, and full-fat milk products in response to warnings that excessive consumption of these foods will lead to heart disease. Other changes that reflect a concern with preventive nutrition are increased sales of vitamin and mineral supplements, nutrition books such as *The Anti-Cancer Diet* and *Sugar Blues,* and shoppers' heightened concern with product ingredients.

Perhaps one of the most widespread systems of belief about preventing and curing illness includes the notion that having bodily fluids, called *humors,* in proper balance is key to good health. Although the notion of humors and balance may have originated in South Asia at least 5,000 years ago, the system was developed and spread to Europe and northern Africa by the Greeks of Hippocrates' time, about 2,500 years ago. In the classical Greek theory, health was believed to result from a balance among the four humors: blood, phlegm, black bile, and yellow bile. Each humor was thought to be either hot or cold and either wet or dry. Blood, for example, was thought to be hot and wet, phlegm to be cold and dry. Illness resulted from a humoral imbalance that caused the body to become too hot or cold, or too dry or wet. From Greece this system spread to the rest of Europe and northern Africa, where Arab physicians further refined it and took it to the Iberian Peninsula at the time of the Moorish occupation of Spain (eighth to fifteenth centuries). The system of **humoral medicine** also traveled eastward to Asia, where it became integrated in the existing ideas of yin and yang (the Chinese belief system of complementary opposites) and was incorporated into classical Chinese medical thought (Anderson 1996).

Humoral medicine was the most advanced form of medical thought in Europe from the time of Hippocrates until the European Renaissance in the sixteenth century. At the time of the European colonization of the Americas, humoral beliefs were orthodox medical beliefs and represented elite medicine. The Spanish and Portuguese brought humoral medicine with them to the Americas. For reasons that are not understood, in the transmission of the system to both the Americas and China, the contrast between wet and dry was lost, and the contrast between hot and cold became the more important (Anderson 1996; Foster 1994). In the Americas, even the idea of humors became less important than the contrast between

humoral medicine

An ancient and widespread system of medical knowledge that holds that the body is composed of bodily fluids called humors. In the healthy body the humors are in balance. Foods, climate, and situations can make the humors fall out of balance. This causes disease. Disease is treated by the use of foods, medicines, other substances, and manipulations that are thought to help get the body back in balance.

 **8.2**

Foods Classified as Hot or Cold in Tzintzuntzan

While there is variation within groups of people on how to classify specific foods, people of Tzintzuntzan, Mexico, studied by George Foster (1994), do substantially agree on the classification of a number of foods. Here are the hot/cold classifications of some common Mexican foods.

HOT	COLD
Chocolate	Mallow (wild vegetable)
Brown sugar	Fava beans
Garlic	Vinegar
Onion	*Verdolaga* (purslane; used as a vegetable)
Rue (a common medicinal herb)	Cucumber
Cedrón tea	Cauliflower
Black pepper	Chayote squash
Clove	Tomato
Tequila	*Jícama* (a root)
Ginger	*Chilacayote* squash (hubbard squash)
Honey	Turnip
Coffee	Cheese, fresh
Peanuts	Beer
Jalapeño chile	Watermelon
Wheat	Cow's milk

hot and cold. In the hot/cold system of belief, illnesses, foods, medicines, and situations are thought to be hot, or cold, or neutral. Hot and cold do not refer to temperature but, rather, to inherent characteristics of these things. For example, in parts of Latin America, unboiled water is cold and boiled water is hot, no matter at what temperature the water is drunk.

Humoral beliefs, in the guise of a hot/cold contrast, continue to exist and to influence diet and other health practices in a large part of the world. Because of contact with Europe and Asia, people as far from there as Latin America, Southeast Asia and Malaysia, and the Philippines think about illness and other aspects of life as hot or cold. Southeast Asian and Latin American migrants to North America have brought the hot/cold system with them. As we noted above, hotness and coldness are not determined by observable characteristics or by the physical temperature of the substance, although, in some places, cooking occasionally may make a cold food hotter. Usually, the quality of a food is determined by its effect on the body, its medical use, or its association with natural elements. For example, "Chili is hot because it produces a burning sensation. . . . Ice is hot because it produces a burning sensation when applied to the skin. Certain medicines, like aspirin, are hot because they make a person sweat. Cabbage is cold because it produces gas or air (aire) in the stomach" (Cosminsky 1975:184).

These notions of hot and cold affect people's views of the causes of illness and are reflected in menu planning and the use of herbal medicines. A Mexican mother, for example, may put a bit of cinnamon (a "hot" spice) into her baby's formula (a cold food) to achieve a better balance. And when her child becomes ill with a cold, she may restrict foods classified as cold and substitute medicines and foods believed to be hot. A person overheated by physical labor in a tropical climate might moderate the coldness of the water she drinks by adding a pinch of salt to the water so that the shock of "cold" water on a hot body will not cause illness.

Foster (1994) argues that the impact of adherence to the hot/cold system on the diets of Mexicans living in the town of Tzintzuntzan is slight (see Box 8.2). Although people do try to maintain balance in their diets, and do respond to illness with diets that are opposite in hot/cold from the illness, for the most part this appears to have little nutritional impact. Anthropologist Carol Laderman (1987) describes the impact of hot/cold beliefs on the diets of Malayan women who have recently given birth. Newly delivered women are thought to be especially vulnerable to illnesses brought on by imbalances between hot and cold. Women avoid "cold" foods and attempt to balance the qualities of foods. Laderman argues, however, that while there are some dietary adjustments, the overall nutrition of women does not suffer. There are enough high-quality foods available to ensure the ability to choose a "balanced" (hot/cold) diet that is nutritious. However, several studies of child nutrition in Malaysia and India have suggested that withholding cold foods, which include some fruits and vegetables, from children may increase their risk of vitamin A deficiency (Foster 1994).

Although the hot/cold theory has been best described in Latin America, where it constitutes the major popular (that is, nonscientific) form of explanation for illness and treatment, traces of the hot/cold humoral model have been found in popular Western health practices as well. "The rationale, or justification, for treatment that is clearly seen in Latin America has in most cases disappeared, but the practices themselves can be accounted for only a 'residual humoral pathology'" (Foster 1979:18).

Examples of the hot/cold model from industrialized societies include the belief that the body is particularly vulnerable to cold through the head, feet, and open pores (e.g., after bathing). Colds and chills may result from failure to maintain an even balance of warmth in the body. Going outside with wet hair; walking barefoot on a cold floor during winter; exposure to cool, night air; going into a cold room right after a bath (when the pores are still open); and exposure to drafts are believed to cause illness. Traditionally, women were advised not to wash their hair, go swimming, or get caught in the rain when menstruating because doing so could cause illness. Treatment practices based on the hot/cold theory include "feed a cold, starve a fever" (reflecting the need to balance the body's temperature in the face of excessive heat from fever or chill from a cold) and special care to avoid cool temperatures, chills, or drafts when sick with a fever (Foster 1979).

Interestingly, many traditional folk remedies involving food and herbs have been shown to be effective from a scientific point of view. Work done by Bernard Ortiz de Montellano (1975) shows that many Aztec herbal remedies have physiological effects—usually the same effects claimed by native healers. For example, wormwood (*Artemisia mexicana*), used as a tonic and de-wormer, contains an **anthelmintic** agent (santonin), as well as a camphor (a mild stimulant and colic reliever). Wormwood is still used today to combat upset stomachs and intestinal worms in rural Mexico. The bark of willow trees (*Salix lasidpelis*) was used to stop rectal bleeding and treat fevers. It contains a salicylic acid, an aspirin-like substance effective in lowering fevers and relieving pain.

Many agents found in herbal remedies are used in drugs today: digitalis (foxglove) for heart conditions, morphine (opium poppy) for pain, rauwolfia for hypertension, chaulmoogra oil for leprosy, and quinine for malaria. Some foods used as medical remedies also contain pharmacologically active ingredients that correspond with their native use. *Epazote*, a spice used to fight intestinal worms, contains *Chenopodium graveolens*, an effective anthelmintic. In Mexico it is classified as "hot" and used with beans ("cold") to provide a more balanced meal and control intestinal worms. Onions, garlic, apples, and radishes contain antibacterial properties. Grain stored in clay pots and mud bins, which allow for mold formation, often develops the antibiotic tetracycline.

Etkin and Ross (1997) examined the antimalarial properties of plants used by the Hausa of Nigeria to treat malaria. The vast majority, 31 of the 35 plants studied, are used as food as well as medicine, and 23 of them are most abundant during the rainy season, when malaria is most common. Some of these plants were shown in Etkin and Ross' laboratory analyses to have therapeutic value. Moreover, the plants' effectiveness may be enhanced when used as food as well as medicine. Thus, the Aztec, Hausa, and many other medical belief systems contain preventive and treatment practices that now are recognized as effective in biomedical terms.

Some of the folk advice common in the United States has been found to be true also. For example, the adage "If you eat just before going to bed, you'll gain weight" has been borne out by research showing that fewer calories are stored as fat if you exercise immediately after eating. The practice of eating homemade chicken soup when suffering from a cold or flu is useful because its steam can clear stuffed nasal passages and the broth contains fluid that is important in times of physical stress.

anthelmintic
A substance that kills parasitic intestinal worms.

Health Care Providers: Shamans, Curers, and Others

In aboriginal times an Eskimo [Inuit] man who continued to feel ill consulted a shaman . . . who "walked about" the patient, examining him from all angles. Then he might touch the patient about the spot where the "pain" lay. He licked his hands, then rubbed them over the painful area. Some shamans blew on the affected part, occasionally sucking it tentatively at first if the case was diagnosed as one of (object) intrusion. When these preparations were completed, the shaman began to sing. (Spencer 1959:306, as cited in Foster and Anderson 1978:102)

In Aritama, a Spanish-speaking mestizo village in Colombia, a curer is called in when illness persists. After a few searching questions concerning the patient's enemies who might have caused the disease, the practitioner asks in detail about the food consumed during recent days or weeks, about any hallucinations, heavy physical efforts, or exposure to sun, rain, wind, or water. The pulse is taken. If it beats rapidly, a "hot" disease is diagnosed, if it beats slowly, a "cold" disease is suspected. Facial expression is studied carefully. . . . Some specialists examine the urine. . . . The pupils of the eyes are examined. . . . Fecal matter, sputum, and vomit are occasionally examined. Only then does the curer prescribe. (Reichel-Dolmatoff 1961:289, as cited in Foster and Anderson 1978:102)

In the United States a . . . man asks for an appointment with his physician after several days of nagging abdominal pain that he suspects is something more than indigestion. The doctor "fits him into" a busy schedule, examines him, presses his stomach, localizes the pain and orders a blood count. If the results are positive, an appendectomy will follow. (Foster and Anderson 1978:102)

Although the behavior of the Inuit shaman, the Colombian curer, and the American physician may appear very different, there are actually many similarities between them. Most ethnomedical systems include roles for people identified as healers. Throughout history, healers have occupied a special role achieved through a selection process (inheritance, visions, dreams, instructions) and training (apprenticeship or medical school). Certification by previous practitioners, concern for professional image, receipt of payment for services rendered, and public belief in their special powers or skills are universal features of medical practitioners (Foster and Anderson 1978).

There is little doubt about the comparative effectiveness of Western medical practitioners and traditional healers in treating obvious pathology. Antibiotics, immunizations, and modern surgery offer today's physician a degree of mastery that far exceeds the sand painting of the Navajo, object extraction by sucking, divination, and spiritual trance. However, as Wood (1979) points out, traditional healers offer many treatment forms with obvious value to the patient: massage, sweat baths, rest, social concern, and reassurance:

There remains much that modern medicine can absorb from traditional healers. The holistic approach to the troubled victim of some malfunction in his or her life, the involvement of family and friends, the sense of social solidarity and support, the expected and delivered explanations for the troubles at hand—all are vital adjuncts to any medical system. As members of the order Primates, we have a heritage millions of years old that renders us in constant need of the reassurance derived from physical as well as social "stroking." (Wood 1979:333)

FOCUS 8.2 Health Belief System of Mainland Puerto Ricans

To gain an understanding of the complexity of health belief systems, let's examine the health belief system of a specific cultural group. An ethnic group whose medical system has received considerable attention in recent years is the Puerto Rican population living on the United States mainland. Our description of their belief system is derived largely from Alan Harwood (1981).

In 1996 an estimated 3.1 million Puerto Ricans lived in the United States. Because Puerto Ricans are U.S. citizens and have been able to come to the mainland with no legal restrictions since 1917, a great deal of movement between the two places has occurred. Puerto Ricans are one of the largest ethnic groups on the mainland United States.

Puerto Rican New Yorkers born in the United States are doing better educationally and financially than those born on the island. Almost half (47%) of Puerto Rican U.S. natives have some college education or are college graduates, whereas only one in four (26%) Puerto Ricans from the island are college-educated. These levels of schooling affect the earning potential of Puerto Ricans; income is higher for U.S.-born Puerto Ricans than for those born on the island.

Today, considerable variation exists among families who live on the mainland. In fact, people of Puerto Rican descent are represented in all educational, occupational, and income categories, and they reside in rural and suburban as well as urban locations. These factors, and the high incidence of families headed by females, have important consequences for health and medical care. Let's take a closer look at some of the aspects of the ethnomedical systems associated with Puerto Ricans.

Perceptions of Disease

Before examining Puerto Rican medical beliefs, it is important to point out the extent to which Puerto Ricans vary in their views. Some Puerto Ricans, especially those who have higher education, subscribe to popularized and orthodox biomedical theories (e.g., the germ theory), whereas some with less education may include more of the aspects of Puerto Rican local and popular ethnomedical knowledge. The length of time spent on the mainland and the degree of exposure to the orthodox medical system also contribute to intra-ethnic variation in health belief systems. Nutritionists and other health care professionals are cautioned, therefore, to approach each Puerto Rican client individually in order to determine the extent to which folk and/or orthodox beliefs are being used.

Typically, traditional peoples combine newly adopted biomedical beliefs with their traditional system. Because most illnesses are believed to respond best to Western medicine, traditional treatment practices and healers are used less and less frequently. Often the traditional healers are consulted primarily for culturally specific illnesses such as *susto* ("fright illness"), for which biomedical practitioners offer little help, or after a disease has been designated as incurable by physicians.

The Hot/Cold Theory

Puerto Rican ethnomedical knowledge includes a version of this widespread disease theory that distinguishes between hot and cold diseases and between hot, cold, and cool foods and medications. According to the Puerto Rican version, cold diseases such as arthritis result from excessive exposure to cold substances or states, as when hands are switched suddenly from hot to cold water. Influenza and colds result from exposure to drafts or going outside without sufficient clothing. Hot diseases are caused by overindulgence in hot foods and medicine or by allowing the body to become overheated.

(continued)

Focus 8.2   (continued from previous page)

Because health is based on bodily balance, hot diseases are treated with cold or cool substances, while cold diseases are treated with medications and food classified as hot. Harwood (1971) found that some Puerto Rican patients in New York would resist physicians' recommendations to drink fruit juices when ill with a cold or flu because these illnesses are seen as cold, like fruits and their juices. Instead, for them, like many Americans, chicken soup seemed the ideal food to feed a cold.

Spiritist Theories

Most Puerto Ricans of various Christian persuasions, like other Latin American and European migrants, conceive of human beings as consisting of two aspects: a finite, physical body and an eternal, nonmaterial spirit. Many Puerto Ricans also believe that the disembodied spirits of both deceased and divine beings play an active role in influencing the circumstances and behavior of the living. Only some Puerto Ricans, however, seek out spiritist mediums to diagnose and influence spiritual interventions in their own and others' lives (Harwood 1981).

In line with this distinction between the material and spiritual aspects of the body, spiritist theories explain disease as resulting from material factors such as germs, a hot/cold imbalance, the activities of disembodied spirits of divine or deceased humans, the activities of living people, or a combination of these. Most conditions are seen as materially caused and referred to medical professionals for care. Because a troubled spirit can complicate a material condition, spiritist mediums are often used to supplement care regimens. Spiritists assist by diagnosing the spiritual problem, recommending ways to rid the body of noxious spiritual influences, improving relations with protective spirits, and counseling on interpersonal problems. Prayers, performance of special rites, use of herbal preparations, and participation in group sessions are common healing methods used by spiritists.

Viruses and Other Popularized Biomedical Concepts

As popularized in the mass media, the term "virus" is used to describe the symptoms of some colds or influenza. Virus colds are distinguished from chill-caused colds, although the criteria used to make this classification vary from person to person. Some Puerto Ricans also use the term to explain diarrhea or other intestinal disorders. In the Puerto Rican ethnomedical system, the notion of "virus" may not correspond to the biological entity understood in biomedicine. It is closer to a more general term for infectious disease.

Other popularized biomedical concepts widely used by Puerto Ricans include the germ theory, parasite infection, and allergy. Like the notion of "virus," these may not be understood in the same way the biomedical professional understands them, as when a behavioral restriction recommended by a physician is interpreted as the cause of the disease. When a doctor recommends that a diabetic sustain from sugar, sugar is seen as the cause of diabetes. Of particular relevance to nutritionists is the Puerto Rican use of the term "rickets" (*raquitis*) and their interpretation of high and low blood pressure (*alta, baja pression*). Some, though certainly not all, Puerto Ricans use rickets to refer to tuberculosis in children, probably because both illnesses are commonly associated with malnutrition. High blood pressure is interpreted by some Puerto Ricans as too much blood and low blood pressure as too little, while weak blood (*sangre debil*) is the term used for iron-deficiency anemia.

Health Care Providers

Puerto Ricans, like most Americans, rely on biomedical professionals for the illnesses they do not treat at home. In certain circumstances, however, depending on symptoms and cost, care is sought from alternative sources: pharmacists, herb shops (botanists), spiritists, curers (*curanderos*), or evangelical faith healers. These traditional healers may be consulted before going to orthodox medical professionals, such as when a woman asks her pharmacist for diagnostic and treatment assistance; they may be consulted secondarily when the medical facility fails to remedy the problem; or they may be used in conjunction with medical professionals. For example, spiritists often help chronically ill and terminal patients maintain spiritual peace while the physician provides biomedical treatment. Finally, some culturally specific conditions require a special type of folk healer. *Empacho,* an indigestion believed to be caused by a bolus of food

(continued)

Focus 8.2 (continued from previous page)

obstructing the intestines, is treated by a *curandero*. These healers administer massages, botanicals, and hot/cold remedies.

By and large, these nonmainstream healers are not used in direct competition with medical services. In some urban areas, medical and mental health clinics have established successful referral networks with these healers, enabling patients to utilize both services simultaneously for maximum effectiveness (Bryant 1975).

Alternative Health Belief Systems in a Plural Society

Perhaps more common in North America than ethnomedical belief systems are the alternative health belief systems associated with the alternative health movement. Although a variety of differing beliefs and practices have gained popularity, several themes are characteristic of the movement as a whole. These include a value placed on natural foods and beverages and other practices such as natural birth control, natural sleeping cycles, and wearing clothing made of natural fibers; a belief in the body's ability to heal itself through proper diet, exercise, relaxation, massage, and other procedures; and the use of proper diets as a major method of preventing as well as curing disease. Some of the more common foods that are to be avoided include meat, dairy products, and foods containing artificial substances.

With the rapid rise of the complementary medicine movement, most large towns and cities now feature a host of naturopathic and osteopathic doctors, chiropractors, energy workers, and nutrition advisors who offer help for a variety of ailments—arthritis, allergies, chronic back pain, fatigue, headache—as well as general medical care to promote longevity and enhance well-being.

Alternative health belief systems are not new in the United States. Their popularity, however, has surged over the last several decades, at least in part because of dissatisfaction with the Western biomedical system. One source of dissatisfaction comes from modern biomedicine's inability to keep pace with consumers' expectations for effectiveness. Keep in mind from our discussion of worldviews that North Americans approach the world as though it can be controlled and as though all problems can be addressed and conquered. Many North Americans feel fundamentally that at some point humans will be able to cure all disease. Biomedicine has been quite effective in improving treatments for illness and providing guidance for prevention. Paradoxically, biomedicine's own dramatic successes also may be partly responsible.

Because modern biomedicine has cured once-fatal diseases, reattached severed hands, transplanted organs, restored vision, and performed a host of other incredible feats, many people now expect cures for all of their health problems. As a result, many consumers are no longer willing to accept biomedicine's incomplete solutions to rheumatoid arthritis, chronic nephritis, cancer, and other diseases.

Figure 8.4

Acupuncture attempts to restore the balance of vital energy in the body through the use of special needles.

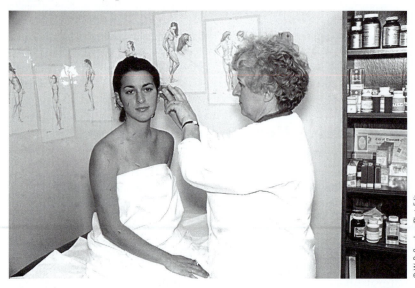

They expect cures—and when the biomedical system doesn't offer them, they look elsewhere for answers.

There are other reasons for dissatisfaction with orthodox medicine. Like all belief systems, the Western biomedical system is influenced by the cultural context in which it has emerged. The overprescription of drugs, frequency of unnecessary surgery, impersonality of the doctor-patient relationship, and restriction of hospital patients' visitors are just a few examples of practices the American health consumer has challenged. Although physicians are typically the target of consumers' criticism, dietitians are also being challenged by a host of alternative nutrition advisors practicing in health food sources and holistic health clinics. When consumers are told by a nutritionist that diet cannot cure arthritis, multiple sclerosis, or other maladies, many turn to alternative nutrition advisors who are willing to promise a dietary solution.

Because most people, like the mainland Puerto Ricans described in Focus 8.2, combine biomedical remedies with other ethnomedical practices, the nutritionist needs to be aware of clients' dietary beliefs and practices. With this awareness, the nutritionist working with people who hold holistic or ethnic folk beliefs can more effectively combine biomedical nutrition beliefs and practices with those of the alternative system.

How important are the components of ideology in influencing diet and acting as barriers to dietary change? It seems clear that many elements of food preferences are based on shared ideas about what constitutes acceptable foods and what makes a meal. We recognize a number of differences in these ideas around the world and among ethnic groups within the United States. Religious belief is perhaps the most powerful of these ideas, with a number of religious traditions either prescribing or proscribing particular foods or types of meals either permanently or periodically. Food is a symbolic representation of ethnic identity, and although research suggests that ethnic identity may not have a great deal of influence on day-to-day diets, it is clear that periodically ethnicity does influence the use of some foods. Finally, health systems tend to use diet in both the treatment and prevention of illness. Followers of particular health systems may alter their diets significantly to accommodate ideas about health and illness.

Summary

- The symbolic meanings of food, worldview, religion, ethnic identity, and health beliefs are among the ideological considerations that affect a person's health care practices and diet.

- Humans create a forest of symbols around themselves by imbuing the world with meanings. Foods and the consumption of foods, so fundamental to human life, become powerful symbols of self and ethnic identity, goodness and naturalness. The symbolic aspects of foods exert influence over diet and nutrition.

- The term "worldview" refers to the way a group defines how the world works. People from two different cultures, looking at the same phenomenon, may describe it very differently. Knowing the history of a person's society and his or her past experiences is helpful in understanding that person's feelings and actions. Ethnocentrism is the belief that one's own worldview and cultural assumptions are the best, the most right, the most natural. Its opposite, cultural relativism, is the position that there are a number of different ways to view the world, with one not necessarily being better or worse, more or less natural, good or bad.

- Religious beliefs can be very powerful aspects of ideology and can exert a great deal of influence on dietary practices in several ways. Religious beliefs sometimes prescribe or restrict certain foods. Through prayer and other ritual practices, people use religion to relieve anxiety about an inadequate food supply. At formal and informal religious meals such as Communion and church suppers, people reaffirm their faith and sense of solidarity with each other. Many religious practices have adaptive value to the group.

- Ethnicity and ethnic identity are important aspects of personal identity that help us understand food use. Beliefs and understandings about foods that make up some of the cultural content of ethnicity may influence diet. More importantly, foods are used to demonstrate solidarity and to reinforce the connections among a group of people and distinguish them from other groups. The use of food in the performance of identity may not have a great influence on the day-to-day diet on many people, but it can be important in seasonal or festival contexts.

- Health belief systems include notions about how the body works and explanations about what makes it healthy and ill. People vary in how they view symptoms and disease, what preventive and treatment measures they use to maintain health, and the kinds of cures they rely on. Food and diet are important components of most preventive and curative health systems. Beliefs about appropriate foods for prevention or treatment of illness may have an important influence on food choice and diet.

- A society's technology, social organization, and ideology influence diet patterns in a synergistic manner: The combined effects are greater than the effect of any single cultural trait.

HIGHLIGHT

Becoming Culturally Competent

Contributions by Arlene Calvo

This section of *The Cultural Feast* has introduced you to the concept of culture and demonstrated the numerous ways in which dietary practices are intertwined with a society's technology, social organization, and ideology. An understanding of cultural influences on dietary practices is increasingly important, as the populations that nutritionists and other health professionals serve grow more culturally diverse. In this Highlight, we discuss ways health professionals are using an appreciation of culture to meet the challenge of serving ethnically diverse populations.

Cultural Competence

Many health professionals and the organizations in which they work have recognized the need to become familiar with the cultural tenets of the people they serve (Holland and Courtney 1998). The ability to recognize and respect the norms, values, and beliefs of other cultural groups, and to respond appropriately to cultural differences in planning, implementing, and evaluating programs and interventions, is called *cultural,* or *multicultural, competence.*

Cultural competence encompasses three major principles: (1) The nutritionist or health provider is knowledgeable about the groups he or she serves; (2) the provider is self-reflective, recognizing the influence culture has on his own nutrition beliefs and practices; and (3) the provider uses this knowledge to skillfully manage the interplay of differing cultural views during health encounters (Cross Cultural Health Care Program 2002; Weaver 1998). Providers do not pretend to become "culturally competent" in the same sense as they may become competent in a language or field of study; rather, they use this label to express their commitment to learn about the culture of others and develop a culturally relative view of cultural differences.

Why Cultural Competence Is Important

Health care organizations that incorporate cultural competence into their missions and operating procedures deliver care more cost efficiently and have better medical outcomes than those that do not (Health Resources and Services Administration 2001). According to the Bureau of Primary Health Care (2000), cultural competence enhances an institution's or organization's health care outcomes in numerous ways:

- Patients feel understood and respected by health care providers and therefore are less likely to delay making or keeping appointments for care.

- Providers can obtain more accurate information and make better diagnoses when they understand and communicate well with their patients.
- Patients are more satisfied with the care they obtain from providers who try to understand and respect them.
- Because they are satisfied, patients are more likely to comply with their providers' recommendations or treatment plans.
- Family members are more likely to support patients' efforts to comply with treatment plans that are compatible with their beliefs and norms.

Implications for Nutritionists: A Case Study

In nutrition counseling, where many therapeutic interventions are on a personal level, sensitivity to the strong influence of culture on an individual's food intake, attitudes, and behaviors is especially imperative. If we are to meet our professional potential in the 21st century, we must strengthen our practice in nutrition counseling and communication. Multicultural competence is not a luxury or a specialty but a requirement for every registered dietitian. (Curry 2000)

To optimize their effectiveness, nutritionists need to not only be able to identify beliefs, values, or norms that influence dietary habits, but also learn to adapt their interventions to accommodate these differences. This approach requires additional time and flexibility. Additional time is required to ask clients about their beliefs and preferences and to discuss nutritionally sound alternatives that are culturally acceptable. It may also require talking with family members who play a more important role in the medical decisions their relatives must make than is customary in North America. More time is also necessary to develop diet plans that contain foods consumed by people from differing backgrounds and to develop other types of culturally appropriate educational materials (e.g., pamphlets, booklets, leaflets, and videotapes). For example, many health care organizations routinely ask people from the communities they serve to help develop or review educational materials to ensure they are easy to understand, culturally acceptable, and effective (Holland and Courtney 1998; Meade, Calvo, and Cuthbertson 2002). Time is also well spent on making clinical facilities more convenient and pleasant for people of differing backgrounds by adding signs, educational posters, and magazines in the appropriate languages.

In addition to time, flexibility is required to meet the linguistic and cultural needs of a multicultural population. Providers may have to adapt their treatment recommendations, and agencies may need to modify policies to accommodate people with differing needs or preferences. In many ways, cultural competence is a mindset, one consistent with cultural relativity, a caring attitude, and a genuine interest in cultural differences.

BOX H8.1

Cultural Food Guide Pyramids

To assist nutritionists in providing general dietary recommendations to people from varying cultural backgrounds, the USDA, the Harvard School of Public Health, Cornell University, and other academic institutions have adapted the USDA Food Guide Pyramid for use by people of differing cultural backgrounds. Food Guide Pyramids are now available for Asian, Arabic, Chinese, Japanese, Russian, and many other populations. For more information on these food pyramids visit the USDA Food and Nutrition Information Center Web site at *www.nal.usda.gov/fnic.*

Figure H8.1
Asian Diet Pyramid.

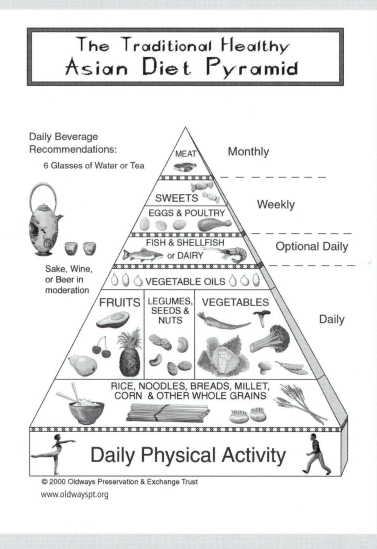

To illustrate the importance of a culturally competent approach to nutrition counseling, consider the case of Mrs. Osorio, an older Mexican American woman who is referred to a nutritionist. The physician recommends that she modify her diet to control her diabetes so that he does not have to prescribe medication. Compare the likely outcomes of ways a health organization can approach this case. In the first scenario, the woman arrives for her appointment 30 minutes late. All signage in the clinic is in English, which Mrs. Osorio is learning to speak but still cannot read, so she spends another 15 minutes trying to find the nutritionist's office. As she approaches the nutritionist's desk, she notices her looking at her watch. Knowing that North Americans are very punctual, Mrs. Osorio begins to feel nervous. She sits down and introduces herself. The nutritionist reads her chart and asks a few questions slowly and clearly. Mrs. Osorio answers several questions in English but does not understand the others well enough to respond. The nutritionist, feeling frustrated with her inability to communicate with Mrs. Osorio, pulls out a brochure, also in English, and slowly reviews the recommendations about counting carbohydrates and limiting saturated fat. When Mrs. Osorio hears bread and pasta mentioned as foods she should limit, she asks about eating tortillas. The nutritionist asks how they are cooked. When Mrs. Osorio explains that she makes them with lard, the nutritionist recommends she greatly limit her intake or eliminate them completely. Mrs. Osorio, wondering how she could possibly survive on a diet without tortillas, leaves with no confidence in her ability to follow these recommendations and decides to ask the physician to prescribe the medication.

In the second scenario, Mrs. Osorio visits a clinic that has adopted a number of measures to become more culturally competent. When she arrives at the clinic, all of the signs are in English and Spanish, enabling her to find the nutritionist easily. After introducing herself to the nutritionist, the nutritionist asks in broken Spanish if she would prefer to speak in English or have the assistance of an interpreter. Mrs. Osorio accepts the offer to have an interpreter, relieved that she will not have to struggle to learn new nutrition information in a new language. While waiting for the interpreter to arrive, Mrs. Osorio notices a breastfeeding promotion poster with a Latina grandmother and mother holding a newborn baby. The message—Da A Madre Apoyo de Amor (Give A Mother Your Loving Support)—makes her smile. She remembers how important her own mother's help was when she was raising her babies. She starts to feel a little more at home in the clinic. After a series of questions about her diet, dietary beliefs, and preferences, the nutritionist gives her a brochure in Spanish. With the help of the interpreter, they review each recommendation, discussing problems Mrs. Osorio might encounter and exploring ways to make the advice easier to follow. When they reach the recommendation to limit breads and pasta, the nutritionist, aware that tortillas are an important staple in the Mexican diet, asks Mrs. Osorio how she prepares them. Mrs. Osorio explains that she uses lard. The nutritionist comments on how much she also likes tortillas and then asks Mrs. Osorio if she can prepare them with a vegetable-based oil, explaining that it is very important to limit saturated fats like lard in her diet. The nutritionist also assures her that it is fine to eat tortillas, but that it would be good to limit the number eaten so that she doesn't take in too many carbohydrates at one time. As they discuss other recommendations, such as the need to limit the amount of sugar consumed, the nutritionist looks for ways to accommodate Mrs. Osorio's food preferences and customs so that she can avoid making unnecessary dietary changes. After being asked if she has any questions, Mrs. Osorio leaves, eager to try out these new ideas so that she can avoid taking new medication.

What Organizations Can Do to Become More Culturally Competent

To help health care organizations become more culturally competent, the Office of Minority Health (OMH 2000) of the U.S. Department of Health and Human Services has issued a set of guidelines: *The National Standards on Culturally and Linguistically Appropriate Services (CLAS) in Health Care* (Meadows 2000; OMH 2000). Essential elements for becoming more culturally competence are to:

1. adopt policies that promote and support attitudes and behaviors necessary for treating clients in a respectful and culturally sensitive manner;
2. recruit and retain a multicultural staff reflecting the community served;
3. provide staff training to enhance the cultural competence of staff;
4. hire and train interpreters or bilingual staff;
5. provide oral and written notices (e.g., signs, forms, educational materials) in clients' languages;
6. hire community members to serve as lay workers or cultural brokers; and
7. seek assistance from community leaders and other members in developing culturally appropriate educational materials and health care interventions. (For a complete copy of the CLAS standards, go to www.omhrc.gov/CLAS.)

What You Can Do to Become More Culturally Competent

The first and most important step you can take to become more culturally competent is to learn about the people with whom you work regularly. Reading books and articles describing the cultural beliefs and practices of others can make you more aware of differences that might be important to explore when working with people from a specific cultural background. Of course, as noted in Chapter 4, not everyone in a social group shares the same dietary beliefs and practices, making it necessary to ask the clients what they believe, do, and consider important rather than relying on generalizations about cultural groups.

Another valuable way to learn about another group is to immerse yourself in their culture. Spending time in their community or country, eating their food, observing their food practices, and talking with them will help you develop a deep respect and fondness for their culture. It is also important to establish ties with community members near where you work and to learn about community resources (e.g., local healers, faith-based organizations, food assistance programs).

Finally, many colleges, universities, and governmental organizations offer training in cross-cultural communication, cultural competence, and related topics simulation. Short courses, workshops, and other types of training are available in most cities and many smaller towns to help health care providers learn to become more sensitive to the cultural differences and the challenges and opportunities they present.

Additional Resources and Readings

Center for Cross Cultural Health
1313 SE 5th Street, Suite 100B
Minneapolis, MN 55414
612-379-3573
www.crosshealth.com

Cross Cultural Health Care Program
1200 12th Ave. S.
Seattle, WA 98144
206-326-4161
www.xculture.org

Health Education Training Centers, Alliance of Texas
The University of Texas
Health Science Center at San Antonio
San Antonio, TX 78284-7787
210-567-7800
www.uthscsa.edu/HETCAT

Indian Health Service
5600 Fishers Lane
Parklawn Building, 6-35
Rockville, MD 20857
301-443-3593
www.his.gov

National Center for Cultural Competence
Georgetown University
Washington, D.C.
800-788-2066
www.georgetown.edu/research/gucdc

Resources for Cross Cultural Health Care
8915 Sudbury Road
Silver Spring, MD 20901
301-588-6051
www.diversityrx.org

Books and Articles

Adams, D. (2000). Making cultural competency work. *Closing the Gap,* January 2000. Washington, D.C.: Office of Minority Health.

American Association for Health Education. (2001). *Report of the 2000 Joint Committee on Health Education and Promotion Terminology.* Reston, VA: American Association for Health Education.

Association for the Advancement of Health Education. (1994). *Cultural awareness and sensitivity: Guidelines for health educators.* Reston, VA: Association for the Advancement of Health Education.

Boston, P. (1999). Culturally responsive cancer care in a cost-contained work classification system: A qualitative study of palliative care nurses. *Journal of Cancer Education,* 14(3), 148–153.

Bracey, J. (2001). Connecting the Dots: Cultural competency in communities of color. *Closing the Gap Newsletter: HIV Impact,* Spring 2001. U.S. Department of Health and Human Services, Office of Minority Health.

Brach, C., & Fraser, I. (2000). Can cultural competency reduce racial and ethnic health disparities? A review and conceptual model. *Medical Care Research and Review, 57,* S(1), 181–217.

Bureau of Primary Health Care. (2000). *Cultural competence: A journey.* New Orleans, LA: Spectrum Unlimited.

Centers for Disease Control and Prevention. (1997). Update: Blood lead levels—United States, 1991–1994. *JAMA, 277*(13), 1031–1033.

Denboba, D. L., Bragdon, J. L., Epstein, L. G., Garthright, K., & McCann Goldman, T. (1998). Reducing health disparities through cultural competence. *Journal of Health Education, 29*(5), S47–S53.

Goode, T., Jones, W., & Mason, J. (2002). *A guide to planning and implementing cultural competence organization self-assessment.* Washington, D.C.: National Center for Cultural Competence.

Guidry, J. J., & Walker, V. D. (1999). Assessing cultural sensitivity in printed cancer materials. *Cancer Practice, 7*(6), 291–296.

Holland, L., & Courtney, R. (1998). Increasing cultural competence with the Latino community. *Journal of Community Health Nursing, 15*(1), 45–53.

Health Resources and Services Administration, Maternal and Child Health Bureau/Children with Special Health Needs Component of the National Center for Cultural Competence. (1999). *Self-assessment checklist for personnel providing services to children with special health needs and their families.* Washington, D.C.

Juarez, G., Ferrell, B., & Borneman, T. (1999). Cultural considerations in education for cancer pain management. *Journal of Cancer Education, 14*(3), 168–173.

Kumanyika, S. K., Morssink, C. B., & Nestle, M. (2001). Minority women and advocacy for women's health. *American Journal of Public Health, 91*(9), 1383–1392.

Leininger, M. (1989). Transcultural nurse specialists and generalists: New practitioners in nursing. *Journal of Transcultural Nursing, 1*(1), 4–16.

Locke, D. (1992). *Increasing multicultural understanding: A comprehensive model.* Newbury Park, CA: Sage.

López, L., & Masse, B. (1998). Income, body fatness, and fat patterns in Hispanic women from the Hispanic Health and Nutrition Examination Survey. *Health Care for Women International,* 14, 117–128.

Loustaunau, M.O., & Sobo, E. J. (1997). *The cultural context of health, illness, and medicine.* Westport, CT: Bergin & Garvey.

Majumdar, B. (1995). *Culture and health: Culture-sensitive training manual for the health care provider.* 4th ed. Hamilton, Ontario: McMaster University.

Meade, C. D., Calvo, A., & Cuthbertson, D. (2002). Impact of culturally, linguistically, and literacy relevant cancer information among Hispanic farmworker women. *Journal of Cancer Education, 17*(1), 50–54.

Meadows, M. (2000). Moving toward consensus on cultural competency in health care. *Closing the Gap,* January 2000. Washington, D.C.: Office of Minority Health.

Office of Minority Health. (2000). *National standards on culturally and linguistically appropriate services (CLAS) in health care.* Federal register, 65 (247). Retrieved from *www.omhrc.gov/CLAS*

Oldways Preservation and Exchange Trust. (2002). *The traditional healthy Asian Diet Pyramid.* Retrieved from *www.oldwayspt.org*

Schilder, A. J., Kennedy, C., Goldstone, I. L., Ogden, R. D., Hogg, R. S., & O'Shaughnessy, M. V. (2001). "Being dealt with as a whole person." Care seeking and adherence: The benefits of culturally competent care. *Social Science & Medicine, 52,* 1643–1659.

Stoy, D. B. (2000). Developing intercultural competence: An action plan for health educators. *Journal of Health Education, 31*(1), 16–19.

Tavelli, S., Beerman, K., Shultz, J. E., & Heiss, C. (1998). Sources of error and nutritional adequacy of the Food Guide Pyramid. *Journal of American College Health, 47*(2), 77–83.

Taylor, T., Serrano, E., Anderson, J., & Kendall, P. (2000). Knowledge, skills, and behavior improvements on peer educators and low-income Hispanic participants after a Stage of Change–based bilingual nutrition education program. *Journal of Community Health, 25*(3), 241–262.

Trotter, R. T. 1988. *Orientation to multicultural health care in migrant health programs.* Austin, TX: National Migrant Resource Program, Inc.

PART III

Strategies for Addressing Nutrition Challenges

CHAPTER 9

Hunger in Global Perspective

The world food situation is at a critical point. Globally, humans produce more food than ever before. Food production has continued to rise in most, but not all, of the world's regions, more than keeping up with population growth over the same period of time. However, the world population continues to grow, although at a much lower rate than in the middle of the twentieth century. Under these conditions we can project a scenario in which food production cannot grow sufficiently in a sustainable way to feed all of the people expected to be alive in the future. As we will see, there is currently enough food produced to adequately feed all of the world's people. At the same time, the Food and Agriculture Organization of the United Nations (FAO) estimates that over 800 million people do not have enough food available to them to meet their nutritional needs. A significant proportion of these people exhibit the physical signs of nutritional deficiency. Most undernourished people are found in countries that have low national incomes, the less developed countries. However, over 30 million people in developed, high-income countries are estimated to suffer from undernourishment. In these more affluent regions, undernutrition exists side by side with obesity and the associated diseases of development.

In the end, the answer to the question "Is there enough for all?" rests on an analysis of the balance between population and food production/distribution now and in the future. More importantly, the nature of the question has shifted a bit in recent years. Rather than asking if there is enough for all now, we have come to ask the question of whether providing enough for all can be accomplished in a sustainable way, that is, one in which the global environment is also maintained. In this chapter we will examine the global food supply and the extent of adequate and inadequate access to food for people at risk for undernourishment, both globally and in specific regions. We will also discuss some of the overarching environmental issues that concern analysts.

First, however, a word of caution is in order. Current analyses and future projections of the world food situation must be studied carefully. Statistical analyses of agricultural productivity, food supply levels, poverty, and malnutrition often present very different—even conflicting—pictures of the world situation. There are several reasons for these often confusing discrepancies. First, data collection is extremely difficult. Imagine trying to determine the extent of malnutrition present in the earth's 6+ billion human inhabitants. What nutrient levels would you consider inadequate? How would you measure dietary intake, nutritional status, or access to food supplies? Second, for many variables, a variety of measures can be made. **Agricultural productivity,** for example, can be calculated as the amount of food produced per unit of **land** or per **agricultural worker (labor);** as the kilocalorie of energy expended in production or in all phases of the food system; or, finally, as the ratio between inputs and outputs in monetary terms.

   Future projections also must be viewed cautiously. Projections do not predict what will actually occur but, rather, depict conditions that are *likely* to develop if current trends continue unchecked. As we will see, there have been several recent important revisions of predictions of increasing population and food production in less than a 10-year period. Of course, the accuracy of the data used to generate projections is also critical in terms of how well they can actually predict the future. Data stretching over a very short period of time, especially when that period is atypical, do not forecast future events nearly as well as data that represent changes caused by factors that will continue to influence events in similar ways both today and tomorrow. Unfortunately, statistics about food production, employment, and population growth in some developing nations are not highly accurate and, in a few cases, are completely unavailable. Despite these limitations, projections of the world food supply, population growth, and other conditions abound; not surprisingly, considerable controversy and contradiction surround them.

   A final problem is the tremendous complexity of the issue. The world food situation is a result of a multiplicity of interacting variables: agricultural and food policies, international trade agreements and economic market conditions, resource distribution, agricultural productivity, environmental conditions, and a host of dietary factors. In this chapter and the next, we present an overview of the world food situation, describing the major factors influencing food supplies and their distribution throughout the world. Our discussion is brief and thus we can explore only superficially many of the complex issues surrounding this topic.

Malthus vs. Boserup

Alarm about food shortages, malnutrition, and hunger has been a recurrent theme since 1798, when Reverend Thomas Malthus publicized his "Essay on the Principle of Population" warning that food production could not keep pace with population growth rates indefinitely. Malthus argued that population grows at geometric or exponential rates whereas food production increases only arithmetically. For this reason, he predicted that the population would eventually grow at a much more rapid pace than food production, ultimately leading to starvation. Figure 9.1 presents Malthus' predicted growth rates. Population growth follows a J-curve; it grows by doubling—1, 2, 4, 8, 16, 32, 64, 128, and so on. If we plot the change in numbers, the line rises slowly at first, and then after a series of doublings, it makes a sharp upward curve in the shape of the letter J. This is the exponential or geometric growth discussed by Malthus. Food production, which climbs in a steadier manner, is suddenly left far behind.

agricultural productivity (land)
Amount of production of a particular crop that is produced per unit of land.

agricultural productivity (labor)
Amount of a particular crop that is produced per person/hour worked.

Figure 9.1

Arithmetic and exponential growth. In 1798, Thomas Malthus hypothesized that population would eventually exceed the food supply because population would increase exponentially and food production would increase only arithmetically.

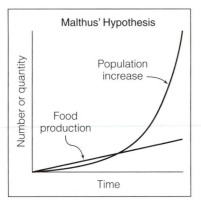

Malthus predicted widespread famine and pestilence as the limits of the ability of the food supply to keep up with population increase became close. In the twentieth century, there were three waves of fear about the world food situation: in the late 1940s, in the mid-1960s, and again in the mid-1970s. In all three cases, the crises resulted from temporary grain shortages created by drought or wars that affected major grain-producing regions (FAO 2000b). Each shortage, however, brought with it concern that world food production was approaching its Malthusian limits and no longer could keep pace with population growth. One book predicted widespread famine by 1975 (Paddock and Paddock 1967).

Neo-Malthusians such as Ehrlich and Ehrlich (1990) argue that, in addition to the potential for the growth in population to outstrip food production, the impact of rapidly rising population has strong negative impacts on natural resources such as forests, water biodiversity, and land. In the long run, the deterioration of these resources would contribute to a situation in which the world's resources would not be able to support the population.

On the other side, analysts following the work of Esther Boserup (1965, 1981) and others argue that the response of households, communities, and societies to pressure from rising population is to intensify production and develop new, more productive technologies to produce more food per unit of land (Pender 2001). Some analysts would argue that technological innovation is almost infinitely substitutable for other factors of production. That is, technological improvements, including the use of technology that protects natural resources, might, in the foreseeable future, continue to be applied so as to replace the need to cultivate more land, cut down more forests, or use more labor in agriculture. From this perspective, agriculture and food production can continue to become increasingly more productive.

Not surprisingly, studies of the impacts of rising population on food production and natural resources over the relatively short term of the last half century or so show conflicting results. A number of studies, including Boserup's own study, have shown that increasing population in a region can result in increased food production, better land management, increased numbers of trees and forested areas, and rising incomes in the region (Templeton and Scherr 1999; Tiffen, Morrison, and Gichuki 1994). At the same time, other studies show that increasing population in rural areas can result in deforestation, overgrazing, soil erosion, and other problems (Kates and Haaramann 1992; Panyotou 1994; Pender 2001; Templeton and Scherr 1999). In large part, the impact of population growth in rural areas is different in different places in the world because the intervening variables can be quite different.

Pender (2001) argues that households and local communities in rural areas react differently to increasing population pressure as a result of other factors and processes taking place in society. For example, in many parts of Latin America, large landholders still control most of the best land. Large landholders may not have any incentive to intensify food production on their large tracts of land. They may find it more economical to use large tracts of land to graze animals, for example. At the same time, small farmers in the same region may have a need to increase production to maintain their families, but not have clear title to land, which is a disincentive to invest in improvements in land. People who cannot be sure that they will be able to use a piece of land for a long time would not be expected to be interested in long-term improvements in the land. Under these sorts of circumstances, small farmers may contribute to deforestation by cutting down trees to cultivate more land. In the terms we used above, they have no incentive to substitute technology for land. On the other hand, in cases where small farmers do own land, there may be great incentives to use land more sustainably and more intensely.

The Factors in the Food Sufficiency Equation

A careful analysis of world food supplies fails to substantiate predictions of widespread famine as a result of food shortage currently or in the very near future. Although debate continues, it is generally concluded that Malthus and other forecasters were wrong, at least in the short term. However, concern is mounting that *sustainable* growth in food production in amounts necessary to sustain the predicted growth in population in the next century may not be possible. In order to see why and to suggest the possible trends for the future, let's take a closer look at the several factors in the global food balance sheet. These include analysis of population increase, the potential for food production, the distribution of food to people who need it, and, finally, the sustainability of the system to continue to provide sufficient food for future populations.

Population

In October 1999 the population of the globe reached 6 billion. This represents a fourfold increase in the number of people over the twentieth century. In 1900 there were about 1.7 billion people in the world. Population reached 2 billion in 1927, 3 billion by 1960, 4 billion by 1974, and 5 billion by 1987 (U.S. Bureau of the Census 1999). Table 9.1 shows the time it took to add a billion people to the global population since 1900. Figure 9.2 represents the increase in global population over the past several centuries, in graphic form. You can see not only that population grew, but also that the rate of population growth accelerated through most of the twentieth century, with an even more rapid rise in growth after World War II. People were added to the global population at an ever increasingly rapid and, to some, frightening rate. However, the period of hyper-population growth appears to be over. The most recent data show that the global rate of population increase is slowing—and slowing even more rapidly than predicted just a decade ago. In large part this

total fertility rate (TFR)
The total number of live children born to a woman during her reproductive life.

Table 9.1 Years Taken to Reach Successive Billion Population Markers in the Twentieth Century

2 billion in 1927	123 years
3 billion in 1960	33 years
4 billion in 1974	14 years
5 billion in 1987	13 years
6 billion in 1999	12 years
7 billion in 2012	13 years
8 billion in 2026	14 years
9 billion in 2050*	24 years

Source: Adapted from U.S. Bureau of the Census, *World Population Profile: 1998* (Washington, D.C.: U.S. Government Printing Office, 1999), p. 9.

* This table has been corrected for the United Nations' revised projections of reaching a global population of 9 billion by 2050, rather than the U.S. Census Bureau's projections of reaching that number in 2043.

is a result of declining rates of fertility. The world's **total fertility rate (TFR),** that is, the total number of children born to a woman, declined from an average of five births per woman in 1960 to fewer than three births per woman in 1998. The TFR is now below 2.1, the level needed to replace the current population, in a number of both developed and developing countries. (As we will note below, even with a TFR below 2.1, population does keep increasing in the short term due to the large number of people of childbearing age currently alive.) In 1998, seventy-nine countries, over a third of all countries, had TFRs below the replacement rate. These included Western and Eastern Europe, the United States, Canada, and China (U.S. Bureau of the Census 1999).

Why has overall population growth slowed so much more than Malthus predicted? In part, Malthus did not anticipate the demographic transition. The demographic transition refers to a change in mortality and fertility from high fertility and high mortality to a situation of low mortality and low fertility. That is, historically countries have reduced TFRs as mortality rates, especially rates of infant and child mortality, decline and income increases (Ray 1998; Williamson 2001). Historically it appears that population growth slows as parents choose to have fewer children as a result of increases in per capita real income, increases in population density, decreased infant and child mortality as a result of improved health and nutrition, and with the realization that less food is available per capita. Declines in mortality, especially infant and child mortality, have historically preceded most of the demographic transitions. The improved survival rate for children allows parents to reduce fertility and still have some children who survive to adulthood (Williamson 2001).

Figure 9.2
World population growth in historical perspective.

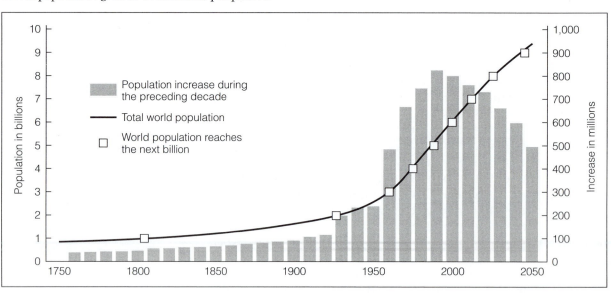

(*Source:* U.S. Census Bureau, *The Official Statistics,* February 2, 1999.)

Conversely, high **infant mortality** and **child mortality** rates contribute to the desire to have more children. In the face of high infant mortality, families in areas with high mortality may have more children in order to ensure that several survive to adulthood. Most families experience the loss of one or more infants or young children, and the uncertainty of their children's survival hangs over most couples. In many poor areas, where large percentages of children die before reaching a productive age, uncertainty and the overcompensation it induces are major factors in the common bearing of seven or eight offspring to ensure two male survivors. Parents, however, do not respond immediately to reduced mortality. There is always a lag time.

The current reduction in fertility is not evenly distributed around the world. While high-income countries and a number of developing countries have an average TFR below replacement levels, some countries of the Near East, northern Africa, Asia, and Latin America continue to show TFRs over replacement levels (U.S. Bureau of the Census 1999). Sub-Saharan Africa continues to show an average TFR of close to six children per woman. Although the U.S. Census Bureau estimates that the average TFR for Latin America will fall below the replacement rate by 2025, the average TFR for sub-Saharan Africa is expected to remain above 2.1 births per woman for some time.

What accounts for the uneven decline in TFR? A number of factors are implicated. Total fertility rates are affected by marriage rates, the average age of marriage of women, infertility, and use of contraceptives. Married women have higher TFRs than unmarried women, and women who marry at an older age have lower TFRs than women who marry earlier. But the most important contributor to a decline in fertility rates in the contemporary world is the use of contraceptives. Contraceptive use is highest in higher-income countries and lowest in lower-income countries, with some interesting exceptions. Of large countries, China, still considered a less developed country, has the highest rate of modern contraceptive use, with over 75% of married women using modern contraception at any one time. In Brazil in 1996, about 70% of married women used modern methods of contraception. The U.S. rate in 1990 was just under 70%. In Nigeria, however, less than 5% of married women used modern forms of contraception in 1990. Modern methods of contraception are just not easily available in a number of poor countries. While some poorer countries have limited availability of family planning services and contraceptives in general, women in urban areas are more likely to have access to and use contraception than women in rural areas. Focus 9.1 addresses some of the social issues in family planning initiatives. Women with higher education are more likely to use contraception than women with less education. In part, the low use of contraceptive use in sub-Saharan Africa can be explained by a largely poor, rural, and less educated population. However, contraceptive use is increasing in all world regions, including Africa.

infant mortality rate
Number of children in a cohort of 1,000 live births who die before their first birthday.

child mortality rate
Number of children in a cohort of 1,000 live births who die before their fifth birthday.

The Effect of HIV/AIDS on World Population

Declining fertility is not the only reason for the current slowing of population growth. As Brown et al. (1999) and others (U.S. Bureau of the Census 1999) have pointed out, some parts of the world, most notably sub-Saharan Africa and parts of Asia, are also experiencing an increase in mortality for the first time since the great famine in China in 1959–60. For sub-Saharan Africa, this may be due, in part, to declining nutritional status as a result of failing agriculture (as we will see below), but an important contributor is the pandemic of HIV/AIDS. In countries in which the HIV/AIDS pandemic is most severe, life expectancy, crude death rates (overall death), infant mortality rates (deaths of children under 1 year per 1,000 live births), and young child mortality rates (deaths of children under 5 years per 1,000 live births) will all be affected. It is estimated that by 2010, the overall life expectancy in Botswana will be reduced from the 66.3 years it would have been without AIDS to 37.8 years due to the effect of the AIDS pandemic. It is also estimated that the national child mortality rate in Botswana will be 119.5 per 1,000 live births, while without AIDS only 38.3 deaths would be expected.

Over 30% of all infants born to mothers with HIV in sub-Saharan Africa will themselves develop an infection. In Cameroon and Côte d'Ivoire, infant mortality rates are already 10% higher as a result of HIV and are expected to be 20% higher in 2010. In eastern and southern Africa, the regions most affected by the HIV/AIDS epidemic, current infant mortality rates are nearly 70% higher than they would be without AIDS. While parts of Africa currently have the highest rates, the infection rate is climbing in parts of Asia, where HIV infection appeared later. It is possible that in absolute numbers, Asia may have the largest number of cases of HIV infection in the next several decades. But even in regions that were a bit less affected in the early years of the HIV/AIDS epidemic, there is likely to be a strong impact. In Brazil, where infant mortality is low, AIDS is expected to raise the infant mortality rate by 56% in 2010. Figure 9.3 shows the crude death rates with and without AIDS for several of the highly affected countries for 1998 and projected to 2010 (U.S. Bureau of the Census 1999). Some neo-Malthusians would argue that increasing rates of infectious diseases, including HIV/AIDS, and declining food production (see below) show that we have entered into the age of disasters predicted by Malthus when the carrying capacity of the world is exceeded.

The United Nations has reduced its estimate of total world population in 2050 downward from a projected 9.3 billion, predicted in the early 1990s, to a new projection of 8.9 billion people. This is due to both declining total fertility rates and increasing mortality from nutritional stress and, particularly, infectious disease. This projection is still close to 50% more people in the world in 2050 than in 2000, but it suggests an even more dramatic reduction in the global growth rate than was anticipated just a few years ago. Current projections are that the global population will stabilize at between 11 and 12 billion people before the end of the twenty-first century.

Figure 9.3

*Life expectancy at birth and deaths in sub-Saharan Africa and
HIV/AIDS-affected countries, 1995–2025.*

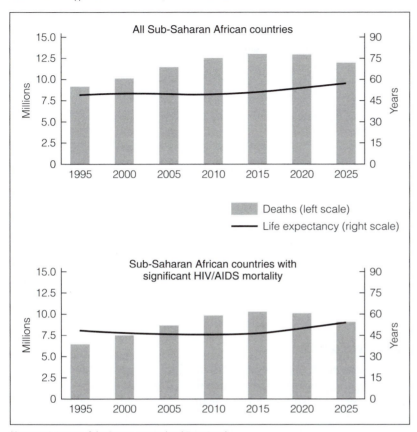

(*Source:* U.S. Bureau of the Census, International Data Base.)

FOCUS 9.1 Family Planning

Why has population growth slowed, proving Malthus and his followers wrong? One factor contributing to declining birthrates is the widespread promotion of family planning programs. Government policy makers in countries inhabited by 90% of the world's population promote some form of family planning. In China, women are encouraged to marry relatively late. It is not considered socially acceptable to marry before the mid to late 20s in towns and early 20s in rural areas. Oral contraceptives are popular, but a variety of fertility measures, including abortion and sterilization, are employed (Allaby 1977).

Family planning programs, especially those in rural areas of developed countries, still encounter many obstacles. First, each of the available contraceptives has significant disadvantages. Vasectomies and tubal ligations are not reversible; the diaphragm, condom, and oral contraceptives require discipline; intrauterine devices are associated with pain, bleeding, and increased mortality; and injectable contraceptives have been shown to cause serious side effects. Second, products must be distributed to areas with bad roads and poorly developed markets. Third, many programs are fraught with administrative, economic, and political problems. For example, the Indian government decided to offer free radios to men willing to have a vasectomy. Large numbers of radios were manufactured at low prices to supply the

(continued)

Focus 9.1   (continued from previous page)

government program, but before the program could succeed, the radios appeared in the open market where men could buy them at reasonable prices and without having to have vasectomies (Allaby 1977).

Finally, family planning programs run into resistance in many rural communities because families want to have many children. As Robert McNamara (1977), president of the World Bank, has said, "It is a mistake to think that the poor have children mindlessly or without purpose—in light of their own personal value systems—irresponsibly. Quite the contrary . . . poor people have large families for many reasons. But the point is, they do have reasons" (p. 37). Among their reasons are the following:

- the need for children to take care of parents in old age and illness
- extra hands to help with agricultural and household labor
- religious beliefs that forbid birth control
- a value on sons as a mark of success or good fortune or virility and the opportunity to pass on the family name and land

Family planning programs do not work until families *want* to restrict the number of children they have, and have access to the means to limit fertility. Health services, economic development, or other changes are needed to bring infant and child death rates down to acceptable levels before family planning can succeed. Because of this, students of population dynamics have come to see overpopulation as a *result* of underdevelopment, rather than a cause. As economic well-being increases, parents are motivated to limit the size of their families. However, even in the face of sometimes strong incentives to have large families, families all over the world are electing to limit their fertility.

Food Production

At the same time the rate of population increase has grown and now declined, the amount of food produced globally has also continued to increase at a faster pace than Malthus, or anyone else of his time, could have predicted. In 1997 there was 24% more food produced per person globally than in 1961, even with an 89% increase in the population of the world (Wood, Sebastian, and Scherr 2001). During this same time, the relative price of food dropped by about 40% as a direct result of increases in agricultural productivity. Decreases in the real price of food represent a significant improvement for poor consumers (Wood et al. 2001). Technological achievements, especially the development of chemical fertilizers and high-yield varieties of wheat, corn, and rice—often referred to as the "green revolution" (see Focus 9.2)—have increased food production more than enough to keep up with population growth, at least on a worldwide basis. Technological advances in irrigation, plant breeding, biotechnology, management of pests, and farming practices enabled people to cultivate new areas at a rate of 0.7% annually between 1950 and 1970 (Knutson, Penn, and Boehn 1983).

agricultural extensification
Increasing production by bringing new land under cultivation.

Extensification   The process of cultivating new lands is called **agricultural extensification.** Extensification was the way in which food production increased throughout most of human agricultural history. Though it has slowed, extensification still continues showing a rate of increase in cultivated land of about 0.25% annually between 1986 and 1996 and has been relatively steady at an average of about 0.3% annually since 1970 (Wood et al. 2001).

FOCUS 9.2   The Green Revolution: Production Benefits and Social Costs

The term "green revolution" is one that we have heard frequently over the last several decades. It is used to refer to several aspects of agricultural change since World War II. Strictly speaking, "green revolution" describes the development and use of new agricultural technology, including high-yield varieties of seeds and the cultivation techniques developed to go with them. Green revolution technology includes chemical fertilizers and pesticides, mechanical preparation and cultivation of the soil, and irrigation. In its broader sense, the term has been used to refer to a hoped-for transformation or modernization of agriculture in less developed countries resulting in an increase in the availability of food, reduction of food shortages, and elimination of hunger and poverty.

It is important to keep both of these uses of the term in mind because they do not necessarily go together. As we will argue in Chapter 10, much of the problem of hunger in poor countries is the result of national and international economic and political policy that results in widespread poverty. Nonetheless, the scientific advances that have produced better seeds and technology for food production have had a tremendous impact on the ability of the world to feed itself, both now and in the future. These advances have not been without costs, however, including damage to the environment and social dislocation. Again, as we have seen, the social costs are closely tied to the political and economic structure within various countries and the world.

The history of plant breeding for desirable food characteristics and higher yields is as old as agriculture itself. The history of domestication has been a process of selecting plants that yield more food than their wild cousins. Domestication has been a process of selecting the best seeds. Before Darwin, the selection of better crops was done by the individual farmer noting the plants and seeds with the most desirable characteristics and replanting them. The scientific basis of plant improvement came with the work of Darwin in the mid-nineteenth century and the discovery of the laws of inheritance by Gregor Mendel in 1865. Mendel's work was overlooked for nearly 40 years and not rediscovered until 1900. Since then, the breeding of food crops using scientific principles has gained increasing momentum. The most important work of conventional plant breeding, however, took place in the second half of the twentieth century.

The first significant increase in potential grain yields due to the development of plants with greater genetic potential came in 1961 when the first semidwarf winter wheat developed in the United States was released. Even more important was the development of two semidwarf spring wheat varieties developed in Mexico by the International Maize and Wheat Improvement Center (CIMMYT). The CIMMYT wheat varieties were adapted to a wide range of climates and could be grown from Canada to Australia. At about the same time, dwarf varieties of rice were being developed at the International Rice Research Institute (IRRI) in the Philippines.

Semidwarf wheat and rice are more productive for several reasons. The plants respond better to the presence of nitrogen. When fertilized, these plants produce more grain-filled heads. Traditional grain plants, which are taller and have weaker stems, will fall over when the heads are very heavy. This is called "lodging." The lodged grain may be lost or eaten by animals. The shorter semidwarf varieties are less likely to fall over and lodge, so the heavy heads of grain are protected. Also, the shorter plants are more efficient producers of grain; that is, they spend less of their energy producing leaves and stems and more of it producing grain.

It must be kept in mind, however, that the improved yield of grain from the new "miracle" seeds produced in the 1960s was dependent on the use of fertilizer and, in most cases, the careful use of water, usually through irrigation. The high-yield varieties (HYVs) did not yield any better than traditional varieties when fertilizer was not applied or sufficient water was not provided. In fact, under these conditions, traditional varieties frequently did better. After all, they are the result of many generations of selection by farmers under the very conditions of poor soil and unpredictable rainfall in which they are grown. The HYVs are most useful when they are cultivated using the entire technological package that includes chemical fertilizer and, frequently, chemical pesticides and herbicides, mechanical plowing, and irrigation. The extra inputs necessary in order to perform the "miracle" can be very expensive in relation to the resources available to small farmers in rural areas of poor countries.

Nevertheless, the adoption of HYVs and the technology they require was rapid in many areas and had dramatic effects on the production of grains in many parts of the world. According to Norman Borlaug (1983), one of the scientists that developed the semidwarf wheat at CIMMYT:

(continued)

Focus 9.2   (continued from previous page)

When high-yielding semidwarf varieties of Mexican wheat were introduced into India during 1966 to 1968, national production stood at roughly 11 million metric tons and average yields were less than 1 ton per hectare. The high-yielding wheat varieties quickly took over, and by 1981 wheat production had increased to 36.5 million metric tons, largely as a result of a 100% improvement in national wheat yields. The 1981 harvest represents sufficient additional grains to provide 186 million people with 65% of the carbohydrate portion of a diet containing 2,350 kilocalories per day. (p. 692)

By the 1980s, the wheat varieties released by the International Maize and Wheat Improvement Center in the 1960s accounted for virtually all of the wheat grown in Mexico (Plucknett and Smith 1982). In Sonora, the Mexican state with the greatest wheat production, green revolution wheat yielded an average of 4 tons per hectare, as compared with 0.7 ton per hectare with the traditional varieties. The area planted with HYV rice also increased dramatically in Asia. By the 1980s, IRRI rice accounted for one-fourth of the rice grown in the world (Plucknett and Smith 1982).

HYVs of maize were less successful in less developed countries. The costs of producing maize using green revolution seeds was high, and the land could more profitably be used for other crops.

The spread of HYVs of rice, wheat, and maize reduced or even eliminated the need for food imports in some lower-income countries. India has been self-sufficient in grains in most years since the introduction of HYVs of wheat and rice. Bangladesh has come closer to feeding itself. The Philippines has only occasionally had to buy rice since 1970, and now it is a net exporter of rice. The most dramatic impact, however, has been the role of HYVs in the lowering of food prices for food consumers. The higher productivity per unit of land and labor increased yields so much that world grain prices fell in real terms. The most important beneficiaries were urban and rural low-income workers. Green revolution varieties and production technologies are in large part responsible for the 40% drop in real food prices globally since 1960.

But there were costs to the green revolution. Because of the early successes with IRRI rice and CIMMYT wheat (and, to some extent, maize), these three crops have received a great deal of emphasis, often to the exclusion of other crops. In fact, monocropping of these cereals has been emphasized even in areas in which legume crops were important dietary staples and were traditionally intercropped with grain. Other crops, such as cassava, sorghum, millet, yams, beans, sweet potatoes, and potatoes—considered minor in much of the developed world but of great importance in many developing countries—were ignored for some time, although interest and cultivation of cassava has increased in recent years.

There has been a loss of crop diversity as HYVs have replaced a wide variety of older varieties. As a result, a great deal of diversity in genetic material could potentially be lost. Genetic diversity in plants has traditionally been an important source of traits that can be useful. For this reason, germ plasm banks for the storage of seeds of many diverse kinds of crop plants are very important. They provide genetic material used by many different countries for developing new and better varieties of seeds.

A second problem is the environmental effects of the chemical fertilizers and pesticides required to grow HYVs. Large-scale use of pesticides in several counties appears to be associated with a rise in malaria as the

Figure 9.4

Rice is a crop in which intensification in the use of irrigation and labor can result in higher yields.

©Jack Fields/CORBIS

(continued)

Focus 9.2 (continued from previous page)

mosquitoes become resistant to the insecticides used. Contamination of food with pesticides is a problem of unknown dimension.

Perhaps the most profound and difficult effects to assess are the broader social consequences of the green revolution. As we saw in Chapter 5, the majority of peasant farmers are poor and unable to afford the more expensive HYV seeds and technology (fertilizers, pesticides, machines) that accompany them. Thus, early in the green revolution, wealthier farmers were able to benefit from the HYVs while the poor farmers lagged behind. This, in turn, led to increasing disparities between the large, rich landholders and the poor small farmers in most areas in which green revolution technology was adopted. In some cases, the new technology allowed large landholders to more profitably cultivate their own fields, pushing tenant farmers and sharecroppers off the land. Moreover, in many instances the adoption of machines to carry out agricultural labor has further reduced the need for laborers and thus increases unemployment for laborers. In this way, then, the green revolution was one of several factors that pushed the poor and the landless to leave the countryside to join the growing ranks of the urban poor (see Figure 9.4). Such migration of people from agriculture need not be a problem if a well-developed industrial sector exists to provide them with jobs. But this is exactly what less developed nations lacked.

The green revolution is an interesting case to think about as agricultural technology moves strongly into the use of biotechnology, including the cultivation of genetically modified crops. What might be the results of the introduction of genetically modified varieties of food crops in less developed countries?

The earth's surface has an estimated 32.9 billion acres of land. Only 30% of this area, approximately 9.8 billion acres, is potentially useful for agriculture. Lack of moisture, severe temperatures, and other environmental conditions limit cultivation of the remaining areas with conventional methods (Ehrlich, Ehrlich, and Holdren 1977). Satellite imagery shows that in the 1990s cropland and managed pastureland occupied about 28% of the global land surface (Wood et al. 2001). Of this land about 31% was planted to crops, with the remaining 69% in managed pastures for animals. However, again, the percentage of land cultivated varies from world region to world region. Western Europe and South Asia show percentages of land under cultivation over 70%, while in sub-Saharan Africa about 26% of land is under cultivation and in North America only about 23% of land is under cultivation. One of the interesting trends noted in the study of global land use is that there has been a decrease in the amount of land devoted to agriculture in some regions. Both Western Europe and the United States have been taking land out of cultivation since 1960. There was also a significant, but perhaps temporary, decrease in cultivated land in Eastern Europe in the 1990s. Wood et al. (2001) suggest that this is the result of temporary economic dislocations following the introduction of a market economy to countries of the former Soviet bloc. Indeed, only West Asia has seen a steady increase of more than 1% per year in land under cultivation. While increases in land undercultivation have slowed, what cultivation does take place, as we will see below, comes primarily at the expense of forests.

Intensification: Irrigation Although extensification continues, albeit more slowly, **agricultural intensification**—that is, coaxing more production out of existing agricultural land through the use of irrigation, better seeds, fertilizers, and pesticides—has been a far more important component of the growth in food production in the twentieth century (Wood et al. 2001). One of the most important aspects

agricultural intensification
Increasing production by increasing yield per unit of land.

of intensification has been the increased use of irrigation. Globally about 17% of cropland is irrigated. The trend for irrigation was upward throughout the second half of the twentieth century, increasing globally between 1966 and 1996 by about 72%. But again, the use of irrigation is much higher in some areas than in others. About 40% of agricultural land in East Asia and South Asia (China and India) is irrigated (see Figure 9.5), whereas less than 1% of agricultural land in Oceania and sub-Saharan Africa is irrigated (Wood et al. 2001). However, the amount of irrigated land in sub-Saharan Africa increased by 80% from 1960 to 1997, and it more than doubled for South America in that same time frame.

With well-designed and well-managed irrigation and drainage, crop production can be increased in both traditional and modern farming systems. The 17% of global cropland that is irrigated produces 30–40% of the world's crops. Clearly, the potential for increasing food productivity through proper water management is great. The major obstacles to realizing this potential are the lack of financial and technical resources for construction of irrigation projects, and poor maintenance. Construction costs alone place large-scale projects outside the reach of many developing countries.

Smaller-scale irrigation projects are becoming increasingly popular. Many Asian governments, for instance, are encouraging small farmers to finance their own tubewells or other small irrigation systems. Pakistani farmers installed 32,000 private tubewells during a 5-year period in the mid-1960s. Most farmers were able to recover the initial investment of $1,000 to $2,500 within 2 years. And for every 5,000 tubewells installed, an additional 1 million acre-feet

Figure 9.5
Development programs often include the mechanization of agriculture. In some places, yields have increased greatly, but small farmers and laborers have been impoverished.

Courtesy U.S. Department of Agriculture

were added to the annual supply of irrigation water (Brown and Finsterbusch 1972).

Many technical factors also constrain the expansion of land through irrigation. If irrigation projects are not designed and maintained correctly, they create several problems. First, water brought in through irrigation channels percolates down through the soil to the groundwater below. As the water accumulates, the water table, or the groundwater level, rises. When the water table reaches the soil's surface, it brings with it salts such as magnesium, carbonates, sulfates, and sodium chloride from ancient subterranean deposits. Water evaporating from above also can leave salts that increase the soil's salinity. These salts can actually sterilize the soil.

A second, but related, problem is waterlogging. As the water table rises to the surface, it reduces the amount of air and aerobic organisms held in the soil. Processes of decomposition are altered and plant roots die from lack of oxygen. Many large-scale irrigation systems have created waterlogged, sterile soils. The ancient irrigation systems in the fertile crescents of Mesopotamia and the Indus Valley, as well as the 1970 Aswan Dam project, are among the more dramatic examples.

Perhaps the most important constraint is that of water itself. Currently, irrigation accounts for 70% of the water withdrawn from freshwater systems for human use. Agricultural use is in direct competition with domestic uses (e.g., drinking water) and industrial uses. There is increasing competition over the use of limited supplies of both ground and surface water. As a result of overuse, the Yellow River in China periodically runs dry well above its mouth. In some areas, underground aquifers are being drawn down at a rate of a meter/decade. While we should be very concerned with the loss of forests and biodiversity due to the expansion of agricultural land, water is likely to be the most limited and limiting natural resource in the twenty-first century.

It does appear that while some additional land may be brought under irrigation, especially using techniques that conserve water and return water to the environment, the potential for widespread additional irrigation is limited.

Intensification: Cropping Intensity   Irrigation, coupled with the use of fertilizers and pesticides, also allows for increased cropping intensity. **Cropping intensity** is a measure of the amount of harvested land as a proportion of total cultivable land available. In the shifting cultivation/horticultural systems that we discussed in Chapter 5, cropping intensity is less than 1. That is, with a high percentage of land left fallow each year, a farmer harvests less than one crop per acre of cultivable land. In some irrigated areas in the tropics, up to three crops per year can be harvested per acre of land. This would yield a cropping intensity of 3. Global cropping intensity has also been increasing and for annual crops was about 0.8 globally in 1997. However, in irrigated areas of South and East Asia, cropping intensity is as high as 1.1–1.2. In areas in which colder temperatures limit cropping, such as Western Europe and North America, cropping intensity

cropping intensity
The proportion of possible cultivable land that is actually under cultivation during an agricultural season.

is closer to 0.6–0.7. Between 1960 and 1997, cropping intensity in sub-Saharan Africa and Southeast Asia increased by 40–50%, but remained steady in areas such as North and South America.

Though extensification and intensification have resulted in increases in food production, these practices cause concern about the use of nonrenewable natural resources. Expansion of agricultural land has clearly come at the expense of forests (Wood et al. 2001). While irrigation has allowed the exploitation of some land previously uncultivated, most of the expansion of cropping area is the result of cutting forests. This represents the replacement of habitats with high biodiversity, with habitats with low biodiversity. As we will note below, however, some cultivation practices can protect biodiversity to some extent. For example, whereas tree cover is relatively low in many agricultural areas, in rain-fed areas of Latin America, sub-Saharan Africa, and South and Southeast Asia, agriculture coexists with significant and, in some cases, increasing tree cover. There is increasing interest in understanding those local farming systems that can provide alternative approaches to developing cultivation techniques that protect biodiversity.

Intensification: Increasing Yields   In the twentieth century, increasing crop yields as a result of improved technology was perhaps the most significant contributor to increasing agricultural production. While the amount of new land being brought under new cultivation has slowed in recent decades, the amount of food produced per unit of land, that is, productivity, has continued to increase, but at a somewhat slower rate than in previous decades. The increase in productivity of the "green revolution" brought the most dramatic increase in world food supplies that has been experienced since the original agricultural revolution 10,000 years ago. The green revolution (see Focus 9.2) refers to the dramatic increase in cereal crop yields as a result of scientific plant breeding for high-yielding varieties and the application of agrochemicals, most importantly chemical fertilizers. Globally, and throughout history, increases in output per acre were scarcely perceptible within any given generation. Only during the twentieth century did more and more regions of the world succeed in achieving rapid, continuing increases in output per acre.

The increase of yields in the production of food crops, most importantly the production of cereal grains, has been the most significant contributor to the increased global availability of calories since 1950. From the late 1940s until the 1990s, the average yield of wheat increased from 1,100 kg/ha (kilograms per hectare; a hectare is 2.47 acres) to 2,600 kg/ha. During the same period, wheat yields in France, where good soils and high use of fertilizer favor high-yielding wheat varieties, wheat yields can attain 10,000 kg/ha (FAO 2000a). The increase in yields as a result of the adoption of high-yielding varieties was first evident in developed countries, such as the United States where agricultural production increased an average of 2.3% per year between 1955 and 1985. However, once less developed countries began to adopt green technology, the increase in yields per unit of land were even more dramatic.

The increase in productivity per agricultural worker is even more dramatic. In 1950 most of the basic grains in the world were produced predominantly by manual farming—that is, the kind of intensive agriculture that characterizes small and peasant farmers in which traditional varieties of plants are cultivated using few chemical fertilizers and pesticides and human or animal labor (see Chapter 5). Using this system a farmer could produce about 1,000 kilograms of grain per worker. In 1950 the most intensive systems of farming, using the best seeds available, mechanized cultivation and harvesting, and agrochemicals, produced about 30,000 kg of grain per worker for a ratio of about 1:30. By 2000 most of the basic grains in the world were still being produced by manual systems of farming, which still produced about 1,000 kg of grain per worker. However, in developed counties and some areas of less developed countries, some farmers who have adopted the highest levels of technology can produce 500,000 kg of grain per worker. That is a ratio of 1:500. Since 1950, then, there has been a 20-fold increase in the productivity per worker of the most intensive systems over the least intensive systems in the world. In large part, as we discussed in Chapter 6, this comes as a result of a huge increase in the use of fossil fuels. It should be clear that the disparity between productivity in more "modern" and less "modern" systems is increasing. Even with the cost of inputs and energy factored in, small farmers in less developed countries are less able to compete in a global grain marketing system every year (FAO 2000a, 2001a).

There is evidence that the ability of conventional breeding and fertilizer used to increase yields in basic grains has reached its limits. The rate of increase in yields in basic grains has declined dramatically. However, crops other than rice, wheat, and corn still show a good deal of potential for increase. Neglected during much of the green revolution, minor cereal grains (minor globally, but important in some regions such as Africa) such as sorghum and millet and root crops such as cassava are now receiving more attention. Average global sorghum yields are only 1,500 kg/ha and millet yields are only 800 kg/ha. Clearly there is room for increases in yields in these grains. Cassava production has grown steadily in recent years rising from just over 100 million tons worldwide in 1975 to a projected production of over 200 million tons in 2005. A number of countries have become net exporters of cassava. Other "neglected" crops such as sweet potatoes and pigeon peas are also being increasingly incorporated into breeding programs and hold a good deal of potential for increased yield and increased production. It appears that there is room for further conventional (non-genetically-modified) improvement in crop yields.

Finally, it is not clear what the full potential of biotechnology, specifically the development of genetically modified crops, will be on global yields of food crops. As we noted in Highlight 6, the development and dissemination of genetically modified (GM) crops has slowed as a result of worldwide consumer concern over the potential health hazards of GM foods. It does appear that GM crops can result in somewhat increased yields as a result of decreased attacks by pests. But more importantly, the adoption of Bt and Ht crops appears to allow for a reduction in the use of chemical pesticides. In many ways, GM crops could be especially effective in less developed countries where agrochemicals are less available and often too expensive for small farmers to use. With some GM crops, the only innovation a farmer needs to make is buying new seeds. However, the adoption of GM crops is likely to be as uneven as was the early adoption of green-revolution technology. Some analysts believe that there is great potential for improving the quality of foods through the use of GM technology. Research is now under way to develop varieties of staple crops with higher content of scarce micronutrients, such as calcium, iron, selenium, and vitamin A, and improved protein quality. Rice with high vitamin A content is already available, although its adoption has been slowed by concerns over the safety of GM crops (Jung 2000).

Declining Gains   In sum, the combination of a modest amount of extensification and a great deal of intensification allowed food production to keep ahead of population increase in the second half of the twentieth century. However, it appears that increases in food production are slowing. In 1999 world agricultural output was estimated to have increased by 2.3%, which was an improvement over the 1.4% growth rate in 1998. Crop production, in particular, expanded more strongly in 1999, but most of the increase was a result of increases in less developed countries. In 1998 crop production in developed countries had declined by 3.4%. In developing countries, agricultural production was also slow. It increased by 2.8% in 1999, which was about the same rate as in 1998, but lower than the 3.2% increase in 1997 and well below the high rates of between 4% and 5% that were recorded from 1993 to 1996. Preliminary estimates for agricultural production for 2000 suggest about a 1% increase (FAO 2001a). Figure 9.6 shows the decline in rate of growth for agricultural production globally.

Increases and declines in agricultural production are not evenly distributed around the globe. As a result of economic and agricul-

Figure 9.6
Decline in growth of agricultural production.

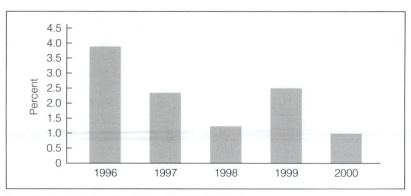

(*Source:* Food and Agriculture Organization of the United Nations.)

tural policy, production in North America and Europe has been stagnant or even declining in some years. However, increases in production continue to be in the range of 4–5% for Latin America and
2–3% for Asia. Food production in sub-Saharan Africa has failed to
keep up with population growth from 1995 to 2000. The decline in
food production in Africa is thought to be the result of a series of
factors, especially drought, war, and the creation of refugees who
have abandoned their farms. Sub-Saharan Africa still has the lowest
use of improved technology for crop production. It is thought that
the HIV/AIDS epidemic has contributed by reducing the number of
adults who can work to produce food. The Food and Agriculture
Organization (FAO 2000b) estimates that between 1985 and 2000
the most affected countries in Africa lost up to 25% of the agricultural labor force as a result of the AIDS epidemic. Finally, continuing desertification and declining soil fertility may also be
contributing to the failure of production to keep up with population
in Africa (FAO 2001a). As we will see below, sub-Saharan Africa is
one of the few regions of the world in which hunger, malnutrition,
and food insecurity are increasing. It is also one of the areas where
the potential for improvement is highest.

Livestock   Livestock production and meat consumption have
become the center of controversy in some discussions of the world
food crisis. On one hand, livestock are depicted as providing an
unnecessary, luxury foodstuff, animals are seen as direct competi-
tors with humans for scarce grain supplies, and because of their inef-
ficiency in converting grain protein to animal protein, they are
considered an important factor contributing to future food shortages. As a result, reduced consumption of animal products is recommended so that more grain can be made available to poor people
in developing countries. Data used to support this viewpoint can be
quite convincing: Three billion head of livestock are maintained to
supply the animal protein consumed in the United States alone.
Although farm animals traditionally lived on forage, wastes, and
surpluses (as when household pigs and chickens consumed spilled
grain and kitchen scraps), today livestock consume 135 million tons
of grain annually. This is 10 times more than the amount consumed
by the entire U.S. population (Pimentel et al. 1980). Further, livestock in general are inefficient converters of protein. Allaby (1977)
compared the efficiency of various livestock forms and shows "quite
clearly, that in terms of food energy and protein no livestock enterprise can compare in efficiency with vegetable crops . . . for every
kilogramme of egg protein the hen must consume more than 4 kg
of grain protein; for each kilogramme of pork protein the pig must
be fed 7.7 kg of grain protein; for every kilogramme of poultry meat
protein the hen must receive 5.9 kg of grain protein; and, if we plan
to produce beef in feedlots, fed on grain, then every kilogramme of
beef protein represents an expenditure of 22 kg of grain protein."
    The effect of animal consumption on grain supplies is obvious
when we compare the amount of grain consumed in the developed
nations that use large quantities of meat with the amount consumed

in developing countries subsisting predominantly on grain. In India and Thailand, less than 200 pounds of grain are used by the average person each year. After setting aside 10% for the next year's seeds, the Indian family has less than 360 pounds, or under 1 pound per day, for actual consumption. Nearly all of this is eaten directly as grain. At the other extreme, Canadians use almost a ton of grain per capita, with most of it used to feed livestock that they consume in the form of meat, milk, and eggs.

On the other side of the controversy are experts who argue that efficient use of grazing animals does not pose a threat for future food supplies. Indeed, livestock production may even increase the yield of high-grade protein and by-products extracted from renewable sources, and one can think of cattle as living storage sites for surplus grain. Livestock, when used in these ways, can actually contribute to the resolution of malnutrition. Central to this view is the unique stomach of ruminants—cattle, sheep, goats, buffalo, deer, antelope, and other meat-producing animals. Ruminants' stomachs allow them to digest the cellulose in grasses and other plant material that humans cannot utilize as food. Horses, rabbits, and many other animals also digest cellulose, though less efficiently than ruminants. These animals convert vegetable matter into animal products, enabling humans to indirectly consume food that otherwise would be unavailable. Thus, when fed grasses and other forage materials, animals become a complementary food source rather than a competitive element.

Because many countries have large quantities of unused forage and large areas of margin land suited only for grazing, important food gains could be realized by switching livestock from grain to forage feeds. In the United States corn belt, cattlemen are mixing corncobs, cornstalks, straw, and even sawdust with molasses (to make it more palatable) and urea (to increase nitrogen content). Gains can also be made by switching pigs and other nonruminants from grain to feeds that do not compete with human needs. In China, pigs are reared in large numbers on vegetable refuse, ground and fermented rice hulls, corn husks, sweet potatoes, soybean vines, water hyacinths, and other wastes.

It is also worth noting that animals provide a full range of nonfood products and services needed in the developing world. Transportation, power to pull machinery, leather, wool, and manure for fertilizer and cooking fuel make animals a valuable part of the food production system. Finally, when markets are open (not always the case as we will see below), livestock exports to developed countries generate funds that can be used to offset the costs of grain and other goods imported by developing nations.

Studies of the nutritional impacts of different diets routinely show that the percentage of calories that come from animal products is one of the best predictors of adequate child growth in poor populations (Allen et al. 1992). It is thought that the growth effect is the result of improved protein quality along with the increased availability of some key micronutrients, such as iron and zinc, in animal products. There are some populations, then, in which increased consumption of animal products is desirable.

While people in developed countries and the affluent in some less developed countries may consume too many animal products for optimal health, the majority of people in the world probably consume too few. Humans appear to be opportunistic omnivores, best suited to a varied diet, which includes some foods of animal origin, and with a propensity to increase the consumption of animal products when that is possible. Economists believe that, as global incomes rise over the next several decades, the worldwide demand for meat and dairy products will also rise. It is not clear that this demand can be met, given the high cost of raising animals that consume grains.

Fisheries In addition to foods produced agriculturally, the global food supply includes fish and seafood either captured wild from the seas and inland waters or farmed through the process of aquaculture (see Focus 9.3). Despite fluctuations in supply and demand, caused by the changing state of fisheries resources, the economic climate, and environmental conditions, fisheries and **aquaculture** are very important as a source of food, employment, and revenue in many countries and communities. Capture fisheries and aquaculture produced about 125 million tons of fish and seafood in 1999. The amount of the fish and shellfish production from capture and aquaculture increased steadily throughout the second half of the twentieth century. For the two decades following 1950, world marine and inland capture fisheries production increased on average by as much as 6% per year, going from about 18 million tons in 1950 to 56 million tons in 1969. However, during the 1970s and

aquaculture
The "farming" of fin- and shellfish in ponds or corrals. Ponds or corrals are "seeded" with immature fish or shellfish, fed formulated feed mixes, and "harvested" when they have attained market size.

FOCUS 9.3 Aquaculture

One area of promise for increased food production that has received a good deal of attention in the last two decades is aquaculture. Aquaculture refers to the "cultivation" of fish and shellfish in a controlled environment. It can be carried out with freshwater species such as catfish, trout, tilapia, and crawfish, predominantly in inland ponds, or with saltwater species such as salmon, shrimp, mussels, and oysters in enclosures constructed in coastal wetland areas. Aquaculture is an ancient practice that has been used for centuries by the Chinese, Japanese, and Egyptians, but it is now spreading rapidly around the world as the demand for fish and shellfish is increasing. In the past 15 years, global production of farmed fish and shellfish has doubled and now accounts for more than a quarter of all fish consumed directly by humans (Naylor et al. 2000).

The cultivation of oysters, mussels, and especially shrimp has become a big business for aquaculturists. Unfortunately, many problems plague ocean ranchers. First and foremost is the lack of basic knowledge about marine science, fish diseases, and construction of safe aquatic environments (Clay 1997). Second, just as agriculture badly practiced can do much harm to the environment, coastal aquaculture has too often resulted in wholesale destruction of ecologically sensitive mangrove swamps, as a result of the pollution of lagoons and estuaries, and with unsustainable use of land and water resources. Third, although coastal areas often appear to be uninhabited and little utilized, they have been important sources of food and fuel for resource-poor people. When large corporations build ponds and other facilities to produce luxury items like shrimp, they often displace local populations, further driving them into poverty (B. R. DeWalt, Vergne, and Hardin 1996). Fourth, aquaculture depends on artificial feeding of the species being cultivated.

(continued)

Focus 9.3   (continued from previous page)

Many aquaculture systems use two to five times more fish protein in fish meal than is supplied by the marketable product (Naylor et al. 2000); paradoxically, then, aquaculture is a contributing factor in the collapse of some worldwide wild fisheries stocks. Freshwater projects are less problematic, but they are still limited to a few species, such as carp and rainbow trout.

Because most of the world's fisheries are already fished at, or beyond, their sustainable yield (National Research Council 1992), meeting demand will undoubtedly result in an increasing expansion of aquaculture. We are already seeing the same kinds of advances and setbacks in aquaculture that have occurred with raising livestock. There are huge conflicts over access to the best places for aquaculture, often pitting resource-poor people against large corporations (e.g., Bailey 1988; B. R. DeWalt et al. 1996; B. R. DeWalt et al. 2001). Poorly designed and implemented projects have resulted in land and sea areas being abandoned after a few years of production (Clay 1997), often accompanied by the destruction of ecologically sensitive environments (Tobey, Clay, and Vergne 1998). Crowding millions of animals into enclosures has resulted in the spread of devastating diseases (Roque 2000) and the overuse of antibiotics and other chemicals in the attempt to arrest production losses (B. R. DeWalt et al. 2001). Those who practice science and implement public policy are scrambling to deal with these and a host of other issues. It is still too soon to tell whether aquaculture will be a "blue revolution" (B. R. DeWalt 1998) that helps in addressing the world's food problems, or a short-term boom that enriches a few while leaving behind ecological and social destruction.

80s, the average rate of increase declined to 2% per year. By the 1990s it was almost zero.

It appears that capture fisheries have reached their maximum potential. Many fisheries experts would argue that we have already gone well beyond the sustainable yield from fishing. It is very unlikely that substantial increases in total catch will be obtained. In contrast, aquaculture production has been growing steadily. Starting from an insignificant amount of total production, primarily from traditional fishponds in Asia, inland and marine aquaculture production grew by about 5% per year between 1950 and 1969 and by about 8% per year during the 1970s and 80s, and it has increased further to 10% per year since 1990. Counting both capture fisheries and aquaculture, the production of fish employs about 36 million people globally.

Consequences of the Agricultural Revolution

Another problem even more widely publicized than declining growth in production concerns the ecological pressures created by modern technology. Evidence of our destructive impact on the environment comes from many parts of our habitat: soil erosion, advancing deserts, air pollution and acid rain, deforestation and severe shortages of firewood, the accumulation of chemical pesticides, and pollution of ponds, lakes, and oceans. The economic and ecological costs of cultivating new lands, as well as the environmental damage associated with some of the newest agricultural practices, have made some scientists question whether we can sustain current production levels. Others fear for our very existence.

Increases in the yield of food crops have a number of important costs that may limit the ability of the globe to feed its inhabitants in the future. Over the past 50 years, cultivated ecosystems have become simpler and more uniform (monocropped) and much less biodiverse. Simplification and specialization result in higher yields, but also in the impoverishment of ecosystems. There are more thistles, wallflowers, poppies, and cornflowers—and fewer insects, birds, and rodents (FAO 2000a). Research by the International Food Policy Research Institute and the World Resources Institute (Wood et al. 2001) suggests that agriculture is taking its toll on the land. Remote sensing suggests that about 40% of the world's agricultural land is seriously degraded, and there is little reason to believe that this situation will improve.

As we noted earlier, without improved water management the global availability of water for irrigation, household, and industrial use is likely to become the key limiting environmental factor of the twenty-first century. There are ways to manage water use in irrigation more effectively and with fewer environmental impacts, but these require expensive investments in infrastructure and technology. More importantly, they require the development of human capital, that is, training and education for large and small farmers globally. As we will see in Chapter 10, the development of human capital—that is, the transfer of knowledge to those that need it—is one of the continuing challenges in the global economy.

Agriculture is also one of the major producers of greenhouse gases. The cutting and burning of forests and the decomposition of crop residues contribute to the production of carbon dioxide, releasing it into the atmosphere. Livestock and some crop residues release significant amounts of methane into the atmosphere. Increasing agricultural production, without control of carbon dioxide and methane, is likely to contribute to the process of global warming.

Decreased Food Aid to Other Countries

Accompanying this increased food production capacity are the revolutions in storage and transportation. Following World War II, there was an increase in the ability to store grain surpluses and move them around the world to areas experiencing food shortages. Improved storage and transportation made it possible to mobilize emergency relief to most parts of the world. Since WWII, food aid has been an important contributor to food availability for regions undergoing food shortages and in the midst of natural disasters. Food aid accounted for a large percentage of food available in some countries. However, in the 1990s, the delivery of food aid to regions in crisis declined dramatically. In 1996 only 7.5 million tons of food aid were delivered. This was less than half of the 16.8 million tons that were delivered in 1992 (Pinstrup-Andersen, Pandya-Lorch, and Rosegrant 1997). Between 1992 and 2000, all regions of the world experienced a reduction in food aid deliveries.

Food aid has historically been driven by surplus production of grains in North America and Western Europe. Both the United States and Western Europe have implemented policies to take land out of production and reduce the production of grains. The result is much lower stockpiles of food. The United States and Western Europe have been channeling the reduced amount of food given as aid through targeted disaster relief efforts. Pinstrup-Andersen et al. (1997) argue that the decline in food aid has put some countries at high risk. Many of these are in sub-Saharan Africa where dislocations as a result of war and drought place a large number of people at risk for starvation. Pinstrup-Andersen et al. also note that at the same time surplus food for food aid was disappearing, the amount of official development assistance has also been declining. They argue that agricultural development and the potential for increasing food production in poor countries will be severely hampered if the decline in food aid and development assistance continues.

Trade Imbalances

In recent years, a number of less developed countries have liberalized policies with respect to foreign trade in food and agricultural products. The liberalization of trade and a commitment to policies favoring the development of export agriculture have been strongly influenced by the structural adjustment policies imposed on less developed countries in order to get help from the International Monetary Fund (IMF) for the restructuring of foreign debt. Less developed countries have been encouraged to produce high-value crops for export and to import less costly food staples such as grains. Unfortunately, developed countries have not opened their markets fully to the commodities that less developed countries have to export. The industrialized countries have been especially reluctant to open up their markets for the import of such commodities as beef, sugar, groundnuts, and dairy products. In addition, many developed countries, such as the United States, so heavily subsidize agricultural production that the price of local products is kept artificially low and other countries can't compete. In theory, open markets in which less developed countries can sell tropical commodities at a high price and import staple grains at a low cost should contribute to the availability of sufficient food in those countries (see Figure 9.7). The asymmetry in trade as a result of closed markets in developed countries makes it very difficult for less developed countries to benefit from the liberalization of their own markets (Pinstrup-Andersen et al. 1997).

What Is Meant by Hunger and Malnutrition?

There is little doubt that malnutrition is the major cause of infant and child mortality in developing nations. One study found that nutritional deficiency was a factor in more than 60% of all childhood deaths from infection and, in the first and second years of life, a major contributor to deaths from all causes (Willett 1983). Pelletier (1994) has demonstrated that even moderate malnutrition contributes to child mortality.

Malnutrition has other long-lasting effects that are not revealed in infant and child mortality rates. First, it reduces the body's abil-

Figure 9.7
The global market allocates food-producing resources on the basis of profit and ability to pay rather than on the need of the poor. Where will the coffee harvested by this Colombian man go, and who will profit from its production and sale?

© Carl Frank/Photo Researchers

ity to resist infection and disease. Protein and energy malnutrition suppresses certain immune responses, leaving the child vulnerable to many childhood illnesses, pneumonia, diphtheria, and other diseases. A deficiency of certain vitamins, such as A and the B-complex, also affects immunity; ascorbic acid (vitamin C) deficiency impairs resistance to bacterial infections; and when malnutrition causes the skin and mucous membranes to bruise and break open, the body is more easily invaded by organisms (Caliendo 1979). Once a malnourished person succumbs to illness, his or her nutritional status is further jeopardized by loss of appetite, diarrhea, and vomiting. Sweat, fever, and intestinal parasites increase nutrient needs and aggravate nutritional deficiencies. This is visible in the comparison of death rates in developed and developing nations.

Second, severe malnutrition is believed to impair brain growth and reduce learning abilities. In an excellent review of existing studies on animals and humans, Caliendo (1979) concludes that more research is needed to sort out the complex relationship between the effects of malnutrition and poor housing, poor sanitation, poor health, low levels of education, ignorance, and a climate of apathy and despair. Clearly malnutrition's strong association with mental development and learning is cause enough for concern.

Third, malnutrition affects productivity and decreases the span of working years. Allan Berg (1973) argues that poorly nourished people fail to achieve their genetic potential in physical size and prowess. In addition, their capacity for work is reduced, as is their life span. Increased medical costs, lost workdays, and reductions in work capacity affect economic growth and serve as obstacles to development in poor countries. The elimination of malnutrition, then, makes good economic sense for developing nations.

Who Are the Hungry and Malnourished?

We have examined several of the factors that enter into the equation of feeding the world population; however, we have not addressed the question of the current state of food security and hunger in the world. As we discuss hunger, nutrition, and food security, we should be clear about some of the terms we use.

Hunger and Undernourishment

Hunger can be defined as "the physiological as well as psychological discomfort caused by the lack of food" (Caliendo 1979:6–7). Put in more descriptive terms:

> *No fear can stand up to hunger, no patience can wear it out, disgust simply does not exist where hunger is; and as to superstition, beliefs and what you may call principles, they are no more than chaff in a breeze. (Joseph Conrad,* Heart of Darkness*)*

The FAO estimates the number of people in the world who are *undernourished,* meaning they don't have access to enough calories or other nutrients to meet their needs (FAO 2001b). In general, this is a calculation made using the total number of calories available in a country as a result of food production, trade (imports), and food aid with an adjustment for inequality in incomes. It is expected that poor people will be able to get fewer calories than the affluent. It is possible for a country to have enough food available at the national level, but still have a number of people who are undernourished because they are poor.

The FAO (2001b) estimates that in 1996–98 there were 817 million people in the world who were undernourished. This includes 779 million in less developed countries, 27 million in countries in transition from the former Soviet bloc, and 11 million in industrialized countries. This represents a decrease of about 39 million people since 1990–92. Although there is a continuing decline in undernourishment, the rate of decline has been slowing since the mid-1980s and is stagnant in some regions. The numbers of undernourished people are not evenly distributed. Twenty-seven percent of the population of sub-Saharan Africa and 23% of the population of South Asia (India, Pakistan, and Bangladesh) are estimated to have been undernourished in 1998. However, only about 10% of the populations of East Asia and Latin America are estimated to have been undernourished in 1998.

Undernourishment estimates the shortfall of energy (calories) available to people. It is not a direct measure of the effects of undernourishment on the nutritional status of people. The term *malnutrition* describes "bad nutrition, or more scientifically, 'a condition in which there is an impairment of health, growth, or physiologic functioning resulting from the failure of a person to obtain all the essential nutrients in proper quantity or balance'" (Caliendo 1979). Malnutrition comes in many forms: vitamin and mineral deficiencies, caloric deficits, and protein or energy malnutrition. In its initial stages, a deficiency may be so mild that physical signs are absent and biochemical methods generally cannot detect the slight changes. As tissue depletion continues, the biochemical changes can be measured in body fluids and tissues. With further depletion the physical signs become apparent, until finally the full-blown deficiency can be easily recognized.

Types of Undernutrition

In general, we are concerned with two types of nutrients when we talk of malnutrition. A shortfall in the intake of energy (calories) and protein is referred to as **protein-energy malnutrition (PEM).** We group energy and protein together because it has been recognized since the 1970s that shortfalls in energy and protein go together. Individuals who don't have enough energy usually don't get enough protein and vice versa. The second form of malnutrition is **micronutrient malnutrition,** sometimes called "hidden hunger" (see Highlight 9). This is caused by a shortfall in the amount of vitamins and minerals that a person consumes. The most easily

protein-energy malnutrition (PEM)
Inadequate intake of both energy (calories) and protein.

micronutrient malnutrition
Inadequate intake of micronutrients (vitamins and minerals).

recognized forms of PEM are kwashiorkor and marasmus, both of which are found predominantly in children under 6 years of age.

Initially recognized by Dr. Cicely Williams in West Africa in the early twentieth century, **kwashiorkor** is the indigenous name used to describe the disease that occurs when another baby displaces a child from the breast. And indeed, the condition clinically recognized as kwashiorkor most frequently develops several months after switching a child from breast milk to a starchy diet providing adequate caloric intakes but insufficient protein. In its more extreme forms, kwashiorkor includes moderate to severe growth failure and muscles that are poorly developed and lack tone. Edema is usually severe, resulting in a large potbelly and swollen legs and face and masking the muscle loss that has occurred. The child has profound apathy and general misery, he or she whimpers, but does not generally cry or scream because to do so would take too much energy.

Marasmus results from a diet lacking in calories, while intakes of protein and other nutrients may be well balanced. Severe growth failure and emaciation are the most striking characteristics of the marasmic infant. The wasting of the muscles and lack of subcutaneous fat are extreme. Marasmus differs from kwashiorkor in several ways. The onset is earlier, usually in the first year of life; growth

kwashiorkor
A nutritional disease characterized by wasting that is masked by edema, depigmentation of hair and skin, and death. It is generally thought to be the result of inadequate protein intake when energy (calorie) intake is adequate. It tends to affect children in the second year of life.

marasmus
A nutritional disease characterized by extreme wasting. It is generally thought to be the result of inadequate intake of both energy and protein. It tends to affect children in the first year of life.

Figure 9.8
left Children suffering from a severe form of protein deficiency called kwashiokor experience swelling in the arms, legs, and stomach area. The swelling hides the devastating wasting that is taking place within their bodies. This child has the characteristic "moon face" (edema), swollen belly, and patchy dermatitis (from zinc deficiency) often seen with kwashiorkor. *right* Individuals with marasmus look starved. People with this condition lack both protein and calories. This child is suffering from the extreme emaciation of marasmus.

© Bob Daemmrich/The Image Works

AP/Wide World Photos

failure is more extreme; there is no edema; and the liver is not infiltrated with fat. The period of recovery is much longer. In both marasmus and kwashiorkor, height and weight-for-height are markedly retarded.

In addition to the extreme types of PEM, there are many combinations of intermediate forms. In fact, for every child with obvious kwashiorkor or marasmus, there are 99 children with more subtle or milder cases that go unnoticed unless the child's growth is measured against a standard reference population of well-nourished children of the same age and sex.

anthropometric data
Measurements of the body such as height, weight, arm circumference, and fat folds.

Measuring PEM

PEM is usually measured by collecting **anthropometric data,** that is, weighing and measuring people. Most commonly, the peo-

Table 9.2   Anthropometric Indicators Commonly Used in Nutritional Surveillance Systems

Anthropometric Indicator	What It Measures	Contexts in Which It Is Used
Children		
Underweight	Underweight (low weight for age) represents both inadequate linear growth and poor body proportions caused by undernutrition.	Underweight is the most common indicator collected in growth monitoring systems.
Stunting	Stunting or "shortness" (low height for age) measures long-term growth faltering as a result of chronic undernutrition.	Stunting is associated with poverty and may be assessed in stable situations to measure changes over time.
Wasting	Wasting or "thinness" (low weight for height) as a result of acute undernutrition.	Wasting is the indicator most commonly assessed in nutrition surveys in natural disasters.
Adults		
Body mass index	"Thinness" (low weight for height) as a result of undernutrition.	BMI is the indicator used to assess adult nutritional status. It is of particular importance where adults may be equally or more vulnerable to undernutrition than children, for example, in natural disasters.
Low birth weight[1]	Babies are measured, but this indicator is associated with poor nutrition in mothers.	Low birth weight is a useful indicator in stable situations, where it can be used to measure changes in maternal malnutrition over time. It is particularly important in Asian countries where maternal undernutrition is common.
Elderly		
Body mass index	"Thinness" (low weight for height) as a result of undernutrition.	Although there are problems with using the BMI to assess undernutrition in the elderly, it is very useful in natural disasters.

Source: From FAO, *The State of Food Insecurity in the World* (Rome: Food and Agriculture Organization, 2001).

[1] Unlike the other indicators, birth weight is measured only once.

ple measured are children. Because children are growing, their nutritional needs are higher than those of adults. Inadequate intake of nutrients, especially protein and energy, has a measurable impact on both the height and weight of children. However, adults are also measured to assess nutritional status. Table 9.2 summarizes the measures most commonly used to assess overall nutritional status. These include underweight, stunting, wasting, and low birth weight for children, and low body mass index (BMI) for adults. Underweight, stunting, and wasting are determined by comparing the height, weight, and weight-for-height of children to average values of a reference population. The reference population is a well-nourished population, primarily the U.S. population. The mean and standard deviation of the population have been calculated for weight, height, and weight-for-height for boys and girls of each age group. In order to assess the adequacy of growth of a particular child, that child's measurements are plotted on a graph, and the extent to which the child falls above or below the mean for his or her age and sex is expressed in standard deviations. A child is considered underweight, stunted, or wasted if his or her measurements fall 2 standard deviations or more below the mean of the reference population. This means that particular child is below the 3rd percentile for the measure. This is a very conservative measure of growth failure.

Table 9.3 gives estimates of the prevalence and numbers of children under 5 years of age who show signs of PEM, as measured by **stunting,** in several regions of the developing world for the years 1980–2000, with projections to 2005. In 1980 approximately 48% of the children under 5 years old in developing countries had a height that was below the 3rd percentile of the reference population. This

stunting (chronic malnutrition)
Shortness (low height for age) as a result of long-term inadequate intake of energy and/or protein. Usually measured by calculating height-for-age.

Table 9.3 Prevalence of Underweight, Stunting, and Wasting by Region, 1999

Region/ Country Group	Percentage of Population That Is:		
	Underweight[1] (weight-for-age)	Wasted[1] (weight-for-height)	Stunted[1] (height-for-age)
Sub-Saharan Africa	31	10	37
Near East and North Africa	17	8	24
South Asia	49	17	48
East Asia and the Pacific	19	6	24
Latin America and the Caribbean	9	2	17
Developing countries	29	10	33
Least-developed countries	40	12	45

Data: ACC/SCN, *Fourth Report on the World Nutrition Situation: Nutrition Throughout the Life Cycle* (Geneva: United Nations, 2000).
[1]Defined as <2 standard deviations (SD) or more below the mean of the reference value.

percentage declined to 32.5% in 2000 and is projected to decline to 29% by 2005. The percentage and the number of stunted children have declined and are projected to decline everywhere but in parts of Africa. The problem is especially bad in eastern Africa where surveys in Ethiopia in 1992 show the rate of stunting to be 64%, and surveys in Malawi and Zambia show rates of 48% and 42% respectively (ACC/SCN 2000). Because of high population growth in these areas, the absolute numbers of children that are stunted are also growing. Western Africa has a lower prevalence of stunting than eastern Africa, but the prevalence has stagnated with little decline in recent years. In South Central Asia, stunting is also widespread, but the percentages have been dropping. In Southeast Asia, about one-third of preschool children are stunted, but this region has had the most dramatic drops in prevalence.

wasting (acute malnutrition)
Thinness (low weight for height) as a result of short-term inadequate intake of energy and/or protein. Usually measured by calculating weight-for-height.

The percentage of children who are underweight, or **wasted,** is lower than the percentage who are stunted. This number has also been declining worldwide, except in western and eastern Africa, where the prevalence of underweight has been increasing by about a half of a percent per year. Of the three measures of child PEM, wasting is the most worrisome. Wasted children are very thin and may also be short (stunted). Whereas stunting and underweight are measures that record chronic undernutrition problems, wasting is a measure of *acute* nutritional deficiency. The wasting rate changes quickly, either increasing or decreasing as short-term food supplies decline or increase. Wasting is not as common as either underweight or stunting. The global prevalence is about 10%. However, wasted children are at high risk of mortality and morbidity.

low birth weight
Generally, birth weight below 5 lb, or 2.5 kg.

Low birth weight is also a measure of nutritional well-being. Low birth weight measures the nutritional status of the fetus, but just as importantly, the nutritional status of the mother. The highest levels of low birth weight in the world are in South Asia (ACC/SCN 2000). Low birth weight in South Asia has been linked directly to the low status of women and girls in this region, which results in their lower nutritional status and impedes their access to health care and education.

In sum, the overall view of nutritional status and malnutrition as measured through the growth of children parallels the view derived from the examination of available food supplies. In general the global nutrition situation has improved dramatically since 1960, but improvement has slowed in the past decade and a half in most regions of the world. However, in sub-Saharan Africa the situation is critical and appears to be deteriorating. Why is the situation in this region of Africa so dire? The causes of malnutrition are complex. However, a study of the situation in sub-Saharan Africa suggests that a number of factors have contributed to the deterioration of food security and nutrition (ACC/SCN 2000). Countries with deteriorating nutrition have had much lower overall economic growth, as measured by a gross domestic product lower than that of countries without deteriorating nutrition. They are poorer and have stagnant economies. They have had declining food production in part as a result of drought and war. Women's social status as measured by women's life expectancy as compared with men's has declined. Women's education levels are lower, and total fertility rate is higher, than in countries without deteriorating nutrition. Low economic growth for these countries is attributed to high debt burdens, high population growth, and the HIV/AIDS epidemic.

Overweight in Children

At the same time that some children are underweight and stunted, others suffer from the opposite form of malnutrition—overnutrition. Overweight in children is increasing around the world. Several studies report increasing levels of obesity in developed countries. In the United States, the rates of overweight and obese children doubled between 1980 and 2000, while the rates for teens tripled during the same time frame. In 2000, 13% of children and 14% of teens were overweight or obese. The percentage of overweight children is also increasing in less developed countries. Just as we use a cutoff point of 2 standard deviations below the mean as a measure of underweight, we use a cutoff of 2 standard deviations *above* the mean as a measure of overweight. Using this criterion, an estimated 17.6 million children were overweight in the developing world in 1995. Northern and southern Africa, East Asia, Central America, and South America had prevalences higher than expected. Not surprisingly, western and eastern Africa did not. In northern Africa and Central America, overweight exists side by side with high prevalences of stunting. Countries in these regions are undergoing a nutritional transition in which some segments of the population are consuming diets high in saturated fats, sugar, and refined foods, while others are undernourished (ACC/SCN 2000).

What Are the Causes of Undernourishment and Malnutrition?

If, as we have noted above, there is enough food produced to meet the nutritional needs of all of the people currently alive, why do we still have hungry and malnourished people in some parts of the world? As Amartya Sen (1981) has pointed out, the problem is not food shortage but, rather, a shortage of entitlement and poor policy. That is, poverty and a lack of resources to secure food, and national and international policy that limits access to available food, are the principal barriers to adequate nutrition. Trade policies, development policies, food aid, and even patented rights to technology are alongside education, sanitation, and access to health care and employment as key factors in the ability of households and individuals to demand food. In Chapter 10, we will review these factors and discuss current approaches to improving food security and nutrition for people at risk.

Projections for the Future— Enough for All?

Projections for the future range from the most dire predictions of widespread famine by the end of the decade to rather optimistic projections of increasing food production and declining population. McCalla and Revoredo (2001) have reviewed over 30 projections and forecasts regarding the world's ability to feed itself made between 1950 and 2000. They find that the most pessimistic forecasts made in the 1960s and 70s (e.g., Brown 1967; Ehrlich 1970; Paddock and Paddock 1964, 1967, 1976) were clustered around what turned out to be temporary crises in world markets. In the 1980s and 90s, pessimistic reviews (e.g., Brown 1995; Ehrlich and Ehrlich 1990; Meadows, Meadows, and Randers 1992; Postel 1992) focused more on the constraints posed by the impact of population and agricultural production on natural resources and environmental degradation. More optimistic predictions were made as the green revolution increased agricultural productivity. At the end of the twentieth century, grain prices were at a 100-year low, food stockpiles were increasing, and some scholars were writing about being able to feed the 10 billion people expected by the end of the twenty-first century (Evans 1998). It was clear by the year 2000 that the massive famines that had been predicted in the 1960s and 70s had not taken place. Amartya Sen received the Nobel Prize in economics, in part, for demonstrating that most food crises and famines have been the result of failures of policy and entitlements (resources necessary for households and individuals to secure enough food), not shortfalls in the food supply.

So, why haven't the most dire predictions come true? In part, starting with Malthus, analysts have overestimated the trajectory of population growth. In the 1960s and 70s, when the rate of population increase was at its historical high, it certainly looked like population increase would never stop. In addition, analysts have also underestimated the potential for growth in agricultural production. McCalla and Revoredo (2001) speculate that, in fact, dire predictions had the effect intended by their authors in that they shocked both the public and policy makers and were incentives for policy and economic change. Their effect may have been to head off the very crises they had predicted. Pessimistic models also rarely take into consideration the process of adjustment that takes place as a good, such as energy, water, or land, gets scarce. Most of us do not change behavior toward conservation until something gets too scarce or expensive.

On the other hand, optimistic predictions have not yet been proven wrong. By the same token, they have not been tested. So far, predictions that science, technology, and policy will continue to keep ahead of population growth and environmental degradation appear to be holding up, but one can never tell if next year will be the year in which the limits of technology and policy are reached.

Now, what are the best current estimates and projections regarding food security in the future? Before the end of the twenty-first century, the food supply will probably have to support over 10 billion people, perhaps as many as 12 billion. One of the conclusions of McCalla and Revoredo is that long-term projections are very likely to be wrong. It probably doesn't make much sense to look out toward 2100. Short-term projections of 5 to 10 years are more likely to be accurate. The International Food Policy Research Institute projects food security and nutrition to 2020 (Pinstrup-Andersen et al. 1997). It argues that slowing population growth and continued growth of agricultural production need to be transferred to less developed countries in order for continuing improvement in global food security and nutrition through 2020. The institute's projection continues the trend of slow improvement that occurred through the 1990s. Pinstrup-Andersen et al. even see some very recent improvement in Africa, where, as noted above, the situation has been deteriorating over several decades.

So, what is the future? It probably falls somewhere between the worst and best case predictions. The issues that will have to be addressed are well outlined:

- Continuing efforts to slow, and even reverse, population growth, and that means continuing to address the problems of infant and child mortality and meeting the existing needs and demand for family planning services and educating women
- Continuing development of improved food production technologies that increase yields and are environmentally sustainable
- Developing technologies and practices that conserve natural resources such as energy, water, land, and forests
- Reducing, or eliminating, poverty

In the next chapter, we will address the specific issues in addressing undernourishment, malnutrition, and food security for the populations at risk.

Summary

- Currently there is enough food produced to feed all of the world's population if it could be equitably distributed.
- Population growth slowed in the second half of the twentieth century and appears to be continuing to slow. In some regions of the world, increased mortality, as a result of infectious disease, especially HIV/AIDS, is also slowing population growth. However, the world is expected to have somewhat more than 10 billion people by the end of the twenty-first century.
- Food production has more than kept up with population growth over the past 100 years. However, there are indications that the growth in food production is slowing and that environmental degradation as a result of agriculture and animal husbandry may further slow growth in food production.

- Projections about the future rest on analyses of the balance between population growth and food production. In the short term, it appears that food production can keep up with population growth. Proponents of the optimistic view of the future argue that slowing population growth and continued technological advances will maintain enough food on a global scale for the foreseeable future.
- The pessimistic view suggests that although there is sufficient food now, the results of population growth, and the environmental degradation which might result from increased food production, will result in famine at some point not too far in the future.
- Despite adequate food production on a global scale, many people experience undernourishment and malnutrition.
- Risk of malnutrition can be measured by estimating the numbers of people who do get an adequate amount of energy and by weighing and measuring people, especially children.
- Malnutrition and undernourishment appear to be the result of the inability to gain access to food that is available.

HIGHLIGHT

Hidden Hunger: Micronutrient Malnutrition

It is evening and a pregnant Nepali woman sits quietly in a corner of the house. When she moves about the hut, her children lead her by the hand. For the past several months, she has been unable to see well after the sun sets. She is experiencing night blindness, one of the early symptoms of vitamin A deficiency. Her diet has always been rather low in vitamin A, but her pregnancy and the vitamin A needs of her unborn child have pushed her over the edge to clinical deficiency. Her night blindness limits her ability to work after dusk and makes caring for her children difficult. If she sits very close to the fire, she can see enough to accomplish a few tasks. She may also experience accidents as a result of poor vision (Christian et al. 1998a, 1998b). Vitamin A deficiency is just one of several micronutrient deficiencies of worldwide significance; night blindness is its mildest form.

The discovery of vitamins and their role in the human body at the end of the nineteenth century resulted in a great deal of interest in vitamin deficiencies. However, in the middle of the twentieth century, attention was turned toward problems with the overall food supply and protein-energy malnutrition (PEM) was seen as the most important area of under-nutrition. Less visible than the growth failure and signs of starvation were deficiencies of vitamins and minerals. During this time, they were seen as secondary problems. Because their symptoms can be subtle, not readily identifiable as a deficiency, in the early phases of deficiency they are sometimes considered "hidden malnutrition." In the last two decades, however, the interest in micronutrient deficiencies has emerged into the center stage of food and nutrition research and action. While there can be problems with any of the 50 or so nutrients essential for humans, three micronutrients have been identified as the ones most likely to be deficient: the minerals iron and iodine, and vitamin A.

Iron Deficiency

Iron is essential in the production of *hemoglobin*, the blood protein that carries oxygen. A deficiency in iron will result in anemia that is a reduced ability of the blood to carry oxygen to the cells of the body. The body's iron balance is affected by iron stores in the body, iron intake and absorption, and iron loss. Most of the iron in the human body is circulating as hemoglobin; however, humans do store some iron, and these stores can be accessed in time of need. Iron intake and absorption are important aspects of iron balance. Iron is widespread in the environment and in a number of foods, including leafy vegetables, grains, and animal products, especially meat. But the iron in many foods is not well absorbed by humans.

The most easily absorbed iron is *heme iron*—iron that is already incorporated into hemoglobin in meats. A number of vegetable sources of iron also contain chemical compounds that bind with iron in the digestive system and make it less available for absorption.

Iron is lost in the body principally as a result of blood loss. Women of childbearing age are at special risk for iron deficiency. Pregnant women must also share their own iron stores with their fetuses. Because of increased need and special loss of iron, women and children are at highest risk for deficiency. Iron-deficiency anemia is probably the most widespread deficiency problem in the world. It is estimated that in less developed countries, 56% of pregnant women, 56% of schoolchildren, and 44% of nonpregnant women are affected by iron-deficiency anemia. Men are also affected with about one-third of adult men estimated to be anemic in less developed countries. It is also a problem for older adults; about 51% of older adults in less developed countries are thought to be anemic (ACC/SCN 2000). Iron-deficiency anemia is also a problem, albeit a smaller one, in industrialized countries, with about 18% of pregnant women and schoolchildren showing deficiency.

Iron deficiency is truly hidden in the sense that without blood tests it is difficult to tell that someone is anemic. However, the impact on lives can be dramatic. Iron deficiency limits the ability of people to do work. More importantly, deficiency in children impairs cognitive development. It is estimated that the global economic impact of lower ability to do work and cognitive deficit is the loss of $4 per person per year. This does not take into account maternal deaths related to anemia.

Figure H9.1
In iodine deficiency, the thyroid gland enlarges—a condition known as simple goiter.

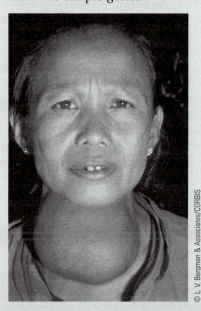

© L. V. Bergman & Associates/CORBIS

Iodine Deficiency

A second important mineral deficiency is iodine deficiency. Iodine is present in the body in infinitesimally small amounts, but its role in human nutrition is key. Iodine is part of the thyroid hormone *thyroxine,* which regulates body temperature, metabolic rate, reproduction, growth, blood cell production, nerve and muscle function, and other bodily functions. Thyroxine enters every cell of the body to control the rate at which cells use oxygen. In other words, thyroxine controls the rate at which energy is released.

The amount of iodine in the body generally reflects the amount of this mineral in the soil in which plants are grown and animals graze. Land that has been under the ocean has soils rich in iodine. Iodine-deficiency diseases (IDDs) are common in areas in which the soils are low in iodine. In areas of high iodine content in soils or where the consumption of seafood is common, IDD is less of a problem. The principal IDDs are goiter, cretinism, and poor growth.

Goiter is an enlargement of the thyroid gland, visible as a swelling in the front of the neck (see Figure H9.1). It occurs when, due to lack of iodine in the diet, the thyroid gland is unable to make sufficient amounts of the hormone thyroxine to meet the metabolic demands of the body. The thyroid gland compensates by enlarging so as to trap as many atoms of iodine as possible. This enlargement usually overcomes mild deficiencies of thyroid hormone. In the United States in the 1930s, when goiter was widespread throughout the Great Lakes, Midwest, and intermountain region (the "goiter belt"), its development was considered a normal part of maturation. Because of the iodization of table salt, goiter due to lack of iodine is now virtually unheard of in the United States. In some countries in Africa, people develop goiters due to an overreliance on foods such as those in the cabbage family that contain antithyroid substances called **goitrogens.**

It is estimated today that 740 million people around the world suffer from goiter, while another 2 billion people live in areas in which the soil is iodine deficient and are therefore at risk. China has come to regard its entire population as at risk. Severe iodine deficiency in pregnancy can cause infants to be born with an extreme and irreversible mental and physical retardation known as *cretinism.* This can be averted by diagnosing and treating iodine deficiency early in pregnancy. It appears that even relatively mild iodine deficiency in children can result in poor growth and cognitive deficits. In the most severe forms, affected individuals are deaf, mute, and severely mentally handicapped.

The most common approach to prevention of IDD is to provide iodized salt. The iodination of salt is a very cheap and effective way to prevent iodine deficiency. Globally about 68% of households in the world consume iodized salt (ACC/SCN 2000). The question becomes "If iodination of salt is easy, cheap, and effective, why isn't it universal?" In part the answer is that many of the most vulnerable populations do not buy commercially produced salt, because either the price is higher or the distribution lines do not reach isolated areas. In other cases, iodized salt is more expensive than non-iodized salt.

Vitamin A Deficiency

Vitamin A is an essential micronutrient that is important for vision, growth, and development. More recently, it has been found to be important in adequate immune function. Vitamin A supplementation reduces morbidity and mortality in children who do not appear to be deficient (Pokhrel et al. 1994; West et al. 1991). However, the most visible sign of vitamin A deficiency is blindness. In early deficiency, people develop night blindness, the inability to see well at night. As the deficiency gets more serious, spots of foamy material appear on the eye. These are called Bitot's spots. **Keratomalacia,** or drying of the cornea, follows; and if vitamin A deficiency (VAD) continues, ulceration, scarring, complete collapse of the cornea, and blindness are the final stages. VAD is the number-one cause of preventable blindness in the world. But even mild deficiency, less than is necessary to cause night blindness, appears to impair immune function.

Night blindness in Nepali is called *ratauni* (Christian et al. 1998a, 1998b). It is so common among pregnant Nepali women that it is considered a normal part of pregnancy. As one Nepali

goitrogens
Chemicals found in some foods that bind with iodine and make it unavailable for use in thyroid hormones. Goitrogens can be important contributing factors to iodine deficiency and goiter.

keratomalacia
A condition associated with vitamin A deficiency and protein-calorie malnutrition, characterized by a hazy, dry cornea. Left untreated, it can lead to ulceration of the cornea and infection.

woman told researchers, *"Don't know what causes it. . . . It happens to some women who are pregnant, like the swelling of hands and feet. It happens on its own, because of the pregnancy"* (Christian et al. 1998a: 882). The cause of night blindness is also attributed to "weakness" of the body during stressful conditions such as pregnancy. Finally, *ratauni* is also incorporated into the hot/cold classification system of health in which the body's humors should be in balance (see Chapter 8). In the Nepali version of humoral medicine, night blindness is considered "hot" and can be caused by eating "hot" foods, especially when the body is already "hot" from pregnancy and the temperature is also "hot." Among the Nepali women interviewed by Christian et al., some treated their night blindness by taking vitamin A supplements, which were seen as "medicine." Others believed that "medicine" was also "hot" and would only make the condition worse. Most of the Nepali women experiencing night blindness during pregnancy simply waited for the pregnancy to be over. Most recovered shortly after delivery. A very small number reported that because it was a "hot" condition, consuming cold foods would be helpful. However, in the ethnomedical system of rural Nepal, night blindness was not seen as a nutritional problem, and for most of the pregnant women the use of supplements was thought to be ineffective. While pregnant women are at special risk, night blindness also affects children and men. Indeed, children are most vulnerable to the more severe forms of vitamin A deficiency, moving quickly from night blindness to keratomalacia and blindness. VAD is also not confined to Nepal. Night blindness, keratomalacia, and corneal ulceration are common among the poor in much of South Asia.

Vitamin A or vitamin A precursors are found widely in green, yellow, and red foods that contain high levels of carotene, a pigment that has a yellow-orange color. Although not all orange foods have high vitamin activity, most do. However, it is precisely these foods that are in the shortest supply and that are the most expensive in the food systems of poor people. Over 250 million children worldwide are estimated to suffer from some form of clinical or subclinical vitamin A deficiency (ACC/SCN 2000; Sommer 2001).

There are several ways to approach improving vitamin A nutrition (Sommer 2001). The best way to prevent VAD is to improve diets, especially of poor children, by diversifying diets and making fruits and vegetables more available. Even poor children with vitamin A rich foods in their diets are less likely to experience keratomalacia than poor children who don't have these in their diets (Shankar et al. 1996). However, vegetable sources provide beta-carotene, not preformed vitamin A. Foods such as eggs and meat products such as liver are better sources for children who need more vitamin A in relation to the overall diet than adults. Programs to increase intake of these foods focus on promoting home gardens in both rural and urban settings, and the marketing of less expensive fruits and vegetables.

Fortification through conventional means is a second way of getting more vitamin A in the diet. Denmark began fortifying margarine with vitamin A in 1900. Fortification of a number of products has been tried in developing countries. These include monosodium glutamate, wheat, maize, noodles, and sugar. Only sugar fortification, primarily in Latin America, seems to have been successful. In order for fortified foods to have an impact, they must have acceptable taste, smell, and look. Also, they must get to the consumers who need them. In some cases the poor, those at greatest risk for deficiency, use very few centrally processed and purchased products. This is similar to the problem of iodine fortification of salt for those who make their own salt or buy locally produced salt.

Foods can be fortified through genetic modification (GM; see Highlight 6). There are several programs to genetically modify some basic grains to increase the vitamin A content. Monsanto has been working with a carotene-rich version of rape seed and mustard seed. Funded by the Rockefeller Foundation, the International Rice Research Institute (IRRI) in the Philippines has developed a "golden rice" variety using GM techniques, but release has been slowed as a result of consumer concerns about the safety of GM foods (Toenniessen 2000). Again, for genetically modified foods to be successful, they must have acceptable flavor and cooking properties. They must also have acceptable agricultural properties and give good yields.

Finally, vitamin A can also be provided through the use of supplementary vitamin capsules or injections. Because vitamin A can be stored in the body for some time, supplementation twice per year is a feasible way to prevent deficiency. In some countries, vitamin A injections are included in vaccination campaigns. As we saw in the Nepali example above, the use of capsules and other supplements may not be perceived as culturally acceptable in all cases.

Of the three most important micronutrients contributing to hidden hunger, iron deficiency is the most difficult to address. Increased consumption of more easily absorbed iron, from animal sources, is likely to be the best approach, but also the most costly and least likely to help the poor, who are at greatest risk. Problems with IDD and VAD are decreasing globally and are expected to continue to do so.

CHAPTER 10

Addressing Global Food Issues

We, the Ministers and Plenipotentiaries representing 159 nations . . . declare our determination to eliminate hunger and to reduce all forms of malnutrition. Hunger and malnutrition are unacceptable in a world that has both the knowledge and the resources to end this human catastrophe. (World Declaration on Nutrition: International Conference on Nutrition, Rome, in December 1992)

In Chapter 9 we reviewed the current global food situation and discussed some of the predictions for the global availability of food in the future. We concluded that there is enough food produced in the world to adequately feed all of the people who are alive today and food supplies will continue to be adequate into the near future. At the same time, there are a large number of hungry, undernourished, and malnourished people in the world. The delegates to the International Conference on Nutrition (ICN) held in Rome in 1992 declared their commitment to eliminate hunger and malnutrition. Following the introduction to the World Declaration on Nutrition quoted above, the representatives continued by stating:

1. . . . We recognize that globally there is enough food for all and that inequitable access is the main problem. Bearing in mind the right to an adequate standard of living, including food, contained in the Universal Declaration of Human Rights, we pledge to act in solidarity to ensure that freedom from hunger becomes a reality. We also declare our firm commitment to work together to ensure sustained nutritional well-being for all people in a peaceful, just and environmentally safe world.

2. Despite appreciable worldwide improvements in life expectancy, adult literacy and nutritional status, we all view with the deepest concern the unacceptable fact that about 780 million people in developing countries—20 percent of their combined population—still do not have access to enough food to meet their basic daily needs for nutritional well-being.

3. We are especially distressed by the high prevalence and increasing numbers of malnourished children less than five years of age in parts of Africa, Asia and Latin America and the Caribbean. Moreover, more than 2,000 million people, mostly women and children, are deficient in one or more micronutrients; babies continue to be born mentally retarded as a result of iodine deficiency; children go blind and die of vitamin A deficiency; and enormous numbers of women and children are adversely affected by iron deficiency. Hundreds of millions of people also suffer from

*communicable and non-communicable diseases caused by con-
taminated food and water. At the same time, chronic non-
communicable diseases related to excessive or unbalanced dietary
intakes often lead to premature deaths in both developed and
developing countries. (ICN 1992, quoted in Latham 1997:1)*

Since the release of the World Declaration on Nutrition, both
the absolute numbers of people estimated to be undernourished and
the percentage of children estimated to suffer from measurable
undernutrition have declined, but as we saw in Chapter 9, the numbers are still unacceptably high. In this chapter we will consider the
range of factors that affect the availability of food and the ability of
people to get the food they need and ways to approach the more
equitable distribution of food to people who need it. Our focus here
will be on the problems of undernutrition and hunger.

Hunger and Malnutrition: The Factors Influencing Adequate Nutrition

We can think about addressing problems of nutrition on a number
of levels of aggregation: the individual, the household, the
region/community, the nation, and the world. At each level of
aggregation, having enough food available is a necessary, but not
sufficient condition for adequate food and nutrition in the lower
levels of aggregation. For example, there is currently enough food in
the world to feed the global population, but that does not mean that
all countries have sufficient food available to feed their populations.
If there is not enough food available in a country, some households
will not have enough food. But even if there is enough food available, some households may still not be able to access that food
because they lack the money or other resources to do so. In the same
way, there may be enough food in a household to meet the nutritional needs of household members, but it may not be equitably distributed within the household. Figure 10.1 summarizes the specific
issues important at different levels of aggregation and the factors
that influence nutrition. We have already addressed some of the
important issues in Chapter 9.

To feed everyone, enough food must be available at global,
national, and regional levels. Agricultural research to improve seeds
and technology, improved transportation, trade policies that allow
equitable access to goods and markets, and food aid in times of crisis
are all part of the equation. However, even when sufficient food is
available, some people cannot get enough food. They simply do not
have the resources to access food. Nobel laureate economist Amartya
Sen (1981, 1990, 1995) refers to these resources as "**entitlements.**"
Economists would call them "effective demand." In an economic
sense, **demand** refers to the extent to which households and individuals can purchase (or trade for) the goods they need. For an
economist, *demand* does not mean "asking strongly for," but is
measured by the degree to which foods can actually be acquired.
There are many households that need food but cannot buy it;

entitlements
Resources necessary in order to
access goods, in this case, food.
Income and access to land and education are examples of entitlements.

economic demand
The ability to purchase goods, in this
case, food. People may need food,
they may ask for food, but there is no
demand for food unless they can pay
for it. The market responds to economic demand.

demand is therefore low. Sen (1990) has argued persuasively that it is entitlement failure—that is, the lack of resources needed in order to demand food—that accounts for the bulk of malnutrition, not a failure of food production: "The starving person who does not have the means to command food is suffering from an entitlement failure" (p. 375).

According to Sen, *poverty* is the most important reason people do not have enough food. But the relationship between cash income and income-in-kind (foods produced by households through either field agriculture or gardening) is not so simple. As Figure 10.1 suggests, even when there is sufficient food or income to procure food in households, appropriate food may not go to the people who need it most. As we noted in Chapters 7 and 8, knowl-

Figure 10.1
Model of the cascade of factors and associated issues influencing nutrition.

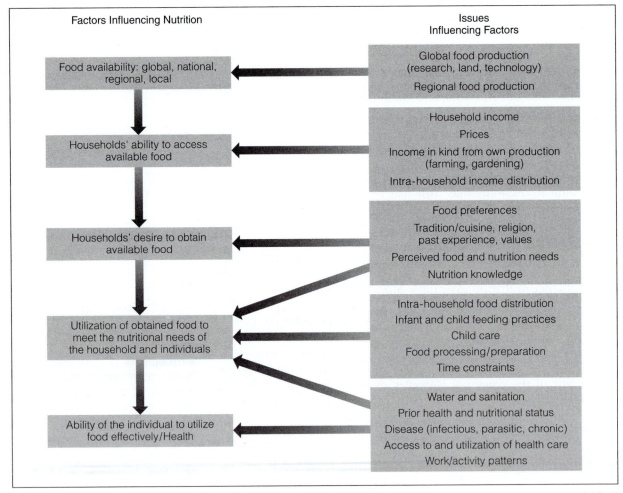

(Adapted from E. Kennedy and P. Pinstrup-Andersen, *Nutrition-Related Policies and Programs: Past Performances and Research Needs* (Washington, D.C.: International Food Policy Research Institute, 1983.)

edge, preferences, beliefs, and work needs may all have an influence on whether appropriate food is consumed by the people who need it. Finally, food is not the only term in the nutrition equation. Health and activity levels interact with food consumption to produce good nutrition. Sick people, especially children, may not eat enough or may not use nutrients effectively enough to be well nourished. Therefore, water quality and sanitation are essential. So is the access to health care, particularly immunizations to prevent childhood diseases. Heavy work demands increase the need for calories and may mean that households make decisions to divert food to those who work the hardest and earn income.

According to nutritionist Michael Latham (1997), data from around the world show that the causes underlying most nutrition problems have not changed very much over the past 50 years. Poverty, ignorance, and disease, coupled with inadequate food supplies, unhealthy environments, social stress, and discrimination, still persist unchanged as a web of interacting factors that combine to create conditions in which malnutrition flourishes.

Recognizing that the issues related to adequate nutrition and food security may differ, the FAO (2000b) defines a series of terms—food adequacy, food security, and malnutrition—differently at different levels of aggregation. Table 10.1 summarizes the definitions of these concepts at different levels. In the following pages, we will look more closely at some of the factors and the types of interventions that may improve them.

Table 10.1 Food Adequacy, Food Security, and Protein-Energy Malnutrition

Individual food adequacy (IFA) is shown, in the short term, by calorie intakes that are sufficient for needs, varying with age, health, work, and adult height; in the medium term, by the absence of acute protein-energy malnutrition (PEM), for example, low child weight-for-age (WA) or low adult weight-for-height (WH); and in the long term, by the absence of chronic PEM in children under 5, of low height-for-age (HA). Inadequate HA (stunting), WA (underweight), or WH (wasting) are often equated with shortfalls of more than 2 standard deviations below the United States median (>2 SD).

Individual food security (IFS) refers to "access to *adequate* safe and nutritious food to maintain a healthy life . . . without undue risk of losing such access" (FAO 1996), that is, IFA as well as the confidence that it can be maintained. Without such confidence, people make hypercautious decisions that forfeit their chances to escape from chronic hunger. A poor person usually obtains 70–80% of his or her calories (and most other nutritional requirements) from one or two of the world's seven main food staples. These are by far the cheapest sources of energy and of most other nutrients. For the poor, access to these staples is the key to achieving IFS.

Household food adequacy/security (HFA/HFS) is necessary for IFA/IFS, but not sufficient, because food may be distributed among household members disproportionately to their individual needs.

National food security (NFS) refers to a nation's capacity to ensure HFS/IFS without undue departure from other policy goals. NFS in a given year is often measured by dietary energy supplies (DES) per person, allowing for the distribution of food and needs among individuals and times; or the ratio of food imports to total exports, although food aid must be allowed for; or staples stocks (publicly controlled or likely to be marketed if prices rise) as a share of normal consumption.

National staples self-sufficiency (NSSS) refers to the production of sufficient staple foods within a country to meet effective demand.

Source: Adapted from FAO, *The State of Food and Agriculture 2000* (Rome: Food and Agriculture Organization, 2000).

Food Availability

Food availability at the national level does not *ensure* household food availability. But does the amount of food available in a country have any impact on overall levels of child malnutrition? Smith and Haddad (2001) argue that national food availability does have an impact. They analyzed the availability of food and indicators of child nutrition for 63 less developed countries between 1970 and 1996. They found that increasing per capita food availability, especially for countries with low availability to start with, did result in better overall child nutrition. This relationship was especially strong for countries in sub-Saharan Africa where per capita food availability is low. But it was much weaker for countries in the Middle East and North Africa where food availability was already high. However, Smith and Haddad also argue that as strong as the relation they found was, education (especially women's education) and improved access to health care and immunizations may be at least as important as national food availability, even in countries with relatively low per capita food availability.

Agricultural Research and Extension

Now let's look at some of the specific programs and outcomes that have an influence on what food is available and where it is available. Agricultural research carried out by international agencies, national research programs, and private companies has an impact on the productivity of agriculture. Policies that promote the production of some kinds of crops over others also have an impact, as do the terms of trade that favor some countries and products over others.

Why aren't the benefits of agricultural research always available to those who need it most? Two problems are key. The first is that agricultural research does not always address the crops that provide food (rather than animal feed or non-nutritive items such as coffee and cotton) to the people at greatest risk for hunger. The second is that the technology produced by research—new seeds, chemicals, and agricultural practices—does not always meet the needs of resource-poor farmers, or women farmers, in less developed countries. In many cases, improved technology, and access to the credit and markets that are necessary to allow small farmers to benefit from technology, have not been made available to these people.

Currently, a large number, indeed the majority, of the hungry are rural, resource-poor farm families—small farmers—and landless workers. But an increasing number of the hungry are poor urban dwellers, many of whom are recent migrants to the city, and this segment of the population is growing, even in Africa. The needs of these groups may actually be quite different from those of small farmers. For urban poor and landless rural workers, the most important benefits of agricultural research would be increases in food availability and the continued lowering of food prices as a result of improved productivity in agriculture.

However, for small farmers, it is important that the prices they receive for their products remain high enough to pay for the costs of production and to provide cash income for food and other goods. Ironically, for many small farmers the global lowering of food prices as a result of improved productivity in industrialized countries makes it less possible for them to sell their agricultural products for enough money to make ends meet. Perhaps more importantly, many industrialized countries, such as the United States, put up trade barriers to importation of food from other countries, or they subsidize their own domestic agriculture to the extent that unsubsidized farmers in developing countries cannot compete. In our postmodern world, poor Mexican corn farmers are competing with subsidized farmers in Iowa. Within developing countries, some governments have placed a higher priority on providing cheap foods for urban populations, rather than improving the economic standing of rural farmers.

Much of the work in agricultural research and improved seeds is carried out by corporations that are most interested in selling their seeds and agrochemicals. Not surprisingly, the crops that get the most attention are those that have the best markets and highest profitability. For this reason, the crops that have had the most improvement are commercial, cash crops: the grains and legumes that provide a lot of food for people, high-profit items (such as vegetables), and feed for animals. These include wheat, rice, corn, and soybeans, tomatoes, peppers, and so on. Grains that have more restricted markets, or are primarily the subsistence crops of small farmers, historically have received little attention from corporations. Small farmers growing traditional food crops, especially for home use, often use their own seed and use minimal amounts of fertilizers and pesticides (see Figure 10.2). These crops, which include sorghum, millet, cassava, sweet potatoes, common beans, and local vegetables, were historically neglected. For this same reason, as we saw in Chapter 9, these crops may hold the highest potential for improvement.

The difficulties that resource-poor farmers, including women farmers, face in adopting some types of technology led to the development of **appropriate technology.** Much of the research work on these "minor" crops is being conducted by the member organizations of the Consultative Group on International Agricultural Research (CGIAR, or CG for short). The CG system currently has 16 member institutions (see Table 10.2). The CG institutions have, as their mandate, to carry out agricultural research on crops that provide food for people at risk of being hungry, with technology that resource-poor farmers can use. They focus on such crops as maize and wheat (CIMMYT), rice (IRRI, WARDA), potatoes, sweet potatoes, and other tubers (CIP, IITA), cassava (CIAT, IITA), sorghum and millet (ICRISAT), beans and other legumes (CIAT), small and large animals (ILRI), and sustainable forest development. CG-generated varieties have had an impact on the productivity of small poor farmers. In Chapter 9 we discussed the impact of the green revolution of the 1960s and 70s on the production of wheat, maize, and rice. While the new varieties were most quickly adopted by larger farmers, who could apply the entire technological package, small farmers also

appropriate technology
Technologies that meet the needs, abilities, and resources of the people who will use them.

Figure 10.2
Sometimes the most appropriate technology for small farmers is not a big tractor.

© Alan Oddie/PhotoEdit

Table 10.2   **International Agricultural Research Centers of the CGIAR System**

CIAT—Centro Internacional de Agricultura Tropical (International Center for Tropical Agriculture), Colombia

CIFOR—Center for International Forestry Research

CIMMYT—Centro Internacional de Mejoramiento de Maíz y Trigo (International Center for Maize and Wheat Breeding), Mexico

CIP—Centro Internacional de la Papa (International Potato Center), Peru

ICARDA—International Center for Agricultural Research in the Dry Areas

ICLARM—International Center for Living Aquatic Resources Management

ICRAF—International Centre for Research in Agroforestry, Kenya

ICRISAT—International Crops Research Institute for the Semi-Arid Tropics, India

IFPRI—International Food Policy Research Institute, Washington, D.C.

IITA—International Institute of Tropical Agriculture, Nigeria

ILRI—International Livestock Research Institute

IPGRI—International Plant Genetic Resources Institute

IRRI—International Rice Research Institute, the Philippines

ISNAR—International Service for National Agricultural Research, the Netherlands

IWMI—International Water Management Institute

WARDA—West Africa Rice Development Association

adopted parts of the technological package and found that their yields also improved. However, newer approaches to improving "minor crops" are more focused both on the needs of small farmers and on the needs of growing urban populations.

Since 1980, Ghana and Nigeria have lowered the prevalence of undernourishment by 30%. In good part, the ability to provide more food was the result of improvements in the production of cassava, a food that is especially important for the poor (FAO 2000b). Cassava roots are high in energy and cassava leaves are high in vitamins A and C, iron, and calcium. The International Institute for Tropical Agriculture (IITA), a CG institution located in Ibadan, Nigeria, spent more than 10 years of research to develop new varieties of cassava and better approaches to marketing. The new cassava varieties yield up to five times more than the older varieties. They mature earlier, are more resistant to diseases, and are suitable for processing into flour and starch. They also have growing characteristics that reduce weeds and make harvesting easier. Finally, they have lower levels of hydrogen cyanide than traditional varieties (see Chapter 5).

Nigerian farmers began adopting the new cassava as a food security crop during several drought years. Resistant to drought, cassava yielded a harvest even when other crops failed. Moreover, because it matures earlier than other crops and can be stored in the ground for up to 3 years, it works well as a "hunger season" crop for small farmers. The development of a market for processed cassava in urban areas made it into a cash crop as well as a subsistence and food security crop. Between 1983 and 1992, annual per capita consumption of cassava in Nigeria increased from 63 kg to 129 kg.

In Ghana the increase came a bit later. The IITA varieties developed at research stations in Nigeria had to be adapted to the local climate and soil conditions of Ghana. Production and consumption of cassava in Ghana began to increase in the 1990s, and by 1998 per capita annual consumption had increased from 126 kg to 232 kg. Cassava is now the largest agricultural product produced in Ghana and accounts for 22% of the value of all the agricultural products in the country.

In both Nigeria and Ghana, introduction of new varieties that were well suited to the needs of small farmers was also accompanied by government policies to support agriculture and to make equipment and planting materials available. Also, urban demand for food was growing, creating an expanded market for the increased production of cassava.

The Role of Biotechnology

Biotechnology and genetically modified (GM) food crops may also be technologies that can help small farmers (see Highlight 6). Reviews of the potential of biotechnology to help small farmers note that the distribution of seeds that are disease, insect, and drought resistant, and tolerant to some pesticides, may be an effective and simple way to provide technology already built into the seed to farmers who have few resources to buy inputs and access to information and new technology (Qaim, Krattiger, and von Braun 2000). GM seeds may allow small farmers to reduce the amount of agrochemicals they use, reduce the need for irrigation, and use land that is currently too high in aluminum content to be effectively cultivated. Indeed, it may be possible to build seeds that contain higher levels of scarce nutrients. The International Rice Research Institute (IRRI) has developed GM rice varieties with significantly higher vitamin A and iron content (Bouis 2000). There is continued controversy over the safety of GM food both in diets and for the wider environment. Until the safety issues are resolved, GM foods are not likely to be made widely available for resources-poor farmers.

Policy Options: Self-Sufficiency vs. Food Security

Over the past several decades there has been a continuing debate between proponents of a strategy and set of policies that favor regional and national food *self-sufficiency* and those who argue that self-sufficiency is less relevant than improved *food security* through the improvement of *entitlements*. It seems axiomatic that self-sufficiency in food production at the national and even household levels should lead to improved national food security and nutrition, but as usual, the relationship is not that simple.

Food Self-Sufficiency and National Food Security

Food self-sufficiency and national food security (NFS) are quite different concepts (see Table 10.1). National food security may be achieved through a number of means. It can include trying to achieve national staples self-sufficiency (NSSS) by promoting policies that increase national production of staple foods or by increasing the capacity to pay for the import of staple foods by exporting agricultural or other products. As we will see below, the importance of NSSS in achieving NFS may be very different for different countries and for the same country at different stages of economic development.

We can think of food self-sufficiency at several levels. For agricultural households, especially small, resource-poor farm households, self-sufficiency refers to the degree to which a household produces the food it needs. As we discussed in Chapter 5, farmers who produce most of what they consume and little else are referred to as *subsistence farmers*. Apart from indigenous peoples living in traditional ways, there are very few fully subsistence farmers in the world. Most farmers, even small farmers, produce a mix of subsistence and commercial (cash) crops. They are *semi*subsistence farmers. At the national level, food self-sufficiency means producing enough food within the country to meet the food *demands* (not necessarily food *needs*) of the population. The green revolution and national policies promoting the production of staple foods allowed a number of countries to become self-sufficient in some staple food products. India went from being a net importer of rice and wheat to being an exporter of these foods. China has met internal demand for rice by internal production in most years since the mid-1980s.

What is the importance of national staples self-sufficiency on food security? It seems that policies promoting NSSS may mean either better or poorer NFS. Achieving NSSS does not ensure food security at the national or household level. India has had NSSS for several decades but still has one of the highest rates of malnutrition. Indeed, it appears that India's ability to achieve NSSS is in part due to improved agricultural productivity as a result of the green revolution, and in part due to continued poverty. Remember that NSSS rests on the ability to meet economic demand for food. Poor people cannot *demand* much food, because they lack the resources to buy it. Even in the face of national self-sufficiency in staples, many of India's households are food insecure, and therefore the country lacks national food security (FAO 2000b).

Other countries, such as Malaysia, import a good deal of their staple foods but are more food secure. Countries such as Malaysia and some Latin American countries have focused on policies that promote export, industrialization, and income growth, rather than staple food production. Despite increases in food importation, Malaysia and much of Latin America show improvements in nutritional status. On the other hand, increasing food imports in some countries of sub-Saharan Africa are symptomatic of devastating failures to achieve food security and of a reliance on food aid.

What Makes NSSS Important?

What are some of the considerations regarding the importance of NSSS in achieving national food security? In some cases, policies aimed at self-sufficiency have been a way to get cheap food from a poor rural population in order to feed a population of urban workers. These kinds of policies may undermine the household food security of farm households and, in the end, discourage farmers from growing more food. This is the case in some countries in Africa with falling food production.

Relying on the production of crops for export, or industrialization to improve national food security, means that the terms of trade must be favorable for the export items relative to the food to be imported. That is, staple foods on the international market have to be cheap, while the goods to be exported must have markets. In fact, the global prices of some staple foods are so low that big producers such as Western Europe and North America have reduced production. However, the terms of trade for exports from developing countries are not always favorable. Many countries, including industrialized countries, put up barriers to trade to protect their own producers of goods, making it hard for less developed countries to export their goods easily. The Cuatro Pinos Project, a relatively successful program to promote the production of vegetables for North American and European markets in Guatemala, has succeeded because the Guatemalan farmers can produce vegetables in a very short window of a few weeks in the winter when the United States allows for the importation of vegetables.

The amount of inequity within countries is also important. A rise in nonstaple exports may not have a positive impact on NFS if the benefits go to only a very small number of capitalists or governments that have priorities other than food security. This is what happened during the oil boom years in the late 1970s in oil-exporting countries such as Nigeria, the Soviet Union, and Indonesia. Rising imports did very little to improve national food security during that time. However, during the same time, Mexico instituted the Mexican Food System, a program that included both price controls in order to keep the price of some foods low and farm subsidies to help support farmers who grew food crops. The Mexican Food System was underwritten by the royalties accrued by the Mexican government from oil sales.

Finally, infrastructure such as transportation, roads, and markets must work well in order to allow both the efficient marketing of nonstaple crops and manufactured goods and the importation and distribution of staple products. In our experience, more projects for producing and selling potentially high-value nonstaple crops have failed as a result of poor access to markets than for any other reason. In 1999 one of the most dramatic effects of the El Niño event in coastal Ecuador was the destruction of the roads needed to get cash crops from farms to markets. Even those crops that benefited from the extra rain, such as rice, were difficult to transport. In fact, the improvement of roads and other means of transportation is now seen as one of the key infrastructure factors in improving food security and nutrition. The State of Food and Agriculture Report for 2000 (FAO 2000a) suggests that countries may go through different stages of development and that NSSS is sometimes important and sometimes not.

Proponents of food self-sufficiency rightly point out that many nations, especially those with poor trading opportunities and poor infrastructure, or those that are not sufficiently developed to move toward industrialization, would probably improve national food security by promoting the production of staple crops, while ensuring that the producers, especially small producers, are fairly compensated for their crops. Those who argue that staples self-sufficiency does not ensure, and may be mostly unrelated to, national food security rightly point out that the key issue in national and household food security is the failure of entitlements. We will discuss the role of poverty in food insecurity and poverty alleviation in the section on entitlements.

Commercialization of Agriculture and Household Food Security of Small Farmers

Issues surrounding the desirability of national food self-sufficiency are mirrored in the discussion of the impact of shifting from subsistence production to production of commercial, cash crops for small farmers. Keep in mind that in many less developed countries, small farmers constitute a large percentage of the poor and those at high risk for household food insecurity. Alleviating poverty among the rural poor may be an important task. In the last two decades there have been several reviews of the impacts of agricultural commercialization, in general, on the food security and nutritional status of rural households (e.g., DeWalt 1993a; Kennedy, Bouis, and von Braun 1992). These reviews reach several conclusions concerning the impacts of the commercialization of agriculture on food security and nutrition. The results of agricultural commercialization are mixed. Empirical research provides examples of both positive and negative impacts on food security and nutritional status (DeWalt 1993a; von Braun and Kennedy 1986). It appears that the impacts are closely tied to specific cultural, social, and policy contexts and to the nature of the commercial crops being produced. However, there are several generalizations that can be drawn, as the remainder of this section will make clear.

Income

The literature examining commercialization of agriculture reinforces the conclusion that the relationship of income generation to food security of small farm households is complex and relies on the profitability of the crop and how income is distributed within households (see Highlight 10). In studies in Guatemala, Kenya, Gambia, New Guinea, Malawi, the Philippines, Sri Lanka, and Rwanda, increases in cash income as a result of growing and selling cash crops are translated into increased household food consumption. Studies that disaggregate the impact of increasing income by income groups show that the strongest effects are in the poorest households where increased income goes to purchasing more calories and often to purchasing a higher-quality diet with more animal products. Although some of these studies also demonstrate improvement in child nutritional status, others show mixed results with some classes of households benefiting and others not, and some show no change (DeWalt 1993a, 1993b).

Increasing income may have differential effects for several reasons. In some cases there may be an increase in cash income, but not in total income; that is, while cash income increases, the loss of substance production (income-in-kind) offsets some or all of the increase in cash. Or a rise in food prices as a result of regional decreases in food production might offset a rise in income. In some cases, income from cash crops may be more "lumpy," that is, come in a few larger payments rather than many smaller payments. Lumpy income is more likely to be spent on nonfood items, such as capital goods, home improvements, and weddings and other social events. Income that comes in smaller amounts in a more regular way is more likely to be spent on food. Shifts from smooth to lumpy income may result in lower food purchases. Furthermore, many shifts to cash cropping result in a shift in income control from women to men. The extent to which increases in family income off-set losses of income to women is unclear. Von Braun (1988) argues that in a rice production scheme in Gambia, increases in family income were able to offset the possible negative impact of loss of income to women. Some researchers found that in Highland Guatemala the shifts in income from women to men that occurred when they adopted winter vegetable production for the North American market may have offset the benefits of increased house-hold income (Barham et al. 1992; Nieves 1987; von Braun, Hotchkiss, and Immink 1989).

Finally, it may be that the degree to which income from cash crops is spent on food may be different from the way in which income from other sources is spent. Kaiser and Dewey (1991) show that, in a Mexican community, income from subsistence crops, sale of cash crops, and remittances from migrant workers are all spent in different ways. Income from cash cropping does not have a signifi-cant impact on food consumption, whereas the percentage of income in remittances is negatively correlated with food consump-tion and the proportion of income that is in the form of home-produced food positively predicts the percentage of income allo-cated to food, even when total income is held constant.

Protecting Food Crop Production

The protection of subsistence production at the household level in the process of commercialization appears to be one of the most important factors in producing a positive impact on food security and nutrition of small farm families. In a review of a series of stud-ies on the nutrition impact of commercialization carried out since 1986 (DeWalt 1993a), it is clear that those projects that specifically included the goal of protecting subsistence production while pro-moting adoption of cash crops generally resulted in improvements in food consumption and/or nutritional status, whereas projects that discouraged subsistence production resulted in negative impacts. The nature of the crop as a food crop, or a nonfood or non-nutritive crop, is an important element in the impact of commer-cialization. Several studies of rice production, for example, show that this crop generates cash income, but at the same time stabilizes subsistence production and household food security as well. The availability of food from gardens in a rubber scheme in Papua New Guinea (Schack, Grivetti, and Dewey 1990) provides the same sta-bility for food consumption with a subsequent added benefit for households producing cash income. Those households that were able to maintain household food production in an irrigated rice scheme in West Kenya (Niemeijer et al. 1988) benefited from cash cropping, while those that were not able to maintain production had the poorest nutritional status. Fleuret and Fleuret (1991), in a study of commercialization in the Taita Hills of Kenya, suggest that the prohibition against intercropping food crops with coffee accounts, in part, for a negative association between coffee produc-tion and nutritional status in one of the communities they studied.

Haaga et al. (1986) suggest that the major factor accounting for poorer nutritional status of children in households adopting new crops in another region of Kenya is the loss of drought-tolerant food crops in a region prone to periodic drought.

At the national level, it appears that promotion of commercial or export crops is also not necessarily at odds with food production. Von Braun and Kennedy (1986) related food production to cash crop production in 78 developing countries and demonstrated that there is a positive correlation between cash crop production and food production. It appears that agricultural and economic policies that benefit agriculture benefit both cash crop and food crop production (von Braun and Kennedy 1986).

Land Tenure

In many cases, the impact of commercialization on land tenure has led to the concentration of land (DeWalt 1993b). This is sometimes the result of increasing accumulation of land by larger landowners or, in many cases, the displacement of tenant farmers and share-croppers. Displacement of tenant farmers is one of the important effects of the cotton and beef booms in Central America. Dewey (1989) argues that land concentration and displacement of marginal farmers is one of the primary mechanisms by which commercialization of agriculture promotes a deterioration in nutritional status.

Bouis and Haddad (1990) find that one of the earliest results of the introduction of a sugar mill in Bukidnon Province, Philippines, was land concentration and displacement of small farmers. They compare the process they observed as akin to land reform—in reverse. Barham et al. (1992) note that the effects of the adoption of winter vegetable production in the Western Highlands of Guatemala resulted in some land concentration, but only by the smallest landholders. Because of diseconomies of scale inherent in winter vegetable production (due to the large amount of hand labor needed, smaller farmers are actually more productive than larger farmers), smaller landholders were able to increase their landholding to closer to the optimal level, whereas there was no incentive for larger farmers to accumulate landholdings. In this case, changes in land tenure appear to benefit the most vulnerable landowners and can be seen as reversing a trend toward increasing fragmentation of landholdings. Under circumstances in which land tenure policy discourages land concentration, commercialization may not have an effect. In a study of shifts from the subsistence production of maize to the commercial production of sorghum in four communities in Mexico, we found the land reform system mitigated against any tendency toward land accumulation (DeWalt et al. 1990).

Health

A failure to improve basic health conditions is likely to offset some or all of the benefit of improved food consumption, especially in children. The IFPRI studies (von Braun et al. 1990) show that increasing income did not result in health improvements in most cases. Safe water and access to immunizations and primary health care are important.

We must keep in mind that food consumption is only part of the nutrition equation. Although it is possible that increasing income from commercialization could be translated into improved sanitation and use of health care, this seems to be less common than impacts on food consumption. Finally, as Bouis and Haddad (1990) point out, increasing demands on the time of women may result in lower-quality care of their children and increased child morbidity. Consequently, the effect of improvements in food availability and consumption on child nutritional status may have been offset by continued high morbidity rates.

Entitlements

While food availability is a necessary precondition, entitlement failures are the most important factors in low levels of household and individual food security. One of the primary reasons that undernourishment and malnutrition continue to be problems in the contemporary world is that there has been failure to alleviate poverty.

Who are the poor? Most of the poor live in poor countries. The *World Development Report 2001* (World Bank 2001) lists over 20 countries with per capita gross national income of $300 or less. Many, but not all, are in sub-Saharan Africa; they also include the Kyrgyz Republic, Tajikistan, Nepal, and Cambodia. Perhaps more importantly, many countries with much higher gross national incomes have income distributions so skewed that most income is in the hands of a very small elite, with many people living in abject poverty. For example, in Mexico and Brazil, people in the top 10% of income distribution have over 40% of national income, while the lowest 10% controls about 1% of income. In Swaziland, the top 10% controls a whopping 50% of income. On the other hand, in poor countries such as Vietnam the top 10% controls 23% of income and the bottom 10% controls about 4%. There is much greater equity in income distribution in Vietnam than in Swaziland. Finally, as economist Phillips Foster (1992) summarizes:

> *Who are the poorest of the poor? Mostly they live in the Third World. By and large they are landless or nearly so. If they do have a bit of land, typically they earn more than half their livelihood working for others. Whether they live packed tightly into city slums or sprinkled across the countryside, they are poorly educated, often illiterate, and commonly superstitious. When employed they accept the most menial jobs. Some are subsistence fishermen. Some live in relative isolation in upland farming areas. Often they are squatters, neither renting nor owning the land on which they put up their hut. Their food larder is usually almost empty. (p. 3)*

As Foster (1992) and others have long pointed out, holding gross national product or gross national income steady, health and nutrition indicators tend to be much worse for countries with highly skewed income distributions. In these countries, the benefits of development tend to go to only wealthy individuals; further, taxes tend to be regressive and are not vehicles for income redistribution through social programming such as health care, foods and nutrition supplementation, or education. As we argue in Focus 10.1, a workforce that is poorly nourished and poorly educated can neither take care of itself adequately nor serve an engine of development.

Also in systems with skewed incomes, the wealthy compete with the poor for food, and because of their greater purchasing power, they win (Foster 1992). Not only do they win when they compete for basic food staples, raising prices, but they also demand foods such as animal products: meat, dairy products, and eggs. The animals that produce these foods compete with the poor for grain. "The more the wealthy consume livestock products, the more dependent the livestock herd becomes on grain as it switches from scavenging and grass to grain. The more the price of grain is bid up to satisfy the wealthy's demand for animal products, the harder it is for the poor to buy the grain they need for minimal nutrition" (Foster 1992:141).

FOCUS 10.1   The Cost of Hunger

We have argued here that poverty is the most important contributor to undernutrition in many families in the world. However, undernutrition may also contribute to both poverty and slower national development by limiting human capital. Human capital includes the skills that people can bring to economic development and their ability to work toward household welfare and economic growth. The FAO (2000a) argues that hunger and undernutrition have measurable economic costs in that they limit human capital in terms of lost work productivity, cognitive impairment, increased morbidity, and, of course, increased mortality especially among children.

There is a direct link between energy (calorie) intake and physical activity. People with lower than optimal energy intake will not be able to perform the same amount of work as people with adequate energy intake. It has long been known that children who are undernourished become lethargic and have lower activity levels. This is also true of economically active adults who are undernourished. Some of the most dramatic impacts may be on the ability of small farm households to feed themselves and others. In research carried out on small farms in Sierra Leone in West Africa, Strauss (1986) calculated that if household members had 50% more calories in their diets, they would be able to put enough extra labor into their farms to yield an additional 379 kilograms of grain. In a study in Ethiopia, Croppenstedt and Muller (2000) showed that adults with a higher weight-for-height or body mass index (BMI; see Table 9.2) had higher output and higher wages. A 10% increase in weight-for-height and BMI resulted in a 23% increase in output and a 27% increase in wages. These researchers also found that taller people—that is, people who probably had better nutrition as children—also had higher wages. A person 7.1 cm (2.8 in.) taller than average would earn about 15% higher wages than average. A similar study carried out in Rwanda showed that those who were poorly fed had to choose to carry out activities that were less physically demanding and less well paid (Bhargava 1997). Although it is hard to estimate the impact of lost labor and productivity on national development, it must be significant. A poorly fed workforce cannot provide the same engine for national development as a well-fed one.

(continued)

Focus 10.1   (continued from previous page)

Protein-energy malnutrition (PEM) is not the only nutritional problem that limits the ability of people to work. Micronutrient malnutrition, or hidden malnutrition, also impairs productivity and income. In Chapter 9 we discussed the prevalence of iron-deficiency anemia. In some countries, as many as three-quarters of women and half of the men have signs of iron deficiency. Research over several decades carried out in Central America, Sri Lanka, Indonesia, and other countries has shown that iron-deficiency anemia reduces the capability of people to work. For example, Basta et al. (1979), in their research in Indonesia, found that workers with anemia produced only 80% of those without anemia.

Lowering of work capacity is not the only effect of PEM and micronutrient malnutrition. As we discussed in Highlight 9, iodine deficiency in extreme forms results in cretinism. But milder forms also cause cognitive impairment (FAO 2000b; Grantham-McGregor, Fernald, and Sethuraman 1999) in children and reduced work capacity in adults.

Nutrition also has an impact on school performance. The relationship between nutrition, school performance, and economic welfare is complex and goes in both directions. "Better nutrition leads to higher cognitive development achievement and increased learning capacity, and thus to higher labor productivity as well as to higher incomes. Higher levels of education lead to higher nutrition" (FAO 2000b:5).

While there has been relatively little research on the impact of malnutrition on cognitive achievement, reviews of the available studies suggest that

- PEM as manifested in stunting is linked to lower development and educational achievement;
- low birth weight is linked to cognitive deficiencies;
- iodine deficiency in pregnant mothers negatively affects the mental development of their children;
- iodine deficiency in children can cause delayed maturation and diminished intellectual capacity;
- iron deficiency can result in impaired learning capacity (FAO 2000b); and
- improved nutrition should further enhance the investments that countries make in education and training.

Finally, undernutrition is clearly linked to health, both morbidity (illness) and mortality (death). Pelletier (1994) found that even mild and moderate malnutrition in children were linked to higher mortality. There are clear synergistic relationships between poor nutrition and a number of infectious diseases including both diarrheal disease and respiratory infections, together the most important causes of illness in children worldwide. Malnutrition in childhood may also impair immune function for life. But not only children are affected. Studies of adults also show that those with lower energy intake are at higher risk for both illness and death (ACC/SCN 2000).

Poor nutritional status in women has an even more insidious long-term, intergenerational impact. In a study in Mexico, Allen et al. (1992) found that women's nutritional status was the best predictor of how effective women were as mothers. Women with unkept houses, poorly groomed children, and poor parenting skills were found to be undernourished themselves. Poorly fed women cannot be as effective as parents as well-fed women (Allen et al. 1992). South Asia is the world area with the highest levels of low-birth-weight babies. Low birth weight is associated with poorer health and lower cognitive performance in children. It is in large part the result of poor nutritional status of pregnant women. Discrimination against women and girls in South Asia would appear to be directly translated into poorer health, work ability, and school performance for the next generation of both girls and boys.

When the impacts of PEM and micronutrient malnutrition are taken together, they have a negative effect not only on the household economic welfare, but also on the economic growth of countries in which a large percentage of people are undernourished. While it is very difficult to establish the exact magnitude of the impact of poor nutrition on economic growth, Arcand (2000) estimates that if the per capita availability of calories were raised to 2,770 calories in countries in which the level is below that level, the per capita gross national product (GNP) would increase between .34 and 1.48 percentage points per year. Using longitudinal data from Europe, Fogel (1994) estimates that improvements in health and nutrition accounted for about half of the economic growth in the United Kingdom and France in the nineteenth century.

This is all strong evidence that investments in improved nutrition will have economic benefits beyond the improved health of the individual and beyond better economic status for households. Improving the nutrition of children and adults at risk may be tied to national economic growth as well.

Alleviating Poverty and Redistributing Income

Some of the ways to alleviate and eliminate poverty are obvious. Creation of jobs along with the institution of minimum wage laws is key. While overall economic growth—that is, a rise in gross national product—does not ensure that the benefits of growth will be equitably distributed among all segments of the population, study after study shows that increasing GNP is associated with improved health and nutrition indicators. The results are more dramatic when the benefits of growth are equitably distributed (ACC/SCN 2000; FAO 2000a, 2001b; Pinstrup-Andersen, Pandya-Lorch, and Rosegrant 1997, 1999). Growth is also more effective when coupled with minimum wage laws. Although some have argued that the institution of minimum wage laws for both rural and urban workers results in a loss of employment, studies of the nutrition impact of these laws suggest that they have an overall positive impact on nutrition.

Land Distribution and Agrarian Reform

Land is the major form of wealth in agrarian communities. As with other resources, at the end of the colonial period agricultural land in most developing countries was concentrated in the hands of relatively few wealthy families. Typically, a few wealthy farmers own large tracts of land that they can farm using the modern techniques associated with the developed nations, while the great majority of peasants own tiny plots cultivated with traditional tools and techniques. In many poor nations, as many as 80% of the farms are 12 acres or less and over 50% are under 2.5 acres. Although many peasants own their small garden plots, others must rent from landlords. Landlords typically have been depicted as harsh, cruel rulers who have little incentive to improve farm production. "It is undeniably true that peasants in many parts of the world have suffered various deprivations, even indignities, at the hands of landlords, and that landlords in general, and absentee landlords in particular, have not been very interested in agricultural development" (Sinha 1976:49). Not all landlords, however, fit this description. Some landlords have made important contributions to agricultural development and have treated their tenants with respect. In India, for instance, landlords have financed major irrigation projects, advanced low-interest loans for construction of wells, contributed to environmental improvement projects, and cooperated with government agricultural experiments (Sinha 1976:50).

In many regions, income disparities between landlords and tenants have been small and partially offset by complex duties and responsibilities of the rich to help the poor. In other areas, income disparities have been accepted as part of tradition. But more recently, especially with the introduction of advanced farming techniques and machines, the traditional status quo has been disrupted. Increased yields have been associated with increased disparities in incomes, farm machinery has caused unemployment to grow among farm laborers, and the poor have become more aware of their poverty. Occasionally, these changes have resulted in outbreaks of violence and heightened interest in land reform (Allaby 1977).

Land tenure policies in many developing countries retard food and agricultural production. Poor tenants and small farmers are unable to maximize use of their land. They do not have the money or credit to purchase the fertilizer or pesticides needed to increase yields with the new seed varieties. Land tenure, then, is seen by many as a serious obstacle to technological change and increased agricultural productivity (Presidential Commission on World Hunger 1980).

Agrarian Reform

"Sometimes argued with ballots, sometimes argued with bullets, and mostly argued with words, the debate about land reform has surfaced time and time again in the 20th century" (Eckholm 1978:13). Land reform is used here to refer to the redistribution of property or land rights for the benefit of small farmers and agricultural workers.

Land reform schemes have varied considerably: Some have occurred gradually (Peru) and peacefully (Japan), others through violent revolutionary changes (China, USSR). Not all land reforms have been successful (Foster 1992). Some governments have failed to back up reforms with credit, farm supplies, or other economic assistance, thus causing small farmers to lose their newly gained possessions. In other countries, opponents of reform have prevented reallocation from occurring. But where reforms have been successful, important gains have been achieved in agricultural production as well as social justice. In Mexico, for example, over 2 million families have been placed on the land. Employment has risen from 50% to 100% over previous levels, and agricultural output has increased from 3.5% to 8% annually over the last 30 years. Taiwan offers another example. The number of families owning land increased from 33% to 59% during the reform period, 1949–53. As a result, the World Bank reported that "the productivity of agriculture has increased, income distribution has become more even, and rural and social stability have been enhanced" (Eckholm 1978:13).

Because traditional land tenure systems act as obstacles to technological change, some experts agree that land reform is essential to increase agricultural productivity and alleviate hunger. Pierre Crosson (1976) questions this notion. He recognizes that a number of countries have made significant progress in agriculture after redistribution of land. Many of these countries, such as Taiwan and Mexico, also have enjoyed other types of technical assistance and progress (e.g., irrigation, foreign aid, and credit) that may have contributed equally to their successes. Crosson also points to countries (India, Pakistan, the Philippines, and Thailand) where substantial technological progress has occurred in the absence of significant land reform. Crosson concludes, therefore, that

we have too little understanding of the relationship between systems of land tenure and technological innovation to predict how important present tenure systems may be as obstacles to future technical progress in LDCs [less developed countries]. That they may impede progress seems certain, but it is equally certain that the land tenure reform is not a generally necessary condition for innovation. This discussion of land tenure systems thus ends on a note of uncertainty. (p. 522)

Foster (1992) also reviews more recent data on the impact of the land reforms carried out in the 1960s, 70s, and 80s. He notes that several of the most dramatic land reforms of the twentieth century—the Soviet Union and China—are being seriously questioned as obstacles to development and were being restructured even before the collapse of the Soviet Union. One of the goals of the Algerian land reform of 1962 was to increase rural employment but, in fact, the redistribution of land reduced rural employment. Land reforms in Peru mostly benefited those who were already wealthy. And as we note in Highlight 10, many land reforms actually disenfranchised women farmers. Finally, the redistribution of land to sharecroppers in several countries resulted in a great reluctance for large and medium-size landowners to allow any sharecropping on their land for fear of subsequent rounds of land reform.

However, in agrarian societies, access to wealth is almost exclusively tied to land use and/or ownership. The landless must work as farm laborers or find nonagricultural employment, both of which are exceedingly scarce in rural communities. Unless industrialization is proceeding rapidly enough in urban areas to absorb the rural poor, land reform represents one of the few means of redistributing wealth and giving the poor a chance to get ahead.

Nutritional well-being of the poor in rural settings has been tied to land tenure systems. For instance, Costa Rican families who farmed small plots of land (under 3 hectares) were 2.3 times more likely to have undernourished children than those with large farms (over 6.5 hectares) (Valverde et al. 1977). (A hectare is a metric unit of land measurement equal to 10,000 square meters, or 2.47 acres.)

Larger landowners typically produce cash products that reap better profits on the international market than crops sold locally. Thus, even though their farms tend to be highly productive, the yields do little to improve the diet in surrounding communities. Also, as DeWalt (1983) has shown for Central America, larger landholders are changing cropland to pasture land, leaving less land available for small farmers who rent from larger landowners. A similar situation has accompanied the green revolution in several developing countries. Land reform is associated with the production of basic grains and other food consumed locally, enabling farmers to better feed their families.

Even though the benefits can be great, the reform schemes are difficult to implement. "The successful change of long-standing rights and rental practices is not something that can be accomplished with the mere passage of a law by some sympathetic politicians. . . . The unspoken dilemma facing many Third World governments is that by promoting the emergence of new peasant groups, they could be endangering their traditional sources of political support—and in some cases the economic interests of their own leaders as well" (Eckholm 1978:16).

Historically, the most successful efforts to eliminate poverty have been rapid governmental changes in distribution of land and all other forms of wealth, combined with the creation of new jobs and expansion of education and health services. These changes in long-standing political and economic systems require an extraordinary degree of political commitment as efforts to meet the basic needs of the poor clash with the individual political freedom valued in Western democracies.

Such measures are, of course, the most politically sensitive that most political leaders can take. In many instances, they threaten the government's very survival. In other cases, however—as in Nicaragua most recently—it is the reluctance to undertake these same reforms that causes governments to fail. At the least, redistributive measures are sure to alienate powerful segments of the society. Therefore, the long-term success of efforts to achieve structural change is largely dependent on firm, unyielding commitment by local political (as well as social, economic, and religious) leaders. Political, financial, and technical support from the outside, as well as supportive trade and investment policy, can often strengthen local resolve and help the process of implementation (Presidential Commission on World Hunger 1980:33).

Credit, Marketing, and Price Systems

Access to land often is not sufficient to ensure an adequate income. Many small farmers cannot survive without support in the form of credit for small loans and markets to sell produce and buy supplies at fair prices. In many developing nations, the lack of these institutions inhibits productivity and fosters rural poverty. It is not uncommon for small landholdings to be mortgaged, sold, or turned over to moneylenders because farmers do not have enough income to purchase supplies or pay high interest rates.

Credit   The small farmer, as any other businessperson, requires money for investments. Long-term investments include land improvements, housing, irrigation, and agricultural tools and machinery. Short-term investments commonly are made for seed, fertilizer, livestock feed, veterinary services, and wages to pay extra help. Without credit to make these investments, many farmers cannot make small landholdings productive.

Credit may come from a variety of sources. Developed nations have an elaborate system of government and private banks, cooperatives, credit unions, and other lending institutions. In some developing countries, these formal credit sources are relatively unavailable to farmers, who turn instead to friends, relatives, neighbors, landlords, shopkeepers, and other informal moneylenders. Some governments offer a limited number of subsidized loans at low interest rates. But if these sources cannot provide farmers with the money needed, bankruptcy may result; at the very least, they may be forced to sell goods, livestock, or other products at low prices. Increasing amounts of credit are essential for rural development and agricultural productivity. Individual governments and regional development banks can greatly enhance the formation of a self-reliant small farm population by expanding sound credit to the rural population.

There are several ways that governments can provide cheap credit to farmers (Foster 1992). They can provide subsidized loans through governmental or quasi-governmental agencies. These agencies loan out money to preferred borrowers at below-market rates. Or governments can require that commercial banks lend some money at low rates to preferred borrowers. In Nigeria, for example, 8% of the loans that banks make to the agricultural sector must be made at half of the going interest rate (Foster 1992). Both of these approaches have costs. Government-run programs can be expensive, and requiring that lenders give low interest rates to some people will mean higher interest rates for others, which can be a constraint on development. An alternative is to promote policies that make saving more attractive to rural people. These include fair return on savings (good interest rates) and ease of putting in and getting out money.

Markets   As farming moves away from the subsistence level, markets become increasingly important as a source of farm supplies and a place to sell produce. In addition to a place for exchanging foods, successful markets require transportation systems, food processing, and storage facilities. Without these services, it is difficult to assemble large quantities of food or make goods available year round. In most parts of the world, postharvest crop losses are great due to difficulties and delays in getting food to the marketplace and inadequate preservation techniques. Losses due to rodents, molds, spillage, and spoilage are particularly high in tropical countries where postharvest losses generally approach 20%. Waste is also high in areas where high-yield varieties are planted uniformly over large areas. This genetic uniformity provides an ideal ecological environment for pests and leads to increased losses (Pimentel and Pimentel 1979). As part of effective marketing, the subsidization of rural roads may be a key investment for resource-poor farmers.

Income Transfers and Subsidies   Progressive taxation and the redistribution of income through targeted food subsidies including price subsidies, supplemental feeding programs, and direct food distribution programs, coupled with education, can have important impacts on poverty. In Focus 10.2 we discuss the Thai approach that used a number of these to reduce malnutrition in a short period of time.

Progressive taxation is taxation that takes a greater percentage of income or wealth from the rich than from the poor. Progressive income taxes are one example. *Regressive* taxation is taxation in which taxes are taken at a higher rate from the poor than from the rich. Sales taxes are regressive because everyone pays the same rate whether they are rich or poor, but the poor are paying a higher

FOCUS 10.2   Thailand: A Success Story

Thailand represents one of the best success stories in the alleviation of undernourishment and malnutrition. The Thai experience demonstrates the potential for significant improvement in nutrition when the will to do so exists. It also demonstrates how addressing several levels and types of factors simultaneously can have a dramatic impact in a relatively short period of time.

The percentage of children showing malnutrition, as measured by underweight, dropped from about 50% in 1982 to under 20% in 1990 and under 10% in 1998. During that same time period, the prevalence of poverty fell from about 32% of households living below the poverty line in 1988 to about 11% in 1996. To achieve these impressive statistics, the government of Thailand focused on community development (FAO 2000b). The 286 districts identified as the poorest were targeted. Poverty elimination was based on five principles:

- Give priority to specific areas where poverty is concentrated.
- Ensure that minimum subsistence-level living standards are met and basic services are available to everyone.
- Encourage people to assume responsibility for their own care.
- Emphasize the use of low-cost technology.
- Support people's participation in decision making and problem solving.

The Thai government decided to focus on the most important types of malnutrition among the poor in rural areas: protein-energy malnutrition (PEM), vitamin A deficiency, and iodine deficiency. The program combined nutritional surveillance (monitoring of children's growth), supplemental feeding of young children, nutrition education, better primary health care, and improved production of nutritious foods. It focused on improving the production of fish, chicken, vegetables, and fruits, promoting appropriate eating patterns, and correcting detrimental food beliefs. Villages produced nutritious local food mixtures of rice, legumes, and sesame or peanuts (capitalizing on protein complementarity) for supplementing the diets of young children. Vegetable gardens, fishponds, and chicken raising were also encouraged to supplement school lunch programs. Much of the program was delivered by volunteers who were selected from local communities and trained.

After the dramatic successes in the 1980s, the Thai government expanded the program to address a wider range of food and nutrition issues. The expanded program included

- production of diversified foods for home consumption;
- skills development and credit schemes for commercially viable food processing and marketing activities;
- fortification of the Thai instant noodle seasoning with vitamin A, iron, and iodine;
- mandatory labeling of food products;
- dissemination and promotion of healthy diets, with special advice for age-specific vulnerable groups such as infants and young children, adolescent girls, and pregnant women;
- free or highly subsidized health services; and
- a monitoring, surveillance, and special feeding program for children under 5 years of age and for those in primary school.

The Thai program highlights the importance of addressing several different levels and sets of factors simultaneously. The program addressed the *availability of food,* particularly on the local and regional level. This was part of a larger, long-standing Thai program to improve economic growth through a strategy of increasing both national staples self-sufficiency and industrialization to develop employment opportunities for urban workers and displaced farmers. The program also addressed the availability of some problematic micronutrients by making them available through fortification of key foods and seasonings.

Elements of the program also addressed the *ability of households to access available food* by increasing income and reducing poverty. Some of the programs focused on providing the resources necessary for small farmers to increase income by producing and processing food for home consumption and sale.

The program also addressed the *desire of households to use available food* by providing appropriate nutrition education and information, simultaneously helping to increase the availability of the foods recommended and the

(continued)

Focus 10.2   (continued from previous page)

formulation of locally based food mixtures for very young children. This also included focusing on improving the nutrition of pregnant women and mothers.

Finally, the program addressed the *health* of individuals, by improving access to both preventive and curative health care through the subsidization of health care services.

percentage of their income as taxes. In much of Latin America, for example, taxes are raised primarily through sales taxes rather than through income taxes. This is common in less developed countries because with poor record keeping and large informal economies, it is hard to collect income taxes and easy to collect sales taxes. Foster (1992) argues that in some rural settings in which much of the economic transactions are through barter, the poor may be able to avoid enough sales tax to make the sales taxation less regressive. However, with increasing urbanization and globalization of markets, we would argue that the ability to avoid sales taxes is rapidly disappearing. Taxing land is another potential way of redistributing income. Progressive taxation will not benefit income redistribution if it is not spent on goods that the poor need. Top among these would be education—especially for women—and health care.

Price Subsidies   One of the potential uses of tax revenues is subsidizing the price of foods. Purchasing power is a combination of income and the cost of goods. It is increased either by increasing income or by decreasing price. Some countries subsidize the price of food to keep the costs low. Price subsidies can be marketwide, the price of a food is lowered to all consumers; or they can be targeted, the price is lowered to some consumers.

There has been a good deal of discussion about the impact of price subsidies on food. In the 1980s a program designed by the Mexican government—the Mexican Food System (SAM)—achieved dramatic results in 2 years by subsidizing the price to consumers of key staple foods such as tortillas and beans, while simultaneously subsidizing farmers to produce food crops. The SAM eventually proved too expensive to sustain and was dismantled after 2 years. The Egyptian government has been subsidizing grain and bread since Joseph. Interpretation of the Pharaoh's dream suggested putting aside the surpluses in 7 fat years in order to subsidize bread in the 7 lean years he foretold would follow (Foster 1992). Egypt has managed the sale of bread, flour, legumes, sugar, tea, and cooking oil within the country, making these items available to consumers at prices well below world prices. At the same time, they supported the prices that farmers received for these goods, so as not to undercut farmers.

Egyptian price subsidies have been credited with keeping undernutrition much lower than one might expect given Egypt's GNP. However, like the SAM, it has been enormously expensive, even though part of the subsidy has been supported by food aid from North America. The heavy use of subsidies is also thought to result in a good deal of wastage; that is, people waste food when it is very cheap, and there is little incentive to develop good storage. While farm prices were also subsidized, the wheat purchased on the international market and sold under cost did have the result of depressing the price that Egyptian farmers received for their wheat, and the inevitable occurred: Looking for higher-value crops, farmers turned away from wheat production, making the government more dependent on international markets.

Both Egypt and Mexico have used market subsidies to distribute food. That is, all of the particular food available in the market is subsidized. This means that both the poor and the rich benefit. Sometimes, food subsidies actually benefit the rich more than the poor. Subsidies on high-value foods such as animal products make them less costly to the wealthy, but they may still be out of reach of the poor. A more efficient way to manage food subsidies is to target the subsidies to foods that benefit the poor more than the rich. Subsidizing foods the poor use more means that less of the subsidy goes to those who do not need it. The consumption of some foods declines as people get richer, because they move away from that food to a preferred food. For example, the poor in Mexico eat more tortillas than the wealthy do, and as people get more income they switch to wheat bread and animal products, replacing less valued grains. In the economic vocabulary, these are "inferior" goods. They may not be nutritionally inferior, but because people prefer other things, they are valued less. In fact, some "inferior" goods, such as maize, sorghum, millet, and sweet potatoes, are more nutritious than the preferred food—rice—in Africa and South Asia. Subsidizing inferior goods means that more of the benefits of the subsidy goes disproportionately to the poor. Even at lower prices, the wealthy will not buy more tortillas, or sorghum, or sweet potatoes, but the poor will. Focus 10.3 describes the diets of some of the poor.

Another way of targeting food subsidies is to limit a program of subsidies geographically, locating subsidized or free food where the poor live. Or people might have to pass a means test, that is, demonstrate low income or existing undernutrition in order to be able to get access to subsidies. The U.S. Food Stamp Program is an example.

Finally, foods can be directly distributed. Most commonly, direct distribution takes place in the context of school lunch programs, soup kitchens, or supplemental feeding programs for women and children, all of which have had some success in alleviating malnutrition. In the United States, school breakfast and lunch programs and the commodities distribution program are examples. (See Highlight 11 on federal food programs for more details.)

Food Aid as Food Subsidy

In the last decade or so of the twentieth century, several net food producers, principally in North America and Europe, reduced production of staple grains and some animal products. The reduction came as a result of increasing food surpluses and decreasing international prices for these commodities. One of the results has been a

FOCUS 10.3 Diets of Hungry People[1]

Undernourishment is typically reported in terms of the percentage of a population with inadequate diets, but what does it mean for an individual to live day in and day out without enough food? These profiles present actual diets of three chronically hungry people—a Bolivian man, a Zambian boy, and a recently married Pakistani adolescent girl. These individuals face continuous nutritional vulnerability because they live in households lacking the resources to produce, buy, and prepare enough food for a healthy diet, although meals may vary by day and by season.

Even when a country has enough to feed its population adequately, some groups within a society may be vulnerable to food insecurity. Within communities, some families are at greater risk of malnutrition than others, and within families some members are at greater risk than others. Young children and pregnant and lactating women are the most vulnerable groups. But undernourishment cuts across all of society: It means a child of school age may not gain the full benefits of education, a laborer may be less productive, and a young woman may risk miscarriage. The diets described here show what these individuals ate during two 24-hour periods and are presented to illustrate common eating patterns; they are not statistically representative of the diets of the larger population.

Zambian Boy

Seven-year-old Mumba Mwansa lives in northern Zambia near a lake on which his father fishes in the employ of another man. Mumba's mother grows vegetables such as sweet potatoes, cowpeas, pumpkins, and groundnuts. His father grows cassava during the rainy season, when the lake is closed to fishing.

During the dry season, Mumba has boiled sweet potato with some roasted groundnuts in the morning. He takes a snack of roasted cassava to eat at school, and in the afternoon he shares a family meal of cassava *nshima* (a thick porridge) and boiled bream fish. At night, his family has a fresh pot of *nshima* with fish relish.

Thanks to fish and cassava flour, Mumba's dry-season diet is above minimum requirements in terms of carbohydrates and protein. However, he does not get enough fat, iron, or calcium, all necessary for growth and health.

The wet season is more difficult. Mumba's parents go to the fields early, so he either roasts cassava himself or goes without breakfast. In the afternoon, Mrs. Mwansa prepares nshima and a relish of vegetables and groundnut flour. The boy snacks on three small mangoes in the morning and again in the afternoon. The Mwansas do not have an evening meal.

At this time of year, Mumba's diet does not provide enough energy for daily activities, health, and growth. A boy of his age needs 1,800 calories each day, yet on this day, he consumed only 1,121 calories. Seasonal hunger and lack of some essential foods throughout the year contribute to the high levels of stunting found in the area where Mumba lives.

The Zambian diet demonstrates how a person can be adequately fed in one season and undernourished in another, especially dangerous for a growing child, as it can lead to stunting and other problems. This diet also shows that important nutrients can be lacking even when dietary energy is sufficient.

Bolivian Man

The Bolivian diet is an example of the importance of indigenous foodstuffs, which often do not appear in official statistics on food supply. Yet even with these nutritious traditional foods, people can be seriously undernourished.

Pedro Quispe, 35, supports four children and his wife by working on a farm near Lake Titicaca in Bolivia. Some mornings he begins the day with a meal of boiled maize, some *chuño* (a preserved potato), and fried broad beans. Two or three times a week, his wife serves him *wallake,* a soup made of *carachi* fish from the lake plus potato, onion, peppers (*ají amarillo*), *koa* (an aromatic herb), lard, and salt.

Mr. Quispe walks one hour to get to the fields. After working for several hours, he has a snack of *chuño* eaten with sauce made of peppers (*ají molido*), onion, and tomatoes.

On his return home in the evening, his wife serves him soup made of rice, potato, onion, carrots, lard, and salt. The soup is eaten with a paste called *quispina* made from quinoa (*Chenopodium quinoa*), a grain indigenous to the Andean Highlands. He enjoys a cold barley drink, called *pito de cebada retostada,* with water and sugar.

(continued)

---

1. Excerpted from *The State of Food Insecurity in the World 2000* (Rome: Food and Agriculture Association), pp. 13–14.

Focus 10.3 (continued from previous page)

Mr. Quispe needs a lot of energy to do agricultural work, walk long distances, and do the heavy tasks that his wife cannot do in the home. It is estimated that a man in this mountainous region needs at least 2,800 calories per day to maintain his level of activity and health. Yet Mr. Quispe's diet provides only 75% of the dietary energy he requires because he consumes too little fat and carbohydrate. Because his diet is poor, he also lacks calcium and vitamin C.

Pakistani Adolescent Girl

The Pakistani diet illustrates concerns about poor nutritional status in adolescent mothers, a factor contributing to high levels of infant and maternal mortality in many parts of the world.

Tahira Khan is a newly married 15-year-old in an isolated hill community in Pakistan. In the morning she fetches water to boil for *chai* (tea), which she drinks with milk and sugar. She and her mother-in-law prepare the family breakfast; after the men leave for the fields, Tahira eats her share, one *paratha* (a type of pancake) made of whole-wheat flour and ghee (butter). Once or twice a week she also has an egg fried in ghee.

She and her mother-in-law spend most of the day on household chores. In the afternoon Tahira eats a *chapati*, a light white bread, with potato and eggplant flavored with tomatoes, onions, and red pepper, cooked in ghee. When the men return from the fields, Tahira serves their evening meal and then eats hers: another *chapati* and mixed vegetables cooked in ghee.

As the village is difficult to reach, the family depends on their gardens for most of their food, so the variety is limited. Tahira's diet contains nearly adequate levels of protein, but it is of low quality because it comes mostly from wheat. Pulses improve the quality of the protein, but she does not get enough of them. Her diet is particularly deficient in fats and also lacks sufficient carbohydrates.

Her limited diet is a concern because she is still growing. In particular, she needs more calcium to nourish her future children and to be stored for her later years. If Tahira becomes pregnant, she and her infant will be at risk because of her diet. Poor nutritional status contributes to high levels of infant and maternal mortality in Pakistan.

decrease in the amount of food aid available for poor countries, especially countries with low food availability, and a stabilization of food prices on the international market, slowing the long-term decline in food prices. Food aid currently goes almost exclusively to disaster relief, including famine as the result of drought, flood, and war, and to refugees (see Figure 10.3). There is little basic food aid as part of ongoing international assistance. Several agencies (ACC/SCN 2000; Pinstrup-Andersen et al. 1999) have called for an increase in international food aid in order to support efforts to jump-start development efforts in some poor countries with food deficits. However, in the past, poorly designed food aid packages have had detrimental effects. Indiscriminate and poorly planned food aid was sometimes siphoned off by corrupt officials and elites. More commonly, it was dumped at low prices on the national market and had the effect of lowering the prices that local farmers, many of whom were struggling small farmers, would receive for their crops. One of the results was the loss of income to poor farmers, but another was the abandoning of staple production by small farmers who had to turn to nonstaple crop production in order to survive.

Even food aid transferred as disaster relief can have negative consequences. If the capacity to provide food internally exists even after a disaster, it may be much more sustainable for governments to support local producers rather than distribute international food aid. Finally, food subsidies tend to benefit urban consumers rather than rural producers.

Figure 10.3
Food aid is of great importance in periods of crisis and for refugees.

© Howard Davies/CORBIS

In sum, there are a number of strategies to reduce poverty and increase household and individual access to foods. These are land redistribution, agrarian reform, credit, markets, and income transfers and subsidies. Most of these strategies have measurable benefits and are appropriate in particular circumstances. However, no one approach alone can solve problems of food security and nutrition, and some have unintended negative consequences. Making choices among alternative policy options to reduce entitlement failure is one of the most important tasks of planners who are attempting to improve nutrition.

Nutritional Quality of Food, Education, and Household Distribution

Having sufficient access to food does not necessarily ensure consumption of an adequate diet within households. In Chapter 8 we discussed a number of the cultural and ideological factors that influence food choice and use. People may not have sufficient information to make the best nutritional choices. The value placed on men and women, girls and boys, older and younger household members may have an influence on the kinds of food that households acquire or the distribution of food within the household. Individuals and households have preferences for foods and combinations of foods. Traditional knowledge about nutrition and diet may be good, neutral, or detrimental to optimal food intake. While we argue that overall food availability and access to food through adequate food entitlements are the most critical aspects of adequate nutrition and food security, there is plenty of evidence to suggest that improved access to information and the introduction of new foods to address specific nutrition problems are also important.

In Kenya, a project to help reduce the prevalence of vitamin A deficiency in children targeted women farmers and provided new varieties of sweet potatoes, rich in beta-carotene, a vitamin A precursor. In this study, some women received planting materials (in this case, seed tubers) and technical assistance in planting and cultivating the new varieties. Other women received nutrition training on the importance of vitamin A for children, and training on food preparation and preservation, in addition to the new sweet potato varieties. Among those children whose mothers received both the planting materials and nutrition education and training, the consumption of vitamin A rich foods increased by 38%. For the children whose mothers received only the planting materials, the consumption of vitamin A rich foods actually declined (Hagenimana et al. 1999). It seems that providing both the means to increase vitamin A consumption and the information about the importance of doing so, exerted a powerful synergistic effect on child nutrition.

As we noted in Chapter 8 and Highlight 9, there are a number of beliefs in different parts of the world that are potentially damaging to the nutrition of children and women. In some cases, nutritional deficiencies such as night blindness (see Highlight 9), and even kwashiorkor and marasmus, are so common that they are considered part of normal childhood development. There are a number of popular medical systems in which diarrheal disease is treated by withholding food, a practice that can potentially push a sick child into malnutrition. In some parts of the world, it is considered inappropriate to push children to eat when their appetites are low as a result of illness or even malnutrition. In other cultures there are a number of food restrictions for women, especially pregnant and newly delivered women.

In the past several decades, nutrition education aimed at improving nutritional status has benefited greatly from improved understanding of how to develop interventions and messages that are culturally appropriate, well understood, and effective. In Chapters 11 and 12, we will address the ways in which new approaches to social marketing have enhanced the effectiveness of nutrition messages.

Some cultural values and attitudes are difficult to change. Several analysts point to the high value placed on boys and men and low value placed on women and girls in South Asia (India, Pakistan, Bangladesh, Nepal) as one of the most important contributors to poor health status and death of girls compared with that of boys. This male preference is also associated with the high prevalence of low birth weight that characterizes South Asia. While boys are more highly valued than girls in a number of cultures, there are few other settings in which there is a strong demonstrable effect of male bias on the growth and health of girls and boys. In India, child mortality is almost 50% higher for girls than for boys (Osmani 1997). More importantly, the very high prevalence of low birth weight (up to 50% of babies in Bangladesh and 33% in India) is attributed to the "woeful condition of maternal nutrition in South Asia" (Osmani 1997:41). However, in a series of studies of childhood growth in Latin America, even in areas in which there is clear male bias, we

Figure 10.4
Safe water supplies are a precious commodity in many parts of the world.

© Jeremy Horner/CORBIS

have failed to find significant differences between boys and girls beyond what one would expect as a result of greater female robustness. Girls grow slightly, but not significantly, better than boys. In most of Asia and Africa the results are similar.

Health and Sanitation

Access to safe drinking water, proper sanitation, and adequate health care are critical to ensuring that the food consumed by individuals actually contributes to improved nutritional status (see Figure 10.4). A series of studies carried out by IFPRI examining the impact of agricultural commercialization on nutrition were able to demonstrate that the food consumption by children was improved in most cases (Kennedy and Cogill 1987; Kennedy and Peters 1992). However, there was little or no impact on nutritional status as a result of poor health. In some cases, increased maternal workload needed in order to grow the new crops took women away from their children and resulted in their having time for important hygienic practices. In other cases, continuing poor health conditions just seemed to dissipate the impact of increased food intake. Basic health care has to be addressed at the same time as nutritional improvement.

Summary

- Although the problems of improving food security and nutrition sometimes seem insurmountable, experience from a number of countries and programs shows that dramatic improvement can be achieved. Strategies and policies for improvement must be directed to several different levels of aggregation and address a number of different sets of factors.

- Food availability can be addressed on the global and national level by continuing attention to agricultural research to improve production especially by small and resource-poor farmers and women farmers. This includes the development of technology appropriate to these farmers, the availability of credit and inputs (e.g., fertilizers and herbicides), and improved access to markets including improved means of transportation.

- Although food ability on global and national levels is necessary for adequate nutrition and food security, the most important reason for food insecurity and malnutrition is the failure of entitlements. Entitlement failure limits the ability of households to access available food. Poverty, then, is the main cause of malnutrition. There are a number of ways to improve entitlements. Redistribution of income and resources such as land reform, subsidized credit, and food subsidies all have some effect, although the impacts may be mixed and different in different settings. Economic growth with equity, creation of employment, and minimum wage laws are all critical in ensuring that people have sufficient income to purchase food.

- The distribution of income within households is also important. Women as farmers and wage earners are often invisible to policy makers. But improving women's education, income, and health may have a much more dramatic impact on food security and nutrition of children than improving men's access to them.

- Sound information about nutrition and health can have an impact on how food is distributed within households and how it is utilized by individuals. Culturally sensitive and appropriate nutrition and health education are also important.

- Improvement in food security may not have full impact unless it is accompanied by improved access to clean water, sanitation, and primary and preventive health care.

- Strategies that address a range of levels and factors are likely to have the best outcomes. Thailand is a country that has done an especially good job of reducing malnutrition. Policies to increase food availability (through improvements in both staples production and importation) and alleviate poverty were coupled with community-based nutrition education programs. Volunteers were used to monitor local families and provide supplementary feeding to undernourished children. Thailand lowered the percentage of children who were underweight from 50% of children in 1980 to less then 10% of children in 1998 (see Focus 10.2).

HIGHLIGHT

Women: A Pivotal Link in the Food Chain

Women are key to improving food security and nutrition globally. In their roles as farmers, workers, household managers, and mothers, women grow food, support families, manage resources, and provide food and care to children and other family members.

Women as Food Producers

In many parts of the world, women are the primary producers of food. In sub-Saharan Africa, where women and men often farm separate plots, women farmers have traditionally been responsible for food production. It has been estimated that women account for about 75% of the food produced in sub-Saharan Africa (see Figure H10.1). Also, women in sub-Saharan Africa carry out 90% of postharvest processing of crops for both subsistence and market (Quisumbing et al. 1995).

In Asia and Latin America, where it is common for men and women to work together on a family farm, women provide between 10% and 50% of the labor in agriculture. In research in indigenous communities of the Ecuadorean Andes, anthropologist Sarah Hamilton (1998) showed that women manage their own pieces of land, spend about 8 hours/day on average in farm work, hire their own agricultural workers, and sometimes even hire their husbands to work on their land. They produce potatoes, quinoa, barley, fava beans, and peas; keep chickens, pigs, dairy cows, and sheep; and produce food crops, eggs, and dairy products for their own households and for sale. They also work side by side with their husbands on family plots of land. When asked to state their primary occupation, over 70% of women responded that they were farmers. (In similar research in other areas of Ecuador, the more common response was "housewife.") When Hamilton asked a woman of nearly 60 years, who was considered "wealthy" by local standards, why she continued to put in full days of work in agriculture, she responded: "I like to work, I am a farmer. What else would I do?" (Hamilton 1998:150). In research in Honduras, we found that women put in less time as agricultural workers, but had the major responsibility for marketing agricultural products (DeWalt and DeWalt 1987).

It is not only in developing countries that women are major food producers. In a study of older adults in rural Kentucky, we found that older women who had been farmwives on small family farms saw themselves as the major food producers for their households. Although they worked with their husbands and hired hands on the family farm, the greatest part of household subsistence production came from extensive "gardens" tended virtually exclusively by women with the help of their husbands and children. They

Figure H10.1
Women produce much or most of the food supply in a great many cultures. These women from Papua New Guinea are fishing.

© Peter Johnson/CORBIS

produced a wide variety of crops, including cabbages, tomatoes, squashes, greens, sweet potatoes, Irish potatoes, turnips, onions, peas, several types of beans (both green beans and dried beans, like pinto beans and kidney beans), and corn, and took major responsibility for animal husbandry and food processing. They used the "butter and egg" money they earned by selling the eggs their chickens produced and the butter they churned to purchase items they could not grow such as coffee, sugar, and sometimes flour and cornmeal. Kentucky farm women not only raised most of their families' food in their gardens, they did all of the postharvest processing and storage as well. They canned, dried, and stored food for use through the year. Their husbands had the responsibility to grow cash crops to generate income and to grow some grains, such as wheat and corn, which were both subsistence and cash crops, but it was the farmwife's job to raise and process the food to feed the family.

Even though women have heavy responsibility for both subsistence and cash food crop production in many areas of the world, women farmers face important constraints not faced to the same extent by men. First, women farmers often have weaker land rights than men. In some countries, religious law restricts women's ownership of land even when civil law allows women to inherit and own land. In some parts of sub-Saharan Africa, where patrilineal inheritance (inheritance through the male line: father to son) and patrilocal residence (living with the husband's family) are practiced, a new wife coming into her husband's community can gain access to land only through allocations made by senior male relatives of her husband. While family heads do understand the importance of allocating land for food production to the women who marry in to the households, women may get smaller amounts of land or land of poorer quality than their husbands. There is some evidence from areas such as Ghana and Sumatra that as a result of religious and other social change, matrilineal inheritance of land and inheritance of land by women are actually shifting to patrilineal inheritance with a preference for inheritance by sons rather than daughters (Quisumbing 1994).

Deere and Leon (Deere 1987; Deere and Leon 2001) have shown that land reform and redistribution programs in a sample of Latin American countries resulted in a weakening of women's land rights compared to men's as, in most cases, land was granted to "household heads," and household heads were almost always defined as the "male household head." This was true even in areas such as the Ecuadorean Andes, where women traditionally owned and inherited land. As land reform took place in Ecuador in the 1960s and 70s, women lost the use rights to land as land was allocated to male household heads. Women were required by the land reform law to get their husbands' permission to apply for membership in agricultural cooperatives, and they could not belong to the same cooperatives as their husbands (Hamilton 1998).

Women farmers also have poorer access to new information and technology. Women farmers tend to have fewer tools than men

(FAO 1996) and the tools that are available are often designed for men, who are larger and have greater upper-body strength. As we noted in Chapter 5, in production systems with very simple technology, such as hoes and digging sticks, women tend to control more of the agriculture. In systems in which plows and animal traction are used, men tend to dominate agriculture. Some of this difference may be as simple as the extra upper-body strength necessary to work with a plow and animals. Technology appropriate to women's size and strength has been slow to develop (the assumption is always that the *farmer* is a man). However, several member organizations of the Consultative Group on International Agricultural Research (CGIAR, or CG) have been working to develop technology appropriate for women. The International Rice Research Institute (IRRI) in the Philippines has developed micro rice mills, direct seeding equipment, and transplanters and threshing machines designed to be used by women. The International Institute for Tropical Agriculture (IITA) in Nigeria has developed cassava-processing technology built for women (Quisumbing et al. 1995).

Historically, women as food producers have been overlooked by agricultural extension workers, development agents, and those who develop technology and bring it to farmers. Most agricultural extension agents are men, and many development agents expect farmers to be men. While small farmers in general get less technical assistance from extension agents, studies in Kenya, Nigeria, Tanzania, Zambia, El Salvador, and Ecuador show that women farmers get even less attention and fewer visits than male farmers (Hamilton 1998; Quisumbing et al. 1995).

Women have less access to extension and information for several reasons. In some cultural settings, it is culturally or religiously inappropriate for male extension workers to interact with women directly. Women often have busier schedules balancing child care and domestic work with farming and sometimes cannot attend demonstrations and talks. Women are often less likely than men to speak the national language. For example, more Ecuadorean Andean women than men speak only an indigenous language. Only a handful of extension agents speak the indigenous language. Finally, as we have noted before, both national extension workers and international consultants and planners *tend to expect farmers to be men.*

Women farmers also have less access to credit than men. With weak land rights, usually having rights to use land, but not having a clear title, women farmers are less likely to have collateral with which to secure credit. Although women farmers, despite all odds, do grow cash crops in addition to subsistence crops, the types of crops they grow tend to be lower-value crops, for which there is less credit available (see Figure H10.2). The amount of paperwork and travel needed to secure credit may be a barrier for very busy women with child-care responsibilities. Women also tend to have less education than men and are less likely to be literate. Poorer education and lower literacy make it more difficult for women to get access to

Figure H10.2
These Indian women are harvesting grain.

UN/DPI Photo

new information or to make successful application for credit (Quisumbing et al. 1995).

At the same time, women have important agricultural knowledge that is frequently overlooked (remember, researchers *tend to expect farmers to be men*) and often not included in the development and testing of new varieties of seeds or technology. In cases in both Latin America and Africa, we have personally observed researchers contracting with men farmers to test new varieties of crops that are, in practice, grown exclusively by women, or to test livestock management practices for animals that only women manage. (The researchers then wonder why these varieties and practices are not adopted by farmers. Perhaps they should have consulted with the *farmers*!)

When women are included in the development and testing of new seeds and techniques, the results can be dramatic. In one case, scientists at the Agricultural Sciences Institute (ISAR) of Rwanda, and the International Center for Tropical Agriculture (CIAT) in Colombia, a CG system institution, were collaborating on the breeding of new bean varieties. In the past, the two or three bean varieties that scientists had thought would be the most productive had actually shown only mildly successful increases in yields when tested under actual field conditions. ISAR and CIAT scientists then invited women farmers (who actually grew the beans in Rwanda) to take a look at the 20 or so new varieties at the research station and take home the ones *they* thought would do the best for *them*. The women planted the varieties they had chosen, using their own methods of experimenting with varieties. Women farmers did not necessarily choose the varieties on the basis of yield. However, under their usual field conditions, their selection out-yielded the scientists' selection by 60–90%. Farmers were still growing the selected varieties 6 years later (Sperling and Ntabomvura 1994).

Hamilton (1998) demonstrated that indigenous Ecuadorean women not only own and manage their own fields, but also have jointly purchased family land with their husbands. She found that women usually manage the household finances and participate equally with their husbands in decision making on the family farm. Hamilton was continually frustrated by the failure of extentionists and development agents to recognize this. She writes:

> For development planners and students of gender and develop-
> ment, resource dynamics within households have even more far
> reaching implications than the gendered division of labor. I have
> been in the company of development professionals, who, observ-
> ing Chanchaleñas [women of the Ecuadorean community of
> Chanchalo] at work in the fields with their husbands, assume
> that the women are merely "unpaid family labor" in market-
> oriented small-scale agriculture. They maintain that, even if
> women do much of the agricultural labor, their husbands make
> all the important production decisions. As development assistance
> will be offered to the household decision-making partner who has
> authority to commit resources or to veto participation in projects,
> these development professionals expect to work with men. (p. 168)

It seems clear that one result of the male bias in reporting agricultural production and agricultural research has been to overlook and devalue women's contribution to food and other agricultural production. It has been argued that agriculture's failure to meet Africa's food needs is in part due to the fact that technology to improve production has been made available to the wrong people. That is, it was given to men and not to women.

Women as Income Earners

Although many farmers, especially food-producing farmers, are women, most women are wage earners, contributing income to households. Women as wage earners are also often invisible to policy makers, for many reasons. Women are more likely to be employed in the informal sector of the economy. That is, they are often self-employed in small entrepreneurial activities such as selling agricultural products in small markets, selling home-prepared foods on the street, producing handicrafts, and taking in laundry. Or they are in jobs that are never registered, such as working as maids or other domestic workers. Policy makers may overlook the needs of women workers, because they do not see them as workers.

A number of studies have now shown that men and women spend the income they earn in different ways. Worldwide, women appear to spend more of their income on food and health care to benefit their families, especially their children. Men spend a somewhat larger proportion of their income on themselves, but they are also generally more likely to spend on "big ticket" items such as housing, appliances, capital goods, and school fees (Dwyer and Bruce 1988; Guyer 1980; Haddad 1992; Kurz and Johnson-Welch 2000; Quisumbing et al. 1995; Tripp 1982).

In a comparison of studies from eastern and southern Africa, Kennedy and Peters (1992) found that women spent a higher proportion of their income on food than men did. This in turn had an overall positive impact on energy intake within households. A study in Kenya (Kennedy and Cogill 1987) found that children in female-headed households had significantly better nutritional status than children in male-headed households. In a study in Chile, Buvinic et al. (1992) found that children in households in which women earned more than half of the family's income, and in which women had greater control over decision making, the prevalence of child stunting (height-for-age) was significantly lower. Quisumbing et al. (1995) summarize some recent studies:

> *Evidence to support the greater impact of women's income on household food security is increasing. In Rwanda, cash income earned by women is positively and significantly associated with household calorie consumption. Although female incomes were lower than total male incomes and men had more than 10 times as much off-farm earnings as women, there were no female-headed households with severely malnourished children and a less than proportional number were found to be calorie-deficient. In Côte d'Ivoire, the share of household cash income earned by women in the household has a positive and significant effect on the budget share for food. Lastly, in the Philippines, after controlling for overall household total expenditures, female income share has been shown to have a positive and significant association with household calorie availability, household budget shares of medical care and child's schooling (important nonfood inputs into children), and preschooler weight-for-age. The probability of preschooler fever and diarrhea is also lower in families where women earn higher incomes. (p. 10)*

Because of these patterns, increasing women's income would have a greater impact on child welfare than increasing men's income. Policies and programs that shift income from women to men, as did several of the agricultural development projects we discussed above, may have a detrimental impact on child nutrition, even though total household income may rise. Studies of the impact of shifting from semisubsistence to cash cropping as a whole suggest that for small producers, entrance of the household into the production of nontraditional export crops (NTAE) tends to shift control of income from women to men (DeWalt 1993b). Both Blumberg (1985) and Nieves (1987) have suggested that this was one of the effects of the shift to NTAE production in the Western Highlands of Guatemala. Barham et al. (1992) speculate that this shift is responsible for the relative lack of impact of rising household incomes on child nutrition in the Cuatro Pinos project of Guatemala, in which households were encouraged to grow vegetable crops for export to the United States. Katz (1994) demonstrates that the contribution of men to food expenses in NTAE-producing households is lower and less responsive to increases in household income than in non-NTAE households in Guatemala. She argues that women are absorbing a larger share of the burden of food expenditures in these households. The impact of these shifts under circumstances in which total household income increases is still unclear.

Women's income also has an impact on their own nutrition (Bisgrove and Popkin 1996; Kurz and Johnson-Welch 2000). In a study conducted in the Philippines, Bisgrove and Popkin found that women who worked and earned income had significantly higher intakes of energy, protein, fat, calcium, and iron. The effect was stronger for women in households with lower incomes or who worked in the informal sector than for women from wealthier households or who worked in the formal sector. Again, better-nourished women may be better able to care for children.

Women as Mothers and Caregivers

Women as mothers are also the primary caretakers of children. The importance of high-quality child care is being increasingly realized. Good child care has a positive impact on child nutrition even when incomes do not change (Engle and Menon 1999; Kurz and Johnson-Welch 2000). Engle and Menon (1999) see care as having several components:

- Care for women, such as providing rest time, or increased food intake during pregnancy
- Breastfeeding, and feeding of young children
- Psychosocial stimulation of children and support for their development
- Food preparation and food storage practices
- Hygiene practices
- Care for children during illness, including diagnosis of illness, and adoption of health-seeking practices

What Factors Are Linked to Providing Good Child Care?

Women's Health and Nutrition

First, women who are poorly nourished themselves may not be able to care adequately for children (Allen et al. 1992; Kurz and Johnson-Welch 2000). Although there has been relatively little research on the linkages between women's nutritional status and their ability to care adequately for children, several studies show that women with low energy intake or low weight-for-height (BMI) have the poorest household cleanliness (Allen et al. 1992). Up to 40% of women in South Asia, and up to 20% of women in Africa, may have low BMIs, suggesting PEM. While low energy intake may affect women's ability to care, iron-deficiency anemia may be the most important nutritional constraint on women's capacity to work and care for children (Engle and Menon 1999; Rahminifar et al. 1992). As we noted in Chapter 9, 60% or more of women in less developed countries suffer from iron-deficiency anemia.

Caregiving and Women's Work Outside the Home

Women's workload is also a major constraint to optimal care. Worldwide, research shows that women spend more time working than men do. They work in agriculture, earn wages, sell street food, make handicrafts, care for children, maintain houses, carry wood, fetch water, and so on. In industrialized countries, women are described as having two work shifts: After work outside the home is completed, the second shift is domestic work. The impact of women's work on child welfare is complex (Leslie 1988). It appears that where women have control over their earnings and can spend their income as they wish, child nutrition is more likely to be positively associated with women's work outside the home. However, in cases in which women turn their income over to others—husbands, fathers, brothers—women's work outside the home may have negative impacts on child nutrition.

An examination of the cross-cultural data reveals several consistent patterns (Blau, Guilkey, and Popkin 1996; Engle and Menon 1999). When women's work is well paid, when women control their income, and when adequate alternative child care is available or the child is over 1 year of age, working outside the home appears to enhance the nutritional status of their children. Other factors also appear to have a positive impact, such as the flexibility of work, closeness of work to home, and the compatibility of work with child care. Women farmers in the Ecuadorean Andes take their infants with them when they go to work in the fields. Women involved in handicraft production such as weaving, knitting, and basketry may be able to work at home. However, income-earning women who are poor, with young infants, inflexible work times, and little control over their income, have children whose growth is impaired (Engle and Menon 1999).

Knowledge and Beliefs About Nutrition and Health

The amount of education that a woman has completed also has a dramatic impact on her ability to be an effective caretaker. As we noted in Chapter 9, local knowledge concerning appropriate child care and child feeding has a strong cultural foundation. Some local knowledge and beliefs are well grounded empirically and effective. The use of herbal teas by Mexican mothers to treat diarrhea in their children has long been seen as an effective way to keep children hydrated during acute illness. At the same time, some aspect of local knowledge may be negative with respect to child nutrition. Katherine Dettwyler (1994) has discussed impact of local knowledge on child nutrition in a peri-urban neighborhood in Mali. Malian women believe that children, even very young children, should be allowed to eat what they want, when they want it. This means that children with anorexia as a result of illness or malnutrition may not be encouraged to eat. Sometimes illnesses, including malnutrition, may be seen as normal stages in child development. In other cases, malnutrition, including severe marasmus, may be seen as a non-food-related illness. Doing research in Brazil, Scheper-Hughes (1992) discusses how women spend money on expensive medicines to treat their children, when the children are clearly suffering from a lack of food. Finally, several studies of child feeding in India show that Indian women do not believe that solid food is appropriate for children until late in their first year or early in their second year of life. The delay in the introduction of solid foods is thought to be an important contributor to child undernutrition in South Asia (Osmani 1997).

Access to adequate information and education is a key component of improving mothers' ability to care for children. In fact, there is a growing consensus that access to education for girls and women may be the single most important noneconomic intervention for the improvement of the lives of both women and their children (ACC/SCN 2000; Sawyer 1997).

Education

The importance of women as farmers, wage earners, and child caretakers highlights the importance of education for women in national strategies to alleviate poverty, hunger, and malnutrition. Worldwide, women's education is correlated with child health and welfare. Even holding income constant, better-educated women have healthier, better-nourished children. It is not entirely clear what the specific effects of education are. They seem to go beyond the benefits of literacy and greater control of information. Levine et al. (1999) argue that the most important impact is the introduction of women to a wider range of possibilities and potential solutions. That is, education shows women that there is a wider world around them and that problems can be thought through and solved using alternatives available in the world.

Higher education for women also translates into greater productivity as farmers (Kurz and Johnson-Welch 2000). Table H10.1 shows that in two studies in Kenya, when women had education and access to resources similar to men's, they were much more productive. Also, better education translates into higher wages in employment for women.

Some countries do much better than others in providing education to girls compared with boys. Some countries, even with the same amount of total resources available as other countries, do a much better job of educating children in general. The same is true for educating girls. Table H10.2 reports the gender gap, that is, the difference in the school enrollment rate of boys and girls for a number of countries with per capita GNPs below $500.

Table H10.1   Returns to Increasing Human and Physical Capital of Women Farmers

Study Site	When Women Have:	Yield Increases
Maize farmers, Kenya (1976)	Age, education, input levels of male farmers	9.0%
Food crop farmers, Kenya (1990)	Primary schooling	24.0%
	Age, education, input levels of male farmers	22.0%
	Same land area as male farmers	10.5%
	Same fertilizer level as male farmers	1.6%

Source: K. M. Kurz and C. Johnson-Welch, *Enhancing Nutrition Results: The Case for a Women's Resources Approach* (Washington, D.C.: International Center for Research on Women, 2000).

Table H10.2   Difference in School Enrollment Between Girls and Boys in Countries with Per Capita GNPs Below $500: Countries with Lowest and Highest Gaps

Countries with Gender Gaps 3 Percentage Points and Under	% Point Gap	Countries with Gender Gaps 15 Percentage Points and Over	% Point Gap
Nicaragua	−3	Benin	36
Haiti	−1	Yemen	34
Bangladesh	0	Chad	28
Malawi	0	Guinea-Bissau	26
Rwanda	0	Togo	22
Zambia	0	Nepal	20
Georgia	1	Gambia	18
Ghana	1	Niger	15
Madagascar	1		
Kenya	2		
Mauritania	2		
Nigeria	2		

Source: H. Sawyer, "Quality Education: One Answer for Many Questions," in UNICEF (Ed.), *The Progress of Nations* (New York: UNICEF, 1997), pp. 33–39.

Note: Negative gender gap indicates a higher enrollment of girls than boys.

Policy Implications of Women's Roles in the Food Chain

The policy implications flowing from this discussion should be fairly clear. They are applicable at all levels of aggregation. At the international level, research and development agencies need to carefully take into consideration the information needs of women and to provide appropriate technology. Needed at the national level are laws and policies securing land rights and fair labor laws protecting women's rights and access to education. Protecting women's health and nutrition for both themselves and their families is also important. The costs of *not* addressing the constraints on women as farmers, wage earners, household managers, and parents are great. The following steps could make a big difference in beginning to remedy the inequities for women that ultimately affect the communities, countries, and world in which they live.

- Increase the access of girls and women to education at all levels.
- Improve the secure access of women farmers to land by making land reform and land titling projects more responsive to women.
- Increase the numbers of agricultural extension workers who are women.
- Increase women's access to credit.
- Develop fair labor laws so that women's wage labor is appropriately compensated.
- Develop effective and appropriate health and nutrition education messages when nutrition education is not available.
- Focus on maternal-child programs that address the food needs of both children and women, not just pregnant women.

CHAPTER 11

Dietary Behavior Change: How People Change Eating Habits

Despite widespread interest in nutrition, most North Americans continue to consume a diet too rich in fat, calories, sodium, and simple carbohydrates, and too low in fruits and vegetables, complex carbohydrates, fiber, and calcium. Consider fat consumption in the United States. Although the proportion of energy derived from fat has declined from 37% to 34% over the past decade in the United States, the actual amount in grams has remained basically the same—80 grams per day. The decreased proportion results from an actual increase of 300 calories per day in the diet's total energy intake (Anand and Basiotis 1998; Hill and Peters 1998).

While many people have increased their consumption of fruits and vegetables, the pace of change is slow (Kantor 1998). As seen in Table 11.1, Americans consume only 59% of the fruits and 65% of the dairy products recommended by the USDA. While vegetable and grain consumption more closely approximates the recommended amounts, most Americans still fail to consume the variety of green leafy vegetables recommended and select grain products made with processed flour, fats, and sugars rather than the more nutritious

Table 11.1 Reported Daily Food Consumption as a Percentage of Food Guide Pyramid Recommendations

Age and Gender	Grain	Vegetable	Fruit	Dairy	Meat
			Percent		
All individuals ages 2 and older	84	89	59	65	82
Males	89	92	54	75	96
Females	80	85	64	55	69
Children ages 2–18	88	70	65	79	68
Males	93	71	63	88	74
Females	84	69	67	71	61
Seniors ages 60 and older	78	94	79	39	78
Males	83	99	75	44	88
Females	75	91	81	35	70

Source: Continuing Survey of Food Intake by Individuals, 1994–1996, Economic Research Service, U.S. Department of Agriculture.

whole-grain products (Economic Research Service 2000). Carbonated soft drinks and other sources of sugar have also increased dramatically in the American diet during the last 10 years (Kantor 1998). (See Figure 11.1.)

These dietary practices are costly. Five of the 10 leading causes of death in the United States—coronary heart disease, certain types of cancer, stroke, diabetes mellitus, and arteriosclerosis—are linked to diet. Diet also plays a role in obesity, hypertension, gastrointestinal disorders, osteoporosis, and neural tube defects. Taken together, these diet-related health conditions may account for over 50% of deaths in the United States and cost society an estimated $250 billion each year in medical costs and reduced productivity—a significant portion of which might be saved by a greater investment in nutrition education and improved nutritional status (O'Halloran et al. 2001).

With the realization that a large proportion of the mortality from chronic diseases is linked to people's behavior, social scientists and public health professionals have increased their efforts to understand preventive health behavior and ways to promote it. Unfortunately, efforts to change dietary and other health-related behaviors have often been disappointing.

Recent studies suggest that nutrition programs are most likely to be effective if they are based on solid theories for changing health-related behaviors (Contento et al. 1995). "Theories are summaries of formal or informal observations, presented in a systematic, structured way" (Goldman and Schmalz 2001). They include a set of interrelated concepts or explanatory factors, definitions of key concepts, and propositions about the relationships among them (Glanz, Lewis, and Rimer 1997). Theories are valuable when designing a program because they help explain why people act as they do and give us important insights into how people change their behavior. By pinpointing the factors that impact behavior, theories enable us to build strategies to help people improve their dietary practices. "Recommendations for changing eating habits are more likely to have an impact if they are based on an understanding of the factors that affect food choices" (Glanz 1993:277).

In this chapter we describe the most widely accepted health behavior theories, review individual and interpersonal factors that impact behavior change, and discuss their implications for designing nutrition interventions. These theories assume that dietary behavior is influenced by "people's knowledge of what to do, what they think about it, and how these merge within their personal worlds" (Glanz 1993:258). We begin the chapter with a look at the behavior change process and then examine theories that describe the impact of knowledge and beliefs, attitudes, intentions, social influences, and self-efficacy on dietary behavior. These theories focus on personal thought processes (knowledge, beliefs, expectations, attitudes, and perceptions), with the belief that people's perceptions of reality are just as, if not more, important than the environment itself. Most of these theories have been developed and tested in industrialized countries, although they may prove useful in

Figure 11.1

Carbonated soft drink consumption has risen dramatically. Beverages are the most popular food items Americans consume outside the home.

© Michael Newman/PhotoEdit

understanding behavior change in developing nations as well. In the next chapter, we describe large-scale nutrition intervention approaches that draw on an understanding of these individual and interpersonal factors and the environmental factors (institutional, community, and public policy) described in Chapter 10. Many of these approaches (e.g., social marketing) have proven successful in bringing about change in international settings as well as in North America.

The Behavior Change Process

Most social scientists now recognize that behavior change is a complex, dynamic process rather than a simple event. People rarely make dramatic, permanent changes in their diets immediately after learning about the need to change. According to many health behavior theorists, people usually move through a series of stages along a change continuum based on their interest, motivation, or readiness to change.

Stages of Change

In an attempt to understand the change process, James Prochaska and his associates (Prochaska and DiClemente 1983; Prochaska, DiClemente, and Norcross 1992) studied thousands of people as they quit smoking or gave up other addictive substances. These studies found that people move through a series of stages based on their readiness to change and that they rely on different intervention strategies to help them depending on the stage they are in at the time. According to Prochaska and DiClemente's Transtheoretical Model of behavior change, people move through the following stages: precontemplation, contemplation, preparation, action, maintenance, and termination. Regardless of whether the person is attempting to change alone or with a professional's advice and support, the characteristics of each stage are similar. These are described below.

- *Precontemplation.* People in this stage do not intend to change in the near future, for example, in the next 6 months. An obese person who is in the precontemplation stage with regard to weight loss would not be planning to follow a weight-loss diet in the next 6 months. He may be unaware of the need to change, underinformed about the benefits of changing, demoralized because he has failed to lose weight many times before, or well informed but still unwilling to try to lose weight in the near future. While in the precontemplation phase, he would tend to avoid activities such as reading or talking with others that require him to think about dieting. Like many people in precontemplation, he may not break out of this stage until encountering a crisis or precipitating event: a 40th birthday, sudden death of an overweight friend, or emergence of medical symptoms related to his condition. Health providers would classify him as "resistant to change," and efforts to enlist him for a weight-loss program would almost certainly fail.

- *Contemplation.* In the contemplation stage, people are seriously thinking about changing. If asked if he planned to lose weight within the next 6 months, the obese person in the contemplation stage would say "yes," even though he did not have a clear-cut plan for how to do so. Compared to the previous stage, he would be more aware of the benefits of losing weight, but he would also be acutely aware of, and perhaps overestimate, the problems or sacrifices he would have to make to change his behavior. While open to feedback from others, he might stay in this stage for a long period of time, weighing the pros and cons of change. In fact, the most distinctive characteristic of this stage is a lack of commitment to change. Efforts to recruit people in the contemplation stage into action-oriented programs, like a weight-loss or exercise program, are also misguided. Many people linger in this stage for long periods until their perceptions of the benefits of losing weight clearly outweigh the perceived costs or disadvantages.

- *Preparation.* Preparation is the stage in which people make a commitment to change, usually defined as within the next month. This intention to change is usually combined with some preliminary activities, such as the purchase of walking shoes, trial of low-fat food substitutes, or collection of information about weight-loss programs. By this phase, the obese person will have begun to formulate a plan of action and will have taken some preliminary steps. He is ready to change and a good candidate for action-oriented programs.

- *Action.* In the action stage, people have initiated specific changes within the last 6 months. Not all diet modifications count as *action;* the person has to have achieved a certain level of success. The obese person, for example, must have reduced caloric intake and/or increased caloric expenditure. The major challenge in the action phase is to prevent relapse. This stage is often defined as the first 6 months after initiating the change.

- *Maintenance.* The maintenance stage is usually defined as the period starting about 6 months after the new behavior has been adopted. In this stage, people are more confident in their ability to maintain the new behavior. Temptation is no longer as strong, and efforts to follow the weight-loss plan require less effort. However, the obese person must still be careful to prevent relapse and maintain the behavioral changes he has adopted. For some behaviors, this stage is the last in the sequence. People either remain in this stage or relapse to an early stage. For other behaviors, such as addictive eating behaviors, this stage may last for many years until the person no longer contemplates resuming previous habits and moves into the next stage.

- *Termination.* In this stage, people are completely confident in their new dietary routine and no longer feel tempted to return to previous dietary practices. This sixth stage is especially important in understanding addictive behaviors. The obese person who was addicted to sweets, for instance, is no longer tempted to binge on desserts, no matter how bored, anxious, or depressed.

Studies show that people may enter or exit at various points on the Stages of Change continuum. They also may remain "stuck" in a stage for long periods of time or cycle back and forth between stages rather than move forward in a steady progression. For example, a person may move quickly from precontemplation through contemplation and preparation immediately after a divorce, but then return to contemplation where he or she languishes for several years.

Processes of Change

While studying how people quit smoking, Prochaska and DiClemente (1983) also found people at different points in the continuum of change benefited from different types of intervention strategies. People in the early "motivational" stages of change were more likely to benefit from educational efforts to raise their awareness of the benefits of change and encouragement to reevaluate their current behavior. In contrast, those in later stages relied more heavily on praise, reinforcement, and coping skills that prevent relapse.

Recognition that intervention activities vary in their effectiveness according to the person's readiness to change has important implications for practice. The Stages of Change Model points to the ineffectiveness of using one intervention strategy to fit everyone and explains why many programs fail to recruit large portions of a population into action-oriented nutrition education programs. Most importantly, the Stages of Change Model helps health professionals identify people who are ready for change and select interventions appropriate for them based on their readiness stage. Prochaska and DiClemente called the intervention strategies people use to progress through the stages of change *processes of change*. A definition of each of these processes and the stages in which each has proven to be most beneficial are given in Table 11.2.

The Stages of Change Model and other "stage theories" have been criticized by some social scientists that question the existence of true, discrete stages. They believe the change process is too multidimensional and complex to be described by the Stages of Change Model. These critics question the claim that people change qualitatively as they move from one stage to another and criticize the use of an arbitrary 6-month cutoff in defining the action and maintenance stages (Bandura 1997). When applied to dietary behaviors, the stages in the change process can be more difficult to define than when studying substance abuse. With smoking cessation, for instance, action is defined as complete abstinence from smoking—a behavior most people can report accurately. But in studying decreased fat intake, people may not know or may be unable to report their dietary intake accurately. In studies of dietary fat reduc-

Table 11.2   Change Processes in the Stages of Change

Processes	Definitions
Stages of Change	
Precontemplation	
Consciousness raising	Finding and learning new information to support the change
Dramatic relief	Experiencing negative emotions that encourage change
Environmental reevaluation	Recognizing the negative impact of the unhealthy behavior and the positive impact of the healthy behavior on the social environment
Contemplation	
Self-reevaluation	Recognizing that the behavioral change is an important part of one's identity
Preparation	
Self-liberation	Making a firm commitment to change
Action and Maintenance	
Contingency management	Increasing the rewards for the healthy behavior and decreasing the rewards for the unhealthy behavior
Helping relationships	Seeking and using support from others to change
Counterconditioning	Substituting healthy alternatives for the unhealthy behavior
Stimulus control	Removing cues that encourage the unhealthy behavior and adding cues and reminders to adopt the healthy behavior

Source: J. O. Prochaska, C. C. DiClemente, and J. C. Norcross, "The Transtheoretical Model and Stages of Change," in K. Glanz (ed.), *Health Behavior and Health Education* (San Francisco: Jossey-Bass, 1997), pp. 60–84.

tion, as many as 46% of people who reported being in the maintenance stage for this dietary behavior change actually consumed more than the recommended amount (30%) of their calories from fat (Ounpuu, Wollcott, and Greene 2000; Povey et al. 1999). As a result, this model is best viewed as a means of classifying people for purposes of program design, and caution should be exercised when developing questions to determine which stage a person is in at a given time.

Despite these limitations, the Stages of Change Model is important because it recognizes that behavior change is a dynamic process with people varying in terms of their readiness to change. It also reminds us that the most appropriate intervention will be different for people in different phases. We should not expect to help everyone using the same program strategies.

Factors That Influence the Change Process

Many nutritionists, health educators, and other health professionals rely almost exclusively on education to bring about behavior change. Yet much of the information imparted is never put into practice. Why? Confusing messages and complicated nutrition information may contribute to the problem, and many people do not understand how to apply nutrition knowledge to daily food choices (Glanz 1993). However, even when people know what they should eat, they often fail to put this knowledge into action. How about you? How often do you fail to eat the foods you know are most nutritious?

In this section, we review numerous factors, in addition to knowledge, that influence the ability to change: health beliefs, attitudes, social influences, intentions, skills and self-efficacy. We also briefly summarize several theoretical models that describe the role these factors are believed to play in the change process.

Before reading further, take the time to answer a few questions about your own experience in attempting to change a habit. If you can understand how behavior change occurs in your life, you will be better able to understand and integrate the discussion that follows.

Choose a health behavior that you have changed at some point in your life. It could be eating less fat, getting more exercise, using sunscreen, reducing stress, or giving up or cutting back on alcohol or tobacco. If you can't think of anything, think of a behavior you would like to change. Write the behavior at the top of a piece of paper. Answer the following questions regarding the behavior.

- What information helped you decide to change the habit? Where did you learn the information? What risks did you associate with not changing your behavior?
- How serious would the results be if you did not change your behavior?
- What were the benefits of changing your behavior?
- What were the "costs" (money, inconvenience, time, etc.) of changing to the positive health behavior?
- What was your attitude about the behavior change?
- Did friends or relatives think you should make the change? Did they support you or hinder you in any way?
- What was your intention? (for example, quit smoking in 2 months, lose 10 pounds by spring break, or get more sleep during finals week)
- Did you have the skills to make the change? (for example, knowing how to read labels to count calories, how to use weights at the gym to increase muscle mass, or how to meditate and do yoga to reduce stress)
- Did you believe that you could make the behavior change?

Health Beliefs

One of the oldest health behavior models was developed by researchers in the public health service to explain why some people

Figure 11.2
Many factors influence behavior change.

"I feel guilty about eating the foods I like."

"Eating right is too expensive."

"I tried eating better, but I didn't stick with it."

"I don't have the time to eat right."

"The vegetables I like aren't available."

"I'm healthy now. Why should I worry about my diet?"

PhotoDisc

use preventive health behaviors while others do not. This theory, called the Health Belief Model, has been applied to a wide variety of health behaviors and is a relatively good predictor of behavior changes. The basic components of this model are the perceived susceptibility and perceived severity of the disease threat, and the perceived benefits and perceived barriers of adopting the preventive behavior (Hochbaum 1958; Rosenstock 1966). As illustrated in Figure 11.2, people are deterred from adopting healthier food practices for a wide variety of reasons.

Perceived susceptibility refers to a person's perception of the risk of getting a particular disease or health condition. It may also be used to describe the perceived risk of developing complications from a health condition, such as diabetes. Perceived risk is believed to motivate people to engage in preventive health behaviors. According to the Health Belief Model, men who believe they are likely to suffer from heart disease when they get older are more likely to take measures to prevent cardiovascular disease than those who do not feel susceptible.

Perceived severity refers to a person's perceptions about the seriousness of the illness. It encompasses the social as well as health consequences of the condition. According to the Health Belief Model, women who believe osteoporosis would be a devastating, crippling disease that will make them look old and feeble are more likely to take steps to prevent it than those who think it is not a particularly serious condition.

Many health professionals attempt to motivate people to change by increasing their perceived threat (a combination of perceived susceptibility and severity). Drug prevention programs often warn young people of the dangers of drug use and attempt to increase their sense of vulnerability to these risks. Unfortunately, fear appeals may often fail because people, especially those who lack confidence in their ability to change, *control their fear* by avoiding the message or discounting its authenticity rather than *control the danger* by changing their behavior. Fear appeals also fail because most people underestimate their susceptibility to a disease threat. To be most effective, fear appeals should meet the following requirements (Stephenson and Witte 2001) for using fear: (1) The onset of the fear should occur before the remedy is offered; (2) the person's susceptibility to the threat should be accentuated by using the word "you" and addressing specific behaviors that place the person at risk; (3) information about a specific person similar to the target audience should be used rather than an average person with whom they will not identify; (4) the recommended health behavior must be demonstrated to be effective in reducing the threat, thereby building a positive attitude toward the preventive behavior and offering immediate relief from the fear. Messages also need to be sufficiently threatening to increase people's sense of vulnerability to the threat, but not so terrifying that people become defensive in an attempt to control their fear (Stephenson and Witte 2001).

A third component of the Health Belief Model is the person's beliefs and expectations about the preventive health behavior's ability to prevent the disease or reduce its impact. Women who believe fruits and vegetables contain properties that greatly reduce the risk of cancer are more likely than those who do not to increase their produce consumption. Perceived benefits may also include nonhealth rewards or advantages the person expects to gain by adopting the recommended behavior. For many people, the driving motive for losing weight is to look better, not to prevent heart disease.

According to the Health Belief Model, and the other theoretical models discussed in this chapter, people do not adopt new behaviors unless they expect to get more than they must give up. People are assumed to weigh or compare the benefits with the costs of changing. In the Health Belief Model, the emotional and tangible costs associated with the new behavior are called *perceived barriers*. Time, embarrassment, inconvenience, loss of pleasure, and the psychic hassle of changing are among the many sacrifices people anticipate when they consider changing. A review of studies that used the Health Belief Model suggests that perceived barriers are one of the most reliable predictors of change for many health-related behaviors. What in your own life did you find that these or other barriers proved difficult to overcome?

An additional component in the Health Belief Model that has not been studied systematically is the concept called *cues to action*. These cues include a wide variety of possible sources of information and persuasion—bodily changes or symptoms, advertising, media campaigns, brochures, medical advice, reminders from friends and relatives—that serve as trigger mechanisms that remind and motivate people to change.

The Health Belief Model (shown in Figure 11.3) is probably most useful when studying behaviors linked to specific disease threats, such as heart disease, diabetes, and cancer. It may also be used to increase adherence to special dietary recommendations that can delay or diminish the impact of a chronic disease. Implications for designing change strategies include the importance of assessing and communicating personal risk for specific diseases, educating people about the consequences of the condition, clearly defining

Figure 11.3
Summary of the key valuables in the Health Belief Mode.

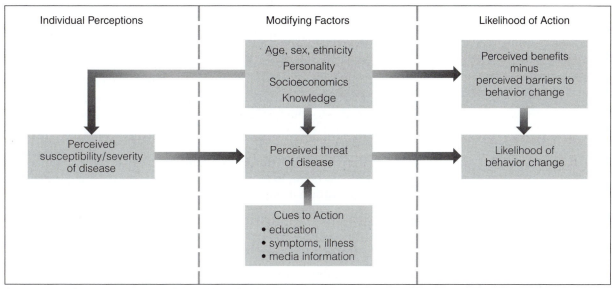

(*Data:* Janz, V. L. Champion, and V. J. Strecher, "The Health Belief Model and Health Behavior," in D. S. Gochman (ed.), *Handbook of Health Behavior Research: Vol. 1, Personal and Social Determinants* (New York: Plenum Press, 2000), p. 52).

the behavior required to prevent or minimize the impact of the disease and other benefits to be gained from changing, and identifying and reducing the barriers through reassurance, correction of misinformation, incentives, and other forms of assistance. Finally, the Health Belief Model points to the value of practical or "how-to" information and use of reminder systems to activate people (Janz, Champion, and Strecher 2000).

Attitudes

Attitudes are people's general evaluations of other people, objects, and issues. In understanding health behavior change, the relevant attitudes are people's positive or negative evaluations of performing a specific behavior. Attitudes are usually measured by asking people to evaluate a behavior, such as eating five servings of fruits and vegetables a day, in terms of bad/good, harmful/beneficial, pleasant/unpleasant, or other bipolar dimensions.

Attitudes are determined by the individual's beliefs about the consequences of behaving in a specific way, weighted by the importance the person places on that outcome. People who hold strong beliefs about the anticipated consequence of a dietary practice that is valued positively will have a positive attitude toward that behavior, whereas people who have strong beliefs about a dietary practice that is valued negatively will have a negative attitude about the same behavior. Many people, for instance, mistakenly believe potatoes are fattening. Those who consider fattening foods to be very bad are likely to develop a negative attitude about eating potatoes, whereas those who share the same belief, but consider fattening foods to be desirable, are likely to develop a positive attitude about eating potatoes.

Attitudes are important in understanding dietary behavior change because they are believed to mediate between the acquisition of new information and behavior change. In other words, new knowledge (e.g., fruits and vegetables prevent cancer) produces new attitudes (e.g., fruits and vegetables are good for me), which motivates people to change (e.g., they eat more fruits and vegetables) (Thomas 1991).

According to early theories about persuasion, new information is transformed into attitude and behavior change via a sequence of five steps: (1) being exposed to new information, (2) paying attention to the information, (3) receiving and storing the information in long-term memory, (4) interpreting or evaluating the information in a way that makes personal sense, and (5) integrating it into an overall attitude that is capable of guiding behavior (Petty and Cacioppo 1984).

Unfortunately for nutrition educators, knowledge does not always translate into practice. Problems can occur at each step along the way. People may not pay attention to information they are given. They may forget what they have heard. They may not believe the information or decide it is not important to them personally. To illustrate how the sequence of steps can fail to lead to behavior change,

> *consider two people who hear that the consumption of oat bran can reduce one's cholesterol level and reduce the risk of heart disease. One person, who is overweight and constantly dieting, responds to this information with the thought that oat bran must be good for one's health. Later that week in the supermarket, the person passes the cereal aisle and selects an oat bran product because it is perceived as a diet food that will help the person to lose weight. In this instance, the original information about the link between the product and heart disease was lost, but the self-generated (and probably mistaken) elaborations or translations of that information (from heart disease prevention to healthful to diet) guides the behavior. Another person hears the same information and responds with the thought that there is no need to be concerned since there is no family history of heart disease. The next week in the supermarket, this person passes over the oat bran products, even though there is a perfect recall of the information that was presented. Since this person does not feel personally vulnerable, learning the new information has no effect on that person's behavior. (Thomas 1991:51–52)*

The fifth step in the sequence, in which information is evaluated and elaborated so that it makes sense, is considered the most important and most vulnerable point in the change process. Studies of this stage suggest that information may be processed in two very different ways (Petty and Cacioppo 1986). First, some information is carefully considered and integrated into a person's beliefs and attitudes. This careful reasoning process is called the *central route* for information processing. When processing information via the central route, people need time, motivation, and opportunity to evaluate the new information in terms of existing beliefs and weigh its merits. This processing route is most likely to be used when the consequences of changing are perceived to be important, as when learning to cope with a recently diagnosed disease.

In contrast, other information may induce attitudinal change by a less conscious or *peripheral route*. This route involves cues that either elicit a feeling (e.g., happiness) associated with a specific food or trigger a simple inference that a behavior is bad (e.g., when a doctor says it will make you sick). Advertisers capitalize on the peripheral processing route when they show their products being consumed by successful role models or when they create an association between their products and fun, fame, or other positive states (Thomas 1991). When behavioral decisions do not seem important, people are more likely to react spontaneously, drawing on attitudes obtained via the peripheral route, rather than carefully weighing costs and benefits associated with the behavior. Have you ever opened the refrigerator door, noticed a piece of your favorite dessert, and eaten it without seriously evaluating the consequences?

Attitudes formed by careful evaluation and consideration of new information (the central route) are believed to have a greater impact on behavior and be more resistant to counterpersuasion than those formed through peripheral association (Petty and Cacioppo 1986). The decision to eat five servings of fruits and vegetables a day is more likely to translate into a permanent dietary change if it is based on extensive reading about, and thorough understanding of, the benefits of increased produce consumption than if it is the response to a physician's recommendation.

Once a person's attitudes have changed, many factors, including old habits and personal experiences, may impact his or her ability to bring about sustained behavior change. You may be convinced by your physician or a magazine article that the omega-3 fatty acids in fish oil will cure your migraines or prevent wrinkles and, as a result, develop a very positive attitude toward increasing the omega-3 fatty acid content of your diet. However, this attitude may sour suddenly if, like one author, you find that taking the recommended dose of fish oil capsules makes your breath smell like the fish the fatty acids came from.

In sum, attitudes have the greatest impact when they are consistent with a person's beliefs and have been formed after careful reasoning and evaluation. Compared with attitudes based on associations created by media messages or other environmental cues, attitudes based on a great deal of knowledge, intensive reasoning, and personal experience are more likely to translate into long-lasting dietary changes. Was this true for you in the health behavior you changed?

Social Influences

As you thought about the factors that helped or hindered your efforts to change your lifestyle, did you consider the contributions made by people important to you? Did anyone provide information and advice, encouragement, reassurance, praise, or tangible help with cooking or shopping? Perhaps you decided to change to avoid their criticism or prevent a conflict. Did any of your friends or relatives make it more difficult to change by pressuring you to continue old habits, ridicule your efforts to change, or tempt you with foods you wanted to avoid?

Social support is used to describe the "aid and assistance exchanged through social relationships and interpersonal transactions" (Heaney and Israel 1997). Four major types of social support have been identified: informational, emotional, appraisal, and instrumental. Informational support includes the provision of new knowledge, advice, or recommendations, as when one shares new recipes or a weight-loss plan with you. Emotional support includes reassurance, love, caring, concern, and empathy. When a friend listens intently to you vent your frustrations or anger, she is providing emotional support. Appraisal support refers to constructive feedback. When someone comments on your weight or affirms your views on how to lose weight, they have provided appraisal support. Finally, instrumental support includes all forms of aid—help with chores, cooking, shopping, lending you money, or giving you a ride to the doctor's office.

Social support has an important impact on health status. It helps people buffer stress and provides access to coping resources in times of need. In a study of economically disadvantaged families in North Carolina, Indu Ahluwalia and her associates (1998) found that people often seek food items, meals for their children, or grocery money from extended family members when they have exhausted their own financial resources. Family members also give valuable emotional support and advice about how to access health care and social services. This support allows families in need to appear "normal" and helps them fulfill their roles as parents and providers. If help is unavailable from family members, people turn to friends, followed by neighbors. Although most people find it stressful to rely on others, in part because of the need to reciprocate, social support is an important way many economically challenged people cope with food insecurity.

People who are isolated socially are more likely to engage in high-risk behaviors than those with many social ties. A longitudinal study of residents living around the Oakland, California, area found that people who lack ties to friends, families, coworkers, and religious and other organizations are more likely to drink alcohol heavily, eat poorly, smoke cigarettes, be overweight, and live a sedentary lifestyle than those with strong network ties (Berkman and Syme 1979).

One explanation for the links between social ties and health status is the access it provides to positive role models and social pressure. Many types of behaviors can be learned through observation (Bandura 1986, 1997). Children learn a great deal about what is good to eat by watching their parents, older siblings, and other important people in their lives. In addition to modeling, parents and other adults can reinforce positive eating behaviors or punish unhealthy ones. Coworkers have also been shown to have an important impact on health behavior. As a result, worksite health promotion programs often include buddy systems, support groups, or other ways to build on existing social norms (Sorensen, Stoddard, and Macario 1998).

While intended to be helpful, some forms of social support may have a negative impact. Reminders can turn into nagging or unwanted cajoling, and friends may model unhealthy behaviors. Elementary school students report that their classmates rarely have anything good to say about vegetables (Cullen et al. 2000). Peer influence has also been shown to impact underage drinking and many other unhealthy behaviors (Sieving, Perry, and Williams 2000).

Most theories of social change recognize the impact other people have on our health behavior. The Theory of Planned Behavior (Ajzen 1988, 1991) asserts that social influences called *subjective norms* are important determinants of behavior change. In this model, social or subjective norms include two components: a person's perceptions of what other people think he or she should do, and the person's motivation to comply with other people's views. To measure people's perceptions of others' opinions, researchers usually ask people to agree or disagree with statements such as "Most people around me think I definitely should eat five servings of fruits and vegetables a day," "My mother thinks I definitely should eat five servings of fruits and vegetables a day," and "My best friend thinks I definitely should eat five servings of fruits and vegetables a day." To assess how motivated someone is to comply with other people's views, researchers ask questions like "With regard to your diet, how much do you want to do what your mother/friends/etc. think you should do?"

Skills and Self-Efficacy

Many dietary behaviors require people to master certain skills—knowledge and the behavioral capacity to carry out specific tasks related to the behavior change. Even the best intentions to decrease fat consumption will fail if a person does not know how to select low-fat foods or prepare them correctly. People also need to have the ability to refuse offers of high-fat deserts when eating at a friend's house and must know how to cope with temptation when depressed, anxious, or bored. For this reason, many nutrition interventions include skills training and provide people with opportunities to practice and obtain feedback to refine performance until successful.

Relapse prevention skills are especially important when dealing with eating addictions or weight-loss regimens. These skills include learning to reinterpret failure in terms of factors that can be overcome (e.g., "This was an especially difficult situation") and successes in terms of personal ability; developing alternative coping mechanisms to deal with high-risk situations, such as boredom and anxiety, or special events like holiday parties; learning to refuse tempting offers; and developing a plan for recovering after a relapse (Conner and Norman 1995).

In addition to acquiring skills, people need to feel confident in their ability to change. A person's confidence in his or her ability to perform a particular behavior is called *self-efficacy*. Introduced by Albert Bandura in 1977, self-efficacy is considered the primary determinant of behavior change for behaviors that have a skills component in Social Cognitive Theory (Bandura 1986). The importance of self-efficacy in bringing about dietary behavior change has been shown in studies of elementary school students' adoption of healthy food choices (Parcel, Edmundson, and Perry 1995), office workers' intentions to engage in eight healthy dietary practices (Sheeshka, Woolcott, and MacKinnon 1993), and adults' adherence to weight-loss plans (Bagozzi and Warshaw 1990; Conner and Norman 1995).

Self-efficacy is assumed to influence behavior in several ways. First, people are more willing to try things they expect they can do well, making self-efficacy an important factor as people move from contemplation to preparation. Second, people set higher goals for themselves and put more effort into activities they believe they can do, making self-efficacy important as people move from preparation to action. Third, people are much more likely to succeed when they think they can (i.e., self-fulfilling prophecy), making self-efficacy an important factor in people's ability to move from action to maintenance. Finally, when encountering setbacks, they are able to cope better and recover more quickly if they have self-efficacy, making self-efficacy important in preventing relapse.

Because self-efficacy has been shown to have a powerful impact throughout the change process, it has been added to the Health Belief Model and many other health behavior theories. In the Theory of Planned Behavior (Ajzen 1988, 1991), it is referred to as *perceived behavioral control*. It is measured by assessing people's confidence in their ability to perform a specific behavior and their perceived ability to overcome any anticipated barriers or external constraints. Researchers ask questions such as "It would be [easy versus hard] for me to eat five servings of fruits and vegetables a day," "When eating out, there is a limited choice of fruit available [likely versus unlikely], and "The limited choice of fruit makes it difficult for me to eat fruit [likely versus unlikely]."

Self-efficacy is derived from several sources. The most obvious source is success in performing a specific task in the change process. Success in carrying out incremental tasks increases the person's perception of mastery. The resulting boost in confidence fuels the process by increasing the amount of effort and persistence devoted to the next task or step.

Therefore, health professionals who are training people with diabetes to self-inject insulin, for example, may divide the self-injection process into a number of small steps, each of which individuals can learn through repetition (for example, filling the syringe with the correct amount of insulin, ensuring that all items remain sterilized . . .). Simplifying each step and allowing individuals to practice each step in isolation, with many repetitions, enables them to build self-efficacy about performing each step. When persons are self-confident about each step, they can progressively put the steps together and build self-efficacy about the entire task. (Baranowski, Perry, and Parcel 2000:173–74)

People may also increase their self-efficacy by comparing themselves to others who have succeeded. A man who sees his wife limit her consumption of sweets may feel more confident in his own ability to do so.

Self-efficacy can also be enhanced by persuasion and support from others whose opinions we value. Our self-efficacy is bolstered when close friends and relatives who know us well tell us they are certain we will succeed. We try to live up to their expectations and feel more confident knowing they believe in our ability to change.

In sum, self-efficacy has a powerful impact on people's decisions to change and their ability to meet the goals they set for themselves. To help bolster people's self-efficacy it is important to (1) break tasks into a number of simple, easy steps, (2) provide them with skills training and adequate time to practice each step, (3) give them guided feedback and support, and (4) enlist support from others whose opinions they value.

Intentions

If you have ever made a New Year's resolution, taken a pledge, or written a personal goal or mission statement, you are familiar with the power of intentions. The Theory of Planned Behavior (Ajzen 1988, 1991) asserts that the most important and direct determinant of behavior change is a person's intentions or assessment of the likelihood he or she will change. In this model, intentions are believed to mediate the impact of attitudes and subjective norms on behavior. This model contains many of the factors we have discussed: health beliefs, attitudes, social norms, and self-efficacy. According to the Theory of Planned Behavior, attitudes—which are determined by health beliefs and the weight placed on their importance—impact intentions, and these intentions lead to behavior change. Subjective norms, formed by what people think others want them to do weighted by their willingness to comply with others' opinions, also impact intentions and thus indirectly influence behavior change. Finally, perceived behavioral control or self-efficacy, which is determined by people's perception of how likely they are to perform a task successfully and their ability to overcome barriers, impacts behavior directly as well as through its influence on intentions.

A review of 56 studies using the Theory of Planned Behavior that was published between 1985 and 1996 revealed that people's intentions were the most important determinant of whether or not they changed. Those who had made a conscious commitment to change were more likely to do so. In about half the studies, the researchers' ability to predict who changed was improved if self-efficacy and intentions were both considered (Godin and Kok 1996). Among those who had decided to change, people with high self-efficacy were even more likely than those with lower self-efficacy to succeed.

The Theory of Planned Behavior assumes that people's decisions reflect careful reasoning or weighing of attitudes, beliefs, and social influences. It also highlights the importance of making commitments or setting goals.

Goal Setting

Goal setting improves performance by helping people focus on an outcome, exert themselves, and persevere (Bartholomew et al. 2001). Setting goals that are difficult but feasible to achieve leads to better performance than setting an easy goal or none at all (Locke and Latham 1991). Of course, goal setting will not work if the task is too complex and/or the person is not given the proper skills to master it.

To be optimally effective, a four-step goal-setting process has been identified: (1) recognizing a need for change, (2) establishing a goal, (3) adopting a goal-directed activity and self-monitoring it, and (4) self-rewarding goal attainment. Because goal setting has been shown to help people lose weight and make other dietary changes, health providers often use it when working with individuals or small groups

Other Factors That Influence Behavior Change

Some behavioral change specialists suggest that personal moral norms, anticipated regret, and concepts of self-identity are important determinants of change (Conner and Norman 1995). Though omitted from most theoretical models, individual's perceptions of the moral or ethical correctness of a behavior can have an important impact on behavior. Moral norms have been found to predict whether or not people will donate blood, donate organs, eat genetically modified food, or follow a vegetarian diet (Conner and Norman 1995). As we have already seen in Chapter 8, many dietary practices are tied to religious or other ideological norms, making these practices difficult to change.

The feelings we expect to experience after doing something can also influence our behavior. We may be deterred from doing something pleasurable now if we anticipate strong feelings of regret or other unpleasant states later. Many people resist tempting foods because they want to avoid the guilt and regret they will feel afterwards.

Finally, people also behave in ways they see as consistent with their personal identity. People who consider themselves to be "green consumers" eat more organically grown vegetables than those who do not share this self-identity (Sparks and Shepherd 1992). A person who sees herself as health conscious or a healthy eater is more likely to adopt new behaviors recommended for their health benefits than one who self-identifies as a "junk food junkie" (Conner and Norman 1995).

Implications for Practice

An important reason to study health behavior theories is to understand the factors that must be addressed to help people make beneficial changes. Review the following recommendations derived from a review of the behavior change literature (Glanz, Rimer, and Lewis 2000) to see if they shed light on your own experience with dietary change or can be incorporated into the intervention you have planned to improve a group's dietary patterns.

- Understand people's beliefs, attitudes, and perceptions regarding their current dietary behavior and the recommended change. Use this understanding to design strategies to (1) change negative beliefs and reinforce positive ones, (2) change how beliefs are evaluated, and (3) introduce new beliefs.
- Make nutrition information personally meaningful. Personalize messages designed to clarify a person's threat from a disease or risk status. Discuss health and other benefits the person finds most attractive, and discuss ways to overcome the barriers he or she perceives as most formidable.
- Clarify the effectiveness of the recommended behavior in preventing health problems.
- Offer incentives to change.
- Correct misinformation about health threats, benefits, and barriers associated with behavior change.
- Provide reminders and other cues to action to promote awareness.
- Provide skills training and practical information about how to perform specific tasks.
- Match interventions to the stage of change a person is in at the time.
- Increase social support. Train members of a person's social network to provide support and model positive behaviors. Use national helpers and form buddy systems and self-help groups.
- Use influential role models to emulate healthy dietary behaviors.
- Provide opportunities for goal setting and self-monitoring.
- Enhance self-efficacy by dividing a task into a number of small, easy steps; providing people with skills training and adequate time to practice each step; giving them guided feedback and support; and enlisting encouragement and reassurance from others.
- Provide relapse prevention training.

Summary

This chapter has examined a variety of individual and interpersonal factors believed to influence people's health behavior and help health providers understand how to promote dietary changes to improve people's health.

- Behavior change is a dynamic process. People move through a sequence of stages based on their readiness to change. In the Stages of Change Model, these stages are called precontemplation, contemplation, preparation, action, maintenance, and termination.
- The most effective intervention strategies or processes of change vary according to the stage of change a person is in at the time. Awareness campaigns and educational activities that stimulate people to evaluate the consequences of change are appropriate during the early motivational stages (precontemplation and contemplation). Social support, skills training, goal setting, and efforts to enhance self-efficacy are most effective as a person moves through preparation into action. Counterconditioning, relapse prevention, helping relationships, and stimulus control enable people to move from action to maintenance and, when appropriate, through termination of the stages of change.
- According to the Health Belief Model, people are motivated to adopt preventive health behaviors by four factors: perceived susceptibility, perceived severity, perceived benefits, and perceived barriers. They are most likely to change when they perceive themselves to be susceptible to a health problem they perceive as severe, expect the behavior to be effective in preventing that threat, and expect the benefits to outweigh the barriers or costs associated with the preventive health behavior. Behavior change is also influenced by cues to action, such as reminders, warnings, and other environmental triggers.
- Attitudes are people's general evaluations of other people, objects, and issues. In understanding health behavior change, the relevant attitudes are people's positive or negative evaluations of performing a specific behavior. Attitudes are an important concept in the Theory of Planned Behavior, which asserts that attitudes are determined by the individual's beliefs about the consequences of behaving in a specific way, weighted by the importance the person places on that outcome. Attitudes can influence dietary and other health-related behavior through their impact on a person's intentions to change. Attitudes that are formed through careful reasoning have a greater impact on long-lasting dietary behavior than those based on simple inferences or environmental cues.
- Social norms also impact people's health behavior. The Theory of Planned Behavior (Ajzen 1988, 1991) asserts that people's intentions to change are influenced by their perceptions of what other people think they should do and their motivation to comply with other people's views. Friends, relatives, and other social ties may impact the change process by providing new information, advice, emotional assistance, feedback, or tangible assistance. They may

also model positive or negative behaviors and pressure us to change or maintain the status quo.

- Self-efficacy, or confidence in one's ability to perform a task, such as follow a weight-loss plan, is another powerful determinant of change for behaviors that have a performance component. Self-efficacy has been shown to impact people's performance, prevent relapse, and help bring about lasting change.
- According to the Theory of Planned Behavior (Ajzen 1988, 1991), the most important determinant of behavior change is a person's intentions. Intentions are determined by a person's attitudes, subjective norms, and perceived behavioral control. Goal setting can be an important way to help people form intentions and increase the likelihood that they will change their behavior.
- Moral norms, anticipated regret, and self-identify are also believed to influence a person's dietary behavior.

Although this chapter has focused exclusively on individual and interpersonal factors, we must be careful not to neglect environmental factors at the institutional, community, and public policy level. When planning a nutrition intervention to improve North American dietary practices, if your list of activities focuses exclusively on teaching people the dangers of high-fat consumption or the benefits of eating more produce, while overlooking environmental factors, such as the price or availability of healthy foods, your efforts may not be effective and you may be guilty of "blaming the victim." Because dietary practices are best understood as part of a holistic, ecological model, nutrition interventions must address factors at multiple levels to be effective. In the next chapter, we examine two models for designing large-scale interventions based on this type of multilevel, ecological model.

HIGHLIGHT

U.S. Federal Food Programs

Karen Miller is 28 years old and the mother of two. Her husband, Rob, also 28, has worked at a local tool and die factory for 8 years. Recently the factory closed, and for the first time in 12 years, Rob Miller was without a job. After 7 months of looking for work, he found a job at a local gas station. It pays one-third what his factory job paid. Karen and Rob were faced with another first: difficulties in being able to feed their children and themselves. After much discussion, they decided to apply for food stamps. Karen describes her experience:

My first visit to the food stamp office took three hours. There was a long wait and the interview took a long time. The whole thing was so new . . . I just didn't know the system. My worker asked me lots of questions about income and expenses. I didn't realize I needed verification—things like pay stubs and utility bills—so the interview could not be completed. I went home and gathered up receipts and took them back to the food stamp office the next day. We received a letter three weeks later saying we qualified for the program. Now every time I go grocery shopping, I just hope I don't run into anyone I know—there's so much stigma with these things. Maybe it's just my imagination, but it seems like people look into my basket at the checkout stand to see what I'm buying. Last week I bought some ice cream as a treat for the girls. You know, I almost felt guilty for buying it, like I didn't deserve it or something. Anyway, we're eating much better since we've gotten the stamps, but I'll be glad when Rob or I can find a job that pays decent wages so we can get off the darned thing.

The Millers are among the 7.4 million households in the United States who receive food stamp benefits. The Food Stamp Program and other federal food programs are the subject of much controversy. Some people argue that these programs drain hard-working, overtaxed citizens in order to support people who are too lazy to work. Others claim the programs are just a "band-aid" keeping us from addressing the basic causes of poverty and hunger. Still others claim that food assistance programs have significantly decreased hunger and malnutrition in the United States and should be expanded. To better understand this controversy, let's take a closer look at U.S. federal food assistance programs—their history, administration, and impact.

The History of Federal Food Programs

Before the Great Depression, the government did little to help farmers or hungry people. Herbert Hoover refused to distribute surplus wheat to the poor in 1931 but expressed the hope "that the hungry and unemployed will be cared for by our sense of voluntary organization and community service" (Kotz 1970). But by 1934, the economic situation had become even more desperate and the government began distributing surplus flour, cornmeal, dried milk, rice, butter, and cheese to millions of people.

The first Food Stamp Program began in 1939. It was a complicated system, difficult to understand and even more difficult to use. In addition, it offered only limited benefits for farmers (whose surplus products were guaranteed a market) and the poor. The program was abandoned from 1943, when World War II created an increased demand for agricultural products, until 1961, when the Kennedy administration reinitiated it on a pilot basis. Recognition of the need for an ongoing program came in the late 1960s as a response to several studies undertaken to assess the extent of hunger and malnutrition in America. The findings of these studies were sobering. A team of physicians sponsored by the Field Foundation reported widespread hunger and malnutrition in the rural South; the Citizen's Advisory Board cited case studies of people going hungry even though they received food stamps; and the Ten State Nutrition Survey found nutrient deficiencies among poor Americans. These studies received widespread media coverage, including the TV documentary *Hunger, USA,* which brought vivid images of hungry American children with swollen bellies and apathetic stares into American living rooms. Americans were shocked to find such conditions right in their own well-fed, diet-conscious society. In response, Congress enacted a series of food program reforms, and the Food Stamp and National School Lunch Programs were expanded. The Special Supplemental Nutrition Program for Women, Infants, and Children (WIC) was also established along with the Title VII feeding program for the elderly. Between 1969 and 1979, the cost of these programs was nearly $50 billion.

To examine the impact of expanded food aid on nutrition among the poor, the Field Foundation sent the physicians who had visited the rural South a decade earlier back to retrace their steps in 1977. In a report to Congress, they concluded that despite the fact that poverty was just as widespread in the South as it had been 10 years before, the nutrition programs were working well and had significantly decreased the prevalence of malnutrition (Sims 1983).

Despite the results of this and other studies showing that food assistance programs increase food security and decrease malnutrition, a conservative political environment in the late 1970s and early 1980s led to widespread cuts in food stamps and other assistance programs.

Throughout the last two decades, participation in nutrition assistance programs has fluctuated like a seesaw, reflecting economic trends and changes in federal policies that impact program eligibility guidelines. In the early 1980s, a politically conservative Congress made deep budget cuts in most nutrition programs. Combined with lower unemployment rates, program participation rates dropped. These rates began to return to their previous levels in the early 1990s as the economy weakened and legislation was passed to ease eligibility requirements for the Food Stamp Program. The most recent fluctuations in participation rates were fueled by declining unemployment rates and the passage of the 1996 Personal Responsibility and Work Opportunity Reconciliation Act (PRWORA). Also known as welfare reform legislation, PRWORA imposed a time limit on a person's eligibility to receive cash assistance from the government or food stamps and required the person to work unless he or she had very young children or met other requirements. PRWORA also made most legal immigrants ineligible for food stamps and some other programs. The highly publicized debate surrounding the passage of PRWORA also heightened the stigma associated with food assistance, making many people reluctant to accept help and confused about their eligibility for governmental assistance (Nord 2001). As shown in Figures H11.1 and H11.2, the impact of these legislative changes and a strong economy resulted in dramatic decreases in program participation: Participation in the Food Stamp Program alone dropped from almost 28 million in March 1994 to 17.3 million in March 2000 (Super 2001).

In this highlight, we look at the three largest governmental programs that provide a nutrition safety net for people in economic need—the Food Stamp Program, the National School Lunch and School Breakfast Programs, and the Special Supplemental Nutrition Program for Women, Infants, and Children (WIC). The goals of these programs are to give poor people access to a more nutritious diet, create markets for farmers' products, and improve children's eating habits.

The Food Stamp Program

The Food Stamp Program is the largest nutrition program in the United States, accounting for over half of government expenditures

Figure H11.1
Food stamp participation and the unemployment rate follow similar trends over time.

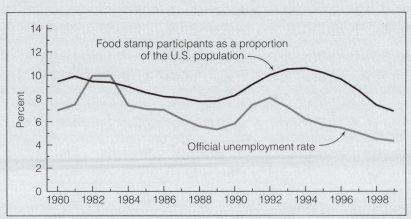

(*Source:* USDA's Food and Nutrition Service, and U.S. Department of Labor, Bureau of Labor Statistics.)

Figure H11.2
Food stamp participation may have been influenced by policy changes.

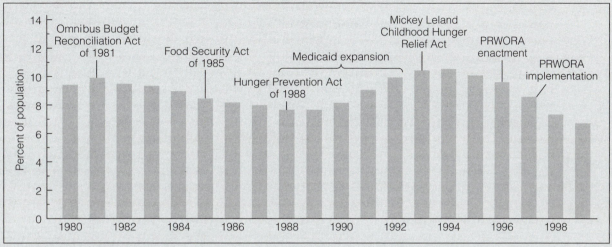

(*Source:* USDA's Food and Nutrition Service.)

on food assistance. Designed to help families improve their diets by increasing their purchasing power, it is the only program available to a broad range of needy households, including those with working members.

Until recently, families were given monthly allotments of paper coupons that looked like money to redeem at grocery stores and some other retail outlets. Now almost all participants use an electronic benefits transfer card that works like a debit card to redeem their food stamp benefits at approximately 155,000 authorized stores in the United States. The total cost of the program in 2000 was $17.1 billion, with $15 billion transferred to families to purchase food and the rest used for program administration and nutrition education (U.S. Department of Agriculture Food and Nutrition Service 2001a).

Program Eligibility

Program eligibility is based on a household's monthly income and assets (i.e., cash on hand, savings accounts, savings certificates, stocks or bonds, vehicles not used for work, and some types of property), minus a deduction for expenses. A household is defined as the people who share a residence and purchase and prepare their food together. With the exception of households that contain an elderly or disabled member, the household's combined gross income must fall at or below 130% of the **federal poverty level** ($1,961 a month for a family of four in 2002). To determine the household's monthly net income, a standard amount is deducted from the gross monthly income to cover expenses and other deductions are made for child care, excess medical bills, and some other costs. To be eligible, the household must have a net monthly income below 100% of the federal poverty guidelines ($1,471 for a family of four in 2001).

federal poverty level
For a discussion of how the federal poverty level is determined each year, see Focus 7.2 in Chapter 7.

Also, households that do not include elderly or disabled members may not have accumulated more than $2,000 in assets, and those that do contain older or disabled members cannot exceed $3,000. Most students, people on strike, and undocumented immigrants are ineligible. Able-bodied adults who are not caring for young children are required to work or participate in training or other work-related activities. To apply, people must appear in person at a Food Stamp Office. Elderly and disabled people may apply by telephone.

Program Benefits

The amount of food stamp benefits a family receives is based on the number of people in their household and the difference between their net monthly income and the cost of the USDA's **Thrifty Food Plan**—a basket of foods that can be used to make an inexpensive, but nutritionally adequate diet. In fiscal year 2002, the maximum monthly benefit for a family of four with no income living in the 48 contiguous states was $465. Because the vast majority of food stamp recipients have some income, they receive less than this amount. The USDA assumes people spend about 30% of their income on food, so a household making a combined net monthly income of $1,000 would receive $300 less than the maximum allowable benefit. In 2002 the average food stamp allocation was $80, or 88 cents per meal per person. Benefits are slightly higher in Hawaii and Alaska where food is more expensive.

The majority of food stamp benefits go largely to people in extreme need. Almost 90% of food stamp households contain a child, an elderly person, or someone who is disabled. About nine out of ten households also have a *gross* income below the poverty line and one-third live equal to or less than half the poverty line (U.S. Department of Agriculture Food and Nutrition Service 2001a). Also, food insecurity rates are higher among households receiving food stamps than low-income households that do not, reflecting the greater propensity of very needy families to seek help from the government (Nord 2001). Focus H11.1 gives more information on food insecurity.

In addition to monetary benefits, the federal Food Stamp Program encourages its state agencies to provide participants with education to help them make healthy food choices consistent with the U.S. Dietary Guidelines and the Food Guide Pyramid. By the year 2000, 48 states were using federal funds to provide their participants in the Food Stamp Program with nutrition education. Most of these states work closely with the Cooperative Extension Service or another nutrition agency to implement their educational activities (Anliker et al. 2000).

Evaluations of the Food Stamp Program's impact on participants' diets have been mixed. Most studies have found participation in the Food Stamp Program increases the amount of money spent on food and the availability of nutrients in the household and improves food security. It is less clear if program participation improves the quality of recipients' diets.

Many studies have examined how people use their food stamp benefits. To what extent do recipients use them to purchase additional

Thrifty Food Plan
See Chapter 7 for a more detailed discussion of the Thrifty Food Plan and how it is calculated.

FOCUS H11.1 Food Insecurity

Food security includes "(1) the ready availability of nutritionally adequate and safe foods, and (2) an ensured ability to acquire acceptable foods in socially acceptable ways (that is, without resorting to emergency food supplies, scavenging, stealing, or other coping strategies)" (Economic Research Service 2001). To assess food security, the U.S. Census Bureau conducts an 18-question survey of a nationally representative sample of about 50,000 households covering topics from worry about running out of food to the number of entire days children go without eating (Nord 2001). Sample questions include:

Was this statement often, sometimes, or never true for you in the last 12 months? "We worried whether our food would run out before we got money to buy more."

Was this statement often, sometimes, or never true for you in the last 12 months? "The food that we bought just didn't last and we didn't have money to get more."

In the last 12 months, were you ever hungry, but didn't eat, because you couldn't afford enough food?

In the last 12 months, did you ever not eat for a whole day because there wasn't enough money for food?

In the last 12 months, did any of the children ever skip a meal because there wasn't enough money for food?

Using these data, households are classified as food secure—they had ensured access at all times to enough food for an active, healthy life; food insecure—they were uncertain of having or unable to acquire adequate food to meet basic needs because there was not enough money for food; and food insecure with hunger—one or more household members were hungry at some time during the year because they could not afford enough food. In 1999 approximately 10% of households were classified as food insecure, including 3%, or 3.1 million households, in which people were hungry at times during the year because they could not afford enough food (see Figure H11.3). (For a more complete description of how food security is measured, visit the Economic Research Service Web site at *http://www.ers.usda.gov/briefing/foodsecurity/*.)

Figure H11.3
Each month, at least 26 million people are hungry in the United States.

© Robert Bremner/PhotoEdit

food as opposed to freeing up money previously allocated for food purchases that then can be used for other expenses? Recent estimates of the amount of money spent on food that would not have otherwise been purchased ranges from 17 cents to 49 cents for every food stamp dollar received (Fraker 1990). Most studies have also found that food stamps have a positive impact on the amount of food available in people's homes. Allen and Gadson (1983) and Devaney et al. (1989) found that participation in the Food Stamp Program increases the availability of calories, protein, vitamin A, vitamin C, the vitamin B complex, calcium, phosphorus, magnesium, and iron. The Food Stamp Program's impact on these nutrients was three to nine times larger than the impact cash assistance had on nutrient availability.

The Food Stamp Program's impact on the quality of what people actually eat is unclear. Some studies found no significant differences in the nutrient intake or other measures of dietary quality between food stamp recipients and other low-income people not participating in the program (Aiken et al. 1985; Bialostosky and Briefel 2000; Blaylock, Jayachandran, and Lin 1999; Gleason, Rangarajan, and Olson 2000; Jensen 1996; Oliveira and Gunderson 2000). In other studies, program participation was associated with improved consumption of some but not all nutrients (Rose, Habicht, and Devaney 1998; Wilde, McNamara, and Ranney 1999). The Human Nutrition Information Service (1989), for instance, found that the consumption of food energy, calcium, folate, iron, magnesium, protein, zinc, and B vitamins was higher among poor children ages 1 to 5 who participated in the program than those who did not. When using the Healthy Eating Index to assess dietary quality, one study (Gleason et al. 2000) found no difference in the scores of food stamp recipients and nonrecipients. (See Focus H11.2 for more information on the Healthy Eating Index.) In another study, the results were mixed. The total scores of food stamp recipients were actually lower than those of nonparticipants. But the value of food stamps received had a significant positive effect on households' scores. As the authors of this study (Basiotis et al. 1998) conclude:

> For each additional dollar of food stamps received, the aggregate household HEI score increases by an estimated .22 points. At the average weekly food stamp value of $34.22, the aggregate household HEI increases 7.5 points, on average. . . . A break-even point is estimated at $17.54 per week. That is to say, when weekly household food stamp benefits are at least $17.54 per week or lower, food stamp participants have diet quality inferior to nonparticipants. Thirty-two percent of Food Stamp Program participating households received food stamps valued at less than $17.54 per week. (p. 12)

The methodology used to evaluate the impact of the Food Stamp Program on diet quality has many limitations. Most researchers compare food stamp recipients' nutrient intake or scores on the Healthy Eating Index with those of people who are qualified for the program but do not participate. There are many differences

FOCUS H11.2 Healthy Eating Index

In an effort to measure how well Americans follow recommended nutrition guidelines, the U.S. Department of Agriculture developed an index, called the Healthy Eating Index. Based on various aspects of a healthful diet, the HEI is designed to provide a measure of the overall quality of a diet. The index assesses the balance of foods people eat, the amount of variety in the diet, and compliance with specific dietary guideline recommendations. Ten dietary components are measured and include the degree to which people follow each section of the Food Guide Pyramid, as well as the amount of fat, saturated fat, cholesterol, sodium, and variety a diet contains. The HEI scores range from 0 to 100. Each of the 10 dietary components has a score ranging from 0 to 10. Individuals with an intake at the recommended level receive a maximum score of 10 points. A score of 0 is assigned when no foods in a particular group are eaten. Intermediate scores are calculated proportionately.

between participants and nonparticipants that may impact dietary intake (e.g., access to assets, gifts, and loans from relatives or friends; different levels of motivation to eat well; health status). Because most studies are unable to control for these differences, researchers are cautious in reaching conclusions about the program's impact on dietary intake and recommend additional research to study this issue (Gleason et al. 2000).

The data are clearer with respect to allegations that the Food Stamp Program encourages people to remain unemployed. As we have already seen, the Food Stamp Program requires all adult recipients to register for work unless they are elderly, disabled, or caring for a child under age 6. Food stamp recipients who fail to meet the work requirement may be dropped from the program temporarily—or permanently, if they repeat the offense. To offer an additional work incentive, people's food stamp benefits are reduced by only 24 to 36 cents for every additional dollar earned when they first find work or move to a better-paying job.

Studies of the length of time people participate in the Food Stamp Program also cast doubt on claims the program breeds dependency. Depending on the state of the economy and other factors, between 42% and over 50% of all households entering the program leave within 6 months, and between 57% and 66% leave before the end of the year. Because these studies include the elderly and disabled recipients who typically receive food stamps for longer periods of time, the proportion of families who rely on the program for very short periods of time is even higher. The major reasons for enrolling in the program are loss of employment and decreased earnings, and the major reason for leaving is increased income (Super 2001; Wilde 2001).

National School Lunch and School Breakfast Programs

In 1946 President Harry Truman signed the National School Lunch Act into law. It was originally created to address nutrition-related health problems found in young military recruits during World War II. In addition to feeding America's youth, the program, operated by the United States Department of Agriculture (USDA), was used as a market for surplus agricultural products.

Background and Overview

Today the program operates in more than 97,000 public and non-profit private schools and residential child-care institutions nationwide. Schools that choose to take part in the program receive cash reimbursements and **commodity foods** for each meal they serve. In turn, they agree to follow federal nutrition requirements and must offer free and reduced-price lunches to eligible children. The National School Lunch Program cost $5.56 billion in 2000 and was served to more than 27.4 million children each day.

In 1966 the Child Nutrition Act established the National School Breakfast Program to complement the lunch program. It began as a pilot project in 1966 and was made permanent in 1975. The breakfast program is not as widespread as the lunch program. It is administered in approximately 72,000 schools and institutions. Congress appropriated $1.40 billion for the National School Breakfast Program in 2000, which fed an average of 7.55 million children a day. Of those, 6.4 million received their breakfast free or at a reduced price.

Children from families with incomes at or below 130% of the federal poverty level are eligible for free meals. Those with incomes between 130% and 185% of the federal poverty level are eligible for reduced-price meals, for which children can be charged no more than 40 cents.

Program Impact on Diet Quality

Early studies of the school lunch and breakfast programs showed that the programs filled important nutritional needs. The benefits of the free lunch program were demonstrated over a 14- year period in Baltimore. After introduction of the lunch program, children scored impressive nutritional gains. Children who eat a good breakfast have been shown to be able to pay better attention in school than those who receive no breakfast or an inadequate one. Their overall nutritional status improves as well.

Though nutritional improvements in children were made with the school meals programs, the School Nutrition Dietary Assessment Study (SNDA-I), completed in the 1991–92 school year, found that school lunches were too high in total fat and saturated fat. This was inconsistent with the *Dietary Guidelines for Americans* published by the USDA, the same agency that administers the National School Lunch and School Breakfast Programs. It was a particularly troubling finding since the percentage of children who were overweight had more than doubled since 1970. Childhood obesity was recognized as a national epidemic, with an earlier onset and increased prevalence of disease.

In response, the USDA launched the School Meals Initiative for Healthy Children (SMI) in 1995, the first full-scale reform of the breakfast and lunch programs since they started. The centerpiece of

commodity foods
Agricultural goods that are usually traded or sold, such as flour, milk, and peanuts.

the initiative was an update of nutrition standards so that school meals met the recommendations of the *Dietary Guidelines for Americans,* including the goal of reducing total fat intake to 30% of total calories and total saturated fat intake to 10% of calories. This is in addition to the long-standing requirement that school lunches provide one-third of the Recommended Dietary Allowance (RDA) for protein, calcium, iron, and vitamins A and C, and one-third of the Recommended Energy Intake (REI) for calories over a week's cycle of menus. School breakfasts must provide one-fourth of the RDA and REI. In addition to the new guidelines, SMI provided schools with educational and technical resources to assist food-service personnel in preparing nutritious and appealing meals and to encourage children to eat more healthful meals.

How well is the School Meals Initiative working? That is one of the questions the SNDA-II set out to answer by assessing school meals in the 1998–99 school year. Between 1991 and 1998, there was a statistically significant trend toward lower levels of total fat and saturated fat in school meals. There was also a marked increase in the percentage of schools that offered lunches that were consistent with the *Dietary Guidelines* (see Figure H11.4).

Though gains were made, it was clear that there was more work to be done. Program regulations require that the meals *selected by* students—not just the meals offered to them—be consistent with SMI standards. In other words, the combination of foods students actually select from the lunch line may be much higher in fat than the meal they are supposed to select "in theory." Elementary schools are doing somewhat better than secondary schools in this regard, but still, in fiscal year 1998–99, only one out of five elementary

Figure H11.4
The National School Lunch Program provides meals to over 97,000 schools and child-care centers.

Courtesy U.S. Department of Agriculture

schools met the SMI standard for calories from fat and roughly one in seven met the SMI standard for saturated fat.

Though the nutrition standards of the school meals programs are admirable, the law makes no provision for enforcement and the federal rules require compliance reviews only once every 5 years. In addition, the menus that are analyzed for review are based on what the cafeteria manager plans to serve, not necessarily what the students select and actually eat. A menu may look nutritious on paper, but in practice students can pick combinations that add up to a very high fat meal. And because American children are used to eating fast food, many cafeterias provide fast-food items such as cheeseburgers, French fries, and chicken nuggets on a regular basis. These foods sell and bring in revenue, making it easier for the program to be financially solvent. Cafeteria managers say, given the nutrition environment of the country, that if the food is "too healthy," students won't buy it and it becomes difficult to make the program financially solvent.

Realizing that there was more work to be done, the USDA established Team Nutrition to support the SMI, operating on the principle that improved nutrition education empowers students to make healthy food choices. Team Nutrition brings together public/private partnerships to teach children the importance of making healthy food choices and to give school food-service staff the tools they need to make nutritious and appealing meals. Partnerships were developed with the Walt Disney Company to develop healthy nutrition messages on television for children and with Scholastic, Inc., to deliver age-appropriate nutrition education to school-age children and their parents.

Other innovations have been made to improve school lunches, which have a notorious reputation for being unappealing. One is the pairing of professional chefs and registered dietitians with school food-service personnel to create award-winning recipes for distribution to cafeterias across the country. A recent School Lunch Challenge Cook-Off generated recipes such as Pack-It Vegi and Beef Burrito, Wild West Vegetables, and Super Cherry Tiramisu.

The National School Lunch Program has been criticized for its dual goals of promoting good nutrition to children while providing a market under the Commodity Food Program for farmers. Some public health professionals argue that the two purposes are not always fully compatible. In fact, up until 1976 it was a requirement that butter be served with every school lunch. In 1997 the Commodity Food Program spent more than 70% of its $436 million budget on animal products, though *Dietary Guidelines* recommend that Americans move toward a plant-based diet.

The Commodities Improvement Council was formed by the USDA in 1995 in an effort to improve the quality of commodities donated to the National School Lunch Program. The goal was for schools to get more useful products that were consistent with providing healthy and appealing meals—a goal not easy to achieve when schools were receiving cases of canned pork, high-fat cheese, and fig nuggets (fig pieces shaped into uniform pellets). Based on the council's recommendations, the USDA reduced the fat, sodium, and sugar content of commodities as well as improved the food quality, packaging, and distribution. It also increased the amount of low- and reduced-fat products available to schools—though sometimes the government offers "advanced processing," which turns healthful commodities like flour and blueberries into high-fat products like blueberry turnovers.

Progress has also been made in increasing the amount of fresh fruits and vegetables available to schools. A cooperative project with the Department of Defense (DOD) has allowed the USDA to increase the variety of produce available to schools by using the DOD's buying and distribution system.

The Special Supplemental Nutrition Program for Women, Infants, and Children (WIC)

The WIC program's mission is to safeguard the health of needy women and their children until they reach 5 years of age by providing them with nutritional assessments, supplemental nutritious foods, nutrition education and infant feeding advice, referrals to health, welfare, and social services, and childhood immunizations (see Figure H11.5).

Program services are provided to women who are pregnant, breastfeeding, or have given birth within the past 6 months; infants; and children under the age of 5. Program participants must be economically disadvantaged and at nutritional risk, based on measurements of height, weight, **hemoglobin** or **hematocrit,** dietary intake, or a history of a nutrition-related condition. The income requirement—at or below 185% of the U.S. poverty level—is more liberal than the Food Stamp Program. People who participate in the Food Stamp, Medicaid, or Temporary Assistance for Needy Families (TANF) programs are automatically income-eligible.

Unlike the Food Stamp Program, in which recipients may purchase whatever foods they would like, WIC participants are given vouchers to buy one of seven food packages. These packages are not designed to meet all nutrient needs but, rather, supply specific nutrients, especially protein, calcium, iron, and vitamins A and C, often deficient in diets during these critical periods of life. The most common foods provided are milk, cheese, fruit juices, dried beans, eggs, iron-fortified cereal and infant formula, and peanut butter (Basiotis et al. 1998; Oliveira and Gundersen 2000). In 1992 a WIC Farmer's Market Nutrition Program was created to provide WIC participants with coupons they can redeem at farmer's markets for fresh fruits and vegetables.

Offered in all 50 states, the District of Columbia, 33 Indian tribal organizations, Guam, the U.S. Virgin Islands, American Samoa, and the Commonwealth of Puerto Rico, WIC is regarded as one of the U.S. government's most successful programs (Center for Budget and Policy Priorities 2001). Program participation peaked in 1997 at 7.4 million. In 2001 the program served 7.34 million people, about half of them children ages 1-4. Approximately 50% of all

hemoglobin and hematocrit
Hemoglobin is a large protein molecule in the blood that carries oxygen. Hematocrit is a measure of the volume of red blood cells in a given amount of blood. Hemoglobin and hematocrit tests are quick and inexpensive ways to evaluate iron status and identify iron-deficiency anemia.

Figure H11.5
In April 2000, 7,855,537 women, infants, and children were enrolled in WIC. The program served 45% of all infants born in the United States and 25% of new mothers.

Courtesy U.S. Department of Agriculture

infants and almost 25% of children in the United States participate in the program. WIC accounts for about 12% of government spending on food assistance (Oliveira and Gundersen 2000).

The WIC program is based on the recognition that early intervention programs during critical growth periods can help prevent medical and developmental problems. During pregnancy, for instance, a mother's diet has an important impact on her baby's weight and health at birth. Evaluations of the program have demonstrated that WIC improves the dietary practices and dietary intake of some disadvantaged pregnant women, contributing to increased birth weight and decreased incidence of maturity. In some studies, WIC has been shown to be cost effective, saving between $1.77 and $4.00 in Medicaid costs for participants' newborns for every dollar spent (Brown, Gershoff, and Cook 1992; Mathematica Policy Research 1990).

Childhood is another important developmental period in which diet has been shown to influence brain development, readiness to learn, and many other health indicators. WIC has been shown to improve children's intakes of iron by almost 21% of the RDA, vitamin B-6 by 23% of the RDA, folic acid by 91% of the RDA (Oliveira and Gundersen 2001), and to improve hematocrit levels (Basiotis et al. 1998). WIC has also been shown to have a strong positive effect on household dietary quality as measured by the Healthy Eating Index (Basiotis et al. 1998). A more recent study (Besharov and Germanis 2001) challenges these findings: Though WIC may have an impact on the nutritional intake of some infants, especially those who would not have been breastfed, the program probably has little impact on the diets of most 1- to 4-year-olds and fails to address the most common nutritional problem facing the poor—obesity.

Another important service provided by WIC is breastfeeding promotion. Since 1989, it has (1) established standards for breastfeeding promotion and support, (2) allocated funds to states that they must spend on breastfeeding promotion, (3) authorized the purchase of breast pumps and other aids, and (4) designated a breastfeeding coordinator for each state (Tuttle 2000). WIC also has established partnerships with the Expanded Food and Nutrition Education Program of the Cooperative Extension Service to promote breastfeeding to low-income populations and has worked with Best Start Social Marketing Inc. to develop a national breastfeeding promotion campaign, called "Loving Support Makes Breastfeeding Work" (see Chapter 12 for a description of this project). While breastfeeding rates among WIC participants is still significantly lower than nonparticipants, breastfeeding has grown more rapidly among WIC participants than among any other segment in the United States (Ryan 1997).

Although funding for the program has been increased, many eligible families do not participate in WIC. In one study, over 40% of eligible families did not realize they qualified, usually because they thought the program served only single-parent or unemployed families. Among those who realized they were eligible, many were reluctant to enroll because of the stigma associated with acceptance of government assistance. Others feared they would be treated disrespectfully or had difficulty signing up (Bryant et al. 2001). Studies of WIC participants have shown that long waits to be served, discourteous service by either WIC staff or grocery store clerks, and embarrassment to be seen redeeming vouchers create dissatisfaction and motivate people to leave the program (Bryant et al. 1998b).

Other Food Assistance Programs

The Child and Adult Care Food Program provides healthy meals and snacks to children in nonprofit child-care centers and family and group day-care homes. In these centers, low-income family members are eligible for free or reduced-price meals.

The Summer Food Service Program provides free meals to children 18 years old and younger and handicapped people during school vacations. In areas where at least half of the children are from poor families, there is no income test for eligibility. In other places, families are screened for eligibility and at least half of those participating must fall below 185% of the federal poverty level.

The Special Milk Program provides funds for free or low-cost milk in schools and other child-care settings that serve poor families.

The Commodity Supplemental Food Program provides free nutritious supplemental foods to children 6 years of age and younger who are eligible but not participating in the WIC program. Elderly people whose income falls below 130% of the federal poverty level may also participate. In some states they must also be at nutritional risk to participate.

The Nutrition Program for the Elderly provides food commodities and cash for meals to states for use in feeding people 60 years and older. These funds may be used to deliver meals to homebound senior citizens or feed them in special centers. All senior citizens may participate regardless of income.

The Disaster Feeding Program provides food commodities after major disasters that have limited food supplies.

Funds and food commodities are also made available to soup kitchens and food banks. Food packages are delivered to American Indians who live far from food stores and meet certain income requirements (Oliveira 1999).

Community food security programs, such as farmer's markets and community gardens, food cooperatives, farm-to-school initiatives, and community-supported agriculture, also help families stretch their food dollars (Kantor 2001).

Beyond Federal Food Programs

Despite evidence that federal food programs improve nutritional status and health, these programs are insufficient to solve the problem of hunger and malnutrition in the United States. Food assistance programs were never intended to eliminate poverty or generate employment. Rather, they provide a nutritional safety net, helping families purchase a more nutritious diet in times of need. The only real solution for hunger and malnutrition is to provide everyone with access to employment that pays a living wage so that they can purchase a nutritionally adequate diet for themselves and their families.

CHAPTER 12

Designing Large-Scale Programs to Change Dietary Practices

The last two chapters demonstrated that efforts to improve individuals' dietary practices are most successful when problems are analyzed to understand the factors that must be addressed to bring about behavior change. This book has also introduced you to the numerous factors that influence what you eat, illustrating the importance of designing programs to impact factors at the individual, interpersonal, and environmental levels. For this reason, many program planners now develop large-scale programs to address factors at all levels simultaneously. In this chapter, we examine two comprehensive, multilevel approaches for improving dietary practices: community-based health promotion and social marketing.

Imagine you were responsible for designing a program to improve North American dietary practices. What types of activities would you use to help people eat less fat and more fruits and vegetables? To plan an effective intervention, think of ways to address factors at each of these levels:

- *Interpersonal:* People's knowledge, beliefs, expectations, attitudes, intentions, skills, self-perceptions, and other personal characteristics that impact their dietary behavior.
- *Intrapersonal:* Beliefs and actions of friends, relatives, coworkers, and other important people who influence a person's dietary beliefs and behavior.
- *Institutional:* Rules, regulations, procedures, and other policies and practices of civic, religious, social, political, and other organizations that impact food availability and dietary practices.
- *Community:* Community norms, values, and resources that constrain or promote dietary practices.
- *Public policy:* Local, state, and federal policies, laws, and regulations that impact food choices and people's behavior.

We'll now look at community-based health promotion and social marketing in detail.

Community-Based Health Promotion

Many public health professionals now believe that change strategies directed by community members are far more likely to succeed than those planned and executed exclusively by outsiders (Green and Kreuter 1991; Minkler and Wallerstein 1998). Increased reliance on community-based approaches to health promotion is consistent with the holistic, ecological model on which this book is based and the recognition that people eat, live, work, love, and play in communities (Mittlemark 1999).

Community organization is the process by which community groups are helped to identify common problems, set goals, mobilize resources, and implement strategies for reaching those goals (Glanz 1993; Minkler and Wallerstein 1998). Community-based approaches vary in the degree of control the community has over the change process and the emphasis placed on enhancing community competence. Minimally, community members provide advice, access to information sources, and assistance with program planning, implementation, or evaluation. They often serve on a planning coalition or board that sets goals and oversees or approves key steps in the process used to achieve those goals. Some program planners restrict the term "community organization" to describe efforts in which community members control the process, and they design the intervention primarily to empower the community rather than solve a specific community problem.

Our exploration of community-based health promotion begins with an examination of the concept of community and then describes the distinguishing features of community-based approaches to improving dietary practices and nutritional status among large populations. After a description of a community-based intervention, this section concludes with a brief discussion of the benefits and challenges of evaluating community-based interventions and the benefits of involving community partners in the process.

Community

Community refers to a group of people who share a sense of social identity, common norms, values, goals, and institutions. They may be based on geographic boundaries (a county, city, or neighborhood), social identity and/or interests (e.g., sororities), or shared political responsibilities (e.g., an animal rights group) (Eng and Parker 1994). Not all neighborhoods, interest groups, or political groups are communities. To qualify as a community, the group must be characterized by the following elements: "(1) membership—a sense of identity and belonging; (2) common symbol systems—similar language, rituals, and ceremonies; (3) shared values and norms; (4) mutual influence—community members have influence and are influenced by each other; (5) shared needs and commitment to meeting them; and (6) shared emotional connection—members share common history, experiences, and mutual support" (Israel et al. 1994:151). This shared identity and sense of belonging develops as people come together to share their experience and work together.

Societal trends toward greater geographic mobility and the use of communication technologies now make it possible for communities to be broader based and more heterogeneous than was once possible (Patrick and Wickizer 1995). As a result of technological changes, such as increased reliance on the Internet and computer networking, communities are less likely to develop around geographic units and more likely to reflect other shared interests (Pilisuk, McCallister, and Rothman 1996).

Distinguishing Features of Community Organization

Community-based approaches to nutrition or other types of health promotion share several important distinguishing characteristics: use of a multilevel ecological perspective, community participation, and relevance (Brownson, Baker, and Novick 1999; Minkler and Wallerstein 1998). The goal is to recognize, use, and enhance the community's resources and abilities to solve problems rather than depend on external sources of help (Mittlemark 1999).

First, community-based approaches are guided by an ecological perspective, addressing factors at multiple levels, either simultaneously or sequentially (Brownson et al. 1999). A community-based approach to decrease fat consumption, for example, might address individual factors by offering classes or individual counseling to increase knowledge and build skills. At the same time, lay health advisors might be trained to act as local advisors and referral sources as a way to build social ties. Community members might advocate for the adoption of low-fat options on restaurant menus. Coworkers might form small groups to teach and support each other. Homemakers and other community groups might develop and promote low-fat recipes. A community coalition might also lobby for changes in federal policies that impact the guidelines for school breakfasts and lunches or nutrition labeling (see Figure 12.1).

Community participation, the second hallmark of community-based approaches, refers to the process by which individuals and

Figure 12.1

This community of breast cancer survivors provides each other with support while lobbying for increased funding for medical research.

© Kent Meireis/The Image Works

families take an active part in efforts to improve community life, services, or resources (Bracht and Tsouros 1990). Even if professionals from outside the community play an important role in the program, they work in partnership with community members—community professionals; lay leaders and activists; representatives of local businesses, churches, voluntary organizations; and citizens. This partnership brings people together who share common goals, resources, and problems (McWilliam, Desai, and Greig 1997). Working together, these partners define and critically analyze community problems, set goals, and design, implement, and evaluate interventions aimed at achieving these goals.

Third, community-based approaches to health promotion are designed to "start where the people are" (Minkler and Wallerstein 1998). The principle of relevance has been a fundamental principle of public health for many years (Nyswander 1956)—one that has gained renewed importance in the last three decades.

> *The health education professional who begins with the individual or community's felt needs and concerns, rather than with a personal or agency-dictated agenda, will be far more likely to experience community ownership and success in the change process than if he or she were to impose an agenda from the outside. (Minkler and Wallerstein 1998:43)*

For this reason, community-based approaches typically start with a needs assessment in which community members, often with the help of health professionals, collect data about local problems and the factors that contribute to them.

While all community-based approaches share these three principles, community organizing is distinguished by the control it gives to community members. Community members are given ownership of community organizing efforts. "Ownership means that local people must have a sense of responsibility for and control over programs promoting change so that they will continue to support them after the initial organizing effort" (Thompson and Kinne 1999:30). In true community organizing, community members select the problem they want to address and control each step of the change process.

Although these principles of community self-determination and relevance are central tenets of community organizing, they are often difficult to implement. Most nutrition interventions are funded by national, state, or other agencies that dictate the problem they want addressed. Even when funding is not tied to a specific problem or health category (e.g., cardiovascular disease prevention), health departments or other lead community organizations, which are used to categorical funding, may be unable to make the transition from a strict problem-solving focus to a broader community development project (El-Askari et al. 1998).

Another feature that distinguishes community organization from other community-based approaches is the emphasis on community empowerment and enhanced competence as desired outcomes of the change process. Empowerment refers to "a social action process by which individuals, communities, and organizations gain mastery over their lives in the context of changing their social and political environment to improve equity and quality of life" (Minkler and Wallerstein 1998:40). Community organizing models attempt to empower communities by helping them work together to set goals, develop strategies for achieving those goals, and gain mastery over the social and political environmental factors that influence the health and well-being of their members. Professionals who live outside the community may teach citizens new skills or provide important resources needed to implement change strategies. However, by prioritizing problems, setting goals, designing interventions, and implementing solutions, citizens and other community participants (e.g., organizations) gain a stronger sense of collective efficacy and learn how to engage in effective problem solving. Participation increases community members' sense of mastery and their realization that they can improve their lives and those of others. As citizens develop the ability to work together to influence decisions in the larger social system, they learn to use their collective skills and power to obtain a more equitable share of resources (Israel et al. 1994; Robertson and Minkler 1994).

Other guiding principles of community-based approaches to health promotion are summarized by Bracht and Kingsbury (1990):

- Planning should be based on a thorough understanding of the community, including its history and conditions that inhibit or facilitate change.
- Multiple interventions are used to address factors at multiple levels.
- Existing social structures and values are used to bring about change.
- Many sectors of the community—political and governmental, faith-based, educational, business, and others—work together to address the problem in a comprehensive effort.
- The project has both short- and long-term goals.
- The community shares responsibility for the problem and for its solution.

Stages in the Community Organizing Process

Community organization typically unfolds in five basic stages: community analysis, design and initiation, implementation, maintenance and consolidation, and dissemination and reassessment (Bracht, Kingsbury, and Rissle 1999).

Community analysis consists of data collection and analysis with the purpose of gaining a better understanding of the community's problems and priorities. Community organizers also assess the community's strengths, its readiness to change, barriers to change, and other factors that may influence the change process.

Many community-based projects train community members to assist with needs assessments and/or research to identify the factors that must be addressed to bring about change. Called participatory research, this process draws on community members to define the research problem and set research objectives, design the methodology and data collection instruments, collect and interpret data, and use results to guide program planning and evaluation (Israel et al. 1998; Schulz et al. 1999). Community members are actively involved in creating new knowledge and determining how they will use that information.

The *design and initiation* stage is devoted to mobilizing a planning group and designing an intervention. Typically, a coalition, committee, or board is established to plan and coordinate the change process. Other tasks in this second stage include recruitment of members, establishment of bylaws or procedures to guide the planning process, and identification of goals and objectives. Composition of the community body is important for success. If members do not truly represent the entire community, they will lack the understanding and authority needed to plan or implement an effective intervention. Once a problem has been targeted and objectives set, the group then develops strategies for resolving the issue.

The *implementation* stage includes the generation of broad citizen participation, development of a sequential work plan, use of comprehensive, integrated tactics, and the integration of community values and norms into programs, materials, and messages used to bring about change. In some community organizing approaches, research is conducted among representative samples of potential audience segments to enable program planners to benefit from feedback from people whose time and interest preclude more active participation.

Finally, during *maintenance and consolidation,* the program is implemented and successful elements are more fully incorporated into community institutions, such as schools, and policies. A key element in this phase is the development of cooperation and trust among members and the recruitment of new volunteers or staff.

Finally, the program moves into the *dissemination and reassessment* stage. Program activities are evaluated and revised, and the results are shared with community members. Community groups may update their needs assessment, review changes in the resources available within the community, and assess the program's effectiveness in changing norms or behaviors. Some groups search for new organizations to sponsor successful activities or find other ways to institutionalize the change so that it is sustained while they turn their attention to new issues. As they begin planning new directions for the future, the five-stage process begins again.

Case Study

An interesting example of community organizing comes from community health educator and organizer Meredith Minkler's work with the elderly poor living in San Francisco's Tenderloin District (Minkler 1998b). The Tenderloin District is a 45-block residential area inhabited by large numbers of elderly people living on small, limited incomes, immigrants, and younger people who are poor and/or homeless because of physical or mental disabilities. At the time this project began, the neighborhood was known for its unsafe housing, absence of any major grocery store chain, numerous liquor stores, and the city's highest crime rate. In terms of assets, residents in the area included a multilanguage newspaper, several churches, a local health center, and an active neighborhood planning coalition.

Beginning in the late 1970s, Minkler and her students and associates at the University of California at Berkeley initiated the Tenderloin Senior Outreach Project (TSOP) to improve health outcomes by empowering the community and increasing their ability to solve local problems. Specific goals were to improve the physical and mental health of the elderly by reducing social isolation and providing relevant health education, and to facilitate dialogue and participation that would help them identify common problems and seek solutions to them.

During the initial planning phase, project organizers conducted a community assessment that identified assets as well as needs. While over 40 service organizations were identified, one church appeared to be universally respected. In order to establish a linkage with the community, the students became church volunteers and asked hotel managers if they could host weekly discussion sessions. A weekly group was formed in one hotel. Students served refreshments and facilitated group discussions. As trust built, students encouraged hotel residents to vent their frustrations and examine the root causes of their problems. Soon additional groups were formed, and a grant was obtained to hire a part-time director for the project. A nonprofit organization was formed and a 12-member board of directors assembled to represent community residents, churches, other helping organizations, the local planning coalition, and other groups.

One of the first problems elderly residents decided to tackle was the lack of access to fresh fruits and vegetables. After discussing many alternatives, the residents contacted a food service and started mini-markets one day a week in hotel lobbies. Other ideas that came to fruition included a weekly breakfast program that used supplies from a local food bank and the development of a "no-cook cookbook" containing recipes made from inexpensive, yet nutritious, foods that residents could prepare in their rooms. The cookbook was given to hundreds of residents and sold outside the community to generate funds for other activities.

In addition to the impact these activities had on fruit and vegetable consumption, the project increased residents' sense of control over their lives, reduced their social isolation, and gave them the self-confidence and skills they needed to address more difficult challenges. Once empowered by their success in tackling the first problem, and by strength in numbers, they tackled other problems. Over the next decade, they

- developed eight hotel-based support groups, with residents replacing students as group facilitators or cofacilitators,
- established 14 tenant's associations,
- organized a successful protest against a sudden, illegal rent increase of 30%,

- forced a landlord to modify a harsh eviction policy,
- persuaded another landlord to make bathrooms wheelchair-accessible,
- installed a vending machine with low-cost, nutritious foods in one hotel lobby,
- convinced landlords to improve pest control, upgrade plumbing and wiring, and
- persuaded government officials and other organizations to make significant improvements in the surrounding neighborhood.

Advantages of Community-Based Interventions

Community-based interventions have many advantages. Because they are guided by a multilevel, economical perspective, community-based interventions are usually more effective than narrowly focused interventions (Butterfoss, Goodman, and Wandersman 1993). A single program may simultaneously increase the amount of information available from local sources, provide culturally appropriate educational activities, increase access to healthy foods, and bring about other changes in schools, hospitals, workplaces, restaurants, supermarkets, and other retail outlets that make it easier for people to change their dietary behavior.

Second, community participation has an indirect effect on health by strengthening the social networks of community members who come together to define and solve problems. As we saw in the Tenderloin Senior Outreach Program, community participation enhances participants' sense of social connectedness, control over their lives, and ability to change (Minkler and Wallerstein 1998).

Community-based approaches offer a third benefit: longer-lasting change. Modifying existing policies and increasing access to healthy food choices in multiple settings makes it more likely that change will be sustained after the initial intervention is launched. When planned by community members, program activities, services, and other activities are more likely to be compatible with community norms, values, and beliefs. The endorsement of local leaders or opinion makers also fosters acceptance by other community members. Community members are also more likely to sustain efforts they have had a role in designing and implementing (Bracht 1990).

Fourth, community-based approaches can be very cost effective. Even small successes can reap important economic benefits because of the large number of people reached. Compare the impact of a community-wide program that assists just 5% of its 150,000 members to exercise regularly, with the impact of an exercise class that achieves a 75% success rate among 250 individuals. Economic benefits are especially impressive when community-based interventions bring about permanent changes in community norms or environmental structures that reinforce and thus facilitate long-term change (Bracht 1990; Frankish and Green 1994).

Finally, by focusing the public health professional's attention on multiple environmental and interpersonal factors, community organization avoids an overreliance on education, communications, and other strategies criticized for "blaming the victim" (Ling et al. 1992; Rothschild 1999).

Challenges Associated with Community-Based Interventions

Community-based interventions are challenging because of the additional time and skill needed to build effective partnerships with community members. Health professionals are not always prepared for the patience required to make community members feel comfortable participating in the program or the additional time it takes to plan a program with a community advisory group (Brownson, Riley, and Bruce 1998).

Trust and cooperation between public health professionals and community members is an essential ingredient for success. Many health professionals are used to controlling planning activities and find it difficult to allow the community to be in the driver's seat.

It is often difficult to build community work groups that truly represent the larger community. Some community members may not understand issues that do not pertain directly to them. For this reason, many community-based interventions rely on data collected from a representative sample of the community to give them an understanding of the views of people who do not have time or interest in attending planning meetings (Eisen 1994).

Nutrition experts from outside the community may fail to understand residents' norms, beliefs, and values. Outside experts often underestimate the strengths of communities and lack an "appreciation of the expertise of potential participants" (McWilliam et al. 1997:32). For this reason, many community organizers now assess communities' assets in addition to their needs.

Many disenfranchised populations or communities have considerable mistrust of public health efforts, including public health professionals and experts (Alter and Hage 1993). This is especially problematic when large power and status differences exist between community members and outside experts. To build trust and help community members feel equal, public health professionals must be willing to respect the needs and desires of community representatives and share power (Flynn, Ray, and Rider 1994).

Social Marketing

Social marketing, the use of marketing to design and implement programs to promote socially beneficial behavior change, is another widely accepted approach to promoting healthier lifestyles. In recent years, the Centers for Disease Control and Prevention, U.S. Department of Agriculture, U.S. Department of Health and Human Services, and other government and nonprofit organizations have used social marketing to increase fruit and vegetable consumption, promote breastfeeding and proper introduction of infant weaning foods, decrease fat consumption, and promote physical activity and a wide variety of other preventive health behaviors (Coreil, Bryant, and

Henderson 2000). State and local communities are also using social marketing to increase utilization of the Special Supplemental Nutrition Program for Women, Infants and Children (WIC), prenatal care, family planning (Rangun and Karim 1991), and other health services (Bryant et al. 1998b).

In the remainder of this chapter, we discuss the defining characteristics and key concepts of social marketing, the phases in the social marketing process, and an example of how social marketing has been used in the United States to promote breastfeeding among economically disadvantaged families.

Distinguishing Features

Social marketing is distinguished from other social and behavioral change approaches by its adherence to six basic principles: (1) a focus on the consumer or target audience, (2) the notion of exchange and the marketing mix, (3) assessment of competition, (4) use of formative research to understand consumers' desires and needs, (5) segmentation of populations into distinct subgroups, and (6) careful, continuous monitoring and revision of program activities (Andreasen 1995).

Consumer Orientation A central principle in the social marketing approach is a steadfast commitment to understand the consumer and design products to satisfy their desires and needs. Marketers use the term "consumers" to refer to the people whose behavior they hope to change. The social marketing approach asks us to see the world through the eyes of the consumers. Research is used to understand the individual, interpersonal, community, institutional, and policy factors that must be addressed to bring about behavior change.

Most nutritionists believe strongly in the value of the services they offer and healthy behaviors they promote. They also expect people to seek out their services and follow their advice because of these health benefits. Consumers who fail to enroll in their programs, throw away the educational brochures, or ignore their advice are called "hard to reach." A social marketing approach calls for a very different response: When services are underutilized or recommendations are overlooked, program planners assume something is wrong in what they are offering people. They listen to consumers to find out what they can do to improve program offerings and make their recommendations more helpful. This willingness to modify the product is a central feature of social marketing and distinguishes it from attempts to simply persuade people to use what health professionals think is best.

The Concept of Exchange Exchange is another key concept in marketing. This concept should not be new to you: It is a central component in the Health Belief Model and the Theory of Planned Behavior (see Chapter 11) and in other behavior change theories that assert people's behavioral choices are influenced by the cost/benefit ratio. In commercial marketing, consumers are expected to attempt to maximize their ability to satisfy their desires and needs while minimizing what they must pay (Kotler and Armstrong 1996). To be successful, each party involved in the exchange weighs the costs of changing relative to its benefits and then compares this ratio with that of alternative choices. When you select a new pair of pants, marketers assume you weigh the benefits you expect to gain with their monetary price and compare it with what you can purchase elsewhere. "The lessons for social marketers are that we must offer something perceived to be of value to our target audiences and recognize that consumers must outlay resources such as time, money, physical comfort, lifestyle changes or psychological effort in exchange for the promised benefits" (Egger et al. 1999:28).

The Marketing Mix   In addition to the notion of exchange, social marketing is also distinguished by its reliance on additional concepts called the 4 P's: the product, price, place, and promotion.

The *product* refers to the healthy behavior (exercising regularly, breastfeeding, eating five servings of fruits and vegetables a day) or service (WIC program, nutrition education class) being promoted. In a social marketing approach, the product is viewed as a bundle of benefits or value being offered to the consumer. The goal is to identify the benefits most appealing to consumers. When promoting physical activity, marketers may promote its ability to prevent depression, osteoporosis, and heart disease; promise a more vigorous, youthful appearance; or highlight it as a way to make new friends. The goal is to discover which benefits have the greatest appeal and design a program to deliver them. Often the most appealing benefits have little or nothing to do with health, as when people exercise to look sexier rather than to be healthier.

Social marketing also reminds public health professionals to consider the competition. For many behaviors of interest to nutritionists, consumers must decide between a healthy (low-fat ice cream) and unhealthy (regular ice cream) alternative. When using a social marketing approach, nutritionists ask: "What products [foods, practices] compete with those we are promoting?" "How do consumers compare our product's benefits with those offered by the competing product?" "How can we make our product's benefits more attractive than those of the competition?"

Price refers to the costs or sacrifices exchanged for the product's benefits. Again, nutritionists must consider price from the consumers' point of view. In contrast to the relatively simple monetary costs of commercial products, dietary behavior changes require many sacrifices—time, loss of pleasure, embarrassment, or the psychological discomfort associated with change, especially when modifying ingrained habits.

In setting "the right price," it is important to know if consumers prefer to pay more to obtain extra, "value-added" benefits and if they think products that are given away or priced too low are inferior to more expensive ones. A good example of how important it is for health professionals to understand consumers' perceptions of price comes from a condom promotion program coordinated by Population Services International. Their consumer research revealed that the reason teens did not use condoms given away by health programs was that they assumed the condoms must be of inferior quality because they were free. While many teens were deterred by the high price of condoms in retail outlets, they could easily afford to pay 25 cents for them. This price was determined as sufficient to reassure consumers that the condoms were reliable, but not too high to make them unaffordable.

Place, the third "P," refers to the location where services are provided, where tangible products are distributed, and where consumers receive information about new products or behaviors. In the Population Services International study, Pizza Huts and other fast-food outlets were identified as ideal locations for condom vending machines because many teen couples went there for a snack before going home together for the evening. Place also refers to the location, time of day, days of the week, and appearance of facilities where nutrition services, like the WIC program, are offered.

Promotion refers to the activities used to bring about behavior change: advertising, public relations, media advocacy, policy changes, consumer education, skills training, support groups, or other means. When marketing nutrition services, the promotional strategy also includes improvements in the delivery of the service, the use of incentives, special events like health fairs, and other means of attracting or communicating with consumers. To be effective, promotional strategies must be carefully coordinated with other components of the marketing mix (Lefebvre and Flora 1988). Promotional efforts are carefully integrated into the marketing mix; promotional efforts cannot succeed if the product's benefits, price, and placement are not also in line with the consumer's wants and needs. To develop a well-coordinated marketing plan, as opposed to a communications program, intervention strategies are based on a careful analysis of an audience's perception of each of the 4 P's and how these variables interact. To this end, the social marketing approach relies on consumer research—the next distinguishing feature of the social marketing approach.

Consumer Research   Social marketing uses research to understand the consumer's perception of product benefits, product price, the competition's benefits and costs, and other factors that influence consumer behavior. As with other approaches, it is critical to identify the factors (perceived threat, social influences, self-efficacy, etc.) that motivate and deter people from adopting the recommended behavior or product. Marketing research is also used to identify the spokespersons who will have the greatest credibility, determine how the message strategy should be developed, decide which media should be used, and identify other activities that would be effective in promoting the product (Egger et al. 1999). Finally, research is used to divide the population into distinct subgroups and identify the segments that the intervention should attempt to reach. This process is called *audience segmentation.*

Audience Segmentation Marketers know that one size does not fit all. People differ in terms of the benefits they find most attractive, the price they are willing to pay, the best place to communicate with them or locate services, and their differential responsiveness to promotional tactics. Just consider the numerous types of automobiles sold in North America. Automobile makers recognize that some people are willing to pay a great deal more than others for a car. They also know the public can be divided into segments based on the features or benefits they value most. Some cars emphasize their safety features, others are designed to meet the needs of a large family, while still others promise speed and a sexy image. In nutrition, audience segmentation makes it possible to identify "targets of greatest opportunity"—subgroups most likely to change their behavior—and understand the types of product benefits, price, place, and promotional mix effective in helping them to change.

Many program planners divide their target population into segments based on the Stages of Change Model (Prochaska and DiClemente 1992) discussed in Chapter 11. They select the segments they hope to reach based on their size, their responsiveness to program activities or services they can provide, and likelihood of success. This method of segmenting an audience enables the program planner to select the processes of change most appropriate for reaching the segments they will try to reach. In designing a program to promote bone density screening to prevent osteoporosis in women, planners might find that 20% of the target population are in precontemplation, 60% are in contemplation, 5% are in preparation, 15% are in action, and the rest have committed to regular screenings and/or preventive treatment (e.g., regular exercise, increased calcium intake, and preventive medication). Based on these findings, the program may focus its efforts on contemplators because they comprise a large proportion of women who are seriously thinking about being screened. Program planners would then design activities to reach this segment, relying on the processes of change appropriate for moving women from contemplation to preparation, such as self-reevaluation, social support, modeling, focusing on positive outcomes, and providing them an easy, low-cost way to try out the recommended behavior (Prochaska and Velicer 1997).

Careful Pretesting and Continuous Monitoring Concepts and prototypes for mass media messages and educational materials are carefully pretested with members of each target audience to make sure they are easily comprehended, attention-getting, believable, and persuasive. All educational and promotional activities are monitored to identify unforeseen problems that may require midcourse revisions to improve their effectiveness. Although many nutrition education programs rely on process and impact evaluations at the end of a program to identify the components that have worked, social marketing devotes considerable resources to this activity throughout the program implementation phase. Program administrators are constantly checking with target audiences to gauge their responses to all aspects of an intervention from the broad marketing strategy to specific messages and materials.

Steps in the Social Marketing Process

The social marketing process can be broken into six major steps or tasks:

- Initial planning
- Formative research
- Strategy formation
- Program development and structuring
- Implementation
- Tracking and evaluation

To illustrate these steps, we describe the use of social marketing to promote breastfeeding among economically disadvantaged families participating in the WIC program in the United States. This case study is based on an article published in the *American Journal of Health Behavior* (Lindenberger and Bryant 2000).

Loving Support Makes Breastfeeding Work

Among the WIC program's many offerings is breastfeeding support, education, and promotion. Since 1989, Congress has designated a portion of the WIC allocation for each state to be used specifically to support and promote breastfeeding among program participants. Despite this effort, breastfeeding rates among WIC participants lagged behind those of more affluent segments of the population. When this project was initiated in 1995, 46.6% of women participating in the WIC program breastfed at hospital initiation, compared to 71% of those who did not (Ryan 1997).

To reach the millions of pregnant women enrolled in WIC, Best Start Social Marketing, a nonprofit social marketing organization, worked with the U.S. Department of Agriculture's Food and Nutrition Service (FNS) to develop a comprehensive, national effort to promote breastfeeding for WIC participants.

Initial Planning During the fall of 1995, Best Start staff met with FNS representatives and members of their breastfeeding promotion consortium to finalize program objectives, select states to serve as pilot sites for the project, and gain a better understanding of current breastfeeding promotion activities and challenges within the WIC program. Program goals were to (1) increase breastfeeding initiation rates, (2) increase breastfeeding duration rates among WIC participants, (3) increase referrals to WIC for breastfeeding support, and (4) increase general public support for breastfeeding.

The primary target population for the program was comprised of pregnant European American, African American, and Hispanic American women who were enrolled in the WIC program or who were income-eligible (i.e, income below 185% of the U.S. poverty guidelines). The program was also intended to reach people who influence the primary target population, especially pregnant women's mothers and partners, WIC nutritionists and clerical staff, and prenatal health care providers. The program was also designed to increase support for breastfeeding among the general public, because they also influence women's decisions to breastfeed.

Formative Research   In-depth interviews and focus groups were conducted with WIC participants and employees, health care providers, and participants' mothers and partners in ten pilot states. Approximately 300 WIC participants were also surveyed.

Research with WIC participants showed that they viewed breastfeeding as a way to realize their goals to have a healthy baby and enjoy a special time with their newborns. Compared with its competition—infant formula—breast milk was perceived to provide better nutrition and better protection from illness and promote a closer maternal-infant bond. For these women, breastfeeding's most appealing benefits were the enjoyment and special time they expected breastfeeding to give them with their babies.

Although most women were attracted to breastfeeding's benefits, they were concerned about its "costs." They worried that breastfeeding would create embarrassing moments when they had to nurse in public settings and open them to criticism from friends or relatives who viewed breastfeeding in a sexual light. Some women worried that the additional time it takes to nurse a baby would conflict with work, school, or social life. Some worried that their partners, mothers, and other close relatives would feel "left out" of the feeding experience and fail to bond to their baby. Other "costs" associated with breastfeeding were the pain the women expected to encounter when nursing, changes they would have to make in dietary and health practices, and anxiety about their ability to produce the quality and quantity of breast milk needed to meet their child's nutritional needs. Finally, many women lacked self-efficacy in their ability to breastfeed successfully.

Research with WIC employees showed that most nutritionists and clerical staff felt that breastfeeding support and education are important components of the WIC program. Nutritionists enjoyed the personal interaction that breastfeeding education afforded them, and many expressed a sense of achievement and satisfaction with their roles as breastfeeding advocates. Clerks in some agencies were actively involved in breastfeeding support. In agencies where training had not been offered to clerical staff, some clerks expressed an interest in receiving breastfeeding training so that they could play a more active role in promoting breastfeeding.

WIC employees also talked about numerous problems that offset the benefits to be gained from breastfeeding promotion: time required to promote breastfeeding (a precious commodity in many agencies where staff shortages and other competing demands make it difficult to allocate sufficient time to educational activities) and lack of administrative support for breastfeeding promotion. Employees working in communities with low breastfeeding rates were also frustrated by clients' lack of receptivity to breastfeeding promotion activities. Finally, staff members' efforts to promote breastfeeding were hampered by external factors, including a perceived lack of support from health care providers and hospitals, direct competition from formula companies, and negative portrayal of breastfeeding by the media.

Research conducted with other health care providers was also revealing. They believed breastfeeding promotion efforts had improved greatly, but reported that a consistent effort to encourage and support breastfeeding was lacking in most communities. They also pointed out that breastfeeding instruction and support during the critical postpartum period were lacking in most areas.

Strategy Formation Formative research results were used to develop a marketing plan to guide the development, implementation, and tracking of the project. This plan spelled out specific, measurable objectives and a step-by-step work plan to guide the development, implementation, and tracking of the program. It gave a vivid description of the audience segments being targeted and strategies for addressing the critical factors associated with the health behaviors being promoted. The plan was organized around marketing's conceptual framework of the 4 P's—product, price, place, and promotion. The product strategy called for an emphasis on the close, loving bond and special joy that breastfeeding mothers share with their babies. Although breastfeeding's health advantages would be mentioned in some materials, the product strategy emphasized the emotional benefits breastfeeding offers because these most clearly distinguished the product from its competition: bottle-feeding. Breastfeeding was positioned as a way families can realize their dreams of establishing a special relationship with their children, and the campaign slogan—"Loving Support Makes Breastfeeding Work"—and program materials emphasized the role family members and friends and public information materials play in the mother's ability to breastfeed. This strategy represented a significant departure from more traditional public health approaches to breastfeeding promotion in which breastfeeding was discussed as the best medical choice.

Because so many women were deterred by their perception of the costs associated with breastfeeding, a focus on lowering costs became a central component of the marketing plan. The target population was segmented based on perceived costs and the largest segments—those who perceived breastfeeding as embarrassing, conflicting with work, school, and an active social life, and/or jeopardizing their relationships with loved ones—were selected as targets. The marketing plan focused on ways to lower these "costs" by countering misperceptions and introducing ways to overcome them. A key tactic for achieving this was a three-step counseling strategy designed to teach health providers to identify patients' perceptions of breastfeeding's costs and help mothers find ways to make breastfeeding fit within their work and social lives. Public information and consumer education materials were developed for each target segment to influence the public's attitudes about breastfeeding and correct common misperceptions about the "costs" of breastfeeding.

The placement strategy highlighted the multiple settings in which women and their social network members obtain information about infant feeding. Partnerships were formed with the Baby Friendly Hospital Initiative, an international breastfeeding promotion program developed by the United Nations Children's Fund and the World Health Organization, to create hospital environments that fully support breastfeeding mothers. Public information materials were developed to reach women in their homes where they could discuss breastfeeding with relatives and friends. Media advocacy and policy development were used in many states to promote policies supportive of breastfeeding in workplaces and pass legislation permitting breastfeeding in public settings.

The promotional strategy utilized multiple approaches to facilitate change, including legislative change, policy development and organizational change, professional training, peer counselor programs, curriculum development, consumer education, public relations, direct marketing, advertising, face-to-face communication, media advocacy, and grassroots advocacy. Message design guidelines were prepared to assist the creative team in developing communication materials. These guidelines recommended an emotional appeal with an upbeat, congratulatory tone, a slice-of-life or vignette manner of presentation, and the use of families as spokespersons.

Finally, Best Start staff met with FNS's Breastfeeding Promotion Consortium to build partnerships needed to institutionalize and sustain the program. Assistance from a wide variety of federal agencies, professional organizations, such as the American Academy of Pediatrics and the American College of Obstetricians and Gynecologists, and breastfeeding promotion organizations, such as La Leche League, proved invaluable in producing additional program materials and disseminating the program outside the WIC community. Efforts were also initiated to obtain funding to monitor breastfeeding rates in the pilot states and elsewhere so that program impact could be assessed.

Program Development and Structuring This phase involved the development and revision of all campaign materials and strategies. Best Start worked with an advertising agency to design public information messages and prepare consumer education materials. Concepts and prototype TV ads, posters, and pamphlets were pretested, and revisions were made before the final materials were developed. A copy of one poster is shown in Figure 12.2.

The final campaign theme, "Loving Support Makes Breastfeeding Work," was then used in a coordinated fashion to build a comprehensive campaign for maximum impact. The original media campaign and collateral materials included

- three 30-second television commercials in English and one spot in Spanish;
- three 60-second radio commercials, two in English and one in Spanish;
- outdoor advertising boards, in English and Spanish;

- nine posters, targeting the primary ethnic groups in WIC (English and Spanish);
- nine educational pamphlets, targeting the primary ethnic groups in WIC (English and Spanish);
- a motivational/information booklet for WIC staff;
- a breastfeeding resource guide;
- a breastfeeding promotion guide; and
- a WIC staff support kit pocket folder to hold breastfeeding resource guide, breastfeeding promotion guide, and motivational/informational booklet for WIC staff.

Program Implementation A key element of the National WIC Breastfeeding Promotion project was training of WIC staff members who would implement the program at the state and local level. A training conference was held for teams from the 10 pilot sites and other interested WIC programs. The program was launched in August 1997 as part of World Breastfeeding Week.

Monitoring and Evaluation State WIC programs monitored the program and reported two problems to Best Start. As a result, additional materials were developed to reach health care providers who worked outside of the public health system and a set of materials was designed for Native American Indians.

An evaluation of the program in one state revealed that after just 2 years, the program had increased WIC employees' knowledge about breastfeeding and their confidence as breastfeeding counselors. WIC participants' perceptions of breastfeeding's benefits and costs also improved significantly. Most importantly, despite similar rates before the program was implemented, breastfeeding rates were higher in WIC programs that implemented the program for 2 years (called intervention sites) than in WIC programs that did not (called control sites) (Khoury et al. 2001). (See Table 12.1.)

This case study is just one example of how social marketing addresses factors at multiple levels, addressing individual beliefs and attitudes, social support, and social norms, and modifies the environment by changing policies and access to information and resources needed to change behavior. Social marketing relies on research and commercial marketing principles and techniques to

Figure 12.2
A "Loving Support" campaign poster designed to elicit social support for breastfeeding women.

Courtesy Best Start, Inc.

Table 12.1 Evaluation of the Mississippi Breastfeeding Promotion Program

	Breastfeeding Rates at Hospital Discharge	Breastfeeding Rates at 4 Months After the Baby's Birth
Intervention sites	49%	31%
Control sites	31%	19%

Source: A. J. Khoury et al., Evaluation of Mississippi's Breastfeeding Promotion Program, 2001.

design programs to understand and satisfy consumers wants and needs in an effort to bring about voluntary behavior change that benefits them and the society in which they live (Andreasen 1995). Another social marketing program is described in Focus 12.1.

FOCUS 12.1   A Social Marketing Approach to Increasing Utilization in the Texas WIC Program

During the early 1990s, many families who could have benefited from WIC were not participating in the program. The Texas Department of Health contracted with Best Start Social Marketing to develop a marketing program and outreach strategies for attracting new customers and retaining those already enrolled.

Target Audience

Extensive formative research was conducted with the three primary target populations: eligible families who were not currently participating in the WIC program (eligible); families currently enrolled in the program (participant); and employees working in 88 local WIC clinics and state administrative offices (staff).

Research

A mix of qualitative and quantitative research methods were used to segment audiences and identify the factors that influence satisfaction with and participation in the WIC program (Bryant et al. 1998b).

The study of *eligible* families was conducted using focus groups, telephone interviews, and a mail survey of a random sample of pregnant Medicaid recipients to identify perceived benefits, costs, and other factors that influenced families' decisions to enroll in WIC.

The *participant* study relied on qualitative data from focus groups, and telephone and structured individual interviews with participants and family members, in rural, midsize cities and metropolitan locales in areas throughout Texas.

A study of WIC *staff members* included in-depth interviews and focus groups with WIC administrators, agency directors, and clinical and clerical staff members, and a survey distributed to employees.

Results

Study of Eligible Families

Research results identified several important attitudes and perceptions that influenced enrollment decisions. Most families were familiar with portions of WIC's food package, but few were aware of additional services that greatly enhanced the program's attractiveness. Major barriers to participation in WIC included

- the mistaken belief they were not eligible if they were married, worked, had a husband who worked, lived with relatives whose incomes were too high, or had been denied food stamps;
- lack of awareness of important program benefits, such as individualized nutritional risk assessment and counseling and health referrals;
- loss of pride and dignity if identified as a recipient of a free food program; and
- apprehension about participating in a program their friends and relatives had complained about, for example, long waits at WIC clinics to obtain food cards and having to listen to educational videotapes, rude treatment by WIC staff, lack of Spanish-speaking staff, and rude treatment while redeeming WIC food vouchers at the grocery store.

Research was used to segment the eligible population into subgroups of women who could benefit from WIC and to identify specific issues to recruit them. For example, women who were employed, completed high school, and earned incomes 135–185% of the federal poverty level were the least likely to know they were income-eligible for WIC. This segment was selected as a target for clarifying eligibility requirements.

(continued)

Focus 12.1 (continued from previous page)

Participant Study

Program participants had positive perceptions of the Texas WIC program, with 64% rating the program "very good," 30% "good," and 6% "fair." Program benefits rated most highly were health and nutrition information (57%), money saved on grocery bills (51%), nutritious foods (39%), and having their children's blood, height, and weight checked (23%). While families valued WIC's many services, they also encountered difficulties that made participation costly. These included

- long waiting times for initial certification, nutrition education, and receiving food cards,
- difficulty securing and paying for supervised child care when visiting WIC clinics,
- disrespectful or rude treatment by clinic staff,
- rude or inconsiderate treatment by grocery cashiers,
- problems redeeming food cards,
- displeasure with specific types or brands of food offered by the WIC program, and
- embarrassment about participating in a government assistance program.

Staff Study

Staff members reported that the primary benefit of working for WIC was satisfying their desire to help others and professional gratification of doing a good job. WIC employees enjoyed serving as WIC team members, saying that this sense of "camaraderie" was key to feeling good about their job. The most significant costs of working in the WIC program were pressure to expand enrollment, poor supervision, racial tension with participants, disrespectful participants, staff shortages, and cramped, noisy work areas that impeded client flow and efficient service.

Marketing Plan

A comprehensive social marketing plan was developed to provide a customer-centered approach to program structure and service delivery. It specified policy change recommendations, service delivery improvements, staff and vendor training, internal promotion, public information and communications, client education, and community based interventions.

Program Implementation

Using the social marketing plan as a guide, Best Start social marketers, health department personnel, and an advertising agency then developed the messages, outreach materials, training programs, data collection instruments, and other resources needed to implement and evaluate the program. A key element in this phase was ensuring that each component was consistent with the overall marketing strategy. Marketing messages and mass media materials were rigorously pretested and redesigned until they proved to be effective. These materials included a new WIC brand and logo, two television spots and three radio spots (each of which were in English and Spanish), billboards, a set of outreach brochures and posters, a community organizer's kit to facilitate coalition building at the local-agency level, several training modules, and data collection instruments for tracking and evaluating the program. Examples of program materials are displayed in Figures 12.3 and 12.4.

Finally, although funds were not released for a formal evaluation of the program's impact, data collected by the Texas WIC program point to significant increases in the number of families who called the toll-free number for more information after the program was launched and, more importantly, in the number of people participating in Texas WIC. The program's caseload grew from 582,819 in October 1993 to 778,558 in October 1998—an increase of almost 200,000 participants.

Figure 12.3
New logo for the Texas WIC program.

Figure 12.4
Texas WIC program posters.

Putting It All Together to Improve Dietary Practices

Most nutritional professionals combine elements of community organizing and social marketing rather than strictly follow the tenets of any single approach. Relying on a framework similar to the one we described in Chapter 1, they design interventions to address each of the multiple components that influence human diets and nutritional status. You may recall the biocultural model we introduced in Chapter 1. This framework describes how nutritional status, biological makeup, human nutrient needs, diet, cuisine, the physical environment, the sociocultural environment, and the economic and political environment all influence how humans eat.

An excellent example of a national nutrition intervention designed to address factors at multiple levels is summarized in the Surgeon General's Call to Action to Prevent and Decrease Overweight and Obesity released in December 2001. Based on extensive study of the obesity epidemic and recommendations from clinicians, researchers, consumers, and advocates, the Surgeon General's report outlined a set of comprehensive, community-based strategies that were subsequently reviewed for their proven scientific effectiveness. In recognition of the multiple factors that contribute to obesity, the report recommends interventions for addressing numerous factors in five key settings: families and communities, schools, health care, media and communications, and worksites. Many of the recommended actions overlap the different settings and can be applied in several or all environments. To give you an idea of how comprehensive a multilevel approach must be, we have selected examples of recommendations for addressing factors in each setting (U.S. Department of Health and Human Services 2001b). As you read through them, you can see that all of the components of the biocultural model are represented in this list. (For a copy of the entire report, go to *http://www.surgeongeneral.gov/topics/obesity/calltoaction/toc.htm*.)

Families and Communities

- Raise policy makers' awareness of the need to develop social and environmental policy that would help communities and families be more physically active and consume a healthier diet.
- Educate individuals, families, and communities about healthy dietary patterns and regular physical activity, based on the *Dietary Guidelines for Americans*.
- Educate parents about the need to serve as good role models by practicing healthy eating habits and engaging in regular physical activity in order to instill lifelong healthy habits in their children.
- Raise consumer awareness about reasonable food and beverage portion sizes.
- Form community coalitions to support the development of increased opportunities to engage in leisure-time physical activity and to encourage food outlets to increase availability of low-calorie, nutritious food items.

- Encourage the food industry to provide reasonable food and beverage portion sizes.
- Evaluate the feasibility of incentives that support healthful dietary and physical activity patterns.
- Identify techniques that can foster community motivation to reduce overweight and obesity.
- Examine the marketing practices of the fast-food industry and the factors determining construction of new food outlets.

School Settings

- Build awareness among teachers, food-service staff, coaches, nurses, and other school staff about the contribution of proper nutrition and physical activity to the maintenance of lifelong healthy weight.
- Educate teachers, staff, and parents about the importance of school physical activity and nutrition programs and policies.
- Educate parents, teachers, coaches, staff, and other adults in the community about the importance they hold as role models for children, and teach them how to be models for healthy eating and regular physical activity.
- Ensure that meals offered through the National School Lunch and School Breakfast Programs meet nutrition standards.
- Adopt policies ensuring that all foods and beverages available on school campuses and at school events contribute toward eating patterns that are consistent with the *Dietary Guidelines for Americans.*
- Provide food options that are low in fat, calories, and added sugars, such as fruits, vegetables, and whole grains, and low-fat or nonfat dairy foods.
- Ensure that healthy snacks and foods are provided in vending machines, school stores, and other venues within the school's control.
- Prohibit student access to vending machines, school stores, and other venues that compete with healthy school meals in elementary schools and restrict access in middle, junior, and high schools.

Health Care

- Inform health care providers and administrators of the tremendous burden of overweight and obesity on the health care system in terms of mortality, morbidity, and cost.
- Inform and educate the health care community about the importance of healthy eating, consistent with the *Dietary Guidelines for Americans,* and physical activity and fitness for the promotion of health.
- Educate health care providers and administrators to identify and reduce the barriers involving patients' lack of access to effective nutrition and physical activity interventions.
- Train health care providers and health profession students in effective prevention and treatment techniques for overweight and obesity.

- Encourage partnerships between health care providers, schools, faith-based groups, and other community organizations in prevention efforts targeted at social and environmental causes of overweight and obesity.
- Establish a dialogue to consider classifying obesity as a disease category for reimbursement coding.
- Explore mechanisms that will partially or fully cover reimbursement, or include as a member benefit, health care services associated with weight management, including nutrition education and physical activity programs.

Media and Communications

- Emphasize to media professionals that the primary concern of overweight and obesity is one of health rather than appearance.
- Emphasize to media professionals the disproportionate burden of overweight and obesity in low-income and racial and ethnic minority populations and the need for culturally sensitive health messages.
- Emphasize to media professionals the need to develop uniform health messages about physical activity and nutrition that are consistent with the *Dietary Guidelines for Americans.*
- Conduct a national campaign to foster public awareness of the health benefits of regular physical activity, healthful dietary choices, and maintaining a healthy weight, based on the *Dietary Guidelines for Americans.*
- Train nutrition and exercise scientists and specialists in media advocacy skills that will empower them to disseminate their knowledge to a broad audience.
- Encourage community-based advertising campaigns to balance messages that may encourage consumption of excess calories and inactivity generated by fast-food industries and by industries that promote sedentary behaviors.
- Encourage media professionals to utilize actors' influences as role models to demonstrate healthy eating and physical activity for health reasons rather than for appearance.

Worksites

- Inform employers of the direct and indirect costs of obesity.
- Communicate to employers the return-on-investment (ROI) data for worksite obesity prevention and treatment strategies.
- Change workflow patterns, including flexible work hours, to create opportunities for regular physical activity during the workday.
- Provide protected time for lunch, and ensure that healthy food options are available.
- Establish worksite exercise facilities or create incentives for employees to join local fitness centers.
- Create incentives for workers to achieve and maintain a healthy body weight.

While this plan demonstrates an awareness of the importance of addressing factors at multiple levels, it is just a start. Many states, with funding from the Centers for Disease Control and Prevention, are working with community coalitions to implement comprehensive initiatives to prevent and control obesity. For a more detailed examination of how a state or county health department can design, implement, and monitor a multilevel intervention, see Highlight 12.

Summary

Social marketing and community-based approaches illustrate the power of comprehensive, multilevel interventions that address the larger social and cultural environments in which we live and eat.

- Many public health professionals now believe that change strategies directed by community members are far more likely to succeed than those planned and executed exclusively by outsiders (Green and Kreuter 1991; Minkler and Wallerstein 1998).
- Community-based approaches are guided by the social ecological perspective of health and the belief that social and physical environments can be modified to augment individual behavior change.
- Community-based efforts use multifaceted strategies to address the environmental, social, behavioral, and political factors that affect health, either simultaneously or sequentially.
- Community-based approaches are designed to address the felt needs of the target community—"starting where the people are."
- Community participation is a hallmark of all community-based interventions. Outside experts work in collaboration with a wide variety of community members from initial planning to the implementation and evaluation phases.
- Community organization differs from other community-based approaches in its emphasis on community empowerment and the level of control the community exerts in defining the issues it will address and planning, implementing, and evaluating strategies designed to meet the goals they set.
- Community-based approaches often involve longer and more complex planning than interventions targeting individuals or groups; however, they offer many benefits. By increasing communities' capacity to solve their own problems, and addressing factors at multiple levels, they increase the likelihood of permanent, large-scale change.
- Social marketing is the application of marketing concepts and techniques to the design and implementation of programs to promote socially beneficial behaviors and services.
- The defining features of the social marketing approach include (1) a focus on the consumer or target audience, (2) the notion of exchange and the marketing mix, (3) assessment of competition, (4) use of formative research to understand consumers' desires and needs, (5) segmentation of populations into distinct subgroups, and (6) careful, continuous monitoring and revision of program activities (Andreasen 1995).
- Social marketing follows a systematic, data-driven process consisting of initial planning, formative research, strategy formation,

program development and structuring, program implementation, and monitoring and evaluation.

• Consumer research is used to help health professionals gain a better perception of how consumers view the 4 P's—product, price, place, and promotion—and then use that understanding to promote services and/or encourage healthier behaviors among those groups.

HIGHLIGHT

Moving from Theory to Practice: A Case Study Using a Multilevel Approach to Changing Dietary Behavior

ARE YOU READY TO TAKE THE 5 A DAY CHALLENGE?

The last two chapters have discussed the importance of using a multilevel approach for changing dietary habits. This highlight illustrates this holistic ecological approach using a case study from the Lexington-Fayette County Health Department in Lexington, Kentucky, a city with a population of approximately a quarter of a million people. It illustrates what a local health department can do with limited funds to promote fruit and vegetable consumption. We'll consider the program from a number of levels: the intrapersonal, interpersonal, institutional, community, and public policy levels.

Background: Why Encourage Increased Fruit and Vegetable Intake?

Large-scale adoption of a diet rich in fruits and vegetables could have a significant effect in reducing the incidence of cancer, heart disease, and stroke, major killers in the developed world. Convincing evidence is also available for the positive role of fruits and vegetables in reducing rates of cataracts, hypertension, **diverticulosis, chronic obstructive pulmonary disease,** and birth defects. The important role of produce in weight management and diabetes also cannot be overlooked (Produce for Better Health Foundation 2000).

Scientists are starting to unravel the reasons produce is so effective in preventing disease. Research suggests that **phytochemicals,** biologically active compounds found in plant foods, work together with nutrients and dietary fiber to protect against disease. To date, over 2000 phytochemicals have been identified (Brown 2002). These compounds have complementary and overlapping mechanisms of action in the body including **antioxidation,** modification of enzymes that break down toxic substances, stimulation of the immune system, modification of hormone metabolism, and antiviral and antibacterial actions (Produce for Better Health Foundation 2000).

Although the evidence supporting the value of eating fruits and vegetables is strong, the practice is not. And what good is brilliant scientific research if it is not put into practice? Only 23% of Americans eat the recommended five servings of fruits and vegetables a day; American children eat an average of only 2.4 servings. Kentucky has a particularly difficult challenge: Only 16% of Kentuckians eat five servings of produce daily. Only two states have a lower rate than Kentucky. According to the U.S. Department of Health and

diverticulosis
A condition created when weak areas or pouches form in the intestine. These pouches, diverticula, can become inflamed.

chronic obstructive pulmonary disease
A permanent obstruction of airflow from the lungs usually caused by smoking or chronic exposure to air pollutants.

phytochemicals
Chemicals in plants that give them their color and flavor and help them to grow. Some of these phytochemicals play important roles in the human body.

antioxidation
Reactions that protect or repair damage to cells caused by exposure to oxygen. Antioxidants are chemical compounds, including nutrients like vitamins C and E, that react with oxygen in ways that prevent it from damaging the cells.

Human Service's report *Healthy People 2010 Objectives,* the key objective for Kentucky is to have 40% of state residents eating "5 A Day" by the year 2010.

A great deal of work has been done around the to world to promote fruit and vegetable consumption. Canada sponsors the 5–10 A Day program, New Zealand promotes 5+ A Day, and Hungary has a more realistic 3 A Day program. In the United States the National Cancer Institute (NCI) and the Produce for Better Health Foundation (PFBH) have partnered to develop a successful comprehensive 5 A Day campaign that has increased produce consumption in America during the past 10 years (Produce for Better Health Foundation 2000). This program has educated consumers, organized health departments around the country, worked with retail grocers, and changed policy including the addition of "Choose a variety of fruits and vegetables daily" to the *Dietary Guidelines for Americans* in the year 2000.

Case Study: The 5 A Day Challenge Contest in Central Kentucky

The Pilot Program: A Small-Scale Employee Wellness Program

To identify problems and correct them before the program was launched on a large scale, a pilot program was implemented with health department employees. This is a crucial phase of program development and can save many headaches and much money before proceeding on a larger scale.

In October 1999, the Employee Wellness Program of the Lexington-Fayette County Health Department (LFCHD) challenged its 310 employees to eat 5 servings of fruits and vegetables a day for 25 out of the 31 days in October and keep track of it on a special calendar. Health Department staff were encouraged to invite their spouses, significant others, and children to participate as well. Those who completed the challenge were eligible for prizes.

A variety of activities were used to make produce readily accessible. On the first day of the contest, employees were greeted with a glass of orange juice in the parking lot. Children dressed in costumes as "fruit fairies" fluttered into offices with baskets of locally grown apples to distribute. The local farmer's market set up business on the front porch of the Health Department and sold its harvest to employees and visitors. Throughout the month, chefs from local restaurants provided food demonstrations that convinced even skeptics that healthy food could taste good. A "Smoothie Morning" and "Juice Hour" were also held. Participating children received letters in the mail and weekly calls from "Fruity Fran Vegetable," an employee who is well loved by children.

Of 310 employees, 142 (or 46%) took the 5 A Day Challenge. Of the people who took the challenge, 73% of employees, 99% of children, 97% of teens, and 53% of spouses successfully completed it. Follow-up interviews with those who participated in the project pointed to a variety of benefits they experienced. Some employees lost unwanted weight, mothers were surprised to see their teenage boys eating vegetables for the first time in years, and parents reported their children started asking them for more fruit. Many people felt a sense of satisfaction at being successful in completing a dietary change. People also liked the contest because it focused on what they *could* eat, rather than focusing on what they *should not* eat. As one woman said, "I get it. If you eat five servings of fruits and vegetables every day, there's not as much room for the junk. But you don't quite notice that you're depriving yourself because the focus is on eating all this delicious fresh fruit."

Because this program was positively received, the health promotion staff of the LFCHD decided to take a variation of the contest region-wide.

The 5 A Day Challenge *Contest of Central Kentucky*

In March 2001, residents of Central Kentucky were challenged to eat five servings of fruits and vegetables a day for seven consecutive days and keep track of it on a special contest entry form. Participants sent their entry forms to the health department to be eligible for a prize drawing. The prizes are listed in Box H12.1.

BOX 12.1

Prizes for the *5 A Day Challenge* Contest

One grand prize:	$500
Two 2nd place prizes	$250
One 3rd place prize	A shopping cart full of fresh fruits and vegetables from Kroger
Ten 4th place prizes	Fruit-of-the-Month Club subscriptions
Ten 5th place prizes	$100

Four corporate sponsors joined the campaign, contributing funds and/or products and services. The four sponsors were Kroger, a retail grocery chain; News Channel 36, a local ABC affiliate television station; area Subway franchisees; and *Lexington Family Magazine*, a local monthly newspaper.

Produce Man, a lettuce-lovin', pepper-packin', watermelon-wearin' messenger for 5 A Day, served as mascot for the campaign. Produce Man, a costume made of 135 servings of fruits and vegetables, is shown in Figure H12.1. Designed by Character Translations, he has hair made of lettuce, bananas for a rib cage, and watermelons for biceps. Produce Man was visible in the community almost every day of the campaign. He could be seen at drive-in bank windows handing out fruit and contest brochures, surprising shoppers at Wal-Mart, and waving to rush-hour drivers from street medians.

Figure H12.1
Billboard from the 5 A Day Challenge *campaign in Kentucky.*

Produce Man hosted the game "Who Wants to Feel Like a Million?" attended by 7,500 students in 14 school assemblies. Surrounded by his entourage, he marched in the St. Patrick's Day parade, attended by thousands of onlookers. And finally, Produce Man appeared in two gigs with Jared, "the Subway Diet Guy."

As is often the case in local health departments, funds were not available for a formal evaluation of the program's impact on behavior using a pre/post survey and comparison group design. It was possible, however, to conduct a follow-up telephone survey with 102 of the 1,022 contest participants. Written surveys and follow-up phone calls were also used with health department nutritionists and school cafeteria managers to document their experience of the campaign. Some of these findings are described in the sections below.

Applying Theory to Practice

The Intrapersonal Level

As you may remember, the intrapersonal level includes intervening at the level of people's knowledge, beliefs, expectations, attitudes, intentions, skills, self-perceptions, and other personal characteristics that impact their behavior. Through billboards, newspaper articles, television and radio commercials, and a Web site, the message of eating 5 A Day was promoted widely at the mass media level. This was reinforced with more personal forms of communication through school assemblies, promotions in school cafeterias, cooking classes, workshops, and clinic counseling sessions that gave more expanded information on the topic. These interventions were designed to change people's *knowledge, beliefs, attitudes, expectations,* and *skills* in relation to eating fruits and vegetables.

The contest encouraged people to set the *intention* of eating five or more servings of produce daily for a week by counting and keeping track on a contest entry form. Carrying out this goal was intended to help them develop the *self-perception* of being someone who was capable of eating an appropriate amount of produce. And indeed, in follow-up interviews assessing the program, 70% of people reported that they felt a sense of self-efficacy in completing the challenge. One man was quoted as saying, "I've never successfully changed an eating habit in my life until I tried this. I always blew it by the second day. Now I know I can do it."

Sixty-six percent of survey respondents reported eating more produce 2 months after the campaign than they had eaten before the campaign started. The average increase was 2.1 servings a day.

The Interpersonal Level

This level of the model refers to the beliefs and actions of friends, relatives, coworkers, and other important people who influence a person's dietary beliefs and behavior. In the pilot project, there was particular effort to include employee's social supports in the contest. Follow-up interviews revealed that this made a difference in some people's ability to follow through with the behavior. One woman said, "At staff meetings, there was fresh fruit instead of doughnuts. My refrigerator was full of produce because my whole family was into it. It sure made it easy for me to eat fruits and vegetables when everyone around me was eating that way."

In the larger 5 A Day campaign, people were encouraged to involve their family members and coworkers in the contest. A number of workplaces promoted the contest among employees. Many teachers encouraged their students to take the challenge. During school assemblies, Produce Man encouraged children to get their families involved. In fact, in the follow-up phone interviews, participants cited family members as the third most common way they heard about the program.

The Institutional Level

The institutional level refers to the rules, regulations, procedures, and other policies and practices of civic, religious, social, political, and other organizations that impact food availability and dietary practices. Managers of 48 school cafeterias received training and agreed to promote the contest in their schools by serving more produce and using various methods to encourage children to eat it. Follow-up interviews found that 50% of cafeterias increased the number of servings of fruits and vegetables served to students by 20–40%.

Sixty-three health department employees in central Kentucky counties agreed to promote the contest in their areas using a variety of educational programs after attending a training session on the campaign. A number of health department clinics encouraged patients to try this new behavior, making it a part of their clinic promotions.

The Community Level

The community level includes the norms, values, and resources that constrain or promote dietary practices. One barrier that constrains people's ability to eat sufficient produce is the cost. Throughout the campaign month, Kroger ran a number of special sales reducing the price of a variety of fruits and vegetables, making them more affordable for consumers.

The Public Policy Level

The public policy level includes the local, state, and federal policies, laws, and regulations that impact food choices and people's behavior. An important public policy that reinforced the program's efforts to increase produce consumption in schools was a USDA cooperative agreement with the Department of Defense (DOD) that allowed the school meals program to increase the variety and quality of affordable produce available to schools by using DOD's buying and distribution system.

In designing health promotion initiatives, chances of success are increased if the interventions are carried out at more than one level of influence. Given the complex sociocultural context within which food habits exist, it is not surprising that the change process often involves a multilevel approach.

REFERENCES

ACC/SCN. (2000). *Fourth report on the world nutrition situation: Nutrition throughout the life cycle.* Geneva: United Nations.

Adams, C. (1990). *The sexual politics of meat: A feminist-vegetarian critical theory.* New York: Continuum.

Ahluwalia, I., Dodds, J., & Baligh, M. (1998). Social support and coping behaviors of low-income families experiencing food insufficiency in North Carolina. *Health Education and Behavior, 25*(5), 599–612.

Aiello, L. C., & Wheeler, P. (1995). The expensive-tissue hypothesis. *Current Anthropology, 36,* 199–221.

Aiken, J., Guilkey, D. K. P., Barry, M., & Smith, K. (1985). The impact of federal transfer programs on the nutrient intake of elderly individuals. *Journal of Human Resources, 20,* 383–404.

Ajzen, I. (1988). *Attitudes, Personality and Behavior.* Philadelphia: Open University Press.

Ajzen, I. (1991). The theory of planned behavior. *Organizational Behavior and Human Decision Processes, 50,* 179–211.

Allaby, M. (1977). *The world food resources: Actual and potential.* London: Applied Science Publishers.

Allen, J. S., & Cheer, S. M. (1996). The non-thrifty genotype. *Current Anthropology, 37*(5), 831–842.

Allen, J. S., Guilkey, D. K., Popkin, B. M., & Smith, K. M. (1983). *Nutrient consumption patterns of low-income households* (Technical Bulletin No. 1685). Washington, DC: Economic Research Service/U.S. Department of Agriculture.

Allen, J., & Gadson, K. E. (1983). *Nutrient consumption patterns of low-income households. Technical Bulletin No. 1685.* Washington, DC: ERS/USDA.

Allen, L. H., Backstrand, J. R., Chavez, A., & Pelto, G. (1992). *People cannot live by tortillas alone: The results of the Mexico nutrition CRSP. Final report.* Washington, DC: USAID.

Alter, C., & Hage, J. (1993). *Organizations working together.* Newbury, CA: Sage Publications.

Amato, P. R., & Partridge, S. (1989). *The new vegetarians.* New York: Plenum Press.

American Dietetics Association. (1998a). *December 1998 highlights from the Journal of the American Dietetics Association: Young adults' breakfasts lack variety, nutrients.* American Dietetics Association. Retrieved December 1998 from http://www.eatright.com/pr/press1298.html

American Dietetics Association. (1998b). *Young adults' breakfasts lack variety, nutrients: December 1998 Journal of the American Dietetic Association highlights.* American Dietetics Association. Retrieved April 4, 2002, from http://www.eatright.org/pr/press1298.html

American Dietetics Association. (1999). *April 1999 highlights from the Journal of the American Dietetics Association.* Retrieved from http://www.eatright.org/pr/1999/press0499.html

American Dietetics Association. (2000). *Nutrition and you: Trends 2000.* Chicago: American Dietetics Association.

Anand, R. S., & Basiotis, P. P. (1998). Is total fat consumption really decreasing? *Family Economics and Nutrition Review, 11*(3), 58–75.

Andersen, A. E., & DiDomenico, L. (1992). Diet vs. shape content of popular male and female magazines. A dose-response relationship to the incidence of eating disorders? *International Journal of Eating Disorders, 11,* 283–287.

Anderson, E. N. (1996). *Ecologies of the heart: Emotion, belief, and the environment.* Oxford: Oxford University Press.

Anderson, M., & Magleby, R. (1997). *Agricultural Resources and Environmental Indicators, 1996–97. Agricultural Handbook No. 712.* Washington, DC: USDA.

Andreason, A. (1995). *Marketing social change: Changing behavior to promote health, social development, and the environment.* San Francisco: Jossey-Bass.

Anliker, J., Bell, L., Miller, C., Harkins, M., Gabor, V., & Bartlett, L. (2002). *Food stamp nutrition education study: Final report* (Technical Report Submitted to U.S. Department of Agriculture).

Washington, DC: Health Systems Research.

Arcand, J. (2000). *Undernourishment and economic growth: The efficiency cost of hunger.* Rome: FAO.

Arhem, K., Homewood, K., & Rodgers, A. (1981). *A pastoral food system: The Ngorongoro Maasai in Tanzania* (Bureau of Resource Assessment and Land Use Planning Research Paper No. 7). Dar es Salaam, Tanzania: University of Dar es Salaam.

Armelagos, G., & Dewey, J. R. (1970). Evolutionary response to human infectious diseases. *Bioscence, 157,* 638.

Arnold, D. (1988). *Famine: Social crisis and historical change.* New York: Basil Blackwell.

Asfaw, B. T., White, T., Lovejoy, B., Latimer, B., Simpson, S., & Suwa, G. (1999). *Australopithecus garhi:* A new species of hominid from Ethiopia. *Nature, 284,* 629–635.

Atkins, P. J., & Bowler, I. (2001). *Food in society: Economy, culture, geography.* London: Edward Arnold.

Axtell, J. (2001). *Natives and newcomers: The cultural origins of North America.* New York: Oxford University Press.

Backwell, L. R., & D'Errico, F. (2001). *Evidence of termite foraging by Swartkrans early hominids.* Paper presented at the National Academy of Sciences, USA.

Bagozzi, R. P., & Warshaw, P. R. (1990). Trying to consume. *Journal of Consumer Research, 17,* 127–140.

Bailey, C. (1988). The social consequences of tropical shrimp mariculture development. *Ocean and Shoreline Management, 11,* 39.

Bak, S. (1998). McDonald's in Seoul: Food choices, identity and nationalism. In J. Watson (Ed.), *Golden Arches East: McDonald's in East Asia* (pp. 136–160). Stanford, CA: Stanford University Press.

Bandura, A. (1982). Self-efficacy mechanism in human agency. *American Psychologist, 37,* 122–147.

Bandura, A. (1986). *Social foundations of thought and action: A cognitive social theory.* Englewood Cliffs, NJ: Prentice Hall.

Bandura, A. (1997). The anatomy of stages of change: An editorial. *American Journal of Health Promotion, 12*(1), 8–10.

Banks, C. G. (1992). Culture in culture-bound syndromes: The case of anorexia nervosa. *Social Sciences and Medicine, 34*(8), 867–884.

Baranowski, T. (1997). Families and health actions. In D. Gochman (Ed.), *Handbook of health behavior research I: Personal and social determinants* (pp. 179–206). New York: Plenum Press.

Baranowski, T., Perry, C. L., & Parcel, G. S. (2000). How individuals, environments, and health behavior interact: Social cognitive theory. In K. Glanz, B. K. Rimer, & F. M. Lewis (Eds.), *Health behavior and health education: Theory, research, and practice* (3rd ed., pp. 165–184). San Francisco: Jossey-Bass.

Barboza, D. (2001, August 2). The power of Roundup; a weed killer is a block for Monsanto to build on. *New York Times.*

Barham, B., Clark, M., Katz, E., & Schurman, R. (1992). Nontraditional agricultural exports in Latin America. *Latin American Research Review, 27*(2).

Barlett, P. (1989). Industrial agriculture. In S. Plattner (Ed.), *Economic anthropology.* Stanford, CA: Stanford University Press.

Bartholomew, L. K., Parcel, G. S., Kok, G., & Gottlieb, N. H. (2001). *Intervention mapping: Designing theory- and evidence-based health promotion programs.* Mountain View, CA: Mayfield.

Bar-Yosef, O. (1995). The lower and middle Palaeolithis in the Mediterranean levant: Chronology and cultural entities. In H. Ullrich (Ed.), *Man and environment in the Palaeolithis.* Liège: Université de Liège.

Bar-Yosef, O., & Meadows, R. H. (1995). The origins of agriculture in the Near East. In T. D. Price & A. B. Gebaruer (Eds.), *Last hunters, first farmers: New perspectives on the prehistoric transition to agriculture* (pp. 39–77). Santa Fe: School of American Research.

Basiotis, P. P., Kramer-LeBlanc, C. S., & Kennedy, E. T. (1998). Maintaining nutrition security and diet quality: The role of the Food Stamp Program and WIC. *Family Economics and Nutrition Review, 11*(1 and 2), 4–17.

Basta, S. S., Soekirman, M. S., & Karyadi, D. (1979). Iron deficiency anemia and the productivity of adult males in Indonesia. *American Journal of Clinical Nutrition, 32,* 916–925.

Beardsworth, A., & Keil, T. (1997). *Sociology on the menu: An invitation to the study of food and society.* New York: Routledge.

Belasco, W., & Scranton, P. (2002). *Food nations: Selling taste in consumer societies.* New York: Routledge.

Bell, D., & Valentine, G. (1997). *Consuming geographies: We are where we eat.* London: Routledge.

Berg, A. (1973). *The nutrition factor: Its role in national development.* Washington, DC: The Brookings Institute.

Berkman, L. F., & Syme, S. L. (1979). Social networks, host resistance, and mortality: A 9-year follow-up study of Alameda County residents. *American Journal of Epidemiology, 109,* 186–204.

Besharov, D. J., & Germanis, P. (2001). *Rethinking WIC: An evaluation of the Women, Infants, and Children Program.* Washington, DC: AEI Press.

Bhargava, A. (1997). Nutritional status and the allocation of time in Rwandese households. *Journal of Econometrics, 77,* 277–295.

Bialostosky, K., & Briefel, R. (2000). *Nutrition and health: What did we know before welfare reform?* Hyattsville, MD: Center for Disease Control, National Center for Health Statistics.

Bisgrove, E., & Popkin, B. (1996). Does women's work improve their nutrition? Evidence from the urban Philippines. *Social Science and Medicine, 43*(10), 1475–1488.

Black, F. (1990). Infectious disease and evolution of human populations: The example of South American forest tribes. In A. C. Swedlund & G. J. Armelagos (Eds.), *Disease in populations in transition: Anthropological and epidemiological perspectives* (pp. 55–74). New York: Bergin and Garvey.

Black, F., et al. (1974). Evidence for persistence of infectious agents in isolated human populations. *American Journal of Epidemiology, 100,* 230.

Blau, D. M., Guilkey, D. K., & Popkin, B. M. (1996). Infant health and the labor supply of mothers. *Journal of Human Resources, 31*(1), 90–139.

Blaylock, J. R., Jayachandran, V. N., & Lin, B. (1999). *Maternal nutrition knowledge and children's diet quality and nutrient intakes* (Food Assistance and Nutrition Research Report No. 1). U.S. Department of Agriculture, Economic Research Service.

Blumberg, R. L. (1985). *Following up on a Guatemalan "Natural Experiment on Women in Development": Gender and the ALCOSA Agribusiness Project in 1985 vs. 1980.* San Diego: University of California.

Bogin, B. (1988). *Patterns of human growth.* Cambridge: Cambridge University Press.

Bogin, B. (2000). The tall and the short of it. In A. H. Goodman, D. Dufour, & G. H. Pelto (Eds.), *Nutritional anthropology* (pp. 192–195). Mountain View, CA: Mayfield.

Bordo, S. (1993). *Unbearable weight: Feminism, Western culture, and the body.* Berkeley: University of California Press.

Boserup, E. (1965). *The conditions of agricultural growth.* New York: Aldine.

Boserup, E. (1970). *Woman's role in economic development.* New York: St. Martin's Press.

Boserup, E. (1981). *Population and technological change: A study of long-term change.* Chicago: University of Chicago Press.

Boster, J. (1983). A comparison of the diversity of Jivaroan gardens with that of the tropical forest. *Human Ecology, 11,* 47–68.

Bouis, H. E. (2000). The role of biotechnology for food consumers in developing countries. In M. Qaim, A. Krattiger, & J. V. Braun (Eds.), *Agricultural biotechnology in developing countries: Towards optimizing benefits for the poor* (pp. 189–213). Boston: Kluwer Academic Publishers.

Bouis, H. E., & Haddad, L. J. (1990). *Effects of agricultural commercialization on land tenure, household resource allocation, and nutrition in the Philippines.*

Washington, DC: International Food Policy Research Institute.

Bowden, B. S., & Zeisz, J. M. (1997). *Supper's on! Adolescent adjustment and frequency of family mealtimes.* Paper presented at the American Psychological Association's 105th Annual Convention, Sheraton Chicago and Towers, River Exhibition Hall.

Bracht, N. (1990). *Health promotion at the community level.* Newbury Park, CA: Sage Publications.

Bracht, N. (1999). *Health promotion at the community level* (2nd ed.). Thousand Oaks, CA: Sage Publications.

Bracht, N., & Kingsbury, L. (1990). In N. Bracht (Ed.), *Health promotion at the community level* (pp. 66–88). Newbury Park, CA: Sage Publications.

Bracht, N., Kingsbury, L., & Rissle, C. (1999). A five-stage community organization model for health promotion: Empowerment and partnership strategies. In N. Bracht (Ed.), *Health promotion at the community level* (2nd ed., pp. 83–104). Newbury Park, CA: Sage Publications.

Bracht, N., & Tsouros, N. (1990). Principles and strategies of effective community participation. *Health Promotion International, 5*(3), 199–208.

Brain, C. K., & Shipman, P. (1993). The Swartkrans bone tools. In C. K. Brain (Ed.), *Swartkrans: A cave's chronicle of early man* (pp. 195–215). Pretoria, South Africa: Transvaal Museum.

Bressani, R. S., & Scrimshaw, N. (1958). Effect of lime treatment on in-vitro availability of essential amino acids and solubility of protein fractions in corn. *Journal of Agricultural and Food Chemistry, 6*(744).

Brink, P. J. (1995). Fertility and fat: The Annang fattening room. In I. de Garine & N. Pollock (Eds.), *Social aspects of obesity* (pp. 71–86). Luxembourg: Gordon and Breach.

Brown, J., Gershoff, S., & Cook, J. (1992). The politics of hunger: When science and ideology clash. *International Journal of Health Services, 22,* 221–237.

Brown, J. E. (1999). *Nutrition now* (2nd ed.). Belmont, CA: West/ Wadsworth.

Brown, J. E. (2002). *Nutrition now* (3rd ed.). Belmont, CA: West/ Wadsworth.

Brown, L. (1967). The world outlook for conventional agriculture. *Science, 158*(3801), 604–611.

Brown, L. (1995). *Who will feed China: A wake-up call for a small planet, 158*(3801), 604–611. New York: W. W. Norton.

Brown, L., Gardener, G., & Halwell, B. (1999). *Beyond Malthus: Nineteen dimensions of the population challenge.* New York: Worldwatch Institute.

Brown, L. R., & Finsterbusch, G. (1972). *Man and his environment: Food.* New York: Harper and Row.

Brownson, R. C., Baker, E. A., & Novick, L. F. (1999). *Community-based prevention: Programs that work.* Gaithersburg, MD: Aspen Publishers.

Brownson, R. C., Riley, P., & Bruce, T. A. (1998). Demonstration projects in community-based prevention. *Journal of Public Health Management and Practice, 4*(2), 66–77.

Brumann, C. (1999). Writing for culture: Why a successful concept should not be discarded. *Cultural Anthropology, 40*(Suppl.), S1–S27.

Bryant, C. A. (1975). The Puerto Rican mental health unit. *Psychiatric Annals, 5,* 8.

Bryant, C. A., Bustillo, M., Kent, E., Unterberger, A., Jeffers, D., & Lindenberger, J. (1998). Determinants of preventive pediatric care and immunization. *Florida Journal of Public Health, 9*(1), 10–17.

Bryant, C. A., Johnson, L., & van Willigen, J. (1981). *Nutrition component report* (Report submitted to the National Institue on Aging). Lexington: University of Kentucky.

Bryant, C. A., Kent, E., Brown, C., Bustillo, M., Blair, C., Lindenberger, J., & Walker, M. (1998, Winter). A social marketing approach to increase customer satisfaction with the Texas WIC Program. *Marketing Health Care Services,* 5–17.

Bryant, C. A., Lindenberger, J. H., Brown, C., Kent, E., Schreiber, J. M., Bustillo, M., & Canright, M. W. (2001). A social marketing approach to increasing enrollment in a public health program—case study of the Texas WIC Program. *Human Organization, 60*(3), 234–246.

Bryant, C. A., & Verbal, M. (1996). *Best Start 3-step counseling strategy.* Tampa: Best Start Social Marketing.

Bulik, C. M., Sullivan, P. F., Wade, T. D., & Kendler, K. S. (2000). Twin studies of eating disorders: A review. *International Journal of Eating Disorders, 27*, 1–20.

Bureau of Primary Health Care. (2002). *Cultural competence: A journey.* New Orleans: Spectrum Unlimited.

Butterfoss, F. D., Goodman, R. M., & Wandersman, A. (1993). Community coalitions for prevention and health promotion. *Health Promotion Research, 8,* 315–330.

Caballero, B., Davis, S., Davis, C. E., Ethelbah, B., Evans, M., Lohman, T., Stephenson, L., Story, M., & White, J. (1998). Pathways: A school-based program for the primary prevention of obesity in American Indian children. *Journal of Nutritional Biochemistry, 9,* 535–543.

Caliendo, M. A. (1979). *Nutrition and the world food crisis.* New York: Macmillan.

Carlson, S. J., Andrews, M. S., & Bickel, G. W. (1999). Measuring food insecurity and hunger in the United States: Development of a national benchmark measure and prevalence estimates. *Journal of Nutrition, 129,* 510–516.

Carpenter, K. J. (2001). *Beriberi, white rice and vitamin B: A disease, a cause and a cure.* Berkeley: University of California Press.

Carroll, D. (2000). Milk doesn't always do a body good. *Emerge, 24.*

Carson, G. (1957). *The cornflake crusade.* New York: Rinehart and Company.

Casper, L. M., & Bryson, K. R. (1998/ 2000). *Co-resident grandparents and their grandchildren: Grandparent maintained families.* (Population Division Working Paper No. 26). Washington, DC: U.S. Census Bureau.

Cassidy, C. M. (Ed.). (1980). *Nutrition and health in agriculturalists and hunters and gatherers.* New York: Redgrave.

Center for Budget and Policy Priorities. (2001). *Quick reference guide to the WIC Program.* Center for Budget and Policy Priorities. Retrieved from http://www. cbpp.org/10-15-01wic-guide.htm

Center for Nutrition Policy and Promotion. (1999). *The Thrifty Food Plan, 1999: Executive summary.* Center for Nutrition Policy and Promotion. United States Department of Agriculture. Retrieved from http://www.usda. gov/cnpp/FoodPlans/TFP99/ TFP99ExecSumm.PDF

Centers for Disease Control and Prevention. (1997). Youth Risk Behavioral Surveillance: National College Health Risk Behavioral Study—United States, 1995. *MMWR, 46*(SS-6), 1–54.

Centers for Disease Control and Prevention. (2000). *School health policies and programs studies: Study overview.* National Center for Chronic Disease Prevention and Health Promotion, Division of Adolescent and School Health. Retrieved from http://www.cdc. gov/nccdphp/dash/shpps/index. htm

Centers for Disease Control and Prevention. (2001, March 16). *Obesity epidemic increases dramatically in the United States: CDC director calls for national prevention effort.* Retrieved January 26, 2002, from http://www.cdc. gov/nccdphp/dnpa/obesity- epidemic.htm

Centers for Disease Control and Prevention. (2002a, May 13). *Obesity trends, 1985 to 2000.* Retrieved June 16, 2002, from http://www.cdc.gov/nccdphp/ dnpa/obesity/trend/index.htm

Centers for Disease Control and Prevention. (2002b, May 13). *Prevalence of obesity among U.S. adults, by characteristics, behavioral risk factor surveillance system (1991–2000).* Retrieved June 13, 2002, from http://www.cdc.gov/ nccdphp/dnpa/obesity/trend/ prev_char.htm

Chagnon, N. A. (1997). *Yanomamo: The fierce people.* New York: Holt, Rinehart and Winston.

Chamorro, R., & Flores-Ortiz, Y. (2000). Acculturation and disordered eating patterns among Mexican American women.

International Journal of Eating Disorders, 28, 125–129.

Chapman, G., & Mclean, H. (1993). "Junk food" and "healthy food": Meanings of food in adolescent women's culture. *Journal of Nutrition Education, 25*(3), 108–113.

Charles, N., & Kerr, M. (1988). *Women, food and families.* Manchester: Manchester University Press.

Cheer, S. M., Allen, J., & Huntsman, J. (2000). Lactose digestion capacity in Tokelauans: A case for the role of gene flow and genetic drift in establishing the lactose absorption allele in a Polynesian population. *American Journal of Physical Anthropology, 113*(1), 119.

Chen, Y., & Akre, R. D. (1994). Ants used as food and medicine in China. *The Food Insects Newsletter, 7*(2), 1–8.

Christian, P., Bentley, M. E., Pradhan, R., & West, K. P. (1998). An ethnographic study of night blindness "ratauni" among women in the Terai of Nepal. *Social Science and Medicine, 46*(7), 879–889.

Christian, P., Thorne-Lyman, A. L., West, K. P., Bentley, M. E., Khatry, S. K., Pradhan, E. K., Leclerq, S. C., & Shrestha, S. R. (1998b). Working after the sun goes down: Exploring how night blindness impairs women's work activities in rural Nepal. *European Journal of Clinical Nutrition, 52*(7), 519–524.

Citro, C. F., & Michael, R. T. (1995). *Measuring poverty: A new approach. Panel on poverty and family assistance: Concepts, information needs, and measurement methods.* Washington, DC: National Research Council, National Academy of Sciences.

Clarke, R. C., & Tobias, P. V. (1995). Sterkfontein Member 2 foot bones of the older South African hominid. *Science, 269,* 521–524.

Clauson, A. (1999). Share of food spending for eating out reaches 47%. *Food Review, 22*(3), 20–22.

Clay, J. (1997, September). Toward sustainable shrimp aquaculture. *World Aquaculture,* 32–37.

Coe, S. D. (1994). *America's first cuisines.* Austin: University of Texas Press.

Cogan, J., & Ernsberger, P. (1999). Dieting, weight, and health: Reconceptualizing research and policy. *Journal of Social Issues, 55*(2), 187–205.

Cohen, M. N. (1989). *Health and the rise of civilization.* New Haven, CT: Yale University Press.

Cohen, Y. A. (1974). Culture as adaptation. In Y. Cohen (Ed.), *Man in adaptation: The cultural present* (2nd ed., pp. 45–68). Chicago: Aldine.

Colbert, D. (2002). *What would Jesus eat?* Nashville, TN: Thomas Nelson.

Colditz, G. (1992). Economics of obesity. *American Journal of Clinical Nutrition, 55,* 503–507.

Conner, M., & Norman, P. (1995). *Predicting health behavior.* Philadelphia: Open University Press.

Contento, I., Balch, G., & Bronner, Y. (1995). The effectiveness of nutrition education and implications for nutrition education policy, programs, and research: A review of research. *Journal of Nutrition Education, 27,* 277–422.

Conway, G. (1997). *The doubly green revolution: Food for all in the 21st century.* Ithaca, NY: Comstock Publishing Associates.

Cook, G. C. (1978). Did persistence of intestinal lactase into adult life originate in the Arabian peninsula? *Man, 13,* 418–427.

Coreil, J., Bryant, C. A., & Henderson, J. N. (2000). *Social and behavioral foundations of public health.* Thousand Oaks, CA: Sage Publications.

Corrigan, P. (1997). *The sociology of consumption.* Thousand Oaks, CA: Sage Publications.

Cosminsky, S. (1975). Changing food and medical beliefs and practices in a Guatemalan community. *Ecology of Food and Nutrition, 4*(3), 183.

Cossins, N. J., & Upton, M. (1988). The impact of climate variation on the Borana pastoral system. *Agricultural Systems, 27,* 117–135.

Costa-Neto, E. M., & Nogueira de Melo, M. (1998). Entomotherapy in the county of Matinha dos Pretos, state of Bahia, northeastern Brazil. *Food Insects Newsletter, 11*(2), 1–2.

Counihan, C. M. (1999). *The anthropology of food and body.* New York: Routledge.

Crapo, R. H. (1993). *Cultural anthropology: Understanding ourselves and others* (3rd ed.). Guilford, CT: Dushkin.

Crews, D. E., & Gerber, L. M. (1994). Chronic degenerative diseases and aging. In D. E. Crews & R. M. Garruto (Eds.), *Biological anthropology and aging: Perspectives on human variation over the life span* (pp. 154–181). New York: Oxford University Press.

Croppenstedt, A., & Muller, C. (2000). The impact of farmers' health and nutritional status on their productivity and efficiency, evidence from Ethiopia. *Economic Development and Cultural Change, 48*(3), 475–502.

Cross Cultural Training Program. (2002). *What is cultural competence?* Retrieved June 10, 2002, from http://www.xculture.org/training/overview/cultural/index.html

Crosson, P. (1976). Institutional obstacles to expansion of the world food production. *Science, 188,* 519.

Crutchfield, S. R., Feather, P. M., & Hellerstein, D. R. (1995). *Benefits of protecting rural water quality: An empirical analysis* (Agricultural Economics Report No. 701). Washington, DC: USDA/ERS.

Cullen, K. W., Baranowski, T., Rittenberry, L., & Olvera, N. (2000). Social-environmental influence on children's diets: Results from focus groups with African-, Euro-, and Mexican-American children and their parents. *Health Education Research, 15*(5), 581–590.

Cummings, C. H. (1999). Entertainment foods: Health of individuals threatened by the spread of fast-food culture. *The Ecologist, 29*(1), 16–20.

Curcio, C. H. (1995). For college students, bulgur beats burgers. *Eating Well, 21.*

Curry, K. R. (2000). Multicultural competence in dietetics and nutrition. *Journal of the American Dietetics Association, 100*(10), 1142.

Curtis, D. J. (1997). *The mongoose lemur (Eulemur mongoz): A study in behaviour and ecology.* University of Zurich.

Davis, S., Going, S. B., Helitzer, D. L., Teufel, N. I., Snyder, P., Gittelsohn, J., Metcalfe, L., Arviso, V., Evans, M., Smyth, M., Brice, R., & Altaha, J. (1999). Pathways: A culturally appropriate obesity-prevention program for American Indian schoolchildren. *American Journal of Clinical Nutrition, 69* (Suppl.), 796S–802S.

Davis, S., & Lambert, L. C. (2000). Body image and weight concerns among Southwestern American Indian preadolescent schoolchildren. *Ethnicity and Disease, 10,* 184–194.

de Garine, I. (1995). Sociocultural aspects of male fattening sessions among the Massa of Northern Cameroon. In I. de Garine & N. Pollock (Eds.), *Interpreting weight: The social management of fatness and thinness* (pp. 11–28). Hawthorne, NY: Aldine de Gruyter.

de Heinzelin, C. J., White, T., Hart, W., Renne, P., Wolde, G., Beyene, Y., & Vrba, E. (1999). Environment and behavior of 2.5-million-year-old Bouri hominids. *Nature, 284,* 625–629.

de Kathen, A. (2000). Managing biosafety capacity development: Technical and political aspects. In M. Qaim, A. Krattiger, & J. V. Braun (Eds.), *Agricultural biotechnology in developing countries: Towards optimizing the benefits to the poor* (pp. 39–67). Boston: Kluwer Academic Publishers.

Deere, C. D. (1987). Latin American agrarian reform experience. In C. D. Deere & M. Leon (Eds.), *Rural women and state policy: Feminist perspectives on Latin American agricultural development.* Boulder, CO: Westview Press.

Deere, C. D., & Leon, M. (2001). *Empowering women.* Pittsburgh: University of Pittsburgh Press.

DeFoliart, G. (1991). Insect fatty acids: Similar to those of poultry and fish in their degree of unsaturation, but higher in the polyunsaturates. *Food Insects Newsletter, 4*(1), 3.

Dettwyler, K. (1994). *Dancing skeletons: Life and death in West Africa.* Waveland Press.

Devaney, B., Haines, P., & Moffitt, R. (1989). *Assessing the dietary effects of the Food Stamp Program—Volumes I and II.* Alexandria, VA: U.S. Department of Agriculture, Food and Nutrition Service.

DeVault, M. (1991). *Feeding the family: The social organization of caring as gendered work.* Chicago: University of Chicago Press.

DeWalt, B. R. (1998). The Ejido reforms and Mexican coastal communities: Fomenting a blue revolution? In W. Cornelius & D. Myhre (Eds.), *The transformation of rural Mexico, reforming the Ejido sector* (pp. 357–379). San Diego: Center for U.S.–Mexican Studies, University of California.

DeWalt, B. R., & Alexander, S. (1983). The dynamics of cropping systems in Pespire, Southern Honduras. *Practicing Anthropology, 5*(3), 11, 13.

DeWalt, B. R., Ramírez, J., Noriega, L., & González, R. (2001). *Shrimp aquaculture, people and the environment in coastal Mexico.* Washington, DC: World Bank, World Wildlife Foundation, Food and Agriculture Organization and Network of Aquaculture Centres in Asia-Pacific.

DeWalt, B. R., Vergne, P., & Hardin, M. (1996). Shrimp aquaculture development and the environment: People, mangroves and fisheries on the Gulf of Fonseca, Honduras. *World Development, 24*(7), 1193–1208.

DeWalt, K. M. (1983). *Nutritional strategies and agricultural change in a Mexican community.* Ann Arbor: University of Michigan Press.

DeWalt, K. M. (1993a). Nutrition and the commercialization of agriculture: Ten years later. *Social Science and Medicine, 36,* 1407–1416.

DeWalt, K. M. (1993b). *Food security and nutrition impacts of nontraditional agricultural exports in Latin America.* Unpublished manuscript. Washington, DC.

DeWalt, K. M., & DeWalt, B. R. (1987). Nutrition and agricultural change in Southern Honduras. *Food and Nutrition Bulletin, 9*(3), 36–45.

DeWalt, K. M., DeWalt, B. R., Escudero, J. C., & Barkin, D. (1990). The nutrition effects of shifts from maize to sorghum production in four Mexican communities. *Food Policy, October 1990,* 395–407.

DeWalt, K. M., Sharma, R., & Documet, P. (1999). *Diet and food security in the El Angel watershed.* University of Pittsburgh, Report to MacAurthur Foundation

DeWalt, K. M., Uquillas, J., & Crissman, C. (1988). *Potato production and consumption in the Sierra of Ecuador.* Washington, DC: USDA.

Dewey, K. G. (1989). Nutrition and the commoditization of food systems in Latin America and the Caribbean. *Social Science and Medicine, 28,* 415–424.

Diamond, J. (1997). *Guns, germs, and steel: The fate of human societies.* New York: W. W. Norton.

Dubos, R. (1980). *Man adapting.* New Haven, CT: Yale University Press.

Dufour, D. (2000a). Use of tropical rainforests by Native Americans. In A. H. Goodman, D. Dufour, & G. H. Pelto (Eds.), *Nutritional anthropology: Biocultural perspectives on food and nutrition* (pp. 77–86). Mountain View, CA: Mayfield.

Dufour, D. (2000b). A closer look at the nutritional implications of bitter cassava use. In A. H. Goodman, D. Dufour, & G. H. Pelto (Eds.), *Nutritional anthropology: Biocultural perspectives on food and nutrition* (pp. 164–172). Mountain View, CA: Mayfield.

Dunkel, F. (1996). Incorporating food insects into undergraduate entomology courses. *Food Insects Newsletter, 9*(2), 1–4.

Dunkel, F. (1997). Food insect festivals of North America. *Food Insects Newsletter, 10*(3), 1–7.

Dunkel, F. (1998). Where can I find information on the efficiency or conversion of rate of traditional livestock and mini livestock (including food insects)? *Food Insects Newsletter, 11*(1), 9.

Dunn, F. (1968). Epidemiologic factors: Health and disease in hunter-gatherers. In R. B. Lee & I. DeVore (Eds.), *Man the hunter.* Chicago: Aldine.

Durham, W. H. (1991). *Coeveolution: Genes, culture and human diversity.* Stanford, CA: Stanford University Press.

Dwyer, D., & Bruce, J. (1988). *A home divided: Women and income in the*

Third World. Stanford, CA: Stanford University Press.

Dyson, L. K. (2000). American cuisine in the 20th century. *Food Review, 23*(1), 2–7.

Eckholm, E. (1978, May). Vanishing firewood. *Human Nature, 1978,* 5–23.

Economic Research Service. (2000). *Diet and health: Food consumption and nutrient intake tables.* USDA. Retrieved July 15, 2002, http://www.ers.usda.gov/briefing/DietAndHealth/data/foods/

Economic Research Service. (2001). *Food security in the United States: Measuring food security.* USDA. Retrieved April 4, 2002, from http://www.ers.usda.gov/briefing/foodsecurity/measurement/index.htm

Egger, G., Spark, R., Lawson, J., & Donovan, R. (1999). *Health promotion strategies and methods.* New York: McGraw-Hill.

Ehrlich, P. R. (1970). *The population bomb.* New York: Sierra Club.

Ehrlich, P. R., & Ehrlich, A. H. (1990). *The population explosion.* New York: Simon and Schuster.

Ehrlich, P., Ehrlich, A. H., & Holdren, J. P. (1977). *Ecoscience: Population, resources, environment.* San Francisco: W. H. Freeman.

Eig, J. (2001). Food industry battles for moms who want to cook— just a little. *Wall Street Journal,* A1.

Eisen, A. (1994). Survey of neighborhood-based, comprehensive community empowerment initiatives. *Health Education Quarterly, 21,* 235–252.

Ekstrom, M. (1991). Class and gender in the kitchen. In E. I. Furst, R. Prattaala, M. Ekstrom, L. Holm, & U. Kjaernes (Eds.), *Palatable worlds: Sociocultural food studies.* Oslo: Solum Forlag.

El-Askari, G., Freestone, J., Irizarry, C., Krowt, K. L., Mashiyama, S. T., Morgan, M. A., & Walton, S. (1998). The healthy neighborhoods project: A local health department's role in catalyzing community development. *Health Education and Behavior, 25,* 146–159.

Eng, E., & Parker, E. (1994). Measuring community competence in the Mississippi Delta: The interface between program evaluation and empowerment. *Health Education Quarterly, 21*(2), 199–220.

Engle, P. L., & Menon, P. (1999). Care and nutrition: Concepts and measurement. *World Development, 27*(8), 1309–1337.

Escobar, A. (1999). Factors influence children's dietary practices: A review. *Food Economics and Nutrition Review, 12*(3 and 4), 45–55.

Estioko-Griffin, A. (1986). Daughters of the forest. *Natural History, 95*(5), 36–43.

Estioko-Griffin, A. (1992). Hunting with the Agta women. In D. Gross (Ed.), *Discovering anthropology* (p. 273). Mountain View, CA: Mayfield.

Etkin, N., & Ross, P. (1997). Malaria, medicine, and meals: A biobehavioral perspective. In L. Romanucci-Ross, D. Moerman, L. Tancredi (Eds.), *The anthropology of medicine: From culture to method* (3rd ed.) Westport, CT: Bergin and Garvey.

Evans, L. T. (1998). *Feeding the ten billion: Plants and population growth.* Cambridge: Cambridge University Press.

FAO. (1996). *Food for all. World food summit.* Rome: Food and Agriculture Organization.

FAO. (2000a). *The state of food and agriculture 2000.* Rome: Food and Agriculture Organization.

FAO. (2000b). *The state of food insecurity in the world 2000.* Rome: Food and Agriculture Organization.

FAO. (2001a). *State of food and agriculture 2001.* Rome: Food and Agriculture Organization.

FAO. (2001b). *The state of food insecurity in the world* 2001. Rome: Food and Agriculture Organization.

Farb, P., & Armelagos, G. (1980). *Consuming passions: The anthropology of eating.* Boston: Houghton Mifflin.

Fasoranti, J. O. (1997). The place of insects in the traditional medicine of southwestern Nigeria. *Food Insects Newsletter, 10*(2), 1–5.

Fernandez-Cornejo, J., & Jans, S. (1999). *Pest management in U.S. agriculture.* Washington, DC: USDA.

Field, C. R., & Simpkin, S. K. (1985). *The importance of camels to subsistence pastoralists in Kenya.* Nairobi: UNESCO.

Fields, J., & Casper, L. M. (2001). *America's families and living arrangements: March 2000* (Current Population Reports, P20-537). Washington, DC: U.S. Census Bureau.

Filer, L. J., & Reynolds, W. A. (1997). Lessons in comparative physiology: Lactose intolerance. *Nutrition Today, 32,* 79–81.

Fitchen, J. M. (1988). Hunger, malnutrition and poverty in contemporary United States: Some observations on their social and cultural context. *Food and Foodways, 2,* 309–333.

Flannery, K. V. (1971). Archeological systems theory and early Mesoamerica. In S. Struven (Ed.), *Prehistoric agriculture.* New York: Natural History Press.

Flatz, G. (1987). Genetics of lactose digestion in humans. *Advanced Human Genetics, 16,* 1–77.

Flegal, K. M., Caroll, M. D., Kuczmaraki, R. J., & Johnson, C. L. (1998). Overweight and obesity in the United States: Prevalence and trends, 1960–1994. *International Journal of Obesity, 22*(1), 39–47.

Fleuret, P., & Fleuret, A. (1991). Social organization, resource management, and child nutrition in the Taita Hills, Kenya. *American Anthropologist, 93,* 91–114.

Flynn, B. C., Ray, D. W., & Rider, M. S. (1994). Empowering communities: Action research through healthy cities. *Health Education Quarterly, 21*(3), 395–405.

Fogel, R. (1994). Economic growth, population theory and physiology: The bearing of long-term processes in the making of economic policy. *American Economic Review, 84*(3), 369–395.

Food Marketing Institute [FMI]. (1997). *Trends in the United States: Consumer attitudes and the supermarket, 1997.* Washington, DC.

Food Marketing Institute [FMI]. (2000). *Trends in the United States: Consumer attitudes and the supermarket, 1999.* Washington, DC.

Food Marketing Institute [FMI]. (2001, April). *Supermarket facts.* Retrieved from http://www.fmi.org/facts_figs/superfact.htm

Fordham, M., DeWalt, B. R., & DeWalt, K. M. (1985). *The economic role of women in a Honduran peasant community.* INTSORMIL, farming systems research in Southern Honduras (Report No. 3) Lexington: University of Kentucky Experiment Station.

Foster, G. M. (1979). Humoral traces in United States folk medicine. *Medical Anthropology Newsletter, 10*(2), 17.

Foster, G. M. (1994). *Hippocrates' Latin American legacy: Humoral medicine in the New World.* Langhorne, PA: Gordon and Breach.

Foster, G. M., & Anderson, B. G. (1978). *Medical anthropology.* New York: John Wiley.

Foster, P. (1992). *The world food problem.* Boulder: Lynn Reiner.

Fox, M. A. (1999). *Deep vegetarianism.* Philadelphia: Temple University Press.

Fraker, T. M. (1990). *The effects of food stamps on food consumption: A review of the literature.* Alexandria, VA: U.S. Department of Agriculture, Food and Nutrition Service.

Frankish, C. J., & Green, L. W. (1994). Organizational and community change as the social scientific basis for disease prevention and health promotion policy. *Advances in Medical Sociology, 4,* 209–233.

Fraser, L. (1997). *Losing it.* New York: Dutton.

Frazao, E. (1999). *America's eating habits: Changes and consequences.* Washington, DC: U.S. Department of Agriculture.

Fresco, L. (2000, August 17–22). *Scientific and ethical challenges in agriculture to meet human needs.* Paper presented at the Third International Crop Science Congress, Hamburg, Germany.

Fussell, B. (1999). Translating maize into corn: The transformation of America's native grain. *Social Research, 66.*

Gabacia, D. R. (1998). *We are what we eat: Ethnic food and the making of Americans.* Cambridge: Harvard University Press.

Gaisford, J. (1978). *Atlas of man.* New York: St. Martin's Press.

Galvin, K. A. (1985). *Food procurement, diet and nutrition of Turkana pastoralists in an ecological and social context*. Unpublished dissertation, State University of New York at Binghamton.

Galvin, K. A. (1992). Nutritional ecology of pastoralists in dry tropical Africa. *American Journal of Human Biology, 4*(2), 209–221.

Galvin, K. A., Coppock, D. L., & Leslie, P. W. (2000). Diet, nutrition and the pastoral strategy. In A. H. Goodman, D. Dufour, & G. H. Pelto (Eds.), *Nutritional anthropology* (pp. 86–96). Mountain View, CA: Mayfield.

Gandhi, Mohandas K. (1954). *How to serve the cow*. Ahmedabad: Navajivan Publishing House.

Gardyn, R. (2000). Retirement redefined. *American Demographics, 22*(11), 52–63.

Gaulin, S. J. C., & Monner, M. (1977). On the natural diet of primates, including humans. In R. J. Wurtman & J. J. Wurtman (Eds.), *Nutrition and the brain* (Vol. 1). New York: Raven Press.

Geraldo, J., Marques, W., & Costa-Neto, E. M. (1997). Insects as folk medicines in the state of Alagoas, Brazil. *Food Insects Newsletter, 10*(1), 7–8.

Germov, J., & Williams, L. (1999). Dieting women: Self-surveillance and the body panopticon. In J. Sobal & D. Maurer (Eds.), *Weighty issues: Fatness and thinness as social problems* (pp. 117–132). Hawthorne, NY: Aldine de Gruyter.

Gerrior, A., & Bente, L. (1997). *Nutrient content of the U.S. food supply, 1909–94* (Home Economics Research Report No. 53). Washington, DC: U.S. Department of Agriculture, Center for Nutrition Policy and Promotion.

Gerrior, S., & Bente, L. (2002). *Nutrient content of the U.S. food supply, 1909–1999: A summary report*. Washington, DC: U.S. Department of Agriculture, Center for Nutrition Policy and Promotion.

Gianessi, L. P., & Carpenter, J. E. (1999). *Agricultural biotechnology: Insect control benefits*. National Center for Food and Agricultural Policy. Retrieved January 20, 2002, from http://www.bio.org/food&ag/ncfap/ag_bio.htm

Glanz, K. (1993). Managing change with nutrition education. In R. Frankel & A. Owen (Eds.), *Nutrition in the community: The art of delivering services*. St. Louis: Mosby.

Glanz, K., Lewis, F. M., & Rimer, B. (1997). *Health behavior and health education: Theory, research, and practice* (2nd ed.). San Francisco: Jossey-Bass.

Glanz, K., Rimer, B. K. & Lewis, F. M. (Eds.). (2000). *Health behavior and health education: Theory, research and practice* (3rd ed.). San Francisco: Jossey-Bass.

Gleason, P., Rangarajan, A., & Olson, C. (2000). *Dietary intake and attitudes among food stamp participants and other low-income individuals* (Technical Report Submitted to U.S. Department of Agriculture, Food and Nutrition Service). Princeton, NJ: Mathematica Policy Research.

Gleason, P., Rangarajan, A., & Olson, C. (2002). *Dietary intake and attitudes among food stamp participants and other low-income individuals* (Technical Report Submitted to U.S. Department of Agriculture, Food and Nutrition Service). Princeton, NJ: Mathematica Policy Research.

Gleason, P., & Suitor, C. (2001). *Children's diets in the mid-1990s: Dietary intake and its relationship with school meal participation* (CN-01-CDC1). Alexandria, VA: U.S. Department of Agriculture, Food and Nutrition Service, Office of Analysis, Nutrition and Evaluation.

Godin, G., & Koch, G. (1996). The theory of planned behavior: A review of its applications to health-related problems. *American Journal of Health Promotion, 12*(11), 87–89.

Goldman, D. (1999). Paradox of pleasure. *American Demographics, 21*(5), 50–51.

Goldman, K., & Schmalz, K. J. (2001). Theoretically speaking: Overview and summary of key health education theories. *Health Promotion Practice, 2*(4), 277–281.

Goldschmidt, W. (1968). Theory and strategy in the study of cultural adaptability. In Y. Cohen (Ed.), *Man in adaptation: The cultural present* (pp. 231–249). Chicago: Aldine.

Goode, J., Curtis, K., & Theophano, J. (1984a). Meal formats, meal cycles, and menu negotiation in the maintenance of an Italian-American community. In M. Douglas (Ed.), *Food in the social order: Studies of food and festivities in three American communities* (pp. 143–218). New York: Russell Sage Foundation.

Goode, J., Theophano, J., & Curtis, K. (1984b). A framework for the analysis of continuity and change in shared sociocultural rules for food use: The Italian-American pattern. In L. K. Brown & K. Mussell (Eds.), *Ethnic and regional foodways in the United States* (pp. 66–88). Knoxville: University of Tennessee Press.

Goodman, A. H., & Armelagos, G. J. (2000). Disease and death at Dr. Dickson's mounds. In A. H. Goodman, D. L. Dufour, & G. H. Pelto (Eds.), *Nutritional anthropology: Biocultural perspectives on food and nutrition.* Mountain View, CA: Mayfield.

Goodman, A. H., Dufour, D. L., & Pelto, G. H. (2000). *Nutritional anthropology: Biocultural perspectives on food and nutrition.* London: Mayfield.

Gordon, B. M. (1983). Why we choose the foods we do. *Nutrition Today, 17.*

Gould, R. S. (1966). Notes on hunting, butchering, and sharing of game among the Ngatajara and their neighbors in the Western Australian Desert. *Kroeber Anthropological Society Papers, 36,* 41–63.

Grantham-McGregor, S., Fernald, L., & Sethuraman, K. (1999). Effects of health and nutrition on cognitive and behavioral development in the first three years of life. Part 2: Infections and micronutrient deficiencies: Iodine, iron and zinc. *Food and Nutrition Bulletin, 20*(1), 75–99.

Green, L. W., & Kreuter, M. W. (1991). *Health promotion planning: An educational and ecological approach.* Mountain View, CA: Mayfield.

Gross, D. (1992). *Discovering anthropology.* Mountain View, CA: Mayfield.

Gross, D., & Underwood, B. (1971). Technological change and calorie costs. *American Anthropologist, 73,* 725.

Gross, D. R., Ettin, G., Flowers, N. M., Leoi, M. F., Ritter, M., & Werner, D. (1979). Ecology and acculturation among native peoples of Brazil. *Science, 206,* 1043–1050.

Guyer, J. (1980). *Household budgets and women's incomes.* (Working Paper No. 28). Boston: African Studies Center, Boston University.

Haaga, J., Mason, J., Omoro, F. Z., Quinn, V., Rafferty, A., Test, K., & Wasonga, L. (1986). Child malnutrition in rural Kenya: A geographic and agricultural classification. *Ecology of Food and Nutrition, 18,* 297–307.

Haddad, L. (1992). The impact of women's employment status on household food security at different income levels in Ghana. *Food and Nutrition Bulletin, 14*(4), 341–344.

Hagenimana, V., Oyunga, M. A., Low, L., Njoroge, S. M., Gichuki, S. T., & Kabira, J. N. (1999). *The effects of women farmers' adoption of orange-fleshed sweet potatoes: Raising vitamin A intake in Kenya. Summary report #3.* Washington, DC: International Center for Research on Women.

Hamilton, S. (1998). *The two-headed household: Gender and rural development in the Ecuadorian Andes.* Pittsburgh: University of Pittsburgh Press.

Harlan, J. (1992). *Crops and man* (2nd ed.). Crop Science Society of America.

Harnack, L. (1998). Guess who's cooking? The role of men in meal planning, shopping, and preparation in U.S. families. *Journal of the American Dietetic Association, 98*(9), 995–1001.

Harris, M. (1974). *Cows, pigs, wars, and witches: The riddles of culture.* New York: Vintage Books.

Harris, M. (1975). *Culture, people, nature: An introduction to general anthropology* (2nd ed.). New York: Thomas Crowell.

Harris, M. (1985). *Good to rat: Riddles of food and culture.* New York: Simon and Schuster.

Harris, M. (2000). India's sacred cow. In A. H. Goodman, D. Dufour, &

G. H. Pelto (Eds.), *Nutritional anthropology: Biocultural perspectives on food and nutrition.* Mountain View, CA: Mayfield.

Harris, M., & Johnson, O. (2002). *Cultural anthropology* (6th ed.). Boston: Allyn & Bacon.

Harris, M. B., Walters, L. C., & Waschull, S. (1991). Gender and ethnic differences in obesity-related behaviors and attitudes in a college sample. *Journal of Applied Social Psychology, 21,* 1545–1566.

Harwood, A. (1971). The hot/cold theory of disease: Implications for treatment of Puerto Rican patients. *Journal of the American Medical Association, 216*(7), 1153.

Harwood, A. (1981). *Ethnicity and disease.* Cambridge, MA: Harvard University Press.

Harwood, J. L., Heifner, R., Coble, K., Perry, J., & Somwaru, A. (1999). *Managing risk in farming: Concepts, research, and analysis.* (Agricultural Economics Report No. 774). Washington, DC: USDA/ERS.

Haworth-Hoeppner, S. (2000). The critical shapes of body image: The role of culture and family in the production of eating disorders. *Journal of Marriage and the Family, 62,* 212–227.

Hayden, B. (1981). Subsistence and ecological adaptations of modern hunter/gatherers. In R. O. Harding & G. Teleki (Eds.), *Omnivorous primates: Gathering and hunting in human evolution.* New York: Columbia University Press.

Hayenga, M. (1998). Structural change in the biotech seed and chemical industrial complex. *AgBioForum, 1*(2), 43–55.

Heaney, R. (2000). The bioavailability of calcium in fortified soy imitation milk with some observations on method. *American Journal of Clinical Nutrition, 71*(5), 1166–1169.

Heaney, C. A., & Israel, B. A. (1997). Social networks and social support. In K. Glanz, B. K. Rimer, & F. M. Lewis (Eds.), *Health behavior and health education: Theory, research, and practice* (2nd ed.). San Francisco: Jossey-Bass.

Heinberg, L. J., & Thompson, J. K. (1995). Body image and televised images of thinness and attractiveness: A controlled laboratory investigation. *Journal of Social and Clinical Psychology, 7,* 335–344.

Hendry, J. (1999). *Other people's worlds: An introduction to cultural and social anthropology.* Washington Square, New York: New York University Press.

Hernandez, M., Hidalgo, C. P., Hernandez, J. R., Madrigal, H., & Chavez, A. (1974). Effect of economic growth on nutrition in a tropical community. *Ecology of Food and Nutrition, 3*(4), 283–291.

Herzog, D. B., Greenwood, D. N., Dorer, D. J., Flores, A. T., Ekeblad, E. R., Richards, A., Blais, M. A., & Keller, M. B. (2000). Mortality in eating disorders: A descriptive study. *International Journal of Eating Disorders, 28,* 20–26.

Hetzel, B. (1978). The changing nutrition of aborigines in the ecosystem of Central Australia. In B. Hetzel & K. Firth (Eds.), *The nutrition of aborigines in relation to the ecosystem of Central Australia.* Melbourne: Commonwealth Scientific and Industrial Research Organization.

Hill, J. O., & Peters, J. C. (1998). Environmental contributions to the obesity epidemic. *Science, 280,* 1371–1374.

Hochbaum, G. M. (1958). *Public participation in medical screening programs: A sociopsychological study* (PHS Publication No. 572). Washington, DC: U.S. Government Printing Office.

Hoebel, E. A., & Frost, E. (1976). *Cultural and social anthropology.* New York: McGraw-Hill.

Holland, L., & Courtney, R. (1998). Increasing cultural competence with the Latino community. *Journal of Community Health Nursing, 15*(1), 45–53.

Honigmann, J. (1959). *The world of man.* New York: Harper and Row.

Hoppe, R. A., Johnson, J., Perry, J. E., Korb, P., Sommer, J. E., Ryan, J. T., Green, R. C., Durst, R., & Monke, J. (2001). *Structural and financial characteristics of U.S. farms: 2001 family farm report* (ERS Agriculture Information Bulletin No. 768). Washington, DC: U.S. Government Printing Office.

Howells, W. W. (1989). Skull shapes and the map: Craniometric analyses in the dispersion of modern Homo. *Papers of the*

Peabody Museum of Archaeology and Ethnology, Harvard University, 79.

Hull, V. (1986). Dietary taboos in Java: Myths, mysteries, and methodology. In L. Manderson (Ed.), *Shared wealth and symbol: Food, culture, and society in Oceania and Southeast Asia.* Cambridge: Cambridge University Press.

Human Nutrition Information Service. (1989). *Low-income women 19–50 years and their children 1–5 years, 4 days* (CSFII Report No. 86-4). Hyattsville, MD: U.S. Department of Agriculture, HNIS.

Hupkens, C. L. H., Knibbe, R. A., Van Otterloo, A. H., & Drop, M. J. (1998). Class differences in the food rules mothers impose on their children. A cross-national study. *Social Science and Medicine, 47*(9), 1331–1339.

ICD-9-CD. (1990). *International classification of disease—Clinical modifications.* Washington, DC: U.S. Department of Health and Human Services.

Inness, S. (2001). *Kitchen culture in America: Popular representations of food, gender, and race.* Philadelphia: University of Pennsylvania Press.

Israel, B. A., Checkoway, B., Schulz, A., & Zimmerman, M. (1994). Health education and community empowerment: Conceptualizing and measuring perceptions of individual, organizational, and community control. *Health Education Quarterly, 21*(2), 149–170.

Israel, B. A., Schulz, A. J., Parker, E. A., & Becker, A. B. (1998). Key principles of community-based research. *Annual Review of Public Health, 19,* 173–202.

James, C. (2000). Transgenic crops worldwide: Current situation and future outlook. In M. Qaim, A. F. Krattiger, & J. V. Braun (Eds.), *Agricultural biotechnology in developing countries: Towards optimizing the benefits to the poor* (pp. 11–23). Boston: Kluwer Academic Publishers.

James, W. P. T., Ferro-Luzzi, A., & Waterlow, J. C. (1988). Definition of chronic energy deficiency in adults. *European Journal of Clinical Nutrition, 42,* 969–981.

Janz, N. K., Champion, V. L., & Strecher, V. J. (2000). The Health Belief Model and health behavior. In D. S. Gochman (Ed.), *Handbook of health behavior research: Vol. 1. Personal and social determinants* (pp. 71–92). New York: Plenum Press.

Jeffrey, W. (2001). Demographic change, economic growth, and inequality. In N. Birdsall, A. C. Kelley, & S. W. Sinding (Eds.), *Population matters: Demographic change, economic growth, and poverty in the developing world* (pp. 106–136). Oxford: Oxford University Press.

Jekanowski, M. D. (1999). Causes and consequences of fast-food sales growth. *Food Review, 22,* 11–16.

Jensen, H. H. (1996). *Factors that affect food choices of food stamp recipients: Interim progress report.* Alexandria, VA: U.S. Department of Agriculture, Food and Consumer Service.

Jerome, N., Pelto, G. H., & Kandel, R. (1980). An ecological approach to nutritional anthropology. In N. Jerome, R. Kandel, & G. Pelto (Eds.), *Nutritional anthropology: Contemporary approaches to diet and culture.* Pleasantville, NJ: Redgrave.

Johns, T. (1991, September/October). Well-grounded diet. *The Sciences,* 39–41.

Johnson, A. O., Buchowski, M. S., Enwonwu, C. O., & Scrimshaw, N. S. (1993). Adaptation of lactose maldigesters to continued milk intakes. *American Journal of Clinical Nutrition, 58,* 879–891.

Johnston, F. E. (1982). *Physical anthropology.* New York: McGraw-Hill.

Jonjuapsong, L. (1996). In Northeast Thailand, insects are supplementary food with a lot of value. *Food Insects Newsletter, 9*(2), 4–7.

Jung, C. (2000). Molecular tools for plant breeding. In M. Qaim, A. F. Krattiger, & J. V. Braun (Eds.), *Agricultural biotechnology in developing countries: Towards optimizing the benefits for the poor* (pp. 25–38). Dordrecht, the Netherlands: Kluwer Academic Press.

Kahn, M. (1986). *Always hungry, never greedy: Food and the expression of gender in a Melanesian society.* New York: Cambridge University Press.

Kahn, M. (1998). Men are taro (they cannot be rice): Political aspects of food choices in Wamira, Papua New Guinea. In C. M. Counihan & S. L. Kaplan (Eds.), *Food and gender: Identity and power* (pp. 29–44). Amsterdam, Netherlands: Harwood Academic Publishers.

Kaiser, L. L., & Dewey, K. G. (1991). Migration, cash cropping and subsistence agriculture: Relationships to household food expenditures in rural Mexico. *Social Science and Medicine, 33,* 1113–1126.

Kalcik, S. (1984). Ethnic food ways in America: Symbol and the performance of identity. In K. Mussell & L. K. Brown (Eds.), *Ethnic and regional foodways in the United States* (pp. 37–65). Knoxville: University of Tennessee Press.

Kantor, L. S. (1998). *A dietary assessment of the U.S. food supply: Comparing per capita food consumption with Food Guide Pyramid serving recommendations* (Agricultural Economic Report No. 772). Washington, DC: Food and Rural Economics Division, Economic Research Service. U.S. Department of Agriculture.

Kantor, L. S. (2001). Community food security programs improve food access. *Food Review, 24*(1), 20–26.

Kaplan, L. (Ed.). (1971). *Archaeology and domestication in American phaseolus (beans).* New York: Natural History Press.

Kates, R., & Haaramann, V. (1992). Where the poor live: Are the assumptions correct? *Environment, 34*(May), 4–28.

Katz, D., & Goodwin, M. T. (1980). The food system: From field to table. In A. Tobias & P. J. Thompson (Eds.), *Issues in nutrition in the 1980s.* Belmont, CA: Wadsworth.

Katz, E. (1994). The nutritional impact of nontraditional export agriculture in Guatemala: An intra-household perspective. *Food and Nutrition Bulletin, 15*(4).

Katz, S. H., Heidger, M. L., & Valleroy, L. A. (1975). Traditional maize processing techniques in the New World. *Science, 184,* 765.

Kaufmann, P. R. (1999). The rural poor's access to supermarkets, large grocery stores. *Rural Development Perspectives, 13*(3), 19–25.

Kay, R. F. (1981). The nut-crackers—a new theory of the adaptations of the Ramapithecinae. *American Journal of Physical Anthropology, 55,* 141–151.

Kennedy, E., Bouis, H., & von Braun, J. (1992). Health and nutrition effects of cash crop production in developing countries—a comparative analysis. *Social Science and Medicine, 35*(5), 689–697.

Kennedy, E., & Cogill, B. (1987). *Income and nutritional effects of the commercialization of agriculture in Southwest Kenya* (IFPRI Research Report No. 63). Washington, DC: International Food Policy Research Institute.

Kennedy, E., & Peters, P. (1992). Household food security and child nutrition: The interaction of income and gender of household head. *World Development, 20*(8), 1077–1085.

Kennedy, E., & Pinstrup-Andersen, P. (1983). *Nutrition-related policies and programs: Past performances and research needs.* Washington, DC: International Food Policy Research Institute.

Khoury, A. J., Hinton, A., Mitra, A., Sheil, H., Moazzem, S. W., & Wakerful, (2001). *Evaluation of Misisissippi's breastfeeding promotion program. Final report to the U.S. Department of Agriculture Food and Nutrition Service and Mississippi's Women, Infants, and Children (WIC) Program.*

Kim, C. S., Taylor, H., Hallahan, C., & Schaible, G. (2001, August). *Economic analysis of the changing structure of the U.S. fertilizer industry.* Paper presented at the 2001 Annual Meeting of the American Agricultural Economics Association.

Kim Estther, H. J., Schroeder, K. M., Houser, R. F., & Dwyer, J. T. (1999). Two small surveys, 25 years apart, investigating motivations of dietary choice in two groups of vegetarians in the Boston area. *Journal of the American Dietetics Association, 99*(5), 598.

Klein, R. G. (1999). *The human career* (2nd ed.). Chicago: University of Chicago Press.

Kluckholn, C. (1968). *Mirror for man: A survey of human behavior and social attitudes.* Greenwich, CT: Fawcett Publications.

Knutson, R. D., Penn, J. B., & Boehn, W. T. (1983). *Agricultural and food policy.* Englewood Cliffs, NJ: Prentice Hall.

Kodicek, E., & Wilson, P. W. (1959). The availability of bound nicotinic acid to the rat. Effect of limestone treatment of maize and subsequent baking into tortilla. *British Journal of Nutrition, 13,* 418.

Kohler, C. L., Grimley, D., & Reynolds, K. (1999). Theoretical approaches guiding the development and implementation of health promotion programs. In J. M. Raczynski & R. J. DiClemente (Eds.), *Handbook of health promotion and disease prevention.* New York: Plenum Press.

Kotler, P., & Armstrong, G. (1996). *Principles of marketing* (7th ed.). Englewood Cliffs, NJ: Prentice Hall.

Kottak, C. P. (2000). *Cultural anthropology* (8th ed.). Boston: McGraw-Hill.

Kotz, N. (1970). *Let them eat promises: The politics of hunger in America.* Garden City, NY: Doubleday Anchor Books.

Kraak, V., & Pelletier, D. L. (1998). The influence of commercialism on the food purchasing behavior of children and teenage youth. *Family Economics and Nutrition, 11*(3), 15–24.

Kurz, K. M., & Johnson-Welch, C. (2000). *Enhancing nutrition results: The case for a women's resources approach.* Washington, DC: International Center for Research on Women.

Laderman, C. (1987). *Wives and midwives: Childbirth and nutrition in rural Malaysia.* Berkeley: University of California Press.

Lalueza, C. A., Perez-Perez, A., & Turbon, D. (1996). Dietary inference through buccal microwear analysis of Middle and Upper Pleistocene human fossils. *American Journal of Physical Anthropology, 100,* 367–387.

Lappe, F. M. (1982). *Diet for a small planet.* New York: Ballantine Books.

Lappe, F. M., & Brady, I. (1986). *World hunger: Twelve myths.* New York: Grove Press.

Last, J. M. (1998). *Public health and human ecology* (2nd ed.). Stamford, CT: Appleton & Lange.

Latham, M. (1997). *Human nutrition in the developing world.* Rome: FAO.

Lee, R. B. (1968). What hunters do for a living, or how to make out on scarce resources. In R. B. Lee & I. DeVore (Eds.), *Man the hunter* (pp. 30–48). Chicago: Aldine.

Lee, R. B. (1972a). The !Kung Bushmen of Botswana. In M. G. Bicchieri (Ed.), *Hunters and gatherers today* (pp. 326–368). New York: Holt, Rinehart, and Winston.

Lee, R. B. (1972b). Population growth and the beginning of sedentary life among the !Kung Bushmen. In B. Spooner (Ed.), *Population growth: Anthropological implications* (pp. 329–342). Cambridge: Cambridge University Press.

Lee, R. B. (1993). *The Dobe Ju/'hoansi* (2nd ed.). Fort Worth, TX: Harcourt, Brace College Publishers.

Lee, R. B. (2000). What hunters do for a living, or how to make out on scarce resources. In A. H. Goodman, D. Dufour, & G. H. Pelto (Eds.), *Nutritional anthropology: Biocultural perspectives on food and nutrition* (Vols. 35–46). Mountain View, CA: Mayfield.

Lee-Thorp, J., & van der Merwe, N. (1993). Stable carbon isotope studies of Swartkrans fossils. In C. K. Brain (Ed.), *Swartkrans: A cave's chronicle of early man* (Vol. 8). Pretoria: Transvaal Museum Monograph.

Lee-Thorp, J. J., Francis Thackeray, J., & van der Merwe, N. (2000). The hunters and the hunted revisited. *Journal of Human Evolution, 38,* 565–576.

Lefebvre, C., & Flora, J. A. (1988). Social marketing and public health intervention. *Health Education Quarterly, 15*(3), 299–315.

Leit, R. A., Pope, H. G., & Gray, J. J. (2000). Cultural expectations of muscularity in men: The evolution of *Playgirl* centerfolds. *International Journal of Eating Disorders, 29*(1), 90–93.

Lenski, G. E. (1970). *Human societies: A macrolevel introduction to sociology.* New York: McGraw-Hill.

Lenten, C. (1999). *Changing food habits: Case studies from Africa, South America, and Europe.* Australia: Harwood Academic Publishers.

Leslie, J. (1988). Women's work and child nutrition in the Third World. *World Development, 16*(11), 1341–1362.

Levine, C. E., Ruel, M. T., Morris, S. S., Maxwell, D. G., & Armar-Klemesu, M. (1999). Working women in an urban setting: Traders, vendors and food security in Accra. *World Development, 27*(11), 1977–1991.

Levi-Strauss, C. (1969). *The raw and the cooked.* London: Jonathan Cape.

Lewin, K. (1935). *A dynamic theory of personality.* New York: McGraw-Hill.

Lin, B., Frazao, E., & Guthrie, J. (1999). *Away-from home foods increasingly important to quality of American diet* (Agriculture Information Bulletin No. 749). Washington, DC: U.S. Department of Agriculture.

Lin, B., Guthrie, J., & Blaylock, J. R. (1996). *The diets of America's children: Influence of dining out, household characteristics, and nutrition knowledge* (Agricultural Economic Report No. 746). Washington, DC: Economic Research Service, U.S. Department of Agriculture.

Lindenberger, H., & Bryant, C. A. (2000). Promoting breastfeeding in the WIC Program: A social marketing case study. *American Journal of Health Behavior, 24*(1), 53–60.

Lindsay, H. (1998). Eating with the dead: The Roman funeral banquet. In I. Nielson and H. Nielson (Eds.), *Meals in a social context: Aspects of the communal meal in the Hellenistic and Roman world.* Oxford, England: Alden Press.

Ling, J. C., Franklin, B. A., Lindsteadt, J. F., & Gearon, S. A. N. (1992). Social marketing: Its place in public health. *Annual Review of Public Health, 13,* 341–362.

Livingstone, F. B. (1958). Anthropological implications of sickle-cell gene distribution in West Africa. *American Anthropologist, 60,* 533.

Locke, E. A., & Latham, G. P. (1991). *A theory of goal setting and task performance.* Englewood Cliffs, NJ: Prentice Hall.

Lunner, K., Werthem, E. H., Thompson, J. K., Paxton, S. J., McDonald, F., & Halvaarson, K. S. (2000). A cross-cultural examination of weight-related teasing, body image, and eating disturbance in Swedish and Australian samples. *International Journal of Eating Disorders, 28,* 430–435.

MacKenzie, D. (1935). *Scottish folklore and folk-life: Studies in race, culture and tradition.* Glasgow: Blackie and Sons.

Maloney, M. J., McGuire, J., & Daniels, S. R. (1988). Reliability testing of children's version of the Eating Attitudes Test. *Journal of the American Academy of Child and Adolescent Psychiatry, 27,* 541–543.

Maloney, M. J., McGuire, J., Daniels, S. R., & Specker, B. (1989). Dieting behavior and eating attitudes in children. *Pediatrics, 84,* 482–489.

Mann, G. V., Spoerry, A., & Gray, J. W. (1971). Atherosclerosis in the Masai. *American Journal of Epidemiology, 96,* 26.

March, K. S. (1998). Hospitality, women, and the efficacy of beer. In C. M. Counihan & S. Kaplan (Eds.), *Food and gender: Identity and power* (pp. 45–80). Amsterdam: Overseas Publishers Association.

Marshall, L. (1976). *The !Kung of the Nyae Nyae.* Cambridge, MA: Harvard University Press.

Mathematica Policy Research. (1990). *The savings in Medicaid costs for newborns and their mothers from prenatal participation in the WIC Program.* Washington, DC: U.S. Department of Agriculture, Food and Nutrition Service, Office of Analysis and Evaluation.

May, J., & McClellan, D. (1972). *The ecology of malnutrition in Mexico and Central America.* New York: Hafner.

Mauss, M. (1967). *The gift: Forms and functions of exchange.* New York: W. W. Norton.

McBean, L. D., & Miller, G. D. (1998). Allaying fears and fallacies about lactose intolerance. *Journal of the American Dietetics Association, 98*(6), 671.

McCalla, A. F., & Revoredo, C. (2001). *Prospects for global food security: A critical appraisal of past projections and predictions.* Washington, DC: International Food Policy Research Institute.

McCracken, R. (1971). Lactase deficiency: An example of dietary evolution. *Current Anthropology, 12,* 479.

McGarvey, S. T. (1991). Obesity in Samoans and a perspective on its etiology in Polynesians. *American Journal of Clinical Nutrition, 53,* 1586S–1594S.

McGee, H. (1985). *On food and cooking: The science and lore of the kitchen.* New York: Scribner.

McIntosh, W. A. (1996). *The sociologies of food and nutrition.* New York: Plenum Press.

McIntosh, W. A. & Shiffleet, P. A. (1984). The impact of religious social support on the dietary adequacy of the aged. *Review of Religious Research.*

McIntosh, W. A., & Zey, M. (1998). Women as gatekeepers. In C. M. Counihan & S. Kaplan (Eds.), *Food and gender: Identity and power* (pp. 125–144). Amsterdam: Overseas Publishers Association.

McKinley, N. M. (1999). Ideal weight/ideal women: Society constructs the female. In J. Sobal & D. Maurer (Eds.), *Weighty issues: Fatness and thinness as social problems* (pp. 97–116). Hawthorne, NY: Aldine de Gruyter.

McNamara, R. (1977). *Accelerating population stabilization through social and economic progress.* Washington, DC: Overseas Development Council Development Paper No. 24.

McWilliam, C. L., Desai, K., & Greig, B. (1997). Bridging town and gown: Building research partnerships between community-based professional providers and academia. *Journal of Professional Nursing, 13*(5), 307–315.

MDConsult. (2002). *Eating disorders.* Clinical Reference Systems. Retrieved July 13, 2002, from http://home.mdconsult.com/das/patient/view/20312068

Meade, B., & Rosen, S. (1996). Income and diet differences greatly affect food spending around the globe. *Food Review, 19*(3), 39–44.

Meade, C. D., Calvo, A., & Cuthbertson, D. (2002). Impact of culturally, linguistically, and literacy relevant cancer information among Hispanic farmworker women. *Journal of Cancer Education, 17*(1), 50–54.

Meadows, D., Meadows, D. L., & Randers, J. (1992). *Beyond the limits: Confronting global collapse, envisioning a sustainable future.* Post Mills, VT: Chelsea Green.

Meadows, M. (2000). *Moving toward consensus on cultural competency in health care. Closing the gap.* Washington, DC: Office of Minority Health.

Mehler, P. S. (1996, January 15). Eating disorders: Part 1. Anorexia nervosa. *Hospital Practice,* 110–117.

Mennell, S., Murcott, A., & van Otterloo, A. H. (1992). *The sociology of food: Eating, diet and culture.* Newbury Park, CA: Sage Publications.

Messina, M., & Messina, V. (1996). *The dietitian's guide to vegetarian diets.* Port Townsend, MD: Aspen Publications.

Meyerhoff, E., & Tobias, A. (1980). The technological impact on the food supply. In A. Tobias & P. J. Thompson (Eds.), *Issues in nutrition for the 1980s.* Belmont, CA: Wadsworth.

Meyer-Rochow, V. B. (1973). Edible insects in three different ethnic groups of Papua and New Guinea. *American Journal of Clinical Nutrition, 26,* 673–677.

Miller, K. J., Gleaves, D. H., Hirsch, T. G., Green, B. A., Snow, A. C., & Corbett, C. C. (2000). Comparison of body image dimensions by race/ethnicity and gender in a university population. *International Journal of Eating Disorders, 27,* 310–316.

Mills, M. (1999). *Journal of the National Medical Association.*

Milton, K. (1999). A hypothesis to explain the role of meat-eating in human evolution. *Evolutionary Anthropology, 8,* 11–21.

Minkler, M. (1998a). *Community organizing & community building for health.* Gaithersburg, MD: Aspen Publishers.

Minkler, M. (Ed.) (1998b). Community organizing among the elderly poor in San Francisco's Tenderloin district. In M. Minkler (Ed.), *Community organizing & community building for health* (pp.

244–258). Gaithersburg, MD: Aspen Publishers.

Minkler, M., & Wallerstein, N. (1998). Improving health through community organization and community building: A health education perspective. In M. Minkler (Ed.), *Community organizing & community building for health.* Gaithersburg, MD: Aspen Publishers.

Mintz, S. (1985). *Sweetness and power: The place of sugar in modern history.* New York: Viking Press.

Mittlemark, M. (1999). Health promotion at the community-wide level: Lessons from diverse perspectives. In N. Bracht (Ed.), *Health promotion at the community level 2: New advances* (pp. 3–27). Thousand Oaks, CA: Sage Publications.

Mogelonsky, M. (1998, January). Food on demand. *American Demographics,* 35–43.

Mokdad, A. H., Serdula, M. K., Dietz, W., Bowman, B. A., Marks, J. S., & Koplan, J. (1999). The spread of the obesity epidemic in the United States, 1991–1998. *Journal of the American Medical Association, 282*(16), 1519–1522.

Mossavar-Rahmani, Y., Pelto, G. H., Ferris, A. M., & Allen, L. H. (1996). Determinants of body size perceptions and dieting behavior in a multiethnic group of hospital staff women. *Journal of the American Dietetics Association, 96*(3), 252–255.

Murcott, A. (1983). "It's a pleasure to cook for him": Food, mealtimes and gender in some South Wales households. In E. Gamarnikow (Ed.), *Public and the private.* London: Heinemann.

Murray, C., Waller, G., & Legg, C. (2000). Family dysfunction and bulimic psychopathology: The mediating role of shame. *International Journal of Eating Disorders, 28,* 84–89.

Muwakkil, S. (2000, June 26). Many who "got milk" get sick. *Chicago Tribune.*

Nadel, D. A., Danin, E., Werker, T., Schick, M. E., Kislev, & Stewart, K. (1994). 19,000-year-old twisted fibers from Ohalo II. *Current Anthropology, 37,* 697–699.

National Center for Chronic Disease Prevention and Health Promotion. (2000). Special focus: Nutrition and physical activity. *Chronic Disease Notes and Reports, 13*(1), 1–19.

National Center for Health Statistics. (1999). *Health, United States, 1999 with health and aging chartbook.* Hyattsville, MD: National Center for Health Statistics, Centers for Disease Control and Prevention.

National Institutes of Health. (1994). *Lactose intolerance* (Publication No. 94-2751). Washington, DC: National Institutes of Health, National Digestive Diseases Information Clearinghouse.

National Marriage Project. (2001). *The state of our unions: The social health of marriage in America, 2001.* Retrieved April 4, 2002, from http://marriage.rutgers.edu/publicat.htm

National Research Council. (1989). *Diet and health: Implications for reducing chronic disease risk.* Washington, DC: National Academy Press.

National Research Council. (1992). *Marine aquaculture: Opportunities for growth.* Washington, DC: National Academy Press.

National Restaurant Association. (1996, November). Home meal replacement finds its place at the table. *Restaurants USA,* 1–6.

National Restaurant Association. (2000, November). Americans' dining out habits. *Restaurants USA,* 1–3.

National Restaurant Association. (2001). *2001 Restaurant industry forecast.* Retrieved February 27, 2001, from http://www.restaurant.org/research/ind_glance.html

Naylor, R., Goldburg, R. J., Primavera, J. H., Kautsky, N., Beveridge, M. C. M., Clay, J., Folke, C., Lubchenco, J., Mooney, H., & Troell, M. (2000). Effect of aquaculture on world fish supplies. *Nature, 405,* 1017–1024.

Neel, J. (1962). Diabetes mellitus: A thrifty genotype rendered detrimental by "progress." *American Journal of Human Genetics, 14,* 353–362.

Neel, J. (1982). The thrifty genotype revisited. In J. Kobberling & R. Tattersall (Eds.), *The genetics of diabetes mellitus* (pp. 283–293). London: Academic Press.

Nelli, H. S. (1976). *The business of crime.* Oxford: Oxford University Press.

Nelson, H., & Jurmain, R. (1979). *Introduction to physical anthropology.* St. Paul, MN: West.

Nestel, P. (1986). A society in transition: Development and seasonal influences on the diet of Maasai women and children. *Food and Nutrition Bulletin, 8,* 2–18.

Neumark-Sztainer, D., Story, M., Hannan, P. J., Beuhring, T., & Resnick, M. D. (2000). Disordered eating among adolescents: Associations with sexual/physical abuse and other familial/psychological factors. *International Journal of Eating Disorders, 28,* 249–258.

Nichter, M., & Nichter, M. (1991). Hype and weight. *Medical Anthropology, 13,* 249–284.

Niemeijer, R., Guens, M., Kliest, T., Ogonda, V., & Hoorweg, J. (1988). Nutrition in agricultural development: The case of irrigated rice cultivation in West Kenya. *Ecology of Food and Nutrition, 22,* 65–81.

Nieves, I. (1987). *Explanatory study of intrahousehold resource allocation in a cash-cropping scheme in Highland Guatemala.* Washington, DC: International Center for Research on Women.

Nord, M. (2001). Food stamp participation and food security. *Food Review, 24*(1), 13–19.

Northwest Territories Bureau of Statistics. (2000, May 23). *National population health survey.* Retrieved February 20, 2002, from http://www.stats.gov.nt.ca/Statinfo/Health/NPHS/NPHS.html

Nyswander, D. (1956). Education for health: Some principles and their applications. *Health Education Monographs, 14,* 65–70.

O'Connell, J. F., Hawkes, K., & Blurton Jones, N. G. (1999). Grandmothering and the evolution of *Homo erectus. Journal of Human Evolution, 36,* 461–485.

Office of Minority Health. (2000). *National standards on culturally and linguistically appropriate services (CLAS) in health car*e. *Federal Register, 65,* 247.

O'Halloran, P., Lazovich, D., Patterson, R. E., Harnack, L., & French, S. (2001). Effect of health lifestyle patterns on dietary change. *American Journal of Health Promotion, 16*(1), 27–33.

Ohnuki-Tierney, E. (1998). McDonald's in Japan: Changing manners and etiquette. In J. Watson (Ed.), *Golden Arches East: McDonald's in East Asia* (pp. 161–182). Stanford, CA: Stanford University Press.

Oliveira, V. (1999). Food-assistance expenditures fall for second year. *Food Review, 22*(1), 38–44.

Oliveira, V., & Gunderson, C. (1996). "Do beliefs, knowledge and perceived norms about diet and cancer predict dietary change?" *American Journal of Public Health, 86*(10), 1394–1400.

Oliveira, V., & Gunderson, C. (2000). *WIC and nutrient intake of children* (Food Assistance and Nutrition Research Report No. 5). Washington, DC: U.S. Department of Agriculture, Economic Research Service.

Olsen, S. (2001, May). Personal communication to K. DeWalt. Pittsburgh.

Ortiz de Montellano, B. (1975). *Science, 188,* 215–220.

Osmani, S. (1997). Poverty and nutrition in South Asia: The Abraham Horwitz lecture. In ACC/SCN (Ed.), *Nutrition policy report #16: Nutrition and poverty: Papers from the ACC/SCN 24th Session Symposium, Katmandu, March 1997.* Geneva: ACC/SCN.

Ounpuu, S., Wollcott, D., & Greene, G. (2000). Defining stages of change for lower-fat eating. *Journal of the American Dietetics Association, 100*(6), 674–677.

Overton, M. (1996). *The agricultural revolution in England: The transformation of agrarian economy, 1500–1850.* New York: Cambridge Press.

Paddock, W., & Paddock, P. (1964). *Hungry nations.* Boston: Little, Brown.

Paddock, W., & Paddock, P. (1967). *Famine 1975! America's decision: Who will survive?* Boston: Little, Brown.

Paddock, W., & Paddock, P. (1976). *Time of famines: American and the world food crisis.* Boston: Little, Brown.

Panyotou, T. (1994). Population, environment and development nexus. In R. Cassen (Ed.), *Population and development: Old debates, new conclusions.*

Washington, DC: Overseas Development Council.

Parcel, G. S., Edmundson, E., & Perry, C. L. (1995). Measurement of self-efficacy for diet-related behaviors among elementary school children. *Journal of School Health, 65*(1), 23–27.

Pares, J. M., & Perez-Gonzalez, A. (1999). Magnetochronology and stratigraphy at Gran Dolina section, Atapuerca (Burgos, Spain). *Journal of Human Evolution, 37,* 313–324.

Parham, E. S. (1999). Meanings of weight among dietitians and nutritionists. In J. Sobal & D. Maurer (Eds.), *Weighty issues: Fatness and thinness as social problems* (pp. 183–205). Hawthorne, NY: Aldine de Gruyter.

Parker, S., Nichter, M., Nichter, M., Vuckovic, N., Sims, C., & Ritenbaugh, C. (1995). Body image and weight concerns among African American and White adolescent females: Differences that make a difference. *Human Organization, 54*(2), 103–114.

Patrick, D. L., & Wickizer, T. M. (1995). Community and health. In B. C. Amick, S. Levine, A. Tarlov, & D. C. Walsh (Eds.), *Society and health* (pp. 46–92). New York: Oxford University Press.

Pelletier, D. (1994). The relationship between child anthropometry and mortality in developing countries: Implications for policy, programs and future research. *Journal of Nutrition, 124*(Suppl. 10), 2047S–2081S.

Pelto, G. H., Goodman, A. H., & Dufour, D. L. (2000). The biocultural perspective in nutritional anthropology. In A. H. Goodman, D. Dufour, & G. H. Pelto (Eds.), *Nutritional anthropology: Biocultural perspectives on food and nutrition* (pp. 1–9). Mountain View, CA: Mayfield.

Pelto, G. H., & Pelto, P. J. (1976). *Food and culture in contemporary society.* Paper presented at Our Daily Bread: Changing Priorities and Human Concerns.

Pender, J. (2001). Rural population growth, agricultural change, and natural resource management in developing countries: A review of hypotheses and some evidence from Honduras. In N. Birdsall, A. C. Kelley, & S. W. Sinding (Eds.), *Population matters: Demographic change, economic growth, and poverty in the developing world* (pp. 325–368). Oxford: Oxford University Press.

Perelman, M. (1976). Efficiency in agriculture: The economics of energy. In R. Mervill (Ed.), *Radical agriculture.* New York: Harper and Row.

Petty, R. E., & Cacioppo, J. T. (1984). Motivational factors in consumer response to advertisements. In R. G. Geen, W. W. Beatty, & R. M. Arkin (Eds.), *Human motivation: Physiological, behavioral, and social approaches.* New York: Allyn and Bacon.

Petty, R. E., & Cacioppo, J. T. (1986). *Communication and persuasion: Central and peripheral routes to attitude change.* New York: Springer-Verlag.

Pilcher, J. M. (1998). *Que viven los tamales: Food and the making of Mexican identity.* Albuquerque: University of New Mexico Press.

Pilisuk, M., McCallister, J., & Rothman, J. (1996). Coming together for action: The challenge of contemporary grassroots community organizing. *Journal of Social Issues, 52*(1), 15–37.

Pimentel, D. (1997). *Livestock production: Energy inputs and the environment.* Paper presented at the Canadian Society of Animal science.

Pimentel, D., Olteracu, M. C., Neshein, M. C., Krummel, J., Allen, M. C., & Chick, S. (1980). Counting kilocalories. *American Geographical Society, 30*(3), 9.

Pimentel, D., & Pimentel, M. (1979). *Food, energy, and society.* London: Arnold.

Pinstrup-Andersen, P., Pandya-Lorch, R., & Rosegrant, M. W. (1997). *The world food situation: Recent developments, emerging issues, and long-term prospects.* Washington, DC: International Food Policy Research Institute.

Pinstrup-Andersen, P., Pandya-Lorch, R., & Rosegrant, M. W. (1999). *World food prospects: Critical issues for the early twenty-first century.*

Washington, DC: International Food Policy Research Institute.

Pipher, M. (1994). *Reviving Ophelia: Saving the selves of adolescent girls.* New York: Ballantine.

Plucknett, D., & Smith, N. J. H. (1982). Agricultural research and Third World food production. *Science, 217,* 215.

Plutarch. (1998). Of eating flesh. In W. Goodwin (Trans.), *Plutarch's miscellanies and essays* (Vol. 5, pp. 3–16). Boston.

Pokhrel, R. P., Khatry, S. K., West, K. P., Jr., Shrestha, S. R., Katz, J., Pradhan, E. K., LeClerq, S. C., & Sommer, A. (1994). Sustained reduction in child mortality with vitamin A in Nepal. *Lancet, 343*(8909), 1368–1369.

Polivy, J., & Herman, C. P. (1995). Dieting and its relation to eating disorders. In K. Brownell & C. Fairburn (Eds.), *Eating disorders and obesity: A comprehensive guide* (pp. 83–91). New York: Guilford Press.

Pollock, D. K. (1998). Food and sexual identity among the Culina. In C. M. Counihan & S. L. Kaplan (Eds.), *Food and gender: Identity and power.* Amsterdam, Netherlands: Harwood Academic Publishers.

Pollock, N. (1995). Social fattening patterns in the Pacific: The positive side of obesity—a Nauru case study. In I. de Garine & N. Pollock (Eds.), *Social aspects of obesity* (pp. 111–126). Luxembourg: Gordon and Breach.

Pope, H. J., Olivardia, R., Gruber, A., & Borowiecki, J. (1999). Evolving ideals of male body image as seen through action toys. *International Journal of Eating Disorders, 26,* 65–72.

Porphyry. (1965). *Porphyry on abstinence from animal food* (T. Taylor, Trans.) London.

Posey, D. (1982). Keepers of the forest. *Garden, 6,* 18–24.

Postel, S. (1992). *Last oasis: Facing waste and scarcity.* New York: W. W. Norton.

Potts, R. (1998). Environmental hypotheses of hominid evolution. *Yearbook of Physical Anthropology, 41,* 93–136.

Povey, R., Conner, M., Sparks, P., James, R., & Shepherd, R. (1999). A critical examination of the application of the Transtheoretical Model's stages of change to dietary behaviors. *Health Education Research, 1*(5), 641–651.

Presidential Commission on World Hunger. (1980). *Overcoming world hunger: The challenge ahead. Report of the Presidential Commission on World Hunger.* Washington, DC: Superintendent of Documents.

Price, C. (2000). Foodservice sales reflect the prosperous, time-pressed 1990s. *Food Review, 23*(3), 23–26.

Prochaska, J., & Velicer, W. (1997). The Transtheoretical Model of health behavior. *American Journal of Health Promotion, 12*(1), 38–48.

Prochaska, J. O., & DiClemente, C. C. (1983). Stages and processes of self-change of smoking: Toward an integrative model of change. *Journal of Consulting and Clinical Psychology, 51,* 390–395.

Prochaska, J. O., & DiClemente, C. C. (1992). Stages of change in the modification of problem behaviors. *Progress in Behavior Modification, 28,* 183–218.

Prochaska, J. O., DiClemente, C. C., & Norcross, J. C. (1992). In search of how people change: Applications to addictive behavior. *Addictive Behaviors, 10,* 395–406.

Prochaska, J. O., Redding, C. A., & Evers, K. E. (1997). The Transtheoretical Model and stages of change. In K. Glanz, et al. (Eds.), *Health behavior and health education* (pp. 60–84). San Francisco: Jossey-Bass.

Putnam, J., & Gerrior, S. (1999). Trends in the U.S. food supply, 1970–97. In E. Frazao (Ed.), *America's eating habits: Changes and consequences* (pp. 133–160). Washington, DC: U.S. Department of Agriculture.

Putnam, J. J., & Allshouse, J. E. (1999). *Food consumption, prices, and expenditures, 1970–97.* Washington, DC: U.S. Government Printing Office.

Qaim, M., Krattiger, A., & von Braun, J. (2000). *Agricultural biotechnology in developing countries: Towards optimizing benefits for the poor.* Boston: Kluwer Academic Publishers.

Quak, S. H. (1994). Lactose intolerance in Asian children.

Journal of Paediatric Child Health, 30, 91–92.

Quandt, S., Arcury, T. A., McDonald, J., Bell, R. A., & Vitolins, M. Z. (1999). *Do food security questions work with rural elderly? Informal support, making do, and trusting in god.* Paper presented at the Annual Meeting of the Society for Applied Anthropology, Tucson, Arizona.

Quisumbing, A., Brown, L. R., Feldstein, H. S., Haddad, L., & Peña, C. (1995). *Women the key to food security.* Washington, DC: IFPRI.

Rabinowicz, H. (1971). Modern views of the dietary laws. *Encyclopedia Judaica Jerusalem, 6,* 26.

Rahminifar, A., Kirksey, A., Wachs, T. D., McCabe, G. P., Galal, O., Harrison, G. G., & Jerome, N. W. (1992). Diet during lactation associated with infant behavior and caregiver interaction in a semirural Egyptian village. *Journal of Nutrition, 123*(21), 164–175.

Rangun, V. K., & Karim, S. (1991). *Teaching note: Focusing the concept of social marketing.* Cambridge: Harvard Business School.

Rappaport, R. (1968). *Pigs for the ancestors: Ritual in the ecology of a New Guinea people.* New Haven, CT: Yale University Press.

Rawlins, N. O. (1980). *Introduction to agribusiness.* Englewood Cliffs, NJ: Prentice Hall.

Ray, D. (1998). *Development economics.* Princeton, NJ: Princeton University Press.

Raymond, J. (2000). Senior living: Beyond the nursing home. *American Demographics, 22*(11), 51–56.

Redon, O. (1998). *The medieval kitchen: Recipes from France and Italy.* Chicago: University of Chicago Press.

Regelson, S. (1976). The bagel: Symbol and ritual at the breakfast table. In W. Arens & S. P. Montague (Eds.), *The American dimension: Cultural myths and social realities* (p. 124). Port Washington, NY: Alfred Publishing Co.

Rhodes, R. (1998). *Deadly feasts: Tracking the secrets of a terrifying new plague.* New York: Simon and Schuster.

Richards, A. (1932). *Hunger and work in a savage tribe.* London: Routledge and Sons.

Richards, M. P., Pettitt, P. B., Stiner, M. C., & Trinkaus, E. (2001). *Stable isotope evidence for increasing dietary breadth in the European Mid-Upper Paleolithic.* Paper presented at the National Academy of Sciences, USA.

Rieger, E., Touyz, S. W., Swain, T., & Beumont, P. J. V. (2001). Cross-cultural research on anorexia nervosa: Assumptions regarding the role of body weight. *International Journal of Eating Disorders, 29,* 205–215.

Robertson, A., & Minkler, M. (1994). New health promotion movement: A critical examination. *Health Education Quarterly, 12*(3), 295–312.

Robinson, C. H., & Lawler, M. R. (1977). *Normal and therapeutic nutrition* (15th ed.). New York: Macmillan.

Robinson, T. N. (1999). Reducing children's television viewing to prevent obesity: A randomized control trial. *Journal of the American Medical Association, 282,* 1561–1567.

Rockefeller Foundation. (2001, January 22). *Press release: International Rice Research Institute begins testing golden rice.* International Rice Research Institute (IRRI).

Roos, E., Lahelma, E., Virtanen, M., Prattala, R., & Pietinen, P. (1998). Gender, socioeconomic status, and family status as determinants of food behavior. *Social Science and Medicine, 46*(12), 1519–1529.

Root, W., & de Rochemont, R. (1976). *Eating in America: A history.* New York: William Morrow.

Roper Starch Worldwide. (2000, October). The 3-D consumer: Nationality, lifestyle, and value. *American Demographic,* S5–8.

Roque, A. M. Trigo de Sousa (2000). *An overview of the shrimp disease problem in Mexico with emphasis on taura syndrome and white spot syndrome.* Mazatlán, Sinaloa, México: Centro de Investigación en Alimentación y Desarrollo.

Rorty, M., Yager, J., Rossotto, E., & Buckwalter, G. (2000). Parental intrusiveness in adolescence recalled by women with a history of bulimia nervosa and comparison women. *International Journal of Eating Disorders, 28*, 202–208.

Rosal, M. C., Ebbeling, C. B., Lofgren, I., Ockene, J. K., Ockene, I. S., & Hebert, J. R. (2001). Facilitating dietary change: The patient-centered counseling model. *Journal of the American Dietetics Association, 101*, 332–341.

Rose, D., Habicht, J., & Devaney, B. (1998). Household participation in the food stamp and WIC programs increases the nutrient intakes of preschool children. *Journal of Nutrition, 128*, 548–555.

Rosenstock, I. M. (1966). Why people use health services. *Milbank Memorial Fund Quarterly, 44*, 94–124.

Ross, E. (1987). An overview of trends in dietary variation from hunter-gatherer to modern capitalist societies. In M. Harris & E. Ross (Eds.), *Food and evolution: Toward a theory of human food habits* (pp. 1–56). Philadelphia: Temple University Press.

Rothschild, M. (1999). Carrots, sticks and promises: A conceptual framework for the management of public health and social issue behaviors. *Journal of Marketing, 63*, 24–37.

Rozin, P. (1988). Social learning about food by humans. In J. T. R. Zentall & B. Golef (Eds.), *Social learning: Psychological and biological perspectives* (pp. 165–187). Hillsdale, NJ: Lawrence Erlbaum.

Ryan, A. (1997). The resurgence of breastfeeding in the United States. *Pediatrics, 99*(4), 1–5.

Salzman, P. (2001). *Understanding culture: An introduction to anthropological theory*. Prospect Heights, IL: Waveland Press.

Sawyer, H. (1997). Quality education: One answer for many questions. In UNICEF (Ed.), *The progress of nations* (pp. 33–39). New York: UNICEF.

Schack, K. W., Grivetti, L. E., & Dewey, K. G. (1990). Cash cropping, subsistence agriculture, and nutritional status among mothers and children in Lowland Papua, New Guinea. *Social Science and Medicine, 31*, 61–68.

Schein, E. H. (1992). Defining organizational culture. In J. M.

Shafritz & J. S. O. Shafritz (Eds.), *Classics in organizational theory* (pp. 490–502). Pacific Grove, CA: Brooks/Cole.

Scheper-Hughes, N. (1992). *Death without weeping*. Berkeley: University of California Press.

Schindewolf, O. (1950). *Basic questions in paleontology: Geologic time, organic evolution, and biological systematics* (J. Schaefer, Trans.). Chicago: University of Chicago Press.

Schlenker, E. D. (1998). *Nutrition in aging* (3rd ed.). St. Louis: Mosby-Year Book.

Schlosser, E. (1999). *Fast-food nation: The dark side of the all-American meal*. Boston: Houghton Mifflin.

School Health Policies and Programs Study, Centers for Disease Control and Prevention. (2000). Study Overview. National Center for Chronic Disease Prevention and Health Promotion, Division of Adolescent and School Health [Web page]. Available http://www.cdc.gov/nccdphp/dash/shpps/index.htm

Schulz, A. J., Parker, E. A., Israel, B. A., Becker, A. B., Maciak, B. J., & Hollis, R. (1999). Conducting a participatory community-based prevention: Programs that work. In R. C. Brownson, E. A. Baker, & L. F. Novick (Eds.), *Community-based prevention: Programs that work* (pp. 84–105). Gaithersburg, MD: Aspen Publishers.

Schwartz, B. M., & Ewald, R. H. (1968). *Culture and society: An introduction to cultural anthropology*. New York: Ronald Press.

Schwartz, J. H. (1995). *Skeleton keys: An introduction to human skeletal morphology, development, and analysis*. New York: Oxford University Press.

Schwartz, J. H. (1999). *Sudden origins: Fossils, genes, and the emergence of species*. New York: John Wiley.

Schwartz, J. H., Brauer, J., & Gordon, P. (1995). Tigaran (Point Hope, Alaska) tooth drilling. *American Journal of Physical Anthropology, 97*, 77–82.

Schwartz, J. H., & Tattersall, I. (1999). *The human fossil record: Vol. 1. Terminology and craniodental morphology of genus* Homo *(Europe)*. New York: Wiley-Liss.

Schwartz, J. H., & Tattersall, I. (2000). The human chin: What is it and who has it? *Journal of Human Evolution, 38,* 367–409.

Schwartz, J. H., Tattersall, I., & Laitman, J. T. (1999). New thoughts on Neanderthal behavior: Evidence from nasal morphology. In H. Ullrich (Ed.), *Hominid evolution: Lifestyles and survival strategies* (pp. 166–186). Schwellm, Germany: Archae.

Sen, A. (1981). *Poverty and famine: An essay on entitlement and deprivation.* New York: Clarendon Press.

Sen, A. (1990). Food entitlements and economic chains. In L. F. Newman (Ed.), *Hunger in history: Food shortage, poverty, and deprivation* (pp. 374–386). Cambridge, MA: Basil Blackwell.

Sen, A. (1995). Food, economics, and entitlements. In J. Dreze, A. Sen, & A. Hussain (Eds.), *The political economy of hunger.* Oxford: Oxford University Press.

Shankar, A. V., West, K. P., Jr., Gittelsohn, J., Katz, J., & Pradhan, P. (1996). Chronic low intakes of vitamin A-rich foods in households with xerophthalmic children: A case-control study in Nepal. *American Journal of Clinical Nutrition, 64*(2), 242–248.

Sheeshka, J. D., Woolcott, J. D., & MacKinnon, N. J. (1993). Social cognitive theory as a framework to explain intentions to practice healthy eating behaviors. *Journal of Applied Social Psychology, 23,* 1547–1573.

Shostack, M. (1981). *Nisa: The life and words of a !Kung woman.* Cambridge, MA: Harvard University Press.

Sieving, R. E., Perry, C. L., & Williams, C. L. (2000). Do friendships change behaviors, or do behaviors change friendships? Examining paths of influence in young adolescents' alcohol use. *Journal of Adolescent Health, 26*(1), 27–35.

Simmons, T., & O'Neill, G. (2001). *Households and families: 2000* (Census 2000 Brief C2KBR/01-8). Washington, DC: U.S. Census Bureau.

Simoons, F. J. (1970). Primary adult lactose intolerance and the milking habit: A problem in biological and cultural interrelations II: A cultural historical hypothesis. *American Journal of Digestive Diseases, 15,* 673–695.

Simoons, F. J. (1971). The antiquity of dairying in Asia and Africa. *Geographical Review, 61,* 431.

Simoons, F. J. (1983). Geography and genetics as factors in the psychobiology of human food selection. In L. M. Baker (Ed.), *The psychobiology of human food selection.* Westport, CT: Avi Co.

Simoons, F. J. (1994). *Eat not this flesh: Food avoidances from prehistory to the present* (2nd ed.). Madison: University of Wisconsin Press.

Sims, L. S. (1983). The ebb and flow of nutrition as a public policy issue. *Journal of Nutrition Education, 15,* 4. Cited in R. T. Frankle & A. L. Owen (1993), *Nutrition in the community: The art of delivering services* (3rd ed., p. 307), St. Louis: Mosby.

Singer, P. (1975). *Animal liberation.* New York: New York Review of Books.

Sinha, R. (1976). *Food and poverty: The political economy of confrontation.* New York: Holmes and Meier Publishers.

Skinner, M. (1991). Bee brood consumption: An alternative explanation for hypervitaminosis A in KNM-ER 1808 *(Homo erectus)* from Koobi Fora. *Journal of Human Evolution, 20,* 493–503.

Smith, L. C., & Haddad, L. (2001). How important is improving food availability for reducing child malnutrition in developing countries? *Agricultural Economics, 26*(3), 191–204.

Sobal, J. (1995a). The medicalization and demedicalization of obesity. In D. Maurer & J. Sobal (Eds.), *Eating agendas* (pp. 67–90). Hawthorne, NY: Aldine de Gruyter.

Sobal, J. (1995b). Social influences on body weight. In K. Brownell & C. Fairburn (Eds.), *Eating disorders and obesity: A comprehensive guide* (pp. 73–77). New York: Guilford Press.

Sobal, J., Khan, L. K., & Bisogni, C. (1998). A conceptual model of the food and nutrition system. *Social Science and Medicine, 47*(4), 853–863.

Sommer, A. (2001, May 17). *Vitamin A deficiency disorders: Origins of the problem and approaches to its control.* AgBiotechnology InfoNet.

Retrieved January, 2002, from http://www.biotech-info.net/disorders.html

Sorensen, G., Stoddard, A., & Macario, E. (1998). Social support and readiness to make dietary changes. *Health Education and Behavior, 25*(5), 586–598.

Sparks, P., & Shepherd, R. (1992). Self-identity and the theory of planned behaviour. In D. R. Rutter & L. Quine (Eds.), *Social psychology and health: European perspectives* (pp. 325–346). Avebury, England: Aldershot.

Sperling, L., & Ntabomvura, B. (1994). Integrating farmer experts into on-station research. In H. S. Feldstein & J. Jiggins (Eds.), *Tools for the field: Methodologies handbook for gender analysis in agriculture*. Hartford, CT: Kumarian Press.

Sponheimer, M., & Lee-Thorp, J. A. (1999). Isotopic evidence for the diet of an early hominid, *Australopithecus africanus. Science, 283,* 368–370.

Steinmetz, K. A., & Potter, J. D. (1990). Vegetables, fruits, and cancer: II. Mechanisms. *Cancer Causes Control,* 427–442.

Stephenson, M. T., & Witte, K. (2001). Creating fear in a risky world: Generating effective health-risk messages. In R. E. Rice & C. K. Atkin (Eds.), *Public communication campaigns* (3rd ed.). Thousand Oaks, CA: Sage Publications.

Stevens, J., Kumanyika, S., & Keil, J. (1994). Attitudes toward body size and dieting: Differences between elderly black and white women. *American Journal of Public Health, 84*(8), 1322–1324.

Stevens, J., Story, M., Becenti, A., French, S. A., Gittlesohn, J., Going, S. B., Juhaeri, L. S., & Murray, D. M. (1999). Weight-related attitudes and behaviors in fourth-grade American Indian children. *Obesity Research, 7*(1), 34–42.

Stinson, S., Bogin, B., Huss-Ashmore, R., & O'Rourke, D. (Eds.). (2000). *Human biology: An evolutionary and biocultural perspective.* New York: Wiley-Liss.

Story, M., Evans, M., Fabsitz, R. R., Clay, T. E., Holy Rock, B., & Broussard, B. (1999). The epidemic of obesity in American Indian communities and the need for childhood obesity-prevention programs. *American Journal of Clinical Nutrition, 69*(Suppl.), 747S–754S.

Story, M., Neumark-Sztainer, D., Sherwood, N., Stang, J., & Murray, D. (1998). Dieting status and its relationship to eating and physical activity behaviors in a representative sample of U.S. adolescents. *Journal of the American Dietetics Association, 98*(4), 1127–1135.

Strasser, S. (1982). *Never done: A history of American housework.* New York: Pantheon Press.

Strauss, J. (1986). Does better nutrition raise farm productivity? *Journal of Political Economy, 94*(2), 297–320.

Strecher, V., & Rosenstock, I. M. (1997). The health belief model. In K. Glanz, F. Lewis, & B. Rimer (Eds.), *Health behavior and health education: Theory, research, and practice* (2nd ed., pp. 41–59). San Francisco: Jossey-Bass.

Strier, K. B. (2000). *Primate behavioral ecology.* Boston: Allyn and Bacon.

Suarez, F. L., & Savaiano, L. (1995). A comparison of symptoms after the consumption of milk or lactose hydrolyzed milk by people with self-reported severe lactose intolerance. *New England Journal of Medicine, 333,* 1–4.

Suarez, F. L., Savaiano, P., Arbisi, P., & Levitt, M. D. (1997). Tolerance to the daily ingestion of two cups of milk by individuals claiming lactose intolerance. *American Journal of Clinical Nutrition, 65,* 1502–1506.

Super, D. (2001). *Background on the Food Stamp Program.* Center on Budget and Policy Priorities. Retrieved April 4, 2002, from http://www.cbpp.org/7-1-01fs.htm

Tannahill, R. (1973). *Food in history.* New York: Stein and Day.

Tannahill, R. (1988). *Food in history* (2nd ed.). New York: Three Rivers Press.

Tattersall, I., & Schwartz, J. H. (2000). *Extinct humans.* Boulder, CO: Westview Press.

Tattersall, I., & Schwartz, J. H. (in press). Diet and neanderthals. *Acta Universitatis Carolinae Medica, 41.*

Tattersall, I., & Sussman, R. W. (1975). Observations on the ecology and behavior of the mongoose lemur *Lemur mongoz mongoz* Linnaeus (Primates, Lemuriformes), at Amijoroa, Madagascar. *Anthropological Papers of the American Museum of Natural History, 52,* 195–216.

Teaford, M. F., & Ungar, P. S. (2000). *Diet and the evolution of the earliest human ancestors.* Paper presented at the National Academy of Sciences, USA.

Templeton, S., & Scherr, S. J. (1999). Effects of demographic and related microeconomic changes on land quality in hills and mountains of developing countries. *World Development, 27*(6), 903–918.

Theophano, J., & Curtis, K. (1991). Sisters, mothers, and daughters: Food exchange and reciprocity in an Italian-American community. In A. Sharman, J. Theophano, K. Curtis, & E. Messer (Eds.), *Diet and domestic life in society* (pp. 147–171). Philadelphia: Temple University Press.

Thieme, H. (1997). Lower Palaeolithic hunting spears from Germany. *Nature, 385,* 807–810.

Thomas, P. R. C. (1991). *Improving America's diet and health: From recommendations to action.* Washington, DC: National Academy of the Press.

Thompson, B., & Kinne, S. (1999). Social change theory: Applications to community health. In N. Bracht (Ed.), *Health promotion at the community level 2: New advances* (pp. 29–46). Thousand Oaks, CA: Sage Publications.

Thompson, J. K., Heinberg, L. J., Altabe, M., & Tantleff-Dunn, S. (1999). *Exacting beauty: Theory, assessment, and treatment of body image disturbance.* Washington, DC: American Psychological Association.

Tiffen, M., Morrison, M., & Gichuki, F. (1994). *More people, less erosion: Environmental recovery in Kenya.* Chichester: John Wiley and Sons.

Tobey, J., Clay, J., & Vergne, P. (1998). *The economic, environmental and social impacts of shrimp farming in Latin America.* Coastal Resources Center, University of Rhode Island.

Toenniessen, G. H. (2000). *Vitamin A deficiency and golden rice: The role of the Rockefeller Foundation.* The Rockefeller Foundation. Retrieved January 17, 2002, from http://www.rockfound.org/

Toops, D. (2000). Hunting and gathering. *Food processing, 61*(11), 14.

Touissaint-Samat, M. (1991). *History of food.* Malden: Blackwell Publishers Ltd.

Townsend, M. S., Peerson, J., Love, B., Achterberg, C., & Murphy, S. P. (2001). Food insecurity is positively related to overweight in women. *Journal of Nutrition, 131* (1738–1745).

Tripp, R. (1982). Farmers and traders: Some economic determinants of nutritional status in Northern Ghana. *Food and Nutrition, 8*(1).

Troiano, R. P., & Flegal, K. M. (1998). Overweight children and adolescents: Description, epidemiology, and demographics. *Pediatrics, 101*(3), 497–504.

Tufts University. (1994). Marketers' milk misconceptions on lactose intolerance. *Tufts University Diet and Nutrition Newsletter,* p. 4.

Turner, V. (1967). *The forest of symbols. Aspects of Ndembu Ritual.* Ithaca, NY: Cornell University Press.

Tuttle, C. R. (2000). An open letter to the WIC Program: The time has come to commit to breastfeeding. *Journal of Human Lactation, 16*(2), 99–103.

Tyron, T. (1983). *The way to health, long life and happiness.* London.

University of Massachusetts. (February 18, 2000). *Community-supported agriculture page.* Retrieved February 6, 2002, from http://www.umass.edu/umext/csa/about.html

U.S. Bureau of the Census. (1998). *United States Department of Commerce News* [Web site]. Census Bureau. Retrieved July 12, 2002, from http://www.census.gov/Press-Release/cb98-56.html

U.S. Bureau of the Census. (1999). *World population profile: 1998.* Washington, DC: U.S. Government Printing Office.

U.S. Bureau of the Census. (2001a). *Profile of selected social characteristics: 2000.* U.S. Census Bureau. Retrieved July 12, 2002, from http://factfinder.census.gov

U.S. Bureau of the Census. (2001b). *Profiles of general demographic characteristics: 2000 census of population and housing.* U.S. Census Bureau, Economics and Statistics Administration, U.S. Department of Commerce. Retrieved from http://www.census.gov/Press-Release/www/2001/2khus.pdf

USDA. (1961). Livestock production units, 1910–1961. *Statistical Bulletin* No. 325.

USDA. (2000). Contracting changes how farms do business. *Rural Conditions and Trends, 10*(2), 50–56.

USDA. (2001a). *Food and agricultural policy: Taking stock of the new century.* Washington, DC: U.S. Government Printing Office.

USDA. (2001b). *Organic agriculture 2001.* Retrieved November 23, 2001, from http://www.ers. usda.gov/Emphases/Harmony/ issues/organic/organic.html

USDA. (2001c). *Food consumption, prices and expenditures 1999–2000.* Economic Research Service, USDA.

USDA/ERS. (1998, August 31). *Results from the 1994–96 continuing survey of food intakes by individuals.* Economic Research Service, U.S. Department of Agriculture. Retrieved July 14, 2002, from http://www.barc.usda.gov/bhnrc/ foodsurvey/96result.html

USDA/ERS. (2001a, February 1). *Agricultural biotechnology: Adoption of biotechnology.* Retrieved February 6, 2002, from http:// www.ers.usda.gov/Briefing/ biotechnology/chapter1.htm

USDA/ERS. (2001b). *Harmony between agriculture and the environment. Current issues: Organic agriculture.* Retrieved, 2001, from http://www. ers.usda.gov/Emphases/Harmony/ issues/organic/organic.html

USDA/ERS. (2002a, April 4). *Food consumption (per capita) data system.* Retrieved June 25, 2002, from http://ers.usda.gov/data/ foodconsumption/

USDA/ERS. (2002b). *Agricultural biotechnology: Adoption of biotechnology and its production impacts.* Retrieved June 24, 2002, from http://www.ers.usda.gov/ Briefing/biotechnology/chapter1. htm

USDA/FNS. (2001a). *Characteristics of food stamp households: Fiscal year 2000.* Retrieved January 30, 2002, from http://www.fns.usda.gov/ oane/MENU/Published/FSP/FILES/ Participation/2000advrpt.pdf

USDA/FNS. (2001b). *How WIC helps.* Retrieved from http://www.fns. usda.gov/wic/CONTENT/ howwichelps.htm

U.S. Department of Health and Human Services. (1990). *International classification of diseases—Clinical modifications.* Washington, DC.

U.S. Department of Health and Human Services. (2001a). *The Surgeon General's call to action to prevent and decrease overweight and obesity.* Retrieved from http:// www.surgeongeneral.gov/topics/ obesity/calltoaction/toc.htm

U.S. Department of Health and Human Services. (2001b). *Healthy People 2010.* Retrieved April 4, 2002, from http://www.cdc.gov/ nchs/hphome.htm

U.S. Department of Health and Human Services. (2002, April 24). *The 2002 HHS poverty guidelines.* Retrieved July 14, 2002, from http://aspe.os.dhhs.gov/poverty/ 02poverty.htm

U.S. Government Accounting Office. (2001). *Long-term care: Baby boom generation increases challenges of financing needed services.* Retrieved May 17, 2001, from http://www. gao.gov/index.html

Valverde, V., Martorell, R., Mejia-Pivaral, V., Delgado, H., Lechtig, A., Teller, C., & Klein, R. (1977). Relationship between land availability and nutritional status. *Ecology of Food and Nutrition, 6,* 1–24.

Vander, A. J. (1981). *Nutrition, stress, and toxic chemicals.* Ann Arbor: University of Michigan Press.

Vickers, W. T. (1988). Game depletion hypothesis of Amazonian adaptation: Data from a native community. *Science, 239,* 1521–1522.

von Braun, J. (1988). Effects of technological change in agriculture on food consumption and nutrition: Rice in a West African setting. *World Development, 16*(9), 1083–1098.

von Braun, J., Hotchkiss, D., & Immink, M. (1989). *Nontraditional export crops in Guatemala: Effects on production, income and nutrition.*

(Research Report No. 73). Washington, DC: International Food Policy Research Institute.

von Braun, J., & Kennedy, E. (1986). *Commercialization of subsistence agriculture: Income and nutritional effects in developing countries.* Washington, DC: International Food Policy Research Institute.

von Braun, J., Kennedy, E., & Bouis, H. (1990). Commercialization of smallholder agriculture: Policy requirements for the malnourished poor. *Food Policy, 15,* 82–87.

von Oppen, A. (1999). "The lazy man's food?" Indigenous agricultural innovation and dietary change in Northwestern Zambia. In C. Lenten (Ed.), *Changing food habits: Case studies from Africa, South America, and Europe* (pp. 43–71). Australia: Harwood Academic Publishers.

Waldmann, E. (1980). The ecology of the nutrition of the Bapedi, Sekhukuniland. In J. R. K. Robson (Ed.), *Food, ecology and culture.* London: Gordon and Breach Science Publishers.

Walker, A. C., & Leakey, R. (Eds.). (1993). *The Nariokotome Homo erectus skeleton.* Cambridge: Harvard University Press.

Wallace, A. F. C. (1966). *Religion: An anthropological view.* New York: Random House.

Walters, K. S., & Portmess, L. (Eds.). (1999). *Ethical vegetarianism: From Pythagoras to Peter Singer.* New York: State University of New York Press.

Warde, A., & Hetherington, K. (1994). English households and routine food practices: A research note. *Sociological Review, 42*(4), 758–778.

Warde, A., & Martens, L. (2000). *Eating out: Social differentiation, consumption, and pleasure.* Cambridge: Cambridge University Press.

Weaver, H. N. (1998). Indigenous people in a multicultural society: Unique issues for human services. *Social Work, 43*(3), 203–209.

Weismantel, M. J. (1988). *Food, gender, and poverty in the Ecuadorian Andes.* Philadelphia: University of Pennsylvania Press.

Weismantel, M. J. (2000). The children cry for bread: Hegemony and the transformation of consumption. In A. H. Goodman, D. L. Dufour, & G. H. Pelto (Eds.), *Nutritional anthropology: Biocultural perspectives of food and nutrition* (pp. 136–144). Mountain View, CA: Mayfield.

Weiss, B. (1980). Nutrition adaptation and cultural maladaptation: An evolutionary view. In N. W. Jerome, R. F. Kandel, & G. H. Pelto (Eds.), *Nutritional anthropology: Contemporary approaches to diet and culture.* Pleasantville, NJ: Redgrave.

Wertheim, E., Koerner, J., & Paxton, S. (2001). Longitudinal predictors of restrictive eating and bulimic tendencies in three different age groups of adolescent girls. *Journal of Youth and Adolescence, 30*(1), 69–81.

West, K. P. J., Pokhrel, R. P., Katz, J., LeClerq, S. C., Khatry, S. K., Shrestha, S. R., Pradhan, E. K., Tielsch, J. M., Pandey, M. R., & Sommer, A. (1991). Efficacy of vitamin A in reducing preschool child mortality in Nepal. *Lancet, 338*(8759), 67–71.

White, R., & Frank, E. (1994). Health effects and prevalence of vegetarianism. *Western Journal of Medicine, 160*(5), 465.

White, R., Seymour, J., & Frank, E. (1999). Vegetarianism among U.S. women physicians. *Journal of the American Dietetics Association, 99*(5), 595.

White, T. D., Suwa, G., & Asfaw, B. (1994). *Australopithecus ramidus,* a new species of early hominid from Aramis, Ethiopia. *Nature, 371,* 306–312.

Whitehead, B. D. (2001, June). *The state of our unions.* The National Marriage Project. Retrieved July 14, 2002, from http://marriage. rutgers.edu/TEXTSOOU2001.htm

Whorton, J. C. (2000). Vegetarianism. In K. Kiple & K. C. Ornelas (Eds.), *The Cambridge world history of food* (pp. 1553–1564). New York: Cambridge University Press.

Wilde, P. E. (2001). Strong economy and welfare reforms contribute to drop in Food Stamp rolls. *Food Review, 24*(1), 2–12.

Wilde, P. E., McNamara, P. E., & Ranney, C. K. (1999). The effect of income and food programs on dietary quality: A seemingly unrelated regression analysis with error components. *American Journal of Agricultural Economics, 81,* 959–971.

Wilfley, D. E., & Rodin, J. (1995). Cultural influences on eating disorders. In K. B. C. Fairburn (Ed.), *Eating disorders and obesity: A comprehensive guide* (pp. 78–82). New York: Guilford Press.

Willett, J. (1983). *Recent developments in the world food situation.* Lexington: University of Kentucky Department of Agricultural Economics.

Williamson, J. (2001). Demographic change, economic growth, and inequality. In N. Birdsall, A. C. Kelley, & S. W. Sinding (Eds.), *Population matters: Demographic change, economic growth, and poverty in the developing world* (pp. 106–136). Oxford: Oxford University Press.

Wolfenbarger, L. L., & Phifer, P. R. (2000). The ecological risks and benefits of genetically engineered plants. *Science, 290,* 2088–2093.

Wood, C. S. (1979). *Human sickness and health: A biocultural view.* Palo Alto, CA: Mayfield.

Wood, S., Sebastian, K., & Scherr, S. J. (2001). *Pilot analysis of global ecosystems: Agroecosystems.* Washington, DC: International Food Policy Research Institute and the World Resources Institute.

World Bank. (1986). *Poverty and hunger: Issues and options for food security in developing countries.* Washington, DC.

World Bank. (2001). *World Development Report 2001.* Washington, DC.

Worthington-Roberts, B., & Williams, S. R. (1989). *Nutrition in pregnancy and lactation* (5th ed.). St. Louis: Mosby.

Zimmerman, P. (2001, June 25). Sara Lee agrees to plea settlement in Bil Mar recall. *Southwest Meat Association/Info Meat, 6*(13), 1.

INDEX

Note: Page numbers in boldface indicate definitions. Page numbers in italics refer to illustrations.